Library of
Davidson College

Memoirs of Anastas Mikoyan

Volume I

Memoirs of Anastas Mikoyan

Volume I

The Path of Struggle

Foreword by
W. Averell Harriman

Preface and Footnotes by
Harrison E. Salisbury

English Edition edited by
Sergo Mikoyan

translated by
Katherine T. O'Connor
Diane L. Burgin

Sphinx Press, Inc.
Madison, Connecticut

Copyright © 1988, Sphinx Press, Inc.
All rights reserved. No part of this book may be reproduced by any means nor translated into a machine language, without the written permission of the publisher.

Library of Congress Cataloging-in-Publication Data

Mikoian, A. I. (Anastas Ivanovich), 1895-1978.
   The path of struggle.

   (Memoirs of Anastas Mikoyan; v. 1)
   Translation of: Dorogoĭ borby.
   1. Mikoian, A. I. (Anastas Ivanovich), 1895-1978.
2. Statesmen—Soviet Union—Biography. 3. Revolutionists
—Transcaucasia—Biography. 4. Transcaucasia—History—
Revolution, 1917-1921. I. Title. II. Series:
Mikoian, A. I. (Anastas Ivanovich), 1895-1978.
Memoirs of Anastas Mikoyan; v. 1.
DK268.M52A3 1988   vol. 1 947.084′092′4 s   87-23324
ISBN 0-043071-04-6   [947.084′092′4]

Manufactured in the United States of America

Contents

Foreword	W. Averell Harriman		vii
Preface	Harrison E. Salisbury		xi
Editor's Preface	Sergo A. Mikoyan		xvii
Author's Preface			1

Part I: **The Formative Years**
- Chapter I — Some Childhood Reminiscences — 7
- Chapter II — My Revolutionary Youth — 25
- Chapter III — My Universities — 39
- Chapter IV — The February Revolution — 59
- Chapter V — Preparing for Armed Revolt — 85

Part II: **The Baku Commune**
- Chapter I — The Victory of Soviet Power in Baku — 103
- Chapter II — The Commune's First Steps — 117
- Chapter III — The Armed Struggle Continues — 135
- Chapter IV — The Fall of the Baku Commune — 149
- Chapter V — We Leave To Return Victorious — 179
- Chapter VI — The Twenty-Six — 193

Part III: **Transcaspian Prisons**
- Chapter I — In Krasnovodsk — 217
- Chapter II — In Ashkhabad — 237

Illustrations — 263

Part IV: **The Baku Underground During the British Occupation**
- Chapter I — The Struggle for the Masses — 301

Chapter II	For the Unification of the Workers of Transcaucasia	329
Chapter III	The Birth of the Slogan "For a Soviet Azerbaijan!"	351
Chapter IV	The First Transcaucasian Party Conference	367
Chapter V	A Letter to Lenin	397
Chapter VI	The Knights of the Revolution	413
Chapter VII	The Tactics of a United Front	435
Chapter VIII	New Arrest	463
Chapter IX	The Unification of the Communist Organizations of Azerbaijan	479
Chapter X	Moscow	497

Part V: The Birth of Soviet Transcaucasia

Chapter I	Return from Moscow	527
Chapter II	Farewell, Baku!	539

Appendix 553

Name Index 577

Foreword

This book on the life of Anastas Mikoyan gives the American reader a personalized insight into the history of the Bolshevik Revolution and the character of the nation that emerged.

I first met Mr. Mikoyan in September 1941 when I visited Moscow as the head of an American mission together with Lord Beaverbrook, who headed the British mission.

President Roosevelt and Prime Minister Churchill had agreed at their meeting at sea in July 1941 that together they would send missions to Moscow to determine whether the Soviet Union could hold out against Hitler's onslaught and what supplies were most urgently needed from their two countries.

Beaverbrook and I dealt largely with Stalin, but we talked on a number of occasions with Mr. Mikoyan, and the members of our missions dealt with Mikoyan to obtain detailed information on the supplies needed. I learned then of the close relationship that existed between Stalin and Mikoyan. In my many visits to Moscow, as well as during my two-and-a-half years service as our wartime ambassador, I had many opportunities to talk with him. He was probably Stalin's closest advisor on foreign trade. Lend-lease came under his supervision. He had a warm personality, but no one was a tougher trader.

On our first evening in Moscow, Stalin was quite cordial when we discussed what supplies might be available to the Soviet Union. However, the next evening Stalin couldn't have been more rude. He said he had talked to his advisors, evidently Mikoyan, and stated among other things that "the paucity of your offers of supplies shows clearly that your governments wish us to be defeated." He said, "A nation such as the United States with an annual steel production of fifty million tons could certainly afford to supply far greater quantities of military equipment than you offer." I rejoined that his figures were inaccurate—the United States had sixty million tons annual steel production, but that was quite different from supplying the tanks and aircraft that he was demanding. We simply had not built the factories and did not have the production capability required.

The third evening Beaverbrook and I agreed that we would give him a detailed list of everything that we could offer him. That evening Stalin's mood was quite different. He accepted our proposals with cordial thanks and invited us to a Kremlin banquet the following evening.

Our associates had spent several days with Mikoyan and his associates. The Soviets gave lists of requirements but would give no information on the reasons for their requirements. I met this same reluctance on the part of Soviet officials to give information of any sort throughout my years as an ambassador. Mr. Mikoyan was always cordial and had a warm manner, but he was adamant on not supplying information that we really needed to convince our people in Washington that they should give up military equipment much needed for our own forces and ship it to Russia. This reluctance made our task of supplying them far more difficult, and I could not fathom the reasons since we were not asking for military secrets. However, I learned that the Soviet principle was to give no information that could possibly leak to the enemy. There was no doubt that Hitler was taken by surprise when the Red Army attacked in great strength at certain points and broke through Hitler's invasion lines. The complete Soviet secrecy was no doubt one of its great assets during the war.

Throughout the years that I knew Mr. Mikoyan, I marveled at his intimacy and frankness with Stalin and yet the fact that he had managed to survive purge after purge of Stalin's other close associates.

I saw Mr. Mikoyan again on several occasions in my visits to Russia in the postwar years. I recall particularly that in 1959 Khrushchev invited me to his dacha for a meal. It started at four in the afternoon and only included a few people—Mikoyan, Koslov, Gromyko, Zhukov, and my companion, Charles Thayer. I sat next to Khrushchev and Mr. Mikoyan was seated opposite. Throughout the meal Khrushchev kept joking with Mikoyan, teasing him and attempting to provoke him by critical remarks. Mikoyan held his own and returned the barbs. The meal lasted until eleven o'clock, and then I was the one who broke it up—with Khrushchev protesting that the evening was still young.

In 1975 President Ford appointed me to head the American delegation to the Thirtieth Anniversary Celebration of VE Day in Moscow. Mikoyan had retired from office then, and yet he came to all the international affairs and ceremonies. I believe he was the first senior official to retire and still retain the privilege of participating in public functions. He came to the American embassy for dinner one evening and sat next to me. He invited me to come to his dacha where we could have an intimate talk of the years we had known each other. He said to me that day that he worked closely with Khrushchev but that Khrushchev was difficult because of his lack of education. In fact, Khrushchev told me himself that his academic education was limited to one year in primary school. Mikoyan said he had opposed Khrushchev's Berlin policy and also his installing the Soviet missiles in Cuba. "When President Kennedy forced Khrushchev to remove the missiles," Mikoyan told me with a smile, "Khrushchev repaid me for my opposition by sending me to tell Castro that the missiles had to be removed."

Unfortunately, something came up on each of my succeeding visits to the Soviet Union that prevented my getting together with Mikoyan. Finally, in 1978, we had a definite engagement to meet in Moscow in early December. Unhappily, Mikoyan died just a few weeks before I arrived in Moscow.

In a final tribute to Mikoyan, I went to his grave and placed a wreath on his tomb. I was happy that a number of his children and grandchildren came to join me in my tribute. It is gratifying that Mikoyan's descendants are playing useful and constructive roles in the life of their country in important positions both at home and abroad.

Anastas Mikoyan had a fascinating life, and the readers will find this book full of interesting insight into the way he was able to participate on the highest level and yet avoid the bitter end that characterized his colleagues. He was an able man and served his country well.

W. Averell Harriman
Washington, DC
1980

Preface

Anastas Ivanovich Mikoyan was not quite twenty-two when the Bolshevik Revolution occurred November 7, 1917. He was a spirited and gifted member of Lenin's party, a dedicated organizer and talented orator, courageous, intelligent, wiry, determined, handsome, witty, and as dogged and driving a young man as ever emerged from the flinty crags of northern Armenia.

He was to live a long, long life, longer than that of any of his fellow Caucasian revolutionaries and any of the men who rose to power with Josef Stalin except for Vyachoslav Molotov and Lazar Kaganovich. In the later years of his life when I had the privilege of knowing him and observing him repeatedly at close range he had lost none of the qualities of his life but had gained the kind of judgment and wisdom which stems only from experience, hardship, and danger.

Armenia is famous for the vitality and talent of her sons, for their energy, obulliance, skill in comprehending complex human events, their talent in government, politics, economics, science, and the broad artistic fields. Mikoyan was a true son of Armenia. In later years his Politburo comrades liked to joke about him. "If Anastas Ivanovich had gone to New York," they said, "he would have become a millionaire." They were probably right. In his capacity as Minister of Foreign Trade he demonstrated a keen nose for a "good deal." But this was beside the point. There is no evidence that the idea of emigration or private enterprise ever entered Mikoyan's mind. He was attracted to the cause of revolution and social change early in boyhood. The inclination grew out of the air he breathed, the hard rock of the mountains, the atmosphere of hardship and poverty, inequality and oppression that surrounded him.

Like many a youngster growing up in the vast "prison house of nations," which imperial Russia had become by the turn of the twentieth century, Mikoyan's mind and soul were pointed toward a new social order that he came to believe might be achieved only through the Social Democratic party and its radical wing, which came to be known as the Bolsheviks, headed by V. I. Lenin. When Mikoyan heard that the party's name was being changed to "Communist" he could not conceal his joy. This sounded just right and for a while he even adopted "Communist" as his party *klichka*, or underground nickname, signing his articles "Communist" or "C-st".

To anyone interested in how the Russian Revolution came about and what forces fired the overthrow of the tsar, Mikoyan's memoirs offer a unique insight. They are the only memoirs written by a leading figure in the Communist party that cover the whole expanse of prerevolutionary, revolutionary, and postrevolutionary days. Revolutionaries often died too young to write memoirs, or were too busy in later days to put their thoughts on paper, or in so many cases failed to survive Stalin's purges.

It is true that Mikoyan's years before 1917 were spent in what Russians call the "periphery," that is, far from the center, deep in the Caucasus, for the most part in Armenia, Georgia, and Azerbaijan. He was a youngster during the prerevolutionary period. But this is not unusual. The leaders of the Revolution were all young. Lenin was the oldest. They called him *starik,* the old man. He was forty-seven years old when the Bolsheviks seized power in the October coup.

The fact that Mikoyan was in the Caucasus when the Revolution broke out does not mean that he was isolated from its main stream. Many of the leading revolutionaries came from the Caucasus. The Baku oil fields represented one of the most advanced industrial regions of Russia, and Baku, Batumi, Tiflis, and other Caucasus cities were important revolutionary centers. Many leading revolutionaries started their careers in this area, particularly Bolsheviks.

Mikoyan's narrative provides a compelling account of the kind of life in prerevolutionary Russia that gave birth to the radical changes of the early twentieth century. In a sense he was especially fortunate in his background as an Armenian. He came from a close and warm family, as his own was to be in his adult life. The Armenian educational system although clerical was strongly nationalist in essence and very liberal by the standards of the day. It provided a hotbed for revolutionary sentiment, and the faculty and teachers within the system for the most part were lenient to their young charges, often sharing their sentiments, sympathizing with them or encouraging them in their more urgent strivings. There was a strong tradition among Armenians of the helping hand and a network of close familial, clannish, or locality ties that proved invaluable in the roles that the young revolutionaries came to play.

If there is a central figure and a central episode in this early passage of the Mikoyan memoirs it is certainly Stepan Shaumian and the episode of the twenty-six commissars. Shaumian was an outstanding Bolshevik in the Caucasus and one of the party's principal leaders. His figure and image have grown with the years especially since the death of Stalin. An Armenian like Mikoyan, he became Mikoyan's personal ideal and was to remain so throughout Mikoyan's life. Shauman died as one of the twenty-six Baku commissars. It was a death Mikoyan could easily have shared for he was arrested along with Shaumian and only by accident was released while Shaumian was held. Had Mikoyan's name been on a list of the party leaders as it properly should have been he would have been held, as was Shaumian, and would have been executed with him—there would have been twenty-seven, not twenty-six commissars executed. By that simple accident Mikoyan escaped and Shaumian did not. All his life Mikoyan was to wonder over this accident, feeling somehow at fault that he had lived while his beloved leader, Shaumian, and his other comrades had died.

The emotional impact of the execution of the twenty-six never left Mikoyan. I personally heard him tell the story more than forty years later and it sounded as though it had occurred only yesterday, so fresh were the details in his memory, so keen his emotion. It was the most dramatic episode of a life that was crowded with dramatic episodes. Nor did Mikoyan ever cease to exclaim over the chance that saved his life—the fact that his name was "not on a list." It was, if you will, a kind of Russian roulette and it echoed a remark that many of his countrymen were later to make concerning the Stalin years, Mikoyan amongst them. Stalin's peril Mikoyan also escaped and equally by chance. His name did not happen to be on a list, although in the last months of Stalin's life when the paranoia deepened he was constructing a final purge trial to end all purge trials, and this time Mikoyan's name *was* on a list along with those of the rest of his surviving Politburo comrades, the list of those whom Stalin had assigned as victims for the final paroxysm of terror.

Mikoyan never tired of talking of Shaumian nor of singing his praises as a remarkable revolutionary leader. He was dedicated to Shaumian's family and children and in later years treated them as if they were his own. In this initial segment of Mikoyan's memoirs we find the picture of the young Armenian growing up; his education, his transformation into a passionate, trigger-ready revolutionary, his progress up the ladder of party leadership and responsibility through the actual fire of revolutionary struggle. We see his character formed; the steel determination, the development of judgment to cure some of his impulsive acts, his steadily broadening responsibilities. Years ahead lie his emergence as one of the major figures in the Soviet leadership, his role under Stalin as a leader in the Caucasus and his transfer into Moscow in 1926 when he became a candidate member of the Politburo. He was to become a full member in the fateful year of 1935.

From 1926 onward Mikoyan was almost always engaged in tasks having to do with trade and production, domestic and foreign. During World War II it was Mikoyan who held the Politburo responsibility for supplying the Red Army with arms and every other need and for seeing that the civilian population did not freeze or starve. It was not the lightest task on the Soviet agenda. Mikoyan and his deputies had the difficult and almost impossible assignment of trying to bring food to blockaded Leningrad, but it was not Mikoyan's fault that more than a million people died there. When he tried to get Leningrad to store additional food supplies early in the war the Leningrad leader, Zhdanov, rebuffed his efforts. No one in Leningrad or Moscow realized that the great northern capital was heading toward a 900-day siege.

For years Mikoyan labored to modernize and improve Soviet production and particularly the supply of goods in the civilian sector. He was an ardent admirer of American production efficiency and the ingenuity of the American distribution systems. After a tour of the United States in the mid-1930s Mikoyan introduced a number of major innovations, such as modern slaughterhouses and the canned food and vitamin industries, to the Soviet homefront. He brought American automatic baking techniques into Russia and built the centralized baking systems that for a long time have supplied all Russian cities, providing twice-daily deliveries of fresh bread and bread products. The system and the machinery could hardly be improved, although many has been the year in which bread has been in short supply in Russia not because of failure of Mikoyan's machinery but because of breakdowns in the supply of grain and flour. He fell in love with New York's Automats, those restaurant marvels of the 1930s where nickels, dimes and quarters inserted in a slot would buy almost any dinner a customer wanted; just put the money in the slot and the apple pie came out. Mikoyan established several Russian *automats* but they did not suit the Russian market—patrons had to buy tickets, not put kopecks in the slot, wait for clerks to supply the *pirogi*, and like as not the variety of foods available was severely curtailed. He had little more success with breakfast cereals. His Soviet *korn flecks* gathered dust on the shelves of the *gastronoms*, or grocery stores, for years. But his Soviet-style tomato juice was a huge success, and Soviet children and adults still enjoy the ice-cream sandwiches he brought over from America. Soviet eskimo pies have been sold for so many years Russians think they are a Soviet invention.

Mikoyan was, as can be seen, an intensely pragmatic man and one of ready imagination. He would introduce anyone's product if he thought it practical and likely to meet a Soviet need. Given those characteristics it is hardly surprising that in the post-Stalin era Mikoyan came into his greatest prominence, actively aiding in efforts to regenerate his country's economic and cultural life after the long grim years of Stalin. He engaged in extensive trade dealings with foreign industrialists, including those of the United States, and often carried out diplomatic missions of great delicacy both in western nations and within the Communist world.

Mikoyan died at the age of eighty-three in 1978. He had retired from active participation in the Soviet leadership more than ten years earlier. He would have been the last to contend that his career was without blemish. He was brought into the Politburo by Stalin, a fellow Caucasian, and became a full member just as the Stalinist purges of the 1930s got underway. He served in the leadership through the purge period, World War II, the postwar purges, the Cold War, the death of Stalin, the rise and fall of Nikita Khrushchev. There was little, indeed, in the Soviet epoch in which he did not participate, and if he did not participate personally he knew what had happened and who was responsible for what happened. He was a shrewd, intelligent man and nothing escaped his quick eye and his remarkable intellect.

By no means all of what Mikoyan saw, knew, and participated in is to be found in these memoirs which, of course, were published first in the Soviet Union and like every word printed there were carefully scrutinized by the official censorship and, in Mikoyan's case, by the high party leadership. This does not mean that Mikoyan would have written more or differently had there been no censor. He was, after all, preeminently a man of the Bolshevik party, had been since his youth, and was dedicated to its principles and practices.

Thus, what he has written here is a view of the Revolution through the eyes not only of a participant but the eyes of one who was totally loyal to its cause and who through the years supported the Soviet cause to the limit of his abilities. It is precisely because Mikoyan has written such a document and has written it in his own shrewd and clear-headed style that his memoirs are worthy of reading and, more than that, worthy of study as a report from inside the Kremlin by the highest-ranking insider who has ever put pen to paper to construct a full study of his life and times.

> Harrison E. Salisbury
> Taconic, Connecticut
> 1980

Editor's Preface

This first English edition of the memoirs of my father Anastas Mikoyan is a translation of the Russian edition which appeared in Moscow in 1971 under the title *The Path of Struggle*.

It is extremely difficult to translate such books into other languages, for both translator and editor must be well acquainted with the history of the country, the language of those years and the places described. Every sentence of this translation has been carefully checked by me with the Russian edition to exclude the possibility of errors. It is to be regretted that this was not done with earlier translations done in Bulgaria, Hungary, Japan and Italy and just recently in Poland. Possibly the fewest mistakes are to be found in the French translation by Progress Publishers in Moscow, and the Armenian edition done in Erevan. This edition contains a name index. Unfortunately, some editions even omitted the author's foreword which I believe to be an indispensable part of the book.

I would like to express my gratitude to Sphinx Press, Inc. for their competent and thorough handling of the English edition, particularly to the President of Sphinx Press, Martin Azarian, and his wife Margaret, who is Vice President and Editor-in-Chief, who gave so much of their time and patience making my cooperation with them deeply satisfying.

I would also like to thank Governor Averell Harriman, author of the foreword to this English edition, and Harrison Salisbury, author of the preface and footnotes, who showed such an interest in seeing this book published in the United States and who expressed their personal observations about a man they knew well.

It has been my work to edit the translation of my father's book, however, I did not supervise the production, which was carried out in the United States.

When writing of the past the author often spans the years to the present by mentioning what one or another of the persons in the book is doing today and where he lives. It must be borne in mind, however, that that present concerns the autumn of 1970 when the Russian edition was being prepared for publication.

My father was seventy-five years old at that time. In 1965, at the age of seventy, he retired from the post of President and in the summer of 1966 he was no longer a member of the Politburo of the Central Committee of the Communist party—the party's supreme executive body—of which he had been a member for exactly forty years. So it was that for the first time in more than half a century, my father finally had what he was always so acutely short of—free time. And he started working on his memoirs with great enthusiasm.

In those years he still retained his phenomenal memory which astonished those who knew him. Absorbed in archives, dictating to his stenographer in his Kremlin office where he was driven every day from his official suburban residence and checking the typed pages of his future book, he obviously immersed himself with great pleasure in the days of his childhood and youth, in the atmosphere of the turbulent prerevolutionary and still more turbulent revolutionary events, reliving again the joys and disappointments of his younger days. In vain were my attempts to dissuade him from spending so much time rereading documents relating to those years and concentrating on certain events in excessive detail, particularly in the early 1920s (the next volume), and to start on the events that crowded the later years of his life. After all, originally, *The Path of Struggle* was to be the first in a seven or eight volume edition. True, the previous year he published another book called *Thoughts and Recollections of Lenin* and after publication of *The Path of Struggle* he worked for four years on the next volume, *In the Beginning of the Twenties* which came out in 1975. Well aware that he had started his race against time rather late in life, in those years he also published magazine articles of his recollections of events, people and the times, including a series of articles on the Second World War and was working on a book covering the period from 1924 to 1939 (the initial plans were for two to three books on this period). Simultaneously, he requested that I prepare material on his missions abroad in the 1950s and 1960s, checked my work, making necessary corrections and adding his own thoughts, and finally saying that I should be the author of the book he planned on this subject.

A great deal had been accomplished. However, a great deal remains to be done in preparing for publication and arranging all of the legacy of Anastas Mikoyan and this I consider to be my obligation to my father.

                Sergo A. Mikoyan
                Moscow
                October 1983

Author's Preface

I have long observed that the older a man becomes the more compelling is his need to sum up his life in some way in order to transmit his knowledge and experience to succeeding generations. This is as it should be. It is especially true of people whose lives have intersected events of major social and political significance.

For more than half a century I have been a witness to and a participant in many of the major events of our national life. I have had the opportunity to meet many outstanding people. I have visited many of the countries of the world, met with their national leaders, their politicians, their leading cultural figures and other prominent people, and I have met with ordinary, simple people, too. For many years I have worked with the leading figures of our party and state. And it goes without saying that I consider my personal meetings with Lenin, when I had occasion to work under his leadership, the greatest good fortune of my life.

It is not without doubt and trepidation that I take up my pen. I am not a writer. And what I am writing is not an artistic work, but a true story about the events I have witnessed and taken part in and the people I have known and worked with in my life. Most of us were still quite young in those turbulent days. As a rule, during periods of extreme social upheaval young people grow up very quickly. Thus, in Russia, during the Great October Socialist Revolution, it was not unusual for young men and women of fourteen or fifteen to join the revolutionary struggle, gather their strength very quickly, and become mature political workers, even leaders, by the time they were eighteen or twenty.

These memoirs are not a history, for a history should be a thorough and profound scientific analysis of the myriad facts and events that constitute the life of a society. I make no pretensions to such a systematic study, nor do I claim that my description of the events which I touch on here is a complete one: that is not the task I have set myself. This book is a memoir of my personal experience of what I have seen and heard. It doesn't pretend to be anything more.

In those distant and stormy revolutionary years none of us, as a rule, kept diaries or even simple notes of the events, meetings, and conversations that took place. We didn't collect or preserve documents "for posterity"; it somehow never occurred to us. The majority of us did not keep copies of the texts of our speeches, nor even outlines of them. They usually weren't even written down: speeches were not read, but delivered extemporaneously, and often improvised on the spot. Only if the speech were a particularly important one would there be a draft or outline, and then it was almost always destroyed immediately afterwards. In the early days we had no stenographers: the minutes were often taken by poorly educated people who wrote down the unimportant things and ignored what was significant. An hour-long report would be summarized on one small page of the minutes, and not always accurately. In any case, it's difficult to make use of these notes now, even though they are, of course, a help in recalling certain details and facts of past experience.

I remember when I first read Lenin's famous book *The State and Revolution*, I took note of the author's Afterword in which he wrote that after publishing the book, he had already conceived a plan for further chapters, but, aside from the title, he never managed to write a single line because a political crisis—the eve of the October Revolution of 1917—"got in his way." "One can only delight," wrote Lenin, "in 'an obstacle such as that.' " Now, he said, "It's more pleasant and more useful to perform an experiment in revolution than to write about it."

It is, of course, a great pity that we will never be able to see the new chapters of this work of Lenin's, but the mood of which he spoke was very understandable to many of us at that time.

My memoirs, like all memoirs, have an inevitably subjective character. Events are described here as I remembered them, and as I perceived them in the distant past. At the same time I also express some of my current attitudes to them as well.

I look at the events described here with the eyes of a participant or an observer, rather than a historian. That is why I devote more attention to events in which I was personally involved than to others which are perhaps more important from the point of view of history. Events which I did not observe or personally participate in made less of an impression on me, and hence are described only briefly and schematically or have been omitted entirely. In the Transcaucasus there were, of course, many remarkable people, outstanding Communists and politically active nonparty workers. I have mentioned many of them by name, while others have not appeared at all although they most certainly deserve to be mentioned. I simply couldn't keep track of the names of all these people for more than half a century.

As a rule, memoirs, if they are true, are valuable in my opinion mainly because they provide the reader with concrete, historical material that has been seen through the eyes of the people who lived at that time: therein lies their chief usefulness.

Insofar as these are personal memoirs, the author's name figures rather frequently in them, although I have made every effort not to exceed the limits of an objective representation of the truth of life. Moreover, I think that the personal element, the personal experience is not without interest for the general reader, who usually wants to know how events were perceived by the people who took part in them, how they interpreted them, what they thought at the time, and what guided them in their actions. All this helps one to gain a better knowledge and deeper understanding of the events themselves as well as an idea of the atmosphere in which they took place and the moods of the people who experienced them, their hopes and expectations, their joys and sorrows.

Should the reader find this to be true of these memoirs, then I shall feel my goal has been achieved.

PART I

THE FORMATIVE YEARS

Chapter I

Some Childhood Reminiscences

*Family and relatives—School—The move to
Tiflis—Enrolling in the seminary.*

    I have only a few memories of my early childhood, and these are fragmentary. It is a strange thing how certain incidents of those far-off years come back to me while others remain forgotten.
    I clearly recall my native village of Sanain, situated in one of the picturesque corners of Armenia. I see acre upon acre of ploughed fields stretching out on our plateau high above steep cliffs on both sides of the gorge. Beyond these fields there rose forest-covered mountains, and beyond the mountains alpine pastures abounding in wildflowers extended as far as the eye could see. Our village nestled against the mountain slope at the edge of the plateau. I can still see the small houses clinging together against the mountainside.
    During my childhood years Sanain was one of the many backward villages on the outlying borders of Tsarist Russia. Its people were downtrodden, lawless, and uneducated. The only two people in our village who could read and write were the monk and the priest from the local monastery. The rest of the population (including my parents) was totally illiterate, with only a vague idea of what a school was, or teachers or textbooks, to say nothing of newspapers or magazines.
    It was in such a place that I was born and spent my childhood. Our family—the Mikoyans—had lived in Sanain for generations, squeezed into extremely "compact" living quarters. Two families, ours (I was one of five children) and Uncle Gevo's, lived in one small house consisting of two rooms, a cellar, and a porch. In another room, built half below ground level and attached to our house, my paternal Grandmother Vartiter lived along with my other uncle, Vartan. In front of our house on the mountain slope there were two more small houses connected by a flat roof, the space beneath which served us as a kind of courtyard. Here Uncle Grisha and Uncle Velikhan lived, while Uncle

Mkrtich lived just below them on the mountainside, and a little beyond him was my Uncle Gigol's house.

For some reason the following incident sticks in my memory. One day I was alone on the roof playing a game I had made up called "Going to the Mill." I lay on my stomach with my head hanging over the edge while I raked up the dried clay from the roof with my fingers and then let the powdery substance dribble over the edge. I suppose I was pretending to be a real mill turning out flour. I was so absorbed in my game that I didn't notice how close I was to the edge and before I knew it I fell head first off the roof. My face banged against the ground when I landed, smashing my nose. Bawling with pain and fear I ran to my mother, who wiped off my nose and comforted me. Evidently the idea for the mill game had come to me from a visit to a real mill with my father a few days earlier. This water mill, which serviced two of the hamlets on our plateau, was down in a ravine, and since it was a small mill the grinding of a sack of grain entailed a stay of at least twelve hours. My father and I would come back for our grain in the evening and then return home the next morning. There were always a number of peasants sitting around on their sacks of grain waiting for their turn.

Once when I was six or seven I was playing in the yard with my cousins. It was autumn, and towards evening when it became cool we went into Uncle Velikhan's house to warm ourselves at his fireside. One of the children accidentally knocked over a kerosene lamp which hung above the fireplace. It fell on Uncle Grisha's son, spilling kerosene all over him. He screamed with terror and ran around the room enveloped in flames, and then he raced outside into the street with the rest of us at his heels screaming at the top of our lungs. Outside, the wind fanned the flames so that it looked as if a huge fiery torch were racing down the street. My cousin was seriously burned. His skin turned completely black, and since there were no doctors in our village he was dead within a few hours.

This incident made a horrible impression on me, the vision of this living torch that was my cousin remained indelibly printed on my mind.

Every fall my father would go into the woods to lay in a stock of firewood for the winter, loading it on the back of the donkey who was our only means of transport. The chance to sit astride this "steed" was for me the height of bliss, so whenever he took me with him on these expeditions he would let me ride the donkey on our way to the forest. The return journey, however, was another story, for with the donkey's back piled high with brushwood, I had to walk like everybody else.

In another woods farther away wild apple and pear trees grew, and my father and I would gather the fruit from early morning until night, loading our filled baskets on the same hardy donkey. Other times I would help my mother pick cornelian cherries by gleefully climbing up the trees, and we also picked a lot of berries in season. In the fall we would go with our neighbors and relatives to a gorge in the forest where the air was warm and where huge walnut trees grew by the riverside. The older boys would climb the trees with great agility and beat down the nuts with sticks for the children to retrieve among the rocks. At the end of the day the entire harvest of nuts would be divided up among the families. After I had begun attending school in Tiflis, leaving home at the end of each August, I was no longer able to take part in these autumn excursions.

We ate the cherries fresh, of course, but we also dried out a large quantity of them for winter use in a fruit soup that was a local favorite. The wild pears and apples were usually pickled in barrels for the winter, and some of them were dried as well. Much of the wild fruit was used to make fruit vodka.

I remember how my father used to get up at dawn while the rest of us were still in bed. At the window ledge he would find the simple breakfast my mother had prepared for him: bread, cheese, and a bit of vodka. He would pour himself a small glass, have a bite, and then leave for work. I don't ever recall seeing him drunk. Even on visits to friends or at weddings he wouldn't drink more than one or two small glasses of vodka. Neither my elder brother or sister nor my mother drank at all and, of course, neither did the children.

In the fall of 1905 at the start of the Revolution a wave of peasant uprisings spread throughout the villages of our region. The seizure of the manor lands had begun. Even our village was affected, but since our local landowners, the Argutinsky princes, were extremely powerful, our peasants refrained from seizing the land. Shortly afterwards the Tsarist authorities sent out a punitive detachment of Cossack infantry to patrol the countryside. The reprisals had begun.

At first these troops appeared only in neighboring villages, such as Akhpat, where the manor lands were restored to the landlords and many peasant-activists arrested as well. Eventually, however, a detachment of Cossack infantry set up camp on the outskirts of our village. The peasants, especially the women, went out of their way to avoid them, but we village boys were burning with curiosity and we finally worked up the courage to go and see what sort of creatures they were. They had already pitched their tents and were preparing their food when we timidly drew near, but to our surprise the Cossacks neither cursed us nor drove us away. We had no way of talking with them, however, since we couldn't speak Russian. The next day the troops broke camp and left.

I remember an incident that took place in 1906, which reveals the temper of the peasants at that time. The Armenian bishop from Tiflis, who stayed during his visits to our district at a dacha on the monastery grounds, talked an Armenian oil magnate from Baku into donating money for a decent road to our village—across a rocky mountain slope—which would be suitable for phaetons. We, the peasants, having no phaetons, had no use for such a road, but the bishop did. Six or seven thousand rubles—a large sum in those days—were set aside for the project, and the peasants of Sanain began work on the road.

The road was finished that summer, and soon after its completion we learned that "our benefactor," the oil magnate, was arriving by train from Baku the next day. It was proposed that the village provide its finest cart for the eminent guest's drive from the station over the new road so that he could inspect it himself.

The next day the boys of Sanain, along with many other villagers, gathered to await the arrival of our guest at the monastery where he was to be put up. When the cart pulled into the yard we caught our first sight of the "benefactor" and broke out into oh's and ah's of surprise at his appearance. He was a man of great height, no longer young, incredibly fat, and a paralytic besides.

Before the guest could be whisked off to his rooms a peasant stepped forward to speak to the crowd which, it seems, had been waiting for this moment. The villagers had been paid only a part of the wages due them for building the road, he said, and since their attempts to obtain payment had met with no success, he proposed that they tie their "benefactor" to the nearest tree until he sent a telegram to Baku requisitioning the money still due them.

The cart had come to a stop near a spring where an old pear tree grew. The guest tried to speak but he was ignored by the half-dozen or so peasants who with great difficulty lifted him from the cart and proceeded to tie him to the pear tree. Only then was he allowed to speak. He explained that he had acted without malice and knew nothing about the debt to the peasants, that on the contrary, he had thought the men had been fully recompensed and had expected the villagers to receive him with gratitude.

His explanation caused an uproar among the peasants, who did not believe one word of it. Finally the victim proposed that his traveling companion send off a telegram to Baku. He was kept tied up until he signed the telegram, and then his captors finally gave in to his pleas to be allowed to lie down somewhere. They unfastened him and settled him in his room but they didn't release him until the next day when the money arrived. The contractor immediately settled accounts in full with the peasants and the benefactor set out for home post-haste.

The peasants were ecstatic over their victory.

The road served our village for the next sixty years. True, it was seldom used, for when automobiles and buses began to appear in our part of the world the impracticality of the road, which skirted a rocky mountain and had many perilous turns, became apparent. It wasn't until 1966 that a decent asphalt road was built to accommodate the buses and automobiles which transported the workers from the plateau down to the copper smelting factory in the gorge.

I recall another incident which provides an insight into peasant psychology of that time. By law, the peasants were obliged to do road construction work without remuneration, and as a consequence, they avoided it in any way possible. When an announcement was posted one Sunday ordering all able-bodied men to turn out to clear that section of the road extending from our village to the railroad station—a distance of some three kilometers—I asked my father to take me with him. I was only nine years old, but he agreed, and we picked up our shovels and set off.

Even at that young age I was struck by the peasants' attitude toward compulsory work, which they grumbled about and performed lazily and grudgingly. Whenever the supervisor's eyes were turned away for a moment, the rocks on the road remained where they were. It was obvious that they were working under the lash, but at that time I couldn't understand their attitude. After all, they were clearing the road for their own benefit, I reasoned; the "authorities" were rarely seen on it. It was only later that I realized their attitude was a natural reaction to prolonged oppression and reflected the peasants' way of expressing their hatred for a hostile regime.

We village children often played in the yard of the ancient monastery of Sanain, which wasn't far from our house, and occasionally we encountered the monk who lived there. He was a tall, black-haired man with a quiet, sensible manner whom everyone respected, including, of course, the children. Once, when I saw him reading a book I became interested and asked him about it. My interest appealed to him and he began to teach me, little by little, until in a few months I was able to read and write in Armenian on a very rudimentary level.

My father had by now acquired a certain facility in several other languages as a result of some six years spent as an apprentice to a master carpenter in Tiflis, where he came into daily contact with the various nationalities of that large, cosmopolitan city. Naturally, his knowledge didn't extend beyond the most simple, everyday conversation, and was limited to the words and phrases of ordinary life situations. However, he took delight in teaching me to count to a hundred in the five different languages he knew. I memorized the numbers without learning how to use them in any mathematical calculations.

One day at a wedding at Uncle Gigol's I had a chance to show off my proficiency. There were a lot of people there, and everyone was standing because there wasn't any place to sit down. My father hoisted me onto his shoulders so that I could see the relatives and guests, and while I was there he told me to count to a hundred "in all languages." I gladly complied, and my father, of course, was very pleased, while the assembled company gaped with astonishment. They thought it a miracle that a child could count in Armenian, Russian, Georgian,

Azerbaijanian, and Greek. They probably were convinced that I really knew all those languages.

After I had learned to read and write I told my father about my new accomplishment, for till then he hadn't known about my studies with the monk. He wanted to give me a formal education but we had no schools in the village.

Around this time a member of the Armenian intelligentsia arrived unexpectedly in Sanain. It was obvious to everyone that he was not from our parts and that he was, no doubt, hiding out in our remote village to escape persecution by the authorities. I think he was an adherent of the Narodniks. After he was settled in the village he made a proposal to the villagers—among them my father—to open a school. There was an empty house at the monastery which could be turned into a schoolroom and he himself agreed to become the teacher for a very modest fee, the equivalent of a subsistence wage.

My father and two other peasants agreed to donate four rubles apiece, stipulating that for those twelve rubles he was to teach not only their children but all the children of the village. About twelve students enrolled in the school, where we were taught reading, writing, arithmetic, and physical education. He devoted a lot of time to our general deportment, teaching us to take care of our appearance, to be neat, to wash our hands before eating and to rinse our mouths afterwards at the little spring next to our school. He taught us to wear clean clothes, to clean our shoes, and to mend holes in our clothes—in short, he instilled in us the rudiments of personal hygiene. Although he lived alone he himself dressed cleanly and neatly, setting us a good example.

He was always thinking up interesting games. I remember how in spring he would take us out into the fields to play with a ball he had made himself. Then he inspired us with the idea of clearing the monastery yard of rocks and planting it with fruit trees. With him working alongside us we enthusiastically fell to work, filled with dreams of our orchard. But the project did not come to fruition. Our parents were indignant when they found out, by accident, that their children were hauling stones instead of studying. My father, for some reason, was especially outraged, and demanded that this "foolish business" be stopped at once. We were disappointed at the loss of our orchard, but our good, kind teacher was apparently most upset of all, for he quit the village soon afterwards, and the school closed down.

We received no schooling for the entire next year, but my ardent love of books had already taken root.

The following summer an Armenian bishop from the Tiflis diocese came to stay in our village. He resided in the monastery in the same little house that had formerly been our school, but which he decided needed renovating and enlarging for his comfort. My father, who was part of the construction crew, was assigned to building a new tile roof and I became his helper. One of my duties was to hand him up the tiles, first one with my right hand, and then one with my left. Once, on becoming momentarily distracted I handed him a tile from the wrong side. My father had automatically lowered his hand for the tile and on feeling none there he shouted, "Where's the tile?" "Right here!" I called, and then saw that I was handing it to him from the wrong side, whereupon my father lost his temper and hit me over the head.

Up until that moment my father had never struck me. It was only a glancing blow, but I was deeply insulted and vowed that I would never work with him again. I never did, for he, too, must have felt he had been unfair and after that he never forced me to work with him again.

I continued to hang around my father while he worked at the monastery, however, and one day the bishop stopped to talk to me. He asked me whether I knew how to read, and I told him that I could read a little and wanted very much to study but that there was no school in our village. "The country isn't the only place one can go to school," he said, and went on to explain that arrangements for schooling could also be made in Tiflis. Never having envisaged such a possibility I was overjoyed and ran to tell my father what the bishop had said. "Bring your son to Tiflis in the fall," the bishop told my father, who had immediately gone to speak to him about the matter, "and I will arrange for him to study at the seminary." Our happiness knew no bounds.

Late in August of 1906 my father and I left for Tiflis. I was ten years old, going on eleven, and this was my first trip on a train. On our arrival we went to Aunt Verginia, my mother's cousin, who was married to Lazar Tumanian. I remember that their house was down in a ravine and had no access road to it so that my father and I had to use the footpath.

Aunt Verginia lived with her husband and children in a house with five small rooms, and when my father asked her to let me board there while I was at school she told him that she was already housing two children—also relatives—who were at school at Tiflis, and had no more room. We had no alternative but to look elsewhere and my father soon found a place for me at six rubles a month with a woman who lived with her son in one room. The son was a year older than I.

My lodgings were now settled but the main objective of enrolling me at the seminary was yet to be accomplished and my father went about it in a roundabout way. Instead of going directly to the bishop he went to the bishop's cook, who was a fellow villager. Martiros Simonian was a good friend of my father's, and moreover he could read and write, which my father could not do, so instead of presenting ourselves at the front door of the bishop's house we went around to the kitchen. There, after an initial exchange of pleasantries, my father asked Martiros if he could dictate to him a petition for my admission to the seminary. Many years later my comrades from Armenia found in the seminary's archives the original copy of my father's petition, written by Martiros Simonian, and I was delighted to receive the photocopy they sent me of this interesting document.

The following is a translation from the Armenian:

To the Highly Esteemed Trustees of the Armenian
Nersesyan Religious Seminary

                         from Ovanes Mikoyan, a resident
                         of the village of Sanain.

A Humble Petition

I shall not take up the esteemed trustees' valuable time with a long letter, but let me say that I am a father and that I have good intentions in regard to my sons. I am a poor, downtrodden peasant, but even I want my son to be able to read and write, and I sincerely believe that he—my son—will one day become a useful citizen both for himself and for his fellow peasants with the gracious help of the trustees. For that reason I have brought him here with great difficulty, but also with sweet hopes for the future, and I ask that the esteemed trustees settle my son Anastas in some corner of the Nersesyan Seminary; he was born on October 12, 1895. I shall provide a birth certificate for him in the near future. My son knows how to read and write and is, I might add, very capable.

Written by Martiros Simonian on account of Ovanes Mikoyan's illiteracy.

September 11, 1906
Tiflis

I should say, by way of explanation, that my father was not in fact a peasant, but a laborer-carpenter. In those days, however, people were classified on official documents in terms of one of the three estates: noble class, inhabitants of cities and towns, or peasantry, rather than according to class and occupation. Since my grandfather had been a serf and my father lived in the country, he usually listed himself as peasant.

As a result of my father's petition I was granted permission to take the entrance examinations. The seminary offered an eleven-year course of studies; four preparatory years in the lower form, and seven primary years in the upper form. On the basis of the knowledge I demonstrated on the exams I was admitted to the second year in the lower form, and so I began my formal studies.

Having accomplished his mission my father then returned home, and I soon found myself suffering from an acute case of homesickness in this strange city. My landlady treated me fairly well but the warm, sustaining relationship I needed to tide me over this first separation from home and family did not develop between us. Her son and his friends looked down on me as a country bumpkin and were often unkind; once they even beat me up. I put up with everything very patiently at first, but after a while I couldn't bear it any longer and decided to leave. I went to Martiros Simonian and told him all my troubles, persuading him to buy me a ticket home.

When I arrived home I told my parents that I wouldn't return to Tiflis. My mother cried, and my father went around looking very gloomy. He felt that once I had started school I ought to see it through. "Let's go back to Aunt Vergush," he suggested, "and I'll try to persuade her to take you in."

I resisted this idea for some time but my father was determined that I have an education and eventually I myself began to realize how important it was for me to go to school. Finally I gave in, but I stipulated that I wouldn't go anywhere unless my mother came too. Preparations for the journey immediately got under way. My parents packed up some cheese, butter, dried cherries, and homemade preserves, and the three of us set out once again for Tiflis.

I remember my parents' conversation with Aunt Vergush as if it took place yesterday. We were on the porch, seated at the table. My father pleaded with her to take me in. "How can I take him?" she said. "I don't even have a place for him to sleep." But my father persisted, pressing his point so energetically that Aunt Vergush finally capitulated, agreeing to give me a place on the ottoman in the corner of her dining room.

As before, however, I didn't want to stay in Tiflis without my parents. My mother postponed her departure as long as she could but soon she too had to leave. This time, however, it worked out all right. I felt at home at my aunt's house, and I wasn't lonely for very long because the kids at Aunt Vergush's made friends with me right away, even though they were younger than I was. I went back to my classes at the seminary and gradually I came to feel at home there, too, as I began to make friends with my classmates. They responded well to me and soon we had an easy and pleasant relationship.

As I've already mentioned, I had a keen desire to learn and my studies came easily to me. I had no luck with singing, however. They began to give us singing lessons the minute I entered the seminary and since I felt I had a good voice and that I sang well, I thoroughly enjoyed them. But three years later when I was promoted from the lower form to the first class of the upper form, an embarrassing thing happened. The singing teacher, who was new to us, suddenly singled me out of the chorus and asked me to sing alone. Never doubting my success I began to sing loudly and confidently (I had in fact a rather strong voice) but I had barely started when the teacher struck the table with his tuning fork and irritably shouted, "Stop!"

I was dumbstruck, and then, blushing with shame and embarrassment, I took my seat. For the first time in my life I learned that I had no ear for music and sang hopelessly out of tune. It had taken our new teacher—whom we later found out was Romanos Melikian, a well-known musician and composer—only a few minutes to see through my musical "talent."

From that moment on I never opened my mouth again in singing class, even avoiding the teacher's glances. Some weeks later Melikian called on me to take dictation, which I did fairly well, but when he asked me to sing I refused. He began to insist. I then announced that although I would do all the assignments in musical theory I would never sing in class again, and despite Melikian's efforts to have me do so I kept my word. At the end of the term he gave me an F in Music and arranged for me to retake the exam. I received a B on my reexamination in music theory, but once again I refused to sing, and I was rewarded with the barely passing grade of C-minus among my usual A's and B's, but at least I was promoted to the next class. When I graduated from the seminary several years later my diploma showed a grade of C in Music. Such was the inglorious end of my musical career.

Soon afterwards I ran into more serious, even scandalous, trouble in another subject—this time a major subject, not a minor one—namely, Scripture.

During my early childhood the entire family went to church on Sundays. Our church was ancient, dating from the tenth to thirteenth centuries, and its walls, floors, and roof were built entirely of huge, dark, well-hewn stones. I remember that when we had to kneel—I imitated everything my father did—my knees got very cold.

At that time, of course, I never gave any thought to the meaning of the liturgy; doubts as to the existence of God did not occur to me until the second year of the upper form at the seminary, when I had a run-in with the priest who was my Scripture teacher.

He was a stern man who never smiled, and since he didn't treat us with any warmth or allow any joking, he didn't win our affection. He was obviously not a very intelligent man, for he used to repeat the same things over and over: God exists, everything happens by God's command, and God is just in everything. He didn't permit any counterarguments, since we had to believe everything to the letter.

Gradually I began to entertain doubts, and I wasn't alone. "If God is so all powerful and so just, then why doesn't he help people when they're sick or starving? Why do some people have everything (even when they're not the best people and sometimes are the worst) and others have nothing?" The number of "why's" began to increase with the years.

The priest's explanations, namely, that God punishes some people for their sins and blesses others for their good deeds, along with other similar "proofs," didn't convince me at all. I began to argue a lot in class. This annoyed the priest, for the other boys in the class were drawn into the argument, too, and many of them were beginning to take my side. Things went on like this for an entire year. I became such a fierce debater on questions relating to God that my classmates started to call me not Anastas, but "Anastvats," which means atheist in Armenian, a nickname that stayed with me throughout my schooldays.

During these arguments and discussions, and especially in my private meditations, I somehow lost faith in the existence of God, although I was as diligent as the others in studying what we called "Holy Scripture," and the teacher had a hard time finding fault with me. Nevertheless, at the end of the year he gave me an F and arranged for a reexamination.

Since I knew the subject I could answer all the questions on the exam, although—luckily for me—there was no question on it concerning the existence of God. I was given a C and promoted to the next class. This C stayed with me to the end and appeared on my diploma.

Some time later, when I was fifteen or so and considered myself an avowed atheist I had an argument with my mother which I later came to regret terribly.

My mother was a devout believer and accordingly she strictly observed all the prescribed fasts. She made sure that we children had an abundance of dairy products but she would not eat cheese or yogurt or any meat products, sustaining herself solely on bread and potatoes, or kasha in vegetable oil. She worked hard all day and when I saw how weak she was getting I tried to persuade her to eat the same way we did, but she refused. I insisted that she was doing herself harm and damaging her health. "There is no God," I said to her. "And even if he did exist, what would your eating habits have to do with him?"

But all my attempts to get her to eat some dairy products had absolutely no effect on her. "God forbid. I never will do it," she would say. I was dismayed to the point of shouting at her—something I had never done before, since I loved my mother very much. But she would not break her fast, and begged me not to go on pestering her about it. I lost my temper then, and grabbing a plate—we had only a few—I smashed it on the floor and stormed out of the house. When I had cooled off I began to realize what a fool I was. After all, my mother believed in these rituals, I told myself, and her stubbornness in adhering to them was a manifestation of her faith. When I returned home my mother greeted me as if nothing had happened between us, and all was well. After that I never again took issue with my mother about her fasting or her faith in God.

Once, however, my mother got me in trouble with my father over the same situation in reverse—by her piling too much food on my plate. It was summer and the whole family was on the porch having a supper of mashed potatoes. I wasn't very hungry, but despite my protests my mother kept adding spoonfuls of potatoes to my plate. On noticing that I hadn't cleaned my plate my father began scolding me. "Why did you take more than you can eat, if you can't finish it?" he said. "You should never refuse God's gifts like that."

I defended myself by explaining that I couldn't eat any more and that it was Mama who had put all those potatoes on my plate. I don't know what came over my father but he suddenly raised his hand and hit me on the cheek. It didn't hurt very much but the insult was devastating. I jumped up from the table and ran off into the wheat fields until I couldn't run anymore, and then I lay stubbornly hidden there while my family searched everywhere for me. When they drew near my hiding place I heard the voices of my father and mother and my sister and my brother, calling, "Artashes, where are you?" It was dark by now and the thought of spending the night in the fields outside the village frightened me. I was terrified of wolves which were still found in our parts in those days but my father had been unfair and pride kept me from answering their calls.

Finally, to my secret delight, my family discovered me. At first for appearance' sake I put on a show of protestation, pretending I didn't want to go home with them, but we were all delighted that I'd been found and in the end they simply took me by the hand and led me home.

After that my father never raised a hand to me. He had probably been reprimanded by my mother, for although she never said anything to him in my presence I suspected that they had had a major row over this incident. On the whole, though, I had an extremely warm and loving relationship with my father.

I never heard my parents arguing with each other. They rarely raised their voices even when they were scolding us children about something. One time, for instance, when I was seven or eight, my mother was sweeping our hard clay floor with a broom when a neighbor dropped by to ask if he could borrow three rubles. My father gave him the money, asking him to return it as soon as possible. When the neighbor left, my mother spoke angrily to my father. "There you go, giving away our last cent—not knowing whether it'll be returned or not. You'll be the ruin of us!"

She continued to upbraid him and I waited for some sort of outburst from my father, but his response was only a mild rebuke. After pacing up and down the room several times he went over to her and making a gesture as if he were going to pinch her on the cheek he said, "Be quiet, you daughter of Otar! [Otar was her father's name.] This is over your head!"

In those days very few peasants had any money, and if they did they had little to spare. It was a self-sufficient economy, with each peasant supplying his family with produce from his own gardenplot. Each family did its own weaving, sewed its own clothes, and made its own shoes and hats. Money was needed only to pay debts and to buy things such as tea, sugar, cloth, kerosene, and matches. Thus, there were only a few households in our village where they drank tea with sugar, that is, they would hold a lump of sugar in their mouths as they drank.

Our social customs were as archaic as our economy. My Grandfather Nerses was no longer alive when I was born, but I remember my Grandmother Vartiter very well. In Russian Vartiter means rose petal, a name that did not suit her, for she was a big tall woman with a severe-looking face who carried a cane and always wore a long black dress.

I remember being home with my mother once when my grandmother came calling. She knocked on the door with her stick, and then came into the house and asked my mother something. According to ancient custom a married woman did not have the right to speak to a man or woman older than she, so my mother whispered her answer in my ear and I loudly repeated it to my grandmother. Nowadays all of this would seem strange, even ridiculous, but in those days tradition was tradition.

It had always seemed to me that my mother was rather afraid of my grandmother, but perhaps having been raised under the old national customs she merely felt shy in her presence. Everything considered, my Grandmother Vartiter was a good woman even though she was very strict. She had worked very hard all her life and had borne and raised eight children, no mean feat in the days of serfdom. It was probably because she was severe, or seemed so, that we, her grandchildren, weren't very attached to her.

During my childhood, until I left for the seminary in Tiflis, I sometimes attended country weddings which took place usually in the fall. Everyone would gather in one room, and if it was warm they would spill over onto the street. It was all very merry. A *zurna* played and drums were beaten. It was fun for us kids and everyone felt free to do what he pleased.

As a rule the girls gathered on one side and the young men on the other, and then they'd look shyly over at each other. Every fellow had a favorite girl whose attention he would try to attract. A kind of friendly rivalry existed between the young men, but if several of them happened to like the same girl, things sometimes ended badly. The young men usually carried daggers with them, and I remember once seeing an argument flare up between two young men who liked the same girl. This led to a quarrel and then to a fight during which one of them was stabbed. He fell to the ground but he wasn't seriously hurt, and being a sturdy young man, he survived. Such things happened more than once.

It was interesting to observe the start of a wedding in which the bride came from another village. The bridal procession was an entertaining spectacle which we usually watched from the flat roofs of our houses. All the bride's relatives rode with her on horseback to "present" her to the people of the bridegroom's village. The procession was accompanied by the shooting of guns, the playing of the *zurna*, and the beating of drums. Once inside the village the horses were slowed to a walk and the young men who rode at the head of the line organized fights and dances throughout the procession.

Numerous relatives of the bridal couple and people from the village would be invited to the wedding. Guests usually brought a ruble or two, or a gift of equivalent value. This was not obligatory, of course, but it was considered improper to come to a wedding without such a gift. Indeed, weddings were very expensive and this gift was viewed as a kind of compensation to the families of the bride and groom for the money they had spent.

My father was more sophisticated than the ordinary villager of Sanain. After having worked for several years in Tiflis he had begun in many respects to look like the Tiflis craftsmen. He dressed neatly and wore a city hat rather than the usual Caucasian fur cap. He had a broad silver belt and wore city shoes, instead of the homemade shoes the villagers wore.

He had his own idiosyncracies, however. Once a long time ago he had got hold of a silver spoon which he always carried in the inside pocket of his caftan. When he went visiting he would eat with his own spoon, then wipe it off thoroughly and put it back in his pocket. Evidently he was squeamish about using the handmade wooden spoons which everyone used at that time.

Although he was illiterate my father nonetheless carried a notebook and a pencil in his pocket. He would make entries in this notebook with certain "hieroglyphs" which only he could decipher: how many days he had worked, how much the boss owed him, etc.

By the time I was eleven I knew a fair amount of arithmetic, and one day my father said to me out of the blue, "Solve a problem for me." I took out a pencil and some paper and took down the problem my father posed me: "I worked for x number of days, I was offered x amount per day. The boss made x number of deductions. How much should I receive in the final tally?" The problem was a simple one but I wasn't prepared for it and I became flustered and couldn't work it out quickly. My father stared at me and finally said in distress, "Forget it. While you've been figuring it out on paper I've already done it in my head. For the life of me I can't understand how your teacher can earn a living if he can't even teach you to solve the simplest problem."

When I was the youngest in the family my father spent a lot of time with me. But five years later, after my younger sister was born and was able to say a few words, my father stopped paying any attention to me and devoted all his free time to my sister. Five years later my sister had the same thing happen to her when a fifth child was born—my youngest brother Artem. We called him Anushavan at home; when he grew up he became an aircraft designer.[1] My father then devoted all his attention to Artem and seemed to forget about me and my little sister, although he always treated us in a warm, fatherly fashion and we knew he loved us very much.

When my younger brother was five or six my father would converse with him in our presence as if he were speaking to an equal. Usually a man of few words, he would become positively loquacious on these occasions. I have never forgotten the intimate chats he used to have with me when I was very young.

When Artem became a little older my father gave him the unenviable task of tending our two goats. I had done it in my day and an exasperating chore it was: the goats would roam all over the steep, rocky slopes that were covered with shrubs and timber, skipping about so quickly that I couldn't catch up with them. At that time we used to wear handmade shoes (a kind of leather moccasin without stiff soles), which did not last long on the rocky slopes nor did they offer protection against the sharp stones. My feet became sore and my fingers were scraped bloody from climbing among the rocks. Even the leather patches I put under the heel and toe of my moccasins were of little help since they slipped out of place when I walked.

I still can't forget an unpleasant incident that happened to me on account of those goats. I had sat down to rest for a while after having chased them all over the mountainside, and when I next looked around for them the goats were nowhere to be seen. I searched all over to the point of exhaustion but couldn't find them. It wasn't easy to go home that evening without them, not knowing what to say to my parents.

Before I reached our house my mother met me and told me bitterly, but without insulting me, that our goats had got into someone's wheat crops, and that they had been seized and locked up. My father had had to pay a fine of twenty kopecks for damages, a rather considerable sum in those days in our village. He said nothing to me but I felt very bad about it and couldn't forgive myself for a long time. So when the responsibilities of goat herding were transferred to my younger brother's shoulders I was vastly relieved, although I couldn't help but sympathize with him.

In any case I was beginning to have entirely new interests by that

[1] See Appendix.

time. I was reading a great deal and was beginning to understand a host of things as well as, and sometimes better than, my father did. He could see that himself.

It was interesting that when I came home from Tiflis for summer vacations my father could not bring himself to ask whether I had been promoted or what my grades had been. For some reason he was embarrassed to ask me about such things, and I never said anything to him. My relations with my mother were much simpler. I would tell her everything about my schooldays and my studies, and she would be delighted.

One night when we were all lying on our ottoman (my elder brother, who was working at the factory, slept on a separate wooden bed), my father, thinking everyone was asleep, asked my mother about my progress in school. My mother sang my praises, telling him I was a good student and that I'd been promoted. This news pleased my father very much and the next morning he was more openly affectionate than usual toward me. But another time when we were all abed, my father said to my mother, "You know, I think we should take Artashes out of school and put him to work in the factory."

My mother was astonished and upset. "But he's such a good student," she objected. "Let him continue his studies and become a man."

To which my father replied, "I'm afraid that in the end he'll go crazy," and he went on to tell her about a prince who lived not far from us who had gone crazy because he read books all the time. "That's just like our Artashes," he said. "He reads from morning to night and in the end I'm afraid he'll go crazy, too, from all those books."

Their argument continued far into the night but I fell asleep before I could find out how it ended.

The next morning my father said to me, "Artashes, what would you say if I were to arrange for you to work in the factory?"

I said that I'd gladly work at the factory during my summer vacations to help my family out, but that I refused to give up my schooling. The reason I spoke so boldly to my father was that by this time my parents weren't spending a ruble on my education. I was a good student and my high marks qualified me for financial aid from an Armenian charitable society which paid the expenses of needy students. They provided me with free meals and paid for the room I shared with three other students. I also earned some money myself by tutoring well-to-do classmates who had fallen behind in their studies, making anywhere from three to nine rubles a month. In those days that wasn't bad if one's needs were modest.

My father didn't argue with me but he took me at my word and decided to arrange for me to work at the factory during my vacation, evidently hoping that a few months there would accustom me to the environment and lead to my permanent employment. His hopes were doomed, however, for when he took me to the copper smelting factory and asked his acquaintance in the personnel office to find me a job, he was turned down. "Even if I had a vacancy at the factory I still wouldn't hire your son," his friend said. "Look at him: what kind of physical labor do you think I could entrust to him?" In those days I was a thin, puny fellow who looked much younger than my age, and I didn't have the appearance of a worker. There was nothing my father could say.

I soon found another way of making some extra money to surprise my mother with, however, for on learning that some of the village kids were picking raspberries in the woods and selling them at the factory to the engineers' families, I decided to profit from their example. The next morning I left the house before daybreak and headed for the edge of a narrow gorge in the mountains where there was a large patch of raspberries. Bears, as everyone knows, are great fanciers of raspberries and I confess I was shuddering with fear at the thought of meeting one, but I managed to stay long enough to fill a basket with the fragrant forest berries. I sold the lot to the engineers' wives for a total of forty kopecks, which was more than my elder brother—a hammerman—made as a daily wage. My mother was very pleased to have the money and I was pleased with myself for being able to contribute to the family larder. I went raspberry picking four or five times until the crop was gone, making between thirty and forty kopecks each time, which I gave to my mother.

That same summer my second cousin, Ashkhen, Aunt Verginia's daughter, came to stay with us while I helped her prepare for her reexaminations. Normally a good student, Ashkhen had done poorly that term. Her teacher had insulted her by reprimanding her coarsely in front of everyone, and since Ashkhen was very proud she had stopped studying, which had led to new reprimands and poor marks. She had failed in four subjects and would be expelled unless she passed reexaminations in these subjects.

I was glad to accede to Aunt Verginia's plea to help Ashkhen, and it so happened that we had an extra room—of sorts—in which to accommodate her. My father had been in the midst of completing an addition to our house when the owner of the plot of land on which our house stood announced that since the land wasn't ours, we had no right to build on it, and he forbade my father to finish the room. Thus one side of the house remained unfinished, but since it was summertime it was possible to live in it. We were grateful that the owner hadn't demanded we take it down altogether.

This room became Ashkhen's summer abode. I lived in the loft, studying and sleeping there, while the rest of the family lived below.

In the morning Ashkhen would do the assignments I gave her and in the afternoon I studied with her for two or three hours. I was very strict with her, avoiding conversations on extraneous topics while

Ashkhen on her part was diligent in studying and doing the assignments. For relaxation from her studies she would usually go into the fields or woods with the girls from the village to gather flowers or edible plants.

My father, I should add, liked Ashkhen very much, conversing freely with her and treating her with politeness and kind attention. Once, though, he inadvertently hurt her feelings over a matter concerning our two goats. In the summer we made good use of the goats' milk, which we drank and made cheese and yogurt out of, but during the winter we boarded the goats in the neighboring village of Akner with a peasant who kept them for us for a fixed amount, returning them to us each spring.

This year, however, the peasant was late in returning the goats and we had begun to feel worried. My father therefore appealed to Ashkhen to write to the goat breeder and ask him to bring them back. She wrote out a note and then read it back to my father. "Dear Mr. Goat-Tender . . . ," she began, but she got no further, for the solemn-sounding title she had given the poor peasant so tickled my father that he couldn't stop laughing, repeating over and over, "Dear Mr. Goat-Tender!"

"Then let your son do it," Ashkhen said in hurt tones, and ran out of the room, but for some reason my father didn't trust me, her tutor, for such a task.

Chapter II

My Revolutionary Youth

Attraction to books—Student strike—Organizing a Marxist circle
First acquaintance with the works of Lenin
Under police surveillance—Journey to the front.

My passion for books began at an early age, and I read whatever I could get my hands on. Until I learned Russian, however, I read only Armenian books during my first years at the seminary in Tiflis.

The story of the Armenian nation's struggle against foreign oppressors fascinated me, and the historical novels of the Armenian writer Raffi—*David-Bek, Samuel,* and others—made a deep impression on me. Shaumian has called Raffi one of the most beloved of Armenian novelists.

I also delved enthusiastically into the works of one of the classic Armenian writers, Hovanes Tumanian, my renowned compatriot whom I later had the good fortune to know personally. He was a great artist whose words vividly depicted the life of the peasants and the landscape of his native land. I was especially drawn to his descriptions of the people and places of my own native region of Lori and I liked his use of the many words and expressions in our Lori dialect which, rather than spoiling the Armenian literary language, enriched it.

I became acquainted with the works of Shirvanzade, Paronian, and others. I read the poetry of the old revolutionary Akop Akopian, who had begun the tradition of Armenian proletarian poetry, and I was greatly excited by the manly verses of Sushanick Kurginian, who celebrated the heroism of the revolutionaries and fiercely condemned the Tsarist system. I was very fond at that time of the poems of the popular Italian poetess, Ada Negri—in Kurginian's translation—whose early works were markedly progressive.

Later, I don't remember under whose influence, I developed a great interest in natural history and science. By that time I was able to read Russian books without difficulty. At first I read popular brochures on natural history; I was particularly interested in Timiriazev's

The Life of Plants and Darwin's *Origin of the Species* and *The Origin of Man and Sexual Selection*. These books marked a real turning point in my conception of the origin and evolution of life on earth, laying the foundation for my consciously formulated atheism. I also began Palladine's *Physiology of Plants*, but it didn't hold my interest.

Next came my fascination with chemistry. It began in school, and then when I discovered Mendeleev I resolved to take his *Foundations of Chemistry* home for the summer and make a thorough study of it. I managed to get through the first 200 pages with some difficulty but the ensuing pages demanded a substantial knowledge of higher mathematics, which I lacked. This really upset me. When one of the seminary teachers who was spending the summer in our village saw me with the book he was surprised and tried to dissuade me from keeping on with it. He explained that it was meant for more advanced students and assured me that no matter how I tried I wouldn't be able to understand it, and he was right.

The social milieu in which I had been raised obviously had an influence on me. Our village was two kilometers from the Alaverdi Copper Chemical Works, the oldest of the copper smelting and copper electrolyte factories still in use in the country. In 1970 it was 200 years old.

In 1801, when the region in which the factory was located passed from Persian to Russian control, the old method of smelting copper was modernized by specialists from the Urals, and then, in 1888, the factory and mine were sold on a concession basis to a French joint stock company. The factory's maximum annual output of copper before the Revolution was 3,800 tons, which comprised one-quarter of all copper smelted in Russia.

My father worked at the factory as a carpenter and my older brother as a hammerman. At one time our whole family had lived in one room in the factory barracks. Other rooms in the barracks slept five to six people on bunk beds. Even as a child I saw how badly off the factory workers were, and how exhausted they were when they returned home after a grueling twelve-hour day.

When I returned to the village on summer vacations after going away to school, I would visit the factory and get to know the workers. On one of these visits I went down into the mine for the first time in my life, to the gallery where the copper ore was extracted. What I saw there made an indelible impression on me. Nor can I forget the sight of my brother climbing the steep hill to the village after work. He was so weak when he got home that he would literally sink down on his cot just to catch his breath. But he was up again at dawn and on his way to work.

About 3,000 people worked at the factory in what were unbelievably bad conditions: approximately 1,000 of them were foreigners —drifters, emigrés from northern Iran who were exploited in the most menial jobs. The miners were, as a rule, Greeks who lived in the countryside near the mine. The work force was multinational, but there was no ethnic strife.

The luxurious living standard of the upper-level management contrasted sharply with the workers' poverty-stricken existence. The former lived in good houses with tennis courts (the first I ever saw), and their wives used to go horseback riding on fine mounts, looking like Amazons. The engineers and upper-level administrators—about fifteen men—were for the most part French, while Armenians, Russians, and other nationalities made up the clerical staff. The overwhelming majority of workers were illiterate.

I told my friends at the seminary everything that I saw there, and I sought explanations for conditions at the factory in books, particularly those on revolutionary history.

In our class there was a student by the name of Eremian. He was two years older than I, had read a lot, and was generally considered to be one of the best-read students at the school, although his marks were quite average. One day I noticed he had a volume of Mommsen's *History of Rome*, and I became interested in it. After talking with Eremian I realized how little history I knew and what an interesting subject it was. At first I decided to read all of Mommsen at once, but then I realized that if I had to read several volumes on Roman history alone, I'd never find enough time to read everything else—particularly modern history, which interested me most of all at that time. When I read Jean Jouresse's *History of the Great French Revolution* I was enthralled by it. I copied the most important facts and dates into separate notebooks, and read over my notes until I learned them by heart.

I also read books about the English bourgeois revolution, but somehow they didn't make as strong an impression on me. It was the French Revolution that imprinted itself indelibly on my mind, because of its vivid events and the colorful figures of its leaders and guiding lights—Marat, Robespierre, and others. I was literally delirious over them.

When the first volume of Dmitri Pisarev's collected works came my way around this time, I gulped down the four remaining volumes that were available. I must say that Pisarev played a prime role in my overall education, in the development of my character and the formation of my world view. Reading his works clarified things for me; I felt I had been cleansed of many of my old prejudices and lifted out of my intellectual rut. I acquired confidence in myself as a result and felt I was beginning to develop a critical attitude toward reality. In a word, Pisarev forced me to grow up.

After Pisarev I began reading Belinsky and Dobroliubov. They stimulated a keen interest in the classics of Russian literature and helped me to overcome the prejudice toward fiction I then held. Up until then I preferred historical works because I felt only they reflected real facts, real events, and real people, while everything in fiction was contrived, divorced from actuality and real life, the fruit alone of the writer's fantasy. I was even more convinced of this after reading books that focused on the problems of love.

Belinsky and Dobroliubov made me realize what a mistake I had made. They opened my eyes to the lofty social purpose of imaginative literature, and I realized that although works of fiction do not always contain reflections of historical events and personages, they are nevertheless based on facts and characters from real life which have been generalized and synthesized by the writer's creative fantasy.

I greedily began the novels of Turgenev and Goncharov, devouring *On the Eve*, *Rudin*, and *Oblomov*. Then came Tolstoy. *Resurrection*, *Hadji Murad*, and *Master and Man* made an enormous impression on me, but I didn't understand the *Kreutzer Sonata*. When I reread it as an adult I realized that I had simply been too inexperienced to understand it.

I remember how excited I was over Voinich's *The Gadfly*. After finishing it I became acquainted with Chernyshevsky's novel, *What Is To Be Done?* Chernyshevsky led me to Fourrier, Thomas More, Saint-Simon, and Robert Owens. Of course, my understanding of much of what they wrote was rather naive and over-simplified, but I found the great Utopians of compelling interest. Everything in their books seemed new and interesting but most importantly they provided answers to questions that were disturbing me at that time.

Finally, I became acquainted with Dostoevsky, but he made me feel gloomy and depressed. Veresaev's *The Notes of a Doctor*, however, interested me greatly. I was also enthusiastic about Garshin. At that time, too, I liked Pushkin best not for his immortal poetry, but for his prose. I was intoxicated by *The History of the Pugachev Rebellion*, *The Captain's Daughter*, and *Dubrovsky*.

Turning now to foreign literature, I read Dickens, Jack London, Victor Hugo, Alexander Dumas, Ibsen, and I still remember Schiller's *The Robbers* very well. Fortunately our Russian teacher, Kapanakian, was progressively inclined and encouraged my interest in the classics to a certain degree.

All the while, my studies at the seminary continued, and as before, I spent summers in the village with my parents where I read away the days, returning to Tiflis each fall.

I remember when we began to have seminary classes in a beautiful, new, rose-colored tufa-stone building. It was built in 1912 and financed by Mantashev, the Baku oil millionaire. The building on the outskirts of Tiflis was on an elevation which offered a splendid view of the city and had large, bright lecture halls and excellently equipped laboratories. The landscaping around the building was done by the students themselves, who planted the saplings which later grew into a beautiful park. When the war came, however, this building, like so many others, was turned into a field hospital, so we had to go back to the old, poorly equipped seminary in the dusty center of Tiflis.

* * *

I recall a certain telling episode that took place in the spring shortly before the war. One of our teachers at the seminary left in the middle of the school year and a former teacher from the Armenian Seminary in Shusha was sent to replace him. His name was Akhnazarian and from the very beginning he incurred our hostility by his rude and arbitrary manner. At first we put up with him without a murmur but our patience began to wear thin until finally one day we rebelled.

Several members of the class stood up and told him that they had had enough of his mockeries and his rudeness and that they didn't plan to attend his classes in the future. At this he immediately left the classroom and went straight to the rector, who, as it happened, wasn't in his office. In a little while a teacher we respected came in to talk to us. He questioned us on the reasons behind our "rebellion" and demanded that we obey the rules, since, as he put it, "teachers are not chosen by the students, but assigned to their jobs by the administration."

We heard him out, but made it clear that although we agreed with him in principle, we refused to be Akhnazarian's pupils any longer and wanted him replaced. The teacher ignored our protests, however, and when he walked out we discussed the situation further and decided to declare the whole class on strike until Akhnazarian was replaced.

We agreed to meet the next day in the Tiflis Botanical Gardens, and at that meeting we took a vote on the question of whether our actions were justified. Everyone agreed that we were in the right and we resolved to stand firm until the directors of the seminary satisfied our demands. Things went on like this for several days. We met each day in the Botanical Gardens for mutual support, to escape parental pressure, and to give encouragement and assistance to the fainter-hearted among us (and there were such).

With each passing day, however, parental pressure grew stronger and the situation became more complicated. Some of the fathers were summoned to the rector's office and told that if their children didn't return to class they would be expelled. Nevertheless, we decided to stand our ground and obtained everyone's promise not to return to the seminary unless everyone agreed on that plan of action.

More than a week went by. As parental pressure intensified, several students were ready to capitulate. Then, one Sunday (we didn't meet on Sundays) when I was out walking, I happened to bump into Andreasian, an underclassman. He asked me if I knew that the strike was over. He had heard the news from Alikhanian, one of the seminary students. That same Sunday morning all the strikers had been summoned to the bishop, who had persuaded them to end the strike and return to classes on Monday. The teacher had been given a reprimand, the bishop said, and would conduct himself properly in the future.

It should be noted that Eremian and I had been the main leaders of the strike, although he had been among the first to break down and exhort the others to return to class on Monday. In any case, I couldn't understand why he had been summoned to the bishop's with the others while I had not.

The next day there was nothing for me to do but to return to class with the others. I wasn't particularly annoyed by the behavior of most of my fellow students; one could more or less understand why they had acted as they did, but Eremian's response bothered me. I went up to him and told him very abruptly that I considered him a traitor and that he was no longer any friend of mine.

To be fair, however, I must add that the teacher who had been hostile to us became "as gentle as a lamb." With time the rift between Eremian and myself began to heal and we became close friends again.

* * *

In 1911–1912 the revolutionary wave which had started to swell in 1910 with the beginnings of the workers' movement surged over the walls of our seminary. I remember the conversations that took place among the upperclassmen about the various political parties. The same talk was heard among my peers. We argued with each other, but finally agreed to make a thorough study of the basic revolutionary literature so that we wouldn't make a mistake in choosing what party to belong to. To this end we formed a political circle and began with a study of Kautsky's book, *The Economic Teachings of Karl Marx*.

There were six of us in the group. We knew and trusted each other and had no desire to let anyone else into our select group; it seemed to us more "orthodox" not to widen our circle.

We met regularly throughout the 1912–1913 school year. We went through Kautsky's book chapter by chapter, discussing everyone's interpretation of it as we went along, a method that facilitated our assimilation of a rather complicated work. The next year our circle widened to include fourteen people. We accepted a small number of lowerclassmen as well as several students, including their organizer Eremian, from a defunct literary circle. Together we reviewed the study of Marx's economic theory, doing so with the help of Bogdanov's book. Meanwhile, we "elders" of the group read other Marxist literature —Bebel's *Woman and Socialism*, and Plekhanov's *On the Question of the Development of a Monistic View of History*.

Eventually we learned that circles similar to ours had been organized in several Tiflis high schools (gymnasia); in the School of Commerce, where students of various nationalities were enrolled in courses taught in Russian; and also in the Gymnasium of the Nobility, where the students were predominantly of Georgian extraction.

We decided to establish contacts with these groups and in the spring of 1914 the first conference of representatives from these various circles was held. Each group sent one representative. The conference took place in the apartment of Liusia Lisinova, a student at the gymnasium, who later became a member of the Moscow Student Commune, and who perished heroically on the barricades in Moscow during the October Revolution. She is buried near the walls of the Kremlin.

An intelligent, well-read young woman, with a good knowledge of theory and compelling good looks, Liusia played a very active role in the conference, and I liked her very much. Later, during that summer of 1914 when I was home in the village, Liusia arrived with her whole family, totally unexpectedly, to spend the vacation. We were usually busy reading during the day but in the evenings we would get together along with her girlfriend and another student from the gymnasium for friendly talks and lively games of croquet (before meeting them I'd had no idea what the game was). It became customary for me to discuss my reading with Liusia and to exchange opinions about political questions of mutual interest. I learned that Liusia had been the favorite of Elena Dmitrievna Stasova when she had taught in Tiflis and that it was under her influence that she'd become a revolutionary.

Another young woman at that conference also had an interesting fate. Her name was Zhenya and she, too, was a gymnasium student. Zhenya became a Bolshevik under the influence of the old Communist, Bograt Borian. After graduation she moved to Baku where she married Borian's son, who died as one of the twenty-six Baku commissars. Zhenya Borian lived in Moscow and died in 1970. These two girls from the conference stand out in my memory, although there were many other interesting young men and women as well.

During that same spring of 1914 I was given a book called *The Development of Capitalism in Russia*, written by a certain Ilin. It was given me by a distant relative of mine, Danush Shaverdian, who was a lawyer in Tiflis (and, I found out later, a member of an underground Bolshevik organization). I had stopped by his apartment one day to renew my acquaintance with him, and while telling him about my life I happened to mention our study group. He already knew there was an underground political circle at the seminary and heartily approved. As I was leaving he handed me a book.

"Here, Anastas," he said, "read this carefully. It's a very useful book. And come and see me as often as you like. I'll be glad to give you advice on studies for your group, and I'll help you get books that aren't in the public library. But be careful with them!"

He then explained to me that the author's real name was Ulianov-Lenin, the theoretician and leader of the Social Democrats (the Bolsheviks), who sometimes wrote under the pseudonym, N. Ilin, for conspiratorial purposes. Thus, thanks mainly to Shaverdian, I discovered Lenin.

I must say that Danush Shaverdian played a rather significant role in the early stages of my party-oriented political life. But more about that later.

Marxism infiltrated our groups via Shaverdian as well as through Stasova and Borian.

A fellow comrade from the underground circle at the seminary and I made an exhaustive study of Lenin's book. I remember how struck we were by his painstaking and elaborate attention to statistical data that reflected major shifts in the Russian economy, and hence confirmed the inevitability of the rise of capitalism in Russia and its subsequent downfall.

During that same spring of 1914, after classes at the seminary had ended, I became involved in an extraordinary brush with the law. To be completely honest I had thought there was no record of the incident but friends of mine accidentally uncovered a reference to it in the archives. Hence I want to relate what actually happened in some detail.

At that time I was sharing a room with two fellow students from the seminary. One day I happened to be delayed at the house of a friend who lived on the outskirts of Tiflis. Suddenly one of my roommates arrived, excited and out of breath from running all the way, to say that the police had come looking for me. They had searched the apartment, confiscating some of my books and notebooks, and on learning where I was they had set out after me, apparently with the intention of arresting me.

"I waited until they left, then ran here the back way to warn you," he concluded. And, in fact, from the window we saw two police officers hurrying down the street. Naturally I didn't wait for them to get there, but ran out the back door and hid in an alley. Fearing a stakeout I spent the night at a friend's house and the next day I bought a railroad ticket and went to the country.

A curious document has been preserved in the archives, an order signed on May 22, 1914, by Colonel Pastriusin, Chief of the Tiflis District Police, mandating "a most meticulous and thorough search" of my house and property.[2]

The search of my living quarters was carried out. Here is the official police report concerning it:

---

[2]See Appendix.

In the pocket of the suspect Mikoyan's trousers were found two bits of a letter written in Russian and German, and six notations in Armenian and Russian, one of which was addressed to Levan Anvazian and signed: "A. Mikoyants, April 18" and read as follows; "Levan, I ask that the next collective meeting be held at my place."

I should explain that "collective meetings" was the name given to meetings of representatives of the student Marxist groups in the various schools in the city. Anvazian was the leader of such an illegal group at the Tiflis Academy of Commerce.

Thus, although they came brandishing a heavy scythe, the police reaped only a paltry "harvest" at my apartment.

I never understood what made the police decide to carry out the search and, more puzzling still, why they intended to arrest me. They had insufficient cause. I hadn't even joined the party yet. I suppose someone had denounced our circle and accused me of reading illegal literature of some sort. Finding nothing very incriminating in my apartment the police apparently backed off.

* * *

When the Imperialist War began in August 1914 I was home on vacation after completing my fifth year at the seminary. In September the new academic year began as usual, but events at the front were disturbing to us all.

The Armenians had a huge stake in the war on the Turkish front, for the western Armenians' hopes for liberation from the Turkish yoke, which affected some two million people, were pinned on a Russian victory over the Turks. Armenian volunteer units were being formed and sent to the Turkish front in support of the Russian army and the Western Armenian Liberation movement.

To go or not to go to the front was the burning issue among the seminary students. Nationalist-minded teachers, who lauded the young people's volunteer movement, nevertheless came out against our quitting school to join the army. They argued that the Armenians were in need of educated people and since there were too few of these, we ought to continue our education in order to become genuinely useful to our nation.

We, on our part, looked on the Armenian volunteer units as an opportunity to take part in the national liberation movement. We cited the example of the Bulgarians, who fought with the Russians against Turkey in their struggle for national liberation, and we also called attention to the exploits of Byron, the famous English poet, who courageously supported the Greek armies in their wars of liberation from the Turkish sultan, and who, in spite of his lameness, went to the front as a volunteer.

In short, the more the teachers tried to dissuade us from going to the front, the more resolved we were to do so. It's true we were afraid that at the last minute the seminary administration, in league with our parents and relatives, would stop us from going to the front because we were underage. That's why we kept our plans from our families.

We signed up as volunteers on a day in November 1914 without saying a word to our parents, and boarded a troop train at Navtlug Station for the border city of Dzhulfa. On the eve of our departure the father of one of my close friends, Aram Shakhgaldian, who lived nearby in Nakhichevan arrived unexpectedly at the camp and took his son home by force. But that was the only incident of its kind. We spent a week at Dzhulfa undergoing a minimal amount of training, and then we were sent to the front as reinforcements for Andranik's troops, who were deployed in Persian territory near the Turkish border.

A middle-aged man, Andranik, had already won fame for his military exploits. His heroism in the war of the Armenian partisans against the Turkish sultan and in the Bulgarian people's liberation struggle was the subject of legend. Bulgarian medals of the highest orders decorated his chest, and among the Armenians an aura of glory enveloped the name of Andranik. Later we saw that his authority was unquestioned among the troops under his command.

It's interesting to note that Andranik had been a member of the Dashnak party for a brief period up until 1907, but he later broke with it and was considered an independent patriot and national hero.[3]

Our arrival at the front coincided with an offensive that was underway in that sector. The surrounding locale was mountainous and lacked roads. Up in the mountains there was snow on the ground. The enemy offered little resistance and our offensive was very successful for several days running. We occupied a series of tiny villages, consisting on the average of ten to fifteen dwellings situated in ravines. The houses were empty. The population had already been evacuated to the rear, leaving everything behind. Overwhelming poverty was in evidence everywhere.

Later we received orders to begin a retreat, although the enemy had not yet forced us to do so. It was said that farther to the south Turkish troops had begun a counterattack. Unpursued by the enemy we fell back slowly to the Russian border. Our regiments halted at Dzhulfa.

My comrades with whom I had arrived at the front decided to return home, complaining of the campaign's difficulty and of their fatigue. The commanders had nothing against their leaving but my

[3]See Appendix.

pride kept me from joining them. I considered it childish to leave after being at the front for less than a month and accomplishing nothing.

After a short rest our regiment marched into Persian Azerbaijan and the area around Khoya. The regiment consisted almost entirely of workers from Baku. They were good, simple people, and they treated me with respect even though I was the youngest person there, being barely nineteen. They looked on me as a member of the intelligentsia, but appreciated the fact that, like any common soldier, I endured the discomforts of military life.

Our regiment soon became involved in a full-scale battle with the Turkish army on Persian territory. In our sector of the front three Armenian regiments saw action together with the Russian divisions. We assumed advantageous defensive positions, entrenching ourselves in the high hills. The Turks had taken up positions in the ravine.

A fierce battle began at early morning. The attacking Turkish soldiers fell under our barrage as we fired at them with rifles from our trenches. In spite of heavy losses, the Turkish forces continued the attack, moving forward in formation. Their artillery bombarded our positions, but for the most part the shells exploded behind our trenches. We incurred significant losses, however, especially from shrapnel, but they couldn't compare with the number of bodies that lay in front of us. By evening the Turkish lines had reached the foothills and halted some hundred meters from our trenches. The firing ceased. Darkness fell and everything became quiet.

A while later, after we had taken advantage of the lull to have a bite to eat, the order was given to prepare for a nighttime bayonet battle: the Russian command suspected that the Turks would not wait till sunrise, but would use the darkness to launch a bayonet attack.

During my first taste of fire I had managed to adjust somewhat to the conditions of war, but now we were facing the possibility of hand-to-hand combat. Like the other soldiers, my morale was high and I didn't feel any particular fear, but I wasn't prepared psychologically for hand-to-hand combat, not having been trained for it. I told my comrades that I had difficulty handling a bayonet and asked one of them to loan me his revolver for the attack. To be completely honest, I was sickened by the thought of sticking a bayonet into a human body. A rifle or a revolver, however—that was a different thing. Once I had a revolver, I calmed down.

We were ordered to sleep in the trenches in shifts so as not to be taken by surprise. I guess my comrades must have taken pity on me because they didn't wake me up until sunrise. There wasn't a Turk to be seen. They had only feigned preparation for a night attack, retreating instead under cover of darkness and carrying off their wounded with them. Only corpses remained on the battlefield.

We descended the hill and advanced through a battlefield strewn with the dead in various poses. I could never have imagined such a sight. As we made our way between the corpses some soldiers frisked the pockets of the slain officers in search of valuables. That really upset me.

That was the biggest battle in which I had so far taken part. After sustaining heavy losses, the Turks began a general retreat, and we soon found ourselves in Turkish territory, but the fighting was relatively light.

The second major battle took place when we were approaching Van, one of the ancient captitals of Armenia. This seemed less of a bloodbath than the first, but the Turks suffered a serious defeat here as well.

By this time my physical condition had begun to deteriorate drastically. The fact is that I had never eaten meat before, and even a tiny piece of meat was enough to make my skin break out in a rash. It wasn't until 1918, when I was in Baku, that I gradually accustomed myself to a meat diet. Before I went to the front, I had eaten dairy products in place of meat, especially cheese, which I usually ate three times a day along with bread. At the front it was impossible to follow such a diet. I lived on bread and kasha, became terribly thin, and began to display symptoms of malnutrition. The long marches finally succeeded in undermining my health. And then a new misfortune struck: as we were approaching Van I fell ill with an acute form of malaria. I had a fever and ran a very high temperature, but it broke in a day's time.

In the beginning of April 1915, a few days after we had entered Van, the sick and wounded, myself included, were evacuated, first to Erevan and then to Tiflis, where we were put in a hospital for wounded Armenian veterans. The city of Van had made a pleasant impression on me because of its tidy little one-story houses, which were surrounded by greenery and orchards. Not far from the city was the deep mirror of Lake Van, flanked on three sides by mountains.

In the hospital they filled me so full of quinine that I almost went deaf. I was weak with continual attacks of dizziness and fainting fits. My malarial attacks abated, however, and eventually stopped entirely. By the middle of May I was released from the hospital and felt completely fit.

When I thought over my experience at the front I realized that, except for a certain maturing and toughening, I had got very little out of it. I recalled one incident during one of the spring marches when we had been caught in a downpour that kept on incessantly. When we finally stopped for the night we were exhausted and soaked to the skin. We dreamed about lighting a fire, warming ourselves, and drying out, if only a bit, but there was no wood or kindling. We wanted to sleep more than anything in the world.

We pitched our tents on the wet grass, covered ourselves with our wet overcoats, and fell asleep. I was certain that I would catch pneumonia but we awoke to bright sunlight and immediately stripped off our soaked uniforms to dry out and get warm. None of us even caught a head cold.

From my readings I had concluded that a man doesn't know his true worth until he has experienced danger. I felt now that I had tested myself in adversity and had survived the test with dignity, which gave me a sense of real satisfaction.

While I was still in the hospital I was overwhelmed by the news of the tragedy that befell the western Armenians in Turkey. The reactionary sultanate had carried out fiendish reprisals by annihilating about a million and a half Armenian civilians. I tried to sort out what had happened, and feeling need of the advice and counsel of a competent person, I went to see Shaverdian, whom I looked upon as my preceptor. He greeted me cordially but asked why I hadn't sought his advice before going to the front. I replied that although the thought had crossed my mind, I was afraid that he, like all the other adults, would try to talk me out of it.

During our conversation he described in detail Lenin's views on the war and the Bolsheviks' tactics in regard to it. Later on that fall he filled me in on the resolutions passed by the conference of the Caucasian organizations (RSDRP [6]), which had taken place in October 1915 in Baku, under Shaumian's leadership. These resolutions had been prepared by Shaumian, in conjunction with a Leninist evaluation of the Imperialist War and the Bolsheviks' attitude toward it. They gave indications of the Caucasian Bolsheviks' disapproval of the national units which were being organized at that time.

He took from a drawer a copy of the newspaper *The Social Democrat*, published in Geneva. It was printed on thin cigarette paper and contained Lenin's article, "The War and Russian Social Democracy." This was no mere article, it was the statement of the Bolsheviks' official political platform on the war.

I asked Shaverdian for the newspaper and for a copy of the *Platform and Rules of the Party*. He gave them to me, along with an unexpected gift of Lenin's pamphlet, *What Is to Be Done?* I immersed myself avidly in a study of these precious documents. No other political books had made such a strong impression on me. This was understandable, for in Lenin's works I found answers to many of the questions that were tormenting me at the time, and *What Is to Be Done?* gave shape and definition to my political views.

The irrefutable logic of Lenin's arguments, his sharp and vivid treatment of the war question and its underlying class implications, and of the Bolsheviks' point of view—all this was near and dear to me and completely comprehensible; it struck a responsive chord in my innermost being and gave me inspiration.

I realized that the correct proletarian stance for all genuinely revolutionary Social Democrats was the struggle to turn the Imperialist War into a civil war.

After making a study of these Leninist materials I discussed with Shaverdian my impressions of what I had read. Danush was sincerely pleased with my "progress," as he put it. Thus encouraged, I decided to tell him my dream. "You know, Danush, I have decided to become a member of the party. And I think that I am to some degree prepared to take this step."

I overcame my shyness and asked if he would help me make this dream come true, a dream which expressed a desire that was sincere, strongly felt, and well thought out. Danush smiled and said that there was no rush, and that I would have to prepare myself for such a decisive step by doing more reading, especially now that I had developed a taste for serious political books. "Also," he said, "I'll introduce you to some party comrades and perhaps in five or six months you'll be able to realize your desires."

He suggested that I read several of Lenin's works, particularly *One Step Forward, Two Steps Back*, and *On the Right of Nations to Self-Determination*, which had been published about a year before in the Bolshevik-oriented Petersburg journal, *Enlightenment*. In addition he recommended the pamphlets of Shaumian and Stalin on the nationality issue, and Plekhanov's book, *Our Divergent Points of View*.

I thanked him and said that I'd soon be going home to the village where I intended to begin my study of the first volume of Marx's *Das Kapital*.

"I don't think that this will prevent me from reading the books you recommended," I said. "I have more than three months to spare. On September first I'm returning to the seminary to make up the time I've lost."

Danush encouraged me in my plan. Taking the books he had given me I bid him good-bye until the fall.

I made one or two more visits to our Marxist study group, which had continued meeting during my absence from Tiflis, and then I left for my village.

Chapter III

My Universities

*Becoming a member of the party—First party duties
Finishing the seminary—About my classmates—Inside academy walls
Our teachers—Debating the agrarian question.*

My mother was overjoyed at my homecoming from the war. My father was happy, too, but he was not one to express his feelings openly. My mother had been afraid I'd be killed at the front, and now that I had returned home safe and sound, she wanted to celebrate the occasion in the traditional way. I resisted the whole idea, but it was impossible to make my mother change her mind. The celebration consisted of slaughtering a lamb in the monastery courtyard on the first Sunday of the month, and then cutting it up into small pieces and distributing them raw to the villagers who gathered at the monastery. All this was prescribed by ancient ritual.

It was a rather productive summer for me. First, I read and outlined the books that Shaverdian had given me. Lenin's *One Step Forward, Two Steps Back* made an indelible impression, clarifying for me the origin and nature of the conflict between the Mensheviks and the Bolsheviks.

As I recall, I had no particular trouble understanding the works on the nationality issue. These books helped me to formulate a correct world view and to understand the principles of proletarian internationalism, how they related to the nationality issue, and how to resolve it.

These studies were a great help to me later on in my everyday party work as well as in speeches I made on the nationality question.

I ran into trouble, however, when I tackled the first volume of *Kapital*. From the first chapter on, it was hard going, but I was determined not to give up. On a second reading I began to grasp some of the things I hadn't understood at first. Then I decided to summarize it in writing, and I read it a third time, taking notes as I went along. But a strange thing happened, for instead of paraphrasing his ideas in my own words, I found I was taking extensive notes verbatim, in Marx's own words. In short, I found it impossible to convey his meaning more concisely than he did.

I decided, nevertheless, to continue with this "summarization" technique because, in the process, Marx was becoming easier and easier for me to understand, and as I gradually became used to his style and his manner of exposition I found reading him less of a chore. In any case, by the end of the summer I had finished reading the first volume of *Kapital* and was very pleased with the results.

The first thing I did on returning to Tiflis at the end of August was to visit Shaverdian and report on what I had learned from my studies over the summer. We also discussed the courses I planned to take at the seminary and the additional Marxist literature I intended to read. Shaverdian gave me some new works on Marxist theory and drew up a rather long list of books he didn't have in his personal library; I was to take the list to the Pushkin Municipal Library and give it to Dzhavaira Ter-Petrosian (a Communist and the sister of the famous Kamo), who worked there. She would get me the books that Shaverdian had recommended. From then on I began to see Ter-Petrosian quite often, and she became my second "benefactor" and supplier of Marxist literature.

I also began to visit Shaverdian more and more frequently as my self-confidence grew. He introduced me to Askanaz Mravian and Sergo Khanoyan, both Communists. I also met several other active party members: Kakhoyan, a worker (who was from a neighboring village of mine), Garagash, and others.

Kakhoyan was one of the leaders of the illegal Bolshevik organization in the Akhpat-Alaverdi District. During that summer whenever I was in my village I would see him at the meetings for workers from the Alaverdi Copper Smelting Factory. Kakhoyan was considerably older than I and had much more practical party experience. He taught me a great deal, and put me in contact with a number of workers. All these meetings helped me to become involved in current politics and to take part in the life of the party organization.

One day I reminded Shaverdian of his promise to help me become a member of the party. "Well, all right," he replied. "I think you've matured enough by now for that step. Our activists know you pretty well and I think it's time to consider you for membership." That November of 1915 I was officially accepted into the party.

I was immediately assigned to do propaganda work amongst students and to prepare the worthiest candidates for party membership. Within our Marxist circle I was instructed to discuss current politics in addition to studying the literature, and to make members aware of the connection between practical politics and revolutionary theory. The war, the struggle for the overthrow of the autocracy, and the nationality issue were the main topics we discussed at our meetings.

The more informed members of the group began writing position papers. I recall that Aram Shakhgaldian wrote one on "The Woman Question and Social Democracy," while mine was on "The Nationality Question and Social Democracy." When we finished our papers we discussed them in the group. The texts of these papers have not been preserved but, as I think back, I must admit that they were indeed the work of students, being essentially attempts to compile and summarize everything we had read. Nevertheless, they represented a real achievement for us and were very much worthwhile.

Before the beginning of the new school year of 1915–1916 I requested permission from the seminary administration to reenter the sixth-year class which I had been in before leaving for the front. I promised to finish that class by the end of the first semester and to pass all the exams so that I could transfer into the seventh, and last, class in the second semester, and graduate along with the others. This meant doing two years' work in one.

It was a very heavy work load. I had no time for any extracurricular activities, nor did I let anything divert me from my schoolwork. I studied during vacations and on holidays, but managed to find time for reading Marxist literature. Our Marxist circle also continued its work. It continued to function after I graduated, right up until the Revolution. It was led first by Andreassian and Alikhanian, and then by Voskanian.

On December 19, 1915, I completed my sixth year successfully and began the seventh. My course work became easier then, allowing me more time for my reading, which absorbed me more and more. Sometimes I would study for days at a time, forgetting about food or rest.

At that time I had a room of my own at the home of Gevorg Alikhanian, a classmate and close friend. His family was relatively well off, although their house was a small one, situated in the courtyard of the Armenian church. Alikhanian gave me his own room and moved in with his parents despite my insistence that we could both share his room if an extra bed were moved in. "You study a lot," he said, "and my being there would take your mind off your work."

Once, when final exams were still a long way off, I stayed away from classes for more than a week to read Marxist literature. No one disturbed me. Once or twice Alikhanian dropped by to tell me what was going on at the seminary.

One day at noon while on my way to the Pushkin Municipal Library to return Engels' *The Origin of the Family, Personal Property and the State*, I unexpectedly ran into our rector, Khununts. He was a tall, broad-shouldered plump man with a proud, dignified bearing. Since he was rich he was one of the city's most influential citizens, but the students didn't like him, and he on his part made no effort to gain our friendship. Immersed in my own thoughts I didn't notice him until he stopped me and asked, rather arrogantly, "What are you doing out here on the street when you should be in class?"

I have no idea how I managed it, but I replied in an equally haughty tone, saying that I was not obligated to answer his question, especially there on the street, and I squared my shoulders and continued on my way. He noticed that I was carrying a book.

That same day, after classes were over, Alikhanian and another close friend of mine came by to tell me that a scandal was brewing over me at the seminary. One of the form-masters, Simak, whom we all knew well, had told them that after his encounter with me, the rector had arrived at the seminary terribly upset, and had headed straight for the Teachers' Room, where he had reported my behavior and declared that I would have to be expelled.

Although I had been absent from the seminary for more than a week, I hadn't spent the entire time reading. Twice a week, in order to earn some money, I tutored students who were behind in their work. With my elder brother now at the front, I didn't want to be an added burden to my family.

Alikhanian and his friend tried to persuade me to go to class the next day. "You know what kind of man our rector is," they warned. "If you don't show up at the seminary tomorrow, things will take a turn for the worse."

I had to agree with them. The next morning when I went to class none of the teachers said anything; my record was "clean," and besides, they knew that if I'd fallen behind I'd catch up quickly. When the class was over, however, the rector came in. He addressed the entire class, relating in agitated tones how I hadn't wanted to speak with him on the street and how I hadn't answered his question. He said that although I was in perfectly good health, I had been absent for ten days for no good reason and was occupied with some unknown project. Furthermore, he had checked the attendance sheet and noticed that I had skipped classes a couple of times before without an explanation. "As rector of this seminary I can no longer tolerate such blatant disregard for the rules," he concluded. "Mikoyan's future as a seminary student is at issue here."

Everyone listened attentively to him. Although I was upset, I controlled myself and kept cool, at least outwardly. I was sitting in the last row at the back of the room where I always sat because it was the best place to read during class without attracting the teacher's attention. When the rector finished his speech a complete silence fell. I rose from my seat and said more or less the following:

"Sir! It's inevitable that you'll have trouble understanding me and my behavior, since you and I come from completely different worlds. You haven't any idea whatsoever of what our life is like. You've said some very insulting and unfair things about me and you've even threatened to expel me from the seminary. No one can deny, of course, that you have power. You can do everything you said, but before you do, hear me out."

I don't know how it happened but as I was talking I decided to exploit the occasion for a bit of revolutionary propaganda.

"We all know," I said, addressing my remarks more to the class than to the rector, "that your son rides to the gymnasium in a fancy carriage, even though he's a bad student, a lazybones who had to repeat a year. You spend a lot of money for tutors, and you probably consider that perfectly normal. We don't know whether you've spoken to your son about this, and, for that matter, we don't really care.

"I shall answer the question you asked me on the street, and thus reply to the insults you've given me. You see, my older brother happens to be fighting in the war at the moment, and my father is no longer young; he has to support the whole family, and I'm ashamed to ask him for money. I have a room in a friend's house, rent free, and I live on what I make as a tutor—it pays for my food. And yet I've managed to complete a year-and-a-half's work in only six months, my grades are good, and my conduct has been irreproachable.

"You want to know what I've been doing while I've been absent from class? Well, I'll tell you—I've been working with the students I coach, along with doing some worthwhile and interesting reading. Nothing I've done has been detrimental to my course work, nor will it keep me from passing my exams. Yet you're threatening to expel me from the seminary, even though I have only three months to go before I graduate. Well, you have the power to do that if you want to. My teachers will tell you that I'm sure to pass all my exams. And I shall come to class regularly in the future if, that is, I'm allowed to remain at the seminary."

I said all this with youthful fervor, but in an even, firm tone of voice that compelled the rector, willingly or not, to hear me to the end. I had the feeling that my words made an impression on him, although he gave no outward indication of it. He made no reply beyond saying that the question would be resolved later. When he left the room my comrades gathered around me, excited and sympathetic.

"Bravo! You hit the nail on the head! You said what needed to be said!" Deep down inside I began to feel better.

The next morning I showed up for class on time.

Two or three days passed while we waited impatiently for the decision of the Educational Advisory Board, but we waited in vain. Several days later, during recess, Simak, the form-master who was sympathetic to us, took me aside privately and said:

"Your case has already been discussed. After a long and stormy debate with the rector, it was tabled. The deciding factor," he said, "was your promise to attend classes faithfully in the future."

I was deeply grateful to Simak for such happy news. I must add that Simak was a member of our party by then and had some influence over us. Later on, he became President of the Supreme Soviet of the Armenian SSR.

As I had predicted, I passed all my final exams. I got my usual C's in the ill-starred courses, Singing and Scripture, but in three other religious subjects I got A's. The only B I received was in Russian—and it was deserved; oral speech was what did me in, for I had almost no conversational practice in Russian, although I received top grades in Russian literature and in my written work.

* * *

As I look back on my days at the seminary, I can't help but be amazed at how many of the students there, particularly my classmates, later became well-known and important Soviet and party figures. There were reasons for this, of course, attributable not to some special distinction of the seminary itself, but rather to certain objective factors.

To begin with, those who attended the seminary were the children of the poor and the less well-off, since well-to-do parents could afford to send their children to tuition-based institutions, either to the gymnasium or to commercial schools. Thus, the students at the seminary constituted a more homogeneous—from the point of view of financial background—and democratic group, particularly receptive to revolutionary ideas.

Secondly, we were in school at a historic moment, between the two Russian Revolutions which witnessed the growth of revolutionary forces. The Revolution rolled toward its crest and drew into its wake everything organically near and dear to it. The Marxist circle we had formed at the seminary in the latter half of 1912 continued in the ensuing years to nourish able young men who were attuned to the spirit of revolution. All of them became members of the Communist party between 1915 and 1918 and worked actively for the cause.[4]

Of significance, too, was the relatively high level of instruction at the seminary. The majority of our teachers had received their higher education in Germany, Switzerland, and France, and many of them held liberal-democratic views.

In general, it should be noted that our school was not, in fact, a seminary in the literal sense of the word. It was not accidental that there was not a single member of the clergy in the administration. The only teacher who was a priest was the Scripture instructor, and besides Scripture, there were only four other religious subjects in the curriculum. These related basically to Church History and Theology, but were taught by lay teachers and took up no more than two hours a week. The rest of the curriculum was devoted to general education: mathematics (including algebra and geometry), geography, literature,

[4]See Appendix.

physics, chemistry, botany, zoology, psychology and physiology. We were required to take only one foreign language; either French or German—the choice was ours (in the gymnasium Greek, Latin, German, and French were required). It's true, however, that we studied Old Armenian in addition to our contemporary native tongue. Finally, we received supplementary instruction in the history and geography of Armenia, and in Pedagogy. The latter was due to the fact that we were being trained as teachers for Armenian schools.

There was an explanation for this. Except for certain church-affiliated schools, religious seminaries and academies, the Tsarist system in Russia forbade schools to teach in the Armenian language. As a way of getting around the Tsarist law, therefore, educational institutions called themselves religious seminaries when they were simply Armenian schools (gymnasia) offering a complete, eleven-year secondary education. It's true that a seminary diploma was not accepted at a Russian institution of higher learning; in Europe, however, they were accepted. As a rule, those who could not afford to study abroad usually became teachers in Armenian schools.

Our seminary was founded in 1824 by Catholicos Nerses, whose name it bore. In Armenian it was called the Tiflis Nersesyan Armenian Religious *School*, not seminary. During the years I was there not a single graduate of our seminary entered the clergy.

* * *

Even before our final exams, I, like my other classmates, spent a lot of time thinking about what I would do after graduation. My first priority was to continue my political education. I wanted to devote myself entirely to political study for two or three more years in order to become a competent Marxist and take part in the revolutionary struggle. This aspiration took complete possession of me.

With this goal in mind I decided to enter the Armenian Religious Academy. This school was located in the ancient town of Echmiadzin and was the only institution of higher learning in Armenia that accepted seminary graduates without entrance examinations. Not only was it tuition-free, but it also provided room and board, a fact of special importance to me, due to my financial situation.

The fact that I had received good grades on my school certificate was a virtual guarantee that I would be accepted at the academy. The majority of my seminary comrades—all members of the Marxist circle—came to the same decision. My petition for admission to the academy, dated July 16, 1916, has been preserved.[5] On September 18, 1916, I was notified that I had been accepted as a first-year student at the academy, and two days later I submitted a petition to the Tiflis Draft

[5]See Appendix.

Board that I be granted a deferment of military service until the completion of my education at the academy. My request was granted.

Conditions for studying Marxist literature were most propitious at the academy. Shaverdian kept me sufficiently supplied with books, which I shared with my friends. We were given three meals a day, and were housed in nice, bright rooms. Although our room contained about twenty beds, everything was clean, neat, and orderly. There was a fixed routine at the academy; everyone arose at six o'clock in the morning, had breakfast at seven, and classes began at eight. We went to bed at nine o'clock in the evening.

Our building accommodated not only the three-year academy program, but a seminary as well, so that, to use the contemporary idiom, the place was an "educational combine" incorporating both middle and higher education. Except for the rector and the Greek instructor, the academy's teachers were lay persons, not clergymen. We studied the history of ancient, medieval, and contemporary Armenia; the historical geography of Armenia; and Armenian language and literature, starting from the earliest period. Essentially, the academy trained specialists on Armenia, not priests.

I quickly became accustomed to my new surroundings, and it took me only a few days to work out a study plan—both for my courses at the academy and for my own Marxist readings. My classmate Aram Shakhgaldian and I got hold of a kerosene lamp and set our own routine of rising each morning at three o'clock when everyone else was still asleep, and going to the lecture hall assigned to our class, where we worked until breakfast at seven o'clock. At that time I was studying the third volume of Marx's *Kapital* (I had read the second during the summer at home). Compared to the first volume, the next two were much easier going. After completing the third volume, I read Marx's two unfinished notebooks which are included in the fourth volume of *Kapital (Theories of Surplus Value)*.

In the meantime Aram was reading the first volume of *Kapital*, and like me before him, was having trouble with it. I tried to help him as much as I could. We studied every day for three or four hours, usually before breakfast. Being young we didn't feel overworked, although I remember that we felt a little tired during our first classes. Aram and I would sit in the last row, at the back of the lecture hall, next to a bookcase where we kept about ten books: Marx, Engels, Lenin, Plekhanov, and others. During class I would sneak in a couple of copies of Kautsky's *Agrarian Question* (in German and Russian) and read them avidly, thus "killing two birds with one stone," by imbibing the contents of the book and learning German as well.

The academy taught two "dead" languages: Latin and ancient Greek, but I also studied German during class. Either the teachers didn't notice, or pretended that they didn't. Our desk, which was in the very last row, was rarely inspected and in any event I managed to get away with it.

Soon after arriving in Echmiadzin we organized a Marxist circle. All my comrades from Tiflis joined the circle, along with two others who had graduated from the seminary in Shusha: one of them, Tovmassian, joined us right away, but Sarkis Danilian waited until several months later. Our circle met regularly. Although the group was rather a motley one as far as the level of theoretical training was concerned, the study sessions were interesting. Each of us would study Marxist literature on his own, and although at meetings our discussions centered primarily on current political issues, we also discussed those questions which had occurred to us during our individual readings.

We were regular subscribers to the weekly newspaper *The Struggle* (*Paikar*) which was published in Tiflis in Armenian, and we discussed each issue in detail. This appealed to me. In December of 1916 I wrote my first article and sent it to *The Struggle* for publication. The article was a sharp polemical attack against a speech by a prominent Dashnak spokesman on the nationality issue which had recently appeared in print. *The Struggle* published my article without any editorial changes, except that the censor's scissors had cut out those places where I had set forth our position, so that it came out with three blank spaces in place of the omitted paragraphs, leaving intact, however, the critical part of the article.

When the members of our circle had got to know each other we wondered how we could extend our influence among the other students of the academy as well as among the upperclassmen of the seminary with whom we were in daily contact. It was soon obvious that the students from Tiflis were much more knowledgeable about politics than the local youth, who knew almost nothing about political economy, either. The one exception was Manukian, a third-year student who had read Adam Smith and Tugan-Baranovsky. He had never read Marx, however, and his views on Marxism were basically those of Tugan-Baranovsky, who, as is well known, distorted Marx's teachings.

At that time I was not familiar with the works of Adam Smith, although I had read Tugan-Baranovsky. After our discussions began, I felt I had to tackle Smith as well. We had heated debates about the labor theory of value, with Manukian zealously defending the views of the bourgeois Austrian economist, Böhm-Bawerk, who rejected Marx's theory of labor and surplus value in favor of his own theory of "marginal utility." For the theoretically unsophisticated, Böhm-Bawerk's theory was easier to grasp than Marx's, which required considerable theoretical training for a proper understanding of it. For that reason, I, who took a Marxist position, often felt that I was less persuasive in my arguments than the other debaters, even though we always gained the upper hand in arguments and discussions on other Marxist positions.

In an effort to extricate myself from this situation I appealed to my professor, Ashot Ionessian, for help in getting hold of works offering serious criticism of Böhm-Bawerk's theories. The next day he gave me the Austrian journal *Marx-Studien* which contained a relevant article. I mastered it with the aid of a German dictionary, and it turned out to be a great help in subsequent discussions.

I should add, by the way, that after the victory of Soviet power Ashot Ionessian became the secretary of the Central Committee of the Communist party of Armenia. Now he is a full member of the Armenian Academy of Sciences and is engaged in a major scientific study of the history of Armenian social thought. At one time or another he assisted all of us, and me in particular. He introduced me to Gevorg Atarbekian—a Bolshevik who lived in Echmiadzin then and who later became famous in the struggle with counterrevolution in 1918–1919 in the northern Caucasus. After that Atarbekian played a leading role in the Transcaucasian Federation, and he died in 1925 in an airplane accident along with Miasnikian and Mogilevsky. Ionessian also introduced us to the Bolshevik Makintsian, an educated and talented writer, who played an important role in Armenia after the establishment of Soviet power. This exposure to older, more experienced and educated Marxists contributed to our growth in a large degree, for we were still very young Communists.

We helped each other, too. One of my classmates, Pogossian, had an excellent command of Georgian, and I asked him to teach me how to read, write, and speak in that language, feeling it would be useful to me in my future revolutionary work. During the next few months I pursued my studies with Pogossian so zealously that I could read the Georgian papers to which he subscribed without too much difficulty. Unfortunately, the facility I acquired in those years has long since vanished after fifty years without spoken practice.

I later made an arrangement with Tovmassian from Shusha, who was striving to perfect his knowledge of Azerbaijanian, to have him teach me that as well. He gladly agreed to help me and ordered some books, after which we began a diligent study of Azerbaijanian. Before very long I could understand it fairly well, and this language subsequently was of great help to me in my practical party work.

* * *

I feel that it might be worthwhile for me to relate some episodes from our student life at the academy during that time.

The rector of the academy was Father Garegin, a man about forty-five, shorter than average, with a nice face and a handsome beard. He was calm and even-tempered, never raising his voice, nor would he tolerate rudeness on the part of the students.

One of the customs at the academy, before the start of morning classes, was to have the students line up in two columns in the corridor, with the teachers flanking them, and all together sing the Lord's Prayer. Then the classes would begin. From the start I avoided this ritual, going instead to the auditorium to read. One day, two weeks or so after I had entered the academy, the rector came into the auditorium and asked me why I was sitting there instead of praying with everyone else. I replied that I was an atheist and therefore couldn't pray. He had evidently heard rumors that I was an atheist, but being an intelligent man he didn't attempt to question me in any way. Rather he said calmly and sternly, "Then you could at least not disrupt the established routine here at the academy. No one is forcing you to sing the prayer, but you are obliged to stand in line with everyone else."

I liked the rector's straightforwardness, and immediately gave my word that I would stand in line in the future. But meanwhile I thought to myself that if he weren't such a broadminded person I would have been expelled from the academy and lost the chance to pursue my study of Marxism.

Some time later the rector summoned me to his office and said, "I've been considering the idea of organizing amateur nights at the academy. My idea was to base these programs on the cultural heritage of the various national groups. Could you and the other Tiflis students start it off? After you, the students from Shusha and from Erevan could perform, and then the local students from Echmiadzin. I'm hoping the project might inject a little spirit into the academy."

I agreed, and said I would discuss the idea further with the Tiflis students. After talking it over with my comrades we came to the conclusion that the rector's proposition was a godsend for us. The idea appealed to us on its own merits, as well as offering us a made-to-order occasion for propaganda work among the student body. So I went to the rector and told him that all my friends were willing to participate in the event, and that we needed about ten days to get ready.

The rector was agreeable to this time schedule and set the date for our performance. He said that all the students at the academy would be invited, along with upperclassmen from the seminary, teachers, and guests from among the monks. Fortunately, he didn't require any formal approval of either the amateur night program or its list of participants.

We set to work with a will on our preparations, having resolved to carry out the performance in the spirit of revolution and internationalism. I happened to have on hand several issues of the Petrograd journal *Enlightenment* for the year 1914, and from these we selected a poem to be read in Russian. I don't remember the name of the poem, but its content was highly militant and revolutionary. It fell to my lot to declaim a sharp, anticlerical poem of Shushanik Kurginian. Ovanes Pogossian, who had a fine voice, was to sing a Georgian folk song, "Gaprindis Shavo," and Tovmassian from Shusha, whom we had included in our "theatrical troupe" was to sing the Azerbaijanian song. Artak Stamboltsian, our resident comic, was chosen to read a story by the Armenian writer Akop Paronian, a master of sociopolitical satire. In this story the author satirizes a certain Armenian who makes himself out to be a great patriot, but who is in fact a coward and a traitor. This "brave fellow" flees the battlefield . . . "so that the numerical strength of the Armenian population will not be diminished." There were to be several additional numbers on our program but I can't recall now what they were.

On the day of our performance the assembly hall of the academy was filled to overflowing. As many as ten prominent churchmen were present in the audience. Every aspect of our performance came as a complete surprise to those present. The show itself, in which four languages were heard, symbolized the idea of internationalism, striking a new note at the academy. Most of all, the content of the various presentations with their revolutionary slant, came as a shock to the entire academy. Everyone listened attentively, but they never ceased to be amazed. The general quality of the performances, judged by the standards of that time, seemed to us, in any case, to be fairly high. However, by the time my turn came near the end of the show to read Shushanik Kurginian's poem "The Workers," the patience of the clergy was exhausted.

Here is a rough approximation of some of the lines that appear in the poem (I'm giving a free translation):

> It's we who are marching—
>   in ragged jackets,
>   in tattered caps,
>   pale and hungry,
>   old before our time . . .
> But we are full of uncontrollable rage,
>   full of contempt and vengeance . . .
> And we firmly believe in a better life for us . . .
>
> It's we who are marching . . .

It's not hard to understand why in those days such words sounded like a bomb exploding within the academy walls. One after another the monks began to leave the hall. But when the show was over, none of the students wanted to leave. Heated arguments and discussions broke out in the corridors. Some of them praised us and others criticized as the halls of the academy buzzed like a disturbed beehive.

We expected the rector to reprimand us because of the show, but even on this occasion he was quite restrained. Without saying a word to us he shelved forever the idea of subsequent amateur nights at the academy. Instead he decided, perhaps as a salutary counterinfluence, to get us interested in the written language and literature of old Armenia. Once he took us on an excursion to the monastery where the ancient manuscripts were preserved. What we saw there overwhelmed us. We hadn't realized that such a valuable collection of old manuscripts existed amongst us—manuscripts written with uncommon beauty and love on papyrus, parchment, and animal hides, and richly illustrated by talented artists. We were struck by the variety and freshness of the colors. There were very thick books bound in richly ornamented leather bindings.

We sincerely thanked the rector for all that he had shown us that day as well as for his detailed and interesting explanations. All these riches are now kept in Matenadaran, the well-known depository for ancient manuscripts in Erevan.

After this successful undertaking the rector hatched a new scheme. He announced that he wanted to invite the students—divided up according to national origin and locale—to visit his home. "We'll begin with the Tiflis group," he added. Naturally, we accepted with pleasure.

I recall how on Sunday after dinner we set out for the rector's apartment. A buffet had been arranged for us there: tea, various hors d'oeuvres, and the local wine. It was soon clear that the rector wanted to involve us in a serious discussion which would stimulate our interest in the old Armenian cultural inheritance. I can't recall now what the conversation was about, but the rector spoke impassionedly. While he spoke he offered us hors d'oeuvres, wine, and tea. At first his remarks seemed very interesting and we listened attentively. I was seated next to him, but after a while some of my comrades seated farther away began to drink the wine until they became quite high, talking and laughing with marked animation. Meanwhile I continued to listen to what the rector was saying, for I found it engrossing; moreover, I wanted to show respect for our host by observing the elementary rules of propriety.

The rector, of course, took note of my comrades' behavior, but pretended not to see them. When he wasn't looking I kept making all sorts of signs to them to try to make them behave, but they were under the influence of wine and continued to carry on. I began to feel highly uncomfortable as it went on for about an hour. The rector finally began to wind up the conversation, seeing that so many were no longer listening to him. Feeling embarrassed, we made our departure, leaving the rector visibly upset.

He didn't invite any of the other national and regional groups to visit him.

I remember how one winter the rector came to tell us that the catholicos himself, George V, would preside over the Sunday service. We all had to attend this service. After the rector left the auditorium we "clansmen" started discussing how we should respond. We hadn't been attending church at all nor had we maintained ties with the clergy. We had, in fact, openly made fun of several students from Shusha who, as staunch Karabachians, had made visits to the cells of monks from Karabachi. Since boycotting the service would mean risking expulsion from the academy we decided to attend the service, but to refrain from kissing the catholicos' hand, as was the custom.

We put in an appearance at the service, but when the time came to kiss the hand of the catholicos, who was seated on his throne, we went up to him, bowed politely, and then, without touching his outstretched hand, retired to the side. Although no one said anything, the faces of those present, especially the monks', reflected astonishment and irritation at our impertinence.

After the service the academy walls buzzed with gossip. Voices were heard in favor of expelling us from the academy. Others, more sober-minded, censured us, but rejected the idea of expulsion. They reasoned that if word got around and the Tsarist police found out that a religious academy was harboring a nest of Bolsheviks, this could result in the school being closed down, so they counseled hushing up the whole affair. And that's just what happened.

Perhaps this is an appropriate time to mention that I met a catholicos of the Armenian church a second time in my life many years later, under completely different circumstances. In 1958 I was in Erevan at a meeting with the electors prior to the elections to the Supreme Soviet of the USSR. The leaders of the Republic arranged a reception attended by about two hundred people.

On noticing several churchmen at the end of the room I asked who they were, and was told that the new All-Armenian catholicos, Vazgen I, was there with his retinue. I was also told that he was an educated and intelligent man, with a positive attitude toward Soviet power who commanded the respect of Armenians both here and abroad. I poured out a glass of wine and, accompanied by the Republic's leaders, strolled to the other end of the room toward the group of churchmen. After greeting the catholicos and his retinue I said in a joking tone that I felt somewhat guilty before the Armenian church since I hadn't justified the hopes and efforts wasted on my education.

"Speaking as an economist," I said, "as a student of the Armenian Religious Academy, I turned out to be a 'defective product.' "

Those around me burst out laughing. The catholicos smiled pleasantly and said, "You're wrong. Not only are we not insulted, we're *proud* that a man like you came out of the academy. We'd like a few more such defective products."

I saw this as the courtesy of a well-brought-up man. After proposing a toast to the flourishing of Soviet Armenia, the catholicos and I clinked glasses and drank up.

* * *

The teacher of ancient Greek at the academy—a tall, thin man around forty years old—was a highly educated monk who was very shy and reserved. He tried to have as little contact with the students as possible, but they often played malicious jokes on him.

I was genuinely sorry for him and would tell the students to stop making fun of him. I pointed out that they didn't act that way toward the other teachers who treated us more strictly, but who had stronger personalities, and in the end I managed to convince them. Whether their consciences began to bother them or they simply got bored, I don't know, but they stopped making fun of him.

This teacher knew that I wasn't studying the Greek language, but I always conducted myself with great seriousness in his classes, with an open book always before me, and he always treated me with marked respect. Moreover, he seemed to trust me, for he once asked a favor of me.

Since the winters in Echmiadzin were generally warm and filled with sunny days I would often get out into the garden on mild mornings and, while regular classes were being held, sit on a bench in the sun and read Marxist literature. No one bothered me there, so I could really concentrate. One day when I was out there the teacher of ancient Greek came over to me and asked me whether I remembered the algebra I'd had at the seminary. I said I remembered it quite well. He explained that he was asking me because he was tutoring the catholicos' son (our catholicos had been a priest, and they had the right to marry) and was having some difficulty with the algebra since his knowledge of it was very rusty, and he wondered whether I could help him with some of the more difficult problems. I agreed, and after that, he came to the park several times with algebra problems with which I was able to help him. He was very grateful to me. He realized, too, that I never told anyone about our meetings—otherwise, there would have been a lot of gossip around the academy.

Another time when I was studying in the park, my close friend Artak ran up to me in a state of great anxiety and told me that a new teacher had arrived at the academy. His name was Navasardian, and he was well-educated, eloquent, and extremely knowledgeable in the social sciences. He had made an excellent impression on the students, who listened to him with great interest. "Now," said Artak, "it'll be difficult for us to maintain our position of superiority in front of the students during political debates." I told Artak not to worry, that I didn't see anything dire about the arrival of this teacher. Nonetheless, deep inside I felt rather alarmed myself.

The next day I went to hear the new prophet. He did indeed speak eloquently and with self-confidence, like an expert. In a word, he made an impression, and the students hung on his every word. In the middle of the lecture he began to extol Catherine II as the "educator of Russia," citing her correspondence with Voltaire and all but calling her "the Russian Voltaire."

Not long before this I had read Shishkin's pamphlet on Catherine, which revealed the real truth about her life and deeds and political policies, which were subservient to the interests of the Russian nobility. The pamphlet exposed everything about Catherine's life—the intrigues, the demagogy, and the wild goings-on. Shishkin told in particular how she had settled huge estates with thousands of serfs on her favorites, particularly on Count Orlov, I remembered, to whom she gave great parcels of land and many serfs.

I asked if I could speak. The teacher interrupted his lecture and let me have my say. I related everything I had read in Shishkin's pamphlet. "What kind of an educator is she?" I asked. "Do a few letters to Voltaire make her worthy of that title?"

My speech made a huge impression on the students. Navasardian himself was flabbergasted. Since he was evidently not prepared to refute the facts I had cited he began to justify Catherine, again referring to her correspondence with Voltaire. But in any event his self-confidence was shaken and he couldn't erase the impression my speech had made on the students.

After two or three classes with him, I was persuaded that Navasardian was indeed a clever man and one with a glib tongue. However, his knowledge was superficial. From what he said, it was clear that he was familiar with Socialist Revolutionary (SR) ideology; he made references to *The History of Russian Social Thought* by Ivanov-Razumnik, who was later to become one of the leading ideologists of the SRs. I too was familiar with this book, but because I had also studied the brilliant works of Plekhanov, who unmasked the ideology of the populists (Narodniks), I knew that our new teacher was not such a dangerous opponent.

At one of our classes Navasardian decided to show off his knowledge of Marx. He quoted by heart some sentence or other, supposedly from Marx, but in essence blatantly anti-Marxist. I shouted out a rejoinder to him from my seat:

"That's not like Marx!"

"It's Marx all right," he said.

"Then tell us, please," said I, not backing down, "in which of his books you read it?"

"In the first volume of *Kapital*," he answered, not batting an eye.

I announced boldly that the quotation he had cited did not come from there. He started to insist, supposing that I would not be able to refute him. At that point I took the first volume of *Kapital* out of the bookcase and asked Navasardian to show us where his quotation came from. Astonished, he walked over to me, took the book, and began to leaf through it. The students waited tensely to see how our "duel" would end.

Finally, not being able to find the quotation—and understandably so—he looked at the book's cover with its inscription, "Translated by I. I. Stepanov-Skvortsov," and he returned it to me, claiming that he had read Marx in another translation, and that his quotation had come from this rare edition of *Kapital*.

That's when I decided to deal the final blow, so I said, "I know that translation, too. It's edited by Struve. Here it is." I took a copy of *Kapital* which belonged to Aram down from the shelf and handed it to Navasardian. He was completely dismayed, but this time he evaded giving a direct answer to my question. Not at a loss for very long, however, he began to ask me where I had gotten such a rare edition, and whether he could borrow it from me. I gave him the book. Several days later he returned it to me in class and thanked me for lending it to him, but he made no reference to our argument. Everyone knew that he had fallen on his face, so I didn't feel it was necessary to bring up the subject of the quotation again. After all, you don't hit a man when he's down.

We never returned to this subject again. He'd had the wind knocked out of him, however, and after this incident he quieted down considerably. Later on, it became clear that Navasardian was an active member of the Dashnak party and considered himself an adherent of the SR ideology. I had another run-in with him, I might add, but it happened later on, after the February Revolution. But more about that later on.

Once, two sixth-form seminarians came to see us to complain about Giandzhetsian, the instructor of old Armenian literature whom we shared with them. They told how during recess they and their comrades had been having a good time singing and dancing. Just at that moment a stout fellow passed by their window, walking arm in arm with his wife, who was also quite fat—it was none other than the Giandzhetsian couple. The seminarians were then in the middle of some Armenian verses from the Azerbaijanian operetta *Arshin mal-allan*, which contains some words roughly like this: "Come here, my love, come here to me . . ." It seems that Giandzhetsian took these words to be an affront to his wife; so he ran into their classroom in a rage and screamed at them, "You're all scoundrels, and so are your fathers!"

I asked him if they were absolutely sure that those were the exact words which Giandzhetsian had uttered. They replied that everyone in their class could testify to the fact that they were indeed the precise words. We told them to go back to their room, but reassured them that we would not tolerate such behavior on the part of a teacher. After they'd gone we discussed the situation, and came to the following decision: the first class we had with Giandzhetsian we would all, in turn, say something like this to him: "I can't attend a class with a teacher who considers our fathers to be scoundrels, as you deigned to call the fathers of the sixth-form seminarians. Our fathers are the same as theirs."

I was to be the first to say this—say it, and then walk out of the classroom. After me, five or six others were going to do the same thing. We assumed that that would be enough.

I remember that morning very well. Giandzhetsian entered the lecture hall, placed two thick books on the rostrum, and sat down. I asked to say a few words, but I hadn't managed to finish the line I had rehearsed when he grabbed his books and shot out of the room like a bullet. We wanted to see what would happen next. The form-master came in and said that the rector wanted to see Mikoyan.

When I appeared in his office the rector greeted me, asked me to sit down, and then began to ask me why I had acted that way, namely, disrupting the class and insulting the teacher. I replied, "Father Garegin, I didn't say anything coarse or insulting to our teacher. I only said that I could no longer attend his class because he had undeservedly called the pupils and their parents 'scoundrels,' and they are just simple people—workers and peasants like our fathers, whom we love and respect. It was not we who insulted him, but he who insulted our parents, and we can't tolerate that. You should understand that."

Being conciliatory by nature, the rector tried to persuade me that Giandzhetsian had probably been in an agitated frame of mind, etc. I replied that this didn't exonerate him in the slightest, and that not only I, but the majority of my classmates, didn't want to study with him in the future.

"If you want us to dismiss him," the rector began, "well, that's out of the question."

"If that's impossible," I replied, "then let him at least apologize to the students and take back what he said!"

Our conversation ended on this note.

Subsequently, Giandzhetsian apologized to the seminarians and our classes with him were resumed.

Late in 1916 Ashot Ionessian informed us that an illegal interparty conference was about to be convened to discuss the agrarian problem in the province of Erevan and how to resolve it, and he advised everyone in our group to be there. I remember that we met in the evening in some dark basement-like room with about thirty people in attendance, among them Mensheviks, SRs, and Bolsheviks.

A report, compiled by someone unknown to me, told of the plight of the Armenian peasants, groaning under the landlords' yoke. At that time the overwhelming majority of landlords in the villages were Azerbaijanian, whereas the peasants were both Armenian and Azerbaijanian; that was the way it had always been.

The speaker took a narrowly nationalistic position on the problem, proposing that the Armenian peasants organize on their own and abolish the landlords' property rights. He linked this suggestion with the recent resolution of the State Duma to form a commission to study the situation of the peasants in the province of Erevan. The commission was headed by Cadet Adzhemov, a member of the Duma and an Armenian by nationality. The speaker proposed that the work of the Duma commission be supported in every way possible, thus revealing himself to be party to the naive illusion that the commission, headed, moreover, by a cadet, would be motivated to resolve the agrarian question to the peasants' advantage. This was Utopian folly, of course, and had nothing to do with the basic tasks imposed by the struggle for the resolution of the agrarian problem. Some spoke out in favor of the proposal, others against. Those in favor were in the majority.

I asked to say a few words, and spoke out sharply against the speaker's position, calling attention to his narrowly nationalistic and hence incorrect approach to the most important problem posed by the class struggle between the landlords and the peasants. I ridiculed his hopes for a solution from Adzhemov's Duma commission and pointed out the significance of the agrarian question not only for the Armenian peasants, but for the Azerbaijanian and Georgian peasants as well. They, too, suffered cruelly under the landlords' yoke.

I therefore urged a consolidation of all the forces in Transcaucasia to fight for the abolition of the landlords' property rights, and I called upon toiling peasants of all nationalities to join in the struggle. I stated that the correct approach to the problem was based on class and not nationality, and that the way to solve it was not through the peasants alone, but through the peasants and the industrial proletariat, united by all the revolutionary parties of Transcaucasia.

Some comrades supported me from the floor, mainly those from our group. When the speaker rose to rebut me he attacked me crudely, instead of addressing himself to what I had said, and stated that the historical materialism which I had used in my arguments had reminded him of "restaurant materialism." An uproar ensued after that. My friends would not, of course, allow me to be insulted in this way so they heatedly objected to the form the discussion had taken. We could come to no agreement whatever, as I recall, and thus ignominiously ended our first and last interparty conference.

Chapter IV

The February Revolution

Work among the soldiers—The first legal meeting of the Tiflis Bolsheviks—Leaving for Baku—My acquaintaince with Shaumian and Dzhaparidze—Work in the Baku Party Committee—The split with the Mensheviks—Back with my family.

The Imperialist War of 1914–1918 exposed and intensified the fundamental contradictions of imperialism. Instead of the promised speedy victory, the Tsarist forces were defeated on many fronts, with the Russian army suffering great losses in manpower. The crucial factor was the regime's inadequate preparation for the war which had left the army in extremely short supply of weapons and ammunition of all kinds.

As the war continued, serious economic difficulties ensued. There was a shortage of foodstuffs, and every day it became harder and harder for the peasants and workers to survive. As the prices of essential goods skyrocketed, the country experienced a general collapse and the people's dissatisfaction with the war became increasingly evident. Meanwhile, a small group of manufacturers and speculators who made fortunes on military supplies continued to get rich at an unprecedented rate. The people's weariness with war and deprivation now began to express itself more sharply. Strikes and mass demonstrations broke out in the country's manufacturing centers, and the Tsarist goverment started to use rifles and machine guns to suppress them, with the result that demonstrations in the cities now ended in bloodshed.

A genuine revolutionary situation was brewing in the country. The Revolution was at the gates.

We heard about this by hearsay, however, because the Tsarist censorship managed to conceal from the general public, especially in the provinces, news of the workers' strikes in Petrograd, the riots involving women on food lines, and the clashes between workers, soldiers, and the police which all resulted in making February 1917 appear a time of civil war. We were taken by surprise, therefore, by the announcement which came over the telegraph one day telling of the overthrow of the autocracy. The tsar had been deposed and a provisional bourgeois government set up with a Soviet of Workers' Deputies.

Our joy knew no limits!

* * *

In Echmiadzin there was no industry and hence no working class. The only fertile field for propaganda work was the depot battalion stationed there, which consisted, by and large, of overage soldiers burdened with families who dreamed of the day when the war would end. We organized a meeting of the soldiers of this battalion at which I spoke to them about the meaning of the fall of the autocracy and the tasks of the Revolution, and called upon them to be done with the war. I explained to them that the people themselves had to resolve the issues of the Revolution, including the question of war and peace.

The soldiers unanimously supported us. Our demands for an end to the war made a special impression on them, and they passed a resolution to form a soldiers' committee for their battalion. However, since they had in effect voted against the war, the soldiers seemed to form the impression that this meant that the war would end immediately, and they proceeded to get joyously drunk.

I should add here that at the start of the Imperialist War the Tsarist government had introduced a dry law which forbade the sale of vodka and wine throughout Russia. This prohibition did not extend to the Caucasus, however, because wine was an integral part of life there, especially among the Georgians and Armenians, and since the Tsar wanted to avoid an open conflict with these peoples, wine was still sold freely there.

One of the drunken soldiers set fire to the trading booths in the bazaar, and looting began as the soldiers ran through the streets bellowing out songs. We tried to reason with them, explaining that they should maintain good relations with the local population, the majority of whom were working people just like themselves. The soldiers understood this very well, and their soldiers' committee promised to restore discipline and to prevent similar excesses in the future. We were convinced that after this everything would be fine.

After we had managed to bring the soldiers back to order, some comrades from our group found out by accident that the Dashnak leadership of Echmiadzin, headed by our teacher Navasardian, had called a meeting of some of the students, along with the local Armenian youth, and had proposed to disarm the soldiers at night when they were asleep. Since such an action would only trigger an inevitable confrontation between the soldiers and the local population and lead to a new wave of nationalist unrest, we had to act fast to prevent it. We decided to give Navasardian an ultimatum; if the Dashnak leaders didn't abandon their plan we would warn the soldiers' committee, and with the soldiers' help, arrest them all.

I went to see Navasardian on behalf of my comrades. He lived in a small, two-story house on the same street as our dormitory and the academy buildings. There was another man in the room with him, but I was alone and unarmed. "If you refuse to cancel your plans," I said, "or decide to detain me here, then within an hour from now my friends will go to the soldiers' committee, rouse the battalion, and arrest your supporters."

Navasardian grew pale at first, then he flew into a rage and began to hurl insults at me, saying that we had no feeling of national pride, that we had allowed the Armenian bazaar to be plundered and torched with impunity, and finally, that we were planning to join hands with the Russian soldiers against the Armenians. I told him that he was completely mistaken. The torching of the bazaar had been a spontaneous phenomenon, which was understandable under the circumstances. I pointed out that we had managed to restore order in the battalion, and took pride in that fact, and that we were having success in improving the relations between the soldiers and the local population.

After long altercations Navasardian agreed to cancel their plans.

The next day when I went to the soldiers' committee I said nothing about the provocation that had been planned, but again explained to them in detail what they had to do to strengthen revolutionary discipline among the soldiers and to restore good relations with the local population.

There was an unexpected outcome to this incident a short time later when, out of the blue, Ashot Ionessian and I received our draft notices into the army. Since we each had student deferments we suspected that the local Dashnak leaders were behind this move, having somehow managed to persuade the military authorities to draft us in order to rid themselves of us Bolsheviks, the most dangerous of their opponents.

In order to spoil their little game Ionessian and I made a trip to the military commander's office in Aleksandropol (now Leninakan). There we protested being drafted, calling attention to the fact that even under the tsar we had obtained a deferment, and that the law on which deferment had been granted was still in effect. Why now, after the Revolution, we asked, were the military authorities violating the law by this arbitrary action? Perhaps the crucial factor was our membership in the Bolshevik party, we went on, but now the Revolution allowed the existence of *all* parties. Therefore, we concluded, we saw no reason to give in so easily to this unfair action.

The military commander was evidently influenced by our energetic pronouncement, for he agreed to cancel our draft order. Much relieved with this outcome, we returned to Echmiadzin.

* * *

Everyone at the academy was preparing for the spring exams when I decided to take a leave of absence from the academy. I wasn't planning in fact, to return, but I didn't want to officially sever relations with the academy because I didn't want to lose my draft deferment. I told the rector that I had to attend to my own affairs in Tiflis and that I would take the exams in the fall. Whether it suited the rector to be free for a while of one of the "rowdies," or because of other considerations, he agreed to my request.

I hurried off to Tiflis where revolutionary events were flaring up, eager to play a more active role in the party. There was, however, another circumstance that made me hurry to Tiflis: during the first ten days of March 1917 a conference was to take place there of delegates from the Transcaucasian Marxist groups (on the model of the first of such conferences which had been convened in December 1916 in Kutaisi).

On my arrival in Tiflis, via Erevan, I immediately went to see Shaverdian to find out the latest news about party affairs. He told me that the first legal meeting of Bolsheviks was to take place the following day at the Zubalov People's Assembly Hall. I attended the meeting, which drew approximately 250 people. Two comrades were seated at the entrance, registering everyone who entered and collecting membership dues (fifty kopecks apiece, as I recall), and then handing out receipts. Three men were chosen to the presidium of the conference: Alyosha Dzhaparidze, Misha Okudzhava, and Amayak Nazaretian.

Alyosha was then passing through Tiflis, en route from Trapezund and the front. A prominent figure in the party, he worked primarily in Baku where he was headed, but he had stayed over in Tiflis to take part in this meeting of Tiflis Bolsheviks.

Misha Okudzhava led the meeting, whose main theme was the unification of the Bolsheviks and the Mensheviks. The reason given for the necessity of union was, as I recall, the new situation which had arisen after the overthrow of the autocracy: a number of tactical differences between the Bolsheviks and the Mensheviks had been automatically resolved by the February Revolution and by the attainment by the people of broad democratic freedoms and political rights. "Under present conditions, the unification of all Social Democrats into one party must necessarily enhance the party's role in the nation's political life," the proponents of unification categorically stated.

As I remember, a sharp speech against unification was delivered by Fyodor Kalantadze, who felt that in a merger the Bolsheviks would inevitably forfeit their political independence and weaken their organizational structure. But the proponents of unification said that a Bolshevik faction would be retained, and that we could also publish our own paper.

After a long debate a resolution for unification was passed by a majority vote. It became clear later on, however, after Lenin's return from Switzerland, that this resolution was ill conceived, both practically speaking and in principle, especially for Tiflis where the Mensheviks were in a decided majority. Practically speaking, the union weakened the Bolsheviks' influence on the working class, a fact which later events confirmed. In the summer of 1917, in any case, the Bolsheviks broke off from the Mensheviks in organization.

The first few days after my arrival in Tiflis we held a conference of representatives from the Marxist student circles. The delegates to this conference consisted of ten to fifteen people each from various schools, in addition to several guests from among the most active members of the different circles. We met on the first floor of the municipal administration building in a conference that lasted two days. Heated arguments sprang up between the Mensheviks and the Bolsheviks, with the most active spokesmen for the Bolsheviks being Dzneladze, Akirtava, Alikhanian, and myself.

I frequently spoke in Armenian at meetings of Tiflis students on the subject of peace and the Russian Revolution, as well as at workers' meetings as a propagandist for the Tiflis Party Committee. I remember my speech on the theme of "Social Democracy and the National Question" since I repeated it on more than one occasion because of the vast interest in the subject. I also gave lectures on "The Self-Extermination of Capitalism," and "Historical Materialism and the Program of Social Democracy."

At that time there were no large scale enterprises in Tiflis except for the railroad shops and yards where several thousand workers were employed. There were also some small tobacco factories, and several small printing houses, shoemaking shops, and garment factories. The largest of these factories was, I think, the Adelkhanov (leather) shoe factory with about 300 employees. We managed some way or other to gather almost all the workers in the factory courtyard. There were two speakers at the meeting: Dzhugeli, who spoke in Georgian, and I, who spoke in Armenian (both Georgians and Armenians worked at the factory).

Dzhugeli spoke very simply and convincingly. His fate was as follows: after the Revolution he, as a Bolshevik, became head of the Red Guards, which came into being at the directive of the Tiflis Soviet. But he went over to the Mensheviks a few months later while the Menshevik government was in control and played the role of executioner in putting down the peasant revolts. I'll have more to say about him later.

During those days I would drop into the editorial offices of the Bolshevik daily, *The Caucasian Worker,* almost every morning to learn the latest news and have a look at the *Petrograd Pravda.* The editor of the *Caucasian Worker* was Sergei Kavtaradze, a highly trained Marxist who was ten years my senior, both in age and length of party service.

At that time, the Bolsheviks in Tiflis met and had their headquarters in the municipal administration building while the Menshevik committee met in the former palace of the governor-general. Every evening after work around seven or eight, a meeting of the Bolshevik Center was held in the municipal administration building. Party activists were also admitted to these sessions, and I attended them almost every evening. For me, still a young Communist, they were an excellent party school which broadened my horizons to a considerable degree.

Usually around seventeen or eighteen people attended these meetings. The most varied questions were discussed: general political topics; organizational questions of local significance; intraparty issues; relations with the Mensheviks; work among the soldiers, employees, and youth. Of those who attended these meetings I would single out the following as the most active: Makharadze (he chaired the meetings), Okudzhava, Toroshevidze, Nazaretian, Kavtaradze, Mravian, Khanoyan, and Dumbadze.

At the end of March 1917 at one of these meetings they discussed a letter from Shaumian in Baku. He reported on the general situation there, and at the end of the letter asked that our old Bolshevik Mravian be sent to him in order to aid in the political work among the Armenian workers. There were many such assistants available, reported Shaumian, but qualified propagandists, particularly those who knew Armenian, were in short supply. The Dashnaks were taking advantage of this circumstance and had managed to bring a lot of Armenian workers under their influence.

Mravian, however, spoke up to say that he couldn't go to Baku. No substitute for him was named, and no one was entrusted with the task of finding one. Accordingly, after the meeting I went to see Shaverdian. "Danush. You know me very well," I said. "Do you think I could handle the work that Shaumian talked about in his letter?"

He thought for a while, and then said, "If you really want to go to Baku, then that's half the battle. Your first experience there might be rather difficult, however, since the situation there is very complex. But I'm sure you'll be able to handle it."

I reaffirmed my desire to go.

"Well, go then," Shaverdian said with conviction. "I'm sure that you'll be all right. Shaumian is a very close friend and I'll write him a note of introduction for you. They'll find some kind of job there for you so that you'll be able to support yourself while you do party work." He wrote a few lines to Shaumian on one of his visiting cards, sealed it in an envelope and handed it to me.

The Bolshevik Center agreed to my trip and issued me a travel warrant, and I left the next day for Baku without going to my village to say good-bye to my parents.

* * *

The people of Baku had greeted the February victory of the Petrograd proletariat with enthusiasm, marking the occasion with a general strike in the oil fields, factories, and shops.

I arrived in Baku at the end of March 1917.

I knew that the Bolsheviks had commanded trust and authority among the Baku workers even in the prerevolutionary period, and that on March 7, 1917, immediately after the February Revolution when a Soviet of Workers' Deputies was set up in Baku, the Bolshevik Stepan Shaumian was chosen as chairman, even though the Bolsheviks were in a minority in the Soviet. Moreover, Shaumian was elected despite his absence, for at the time he was en route home from a term of exile imposed on him under the tsar. Of course, Shaumian's personal prestige played no small role in this, for he had long been a familiar figure among the Baku workers, but his election unquestionably testified to the great confidence the Bolsheviks enjoyed among the Baku proletariat.

"Nonetheless," warned my Tiflis friends, "keep in mind that the situation in Baku is very complex."

To be honest, I didn't have a very clear idea at all of the situation there and it was only much later that I gained any understanding of it.

The complicated international and internal situation of the Baku area was due to many factors: political, socioeconomic, and geographic. Baku oil had attracted the attention of world capital even earlier than Persian oil: before the Revolution about 15 percent of the world's oil came from Baku. By 1917 the largest foreign corporations were in operation here; Nobel, Rothschild, and others, while Russian, Azerbaijanian, and Armenian oil manufacturers were also active.

The interests of capital, constantly in opposition to the working class, dictated that the capitalist forces join together into a "Union of Oil Manufacturers." It should be added, however, that this unity was purely theoretical because of the fiercely competitive struggle in which the oil magnates were engaged—a struggle which intensified during the war that had arisen between the Central Powers (Germany, Austria, and Turkey) and the Entente (England, France, and Russia).

Under these conditions the flexible tactics of mass workers' organizations assumed great importance—within the Bolshevik party, especially, which was seeking newer and better forms of leadership for the working masses. Our party, for instance, encountered great difficulties over the nationality issue, the contradictions of which were ever being sharpened both under tsarism and by the chauvinists of the Armenian and Azerbaijanian bourgeoisie. The nationality question weakened the workers' front in the struggle against capitalism and disrupted unity in their own ranks.

The situation in Baku was further complicated by the fact that immediately after the February Revolution new legions of workers were drawn into political action. Lenin wrote about them thus: "Politically oppressed by the horrible yoke of tsarism and hard labor . . . they suddenly awakened and joined the struggle against the landlords and capitalists." Before, they hadn't been interested in politics, didn't participate in it, and often didn't even know that various political parties existed, each with its own individual program for action. Now at rallies and meetings they began to listen to the speeches of the representatives of the different parties, and in the attempt to win these workers over to their side, speakers from the bourgeois and petit bourgeois parties didn't hesitate to use the most vivid and showy "revolutionary" phrases, making liberal promises of all kinds of freedoms and benefits in the names of their parties of democracy and equality.

The broad working masses were becoming firsthand witnesses, and then participants in the struggle between the Bolsheviks and those affiliated with them—the Gummetists* and the Adaletists,** on the one hand, and the Mensheviks, SRs, Muscavatists,*** and Dashnaks,**** on the other.

It was often very difficult for the theoretically untrained rank-and-file workers of Baku, especially those who had been members of the downtrodden peasantry (there were many like this) and had entered the ranks of the proletariat only recently, to grasp the essence of these political debates and disagreements. Thus they often swam with the tide and were drawn in by archrevolutionary rhetoric, such as that used by the largest party in Baku at this period, the SRs, whose vagueness and the archrevolutionary coloring of their program, cou-

---

*"Gummet" (Energy) was an Azerbaijanian Social Democrat organization which came into being at the end of 1904, under the auspices of the Baku Party Committee, and had as its purpose mass political work among the Azerbaijanian workers. Basically, this organization held the correct Bolshevik positions, even though during some periods the Mensheviks played a prominent role in it, too.

**"Adalet" (Justice)—a Social Democrat organization which worked under the general leadership of the Baku Party Committee of the Bolsheviks, and which recruited Communists among emigrants from Iran.

***"Muscavat" (Equality)—an Azerbaijanian nationalist party which was formed in Baku in 1912.

****"Dashnaktsutiun" (Union)—an Armenian nationalist party which had sprung up in the early 1890s in Transcaucasia.

pled with their ability to accommodate themselves to local moods, accounted for their numerical superiority.

To jump ahead a little, I'd like to say something about the frequent appearance at that time at mass workers' meetings of one of the leaders of the local SRs, a certain Zvonitsky. It should be said that he was a very experienced speaker. It was his custom to deliver loud and vivid speeches that were essentially empty and demogogic, in which he tried to show that the SR program offered a better solution to the land issue than that of other political parties. He even cited certain figures: if the peasants supported the SRs and accepted their programs, each of them would receive thirty-seven *desyatinas* of land apiece. Then he always made his pitch: "Join the Socialist Revolutionary party!" since at the end of every meeting there was usually a drive to register new members. Admittedly, Zvonitsky's "catch" was often fairly large.

The Dashnaks and Muscavatists were also often very effective at meetings, playing very subtly and cleverly on the backward nationalist moods and prejudices which filled the heads of many Azerbaijanian and Armenian workers.

The Baku workers had to make their way among these diverse and opposing political programs and platforms. But the only party which could help them do this—the Bolshevik party—was, numerically, rather small at that time, since the party was then very strict in its recruitment policies, accepting, on the whole, only the activists among the workers as members. Later on, when we were confronted by great trials, this strict approach to recruitment more than justified itself, but in the meantime the Bolsheviks in Baku had to work under very difficult, complicated, and unequal conditions in regard to the other parties.

The diversity of the political parties which existed in Baku at that time was reflected in the early staffing of the Baku Soviet. For although its first president was a Bolshevik, the majority of those in the Soviet were representatives of the other parties. One shouldn't forget, either, that at first the Baku Soviet existed side by side with an elected Duma, in which the majority of seats had been seized by capitalists, representatives of right-wing socialist parties, as well as Dashnaks and Muscavatists. Soon after the fall of the provisional government this Duma began to aspire to more than the fulfillment of municipal functions and strove instead to achieve political power, thus setting itself up in opposition to the Baku Soviet. The situation was so complex that the Bolsheviks in the Baku Soviet, who did not have absolute power, found themselves poised atop a powder keg for several months running.

Up until then I had never been in Baku. I knew, however, that it was the biggest industrial center in Transcaucasia and that it was renowned for its rich oil resources, its developed industry, and its strong working-class vanguard which had had a long revolutionary tradition.

Modern day Baku is a large industrial seaport, often said to be as beautiful as Naples or Marseilles. I don't know whether these comparisons are justified, but I can say with total assurance that back in 1917–1918 we couldn't have imagined even in our dreams the majestic and modern city of Baku was to become during the years of Soviet rule.

Now, for example, it's almost impossible to imagine Baku as a dirty, trash-filled, litter-strewn city which was always covered with a fine—almost microscopic—coating of sand from the hurricane winds blowing in from the north. It's hard to imagine Baku without buses and trolleys and cars, but instead with antediluvian horse-drawn trams to which the passengers themselves were often harnessed in order to help the horses pull the car up the hill. It's hard to imagine the depressing tinkling of camel bells from the caravans that moved lazily along the then vacant stretches of land where you now see housing units and trees. It's hard to picture Baku without the greenery of its parks and gardens, without its cable cars, its modern sewage system and water supply. But in the far-off years of my youth when I first became acquainted with Baku it was just that—a dusty, impoverished city—and as I walked along its streets and threaded its confusing array of narrow lanes, I could not know that this was the city where I would soon experience one of the most colorful revolutionary periods of my life.

I headed straight from the station to Merkurevsky Street (now Shaumian Street) where the Baku committee of the Russian Social Democratic Workers' party (RSDWP) was located. There I introduced myself to Kote Tsintsadze, the Secretary of the committee, gave him my documents, and told him that I wanted to meet Shaumian, for whom I had a letter from Tiflis. He told me that I'd have to wait to see Shaumian because he usually worked at the Soviet during the day and came to the party committee only toward evening.

The room where we had our conversation was a large one, with people constantly coming in. It was interesting for me to take them in and listen to what they were saying. In a little while a middle-aged man carrying a briefcase literally burst into the room and headed straight for Tsintsadze. He was short and stocky and very animated. I recognized him right away—it was Alyosha Dzhaparidze, whom I had seen recently for the first time at the legal meeting of Bolsheviks in Tiflis. I can still recall his masculine, strong-looking face with its neat beard and moustache. He spoke in a rapid patter, forcefully and expressively, gesturing all the while with his hands. He looked as if he were about to rush off somewhere, and that's the way I always remembered him.

He exchanged greetings with Tsintsadze and began to talk to him.

"You know, Kotè, it's getting really absurd. Out in the Mantashev oil fields an organizational meeting of Armenian workers is being held tonight to bring them into the union—and there's no one who can speak to them in Armenian. Can't the Baku Party Committee find someone who can do the job? Otherwise this important meeting will be ruined. The comrade who's organizing the meeting on the site is a good party member, Mukhtadir, a Lezghin, but he's semi-literate, hasn't any experience, doesn't know Armenian to boot, and can't explain anything clearly to the workers." (I found out later that Alyosha Dzhaparidze was currently the President of the Union of Oil-Industry Workers and had charge of the recruitment drive to get workers into the union.)

Tsintsadze thought it over and then replied that they didn't have anyone qualified for the job at the moment. Then he happened to look over at me and said unexpectedly, "Here's a comrade who just arrived from Tiflis. Maybe he can help."

Dzhaparidze walked over to me, introduced himself, asked me who I was, where I came from, and why I was there. Then he said, "Well, why shouldn't he be able to help? He'll do just fine!"

I began to protest that I was new in Baku, didn't know the situation there, and could hardly manage such an assignment, especially since I had never done any work in the unions.

But Alyosha insisted. "You'll manage," he said. "You'll give a speech and talk about the general tasks of the Revolution, the need to organize the working class, and the role of the unions. All the rest, the purely organizational work, will be taken care of by Mukhtadir. After the meeting he'll be the one to sign up the workers for membership in the union."

I still couldn't make up my mind, but Dzhaparidze was insistent. "Go, go! I'm sure it'll work out fine. This'll be a good start for you here in Baku." He went on to tell me that the meeting was to take place in Zabraty in the courtyard of the school of the Union of Oil Manufacturers. "When you get to the school," he said, "find Mukhtadir and tell him that Dzhaparidze sent you, and he'll do everything that's necessary."

Seeing that further objections were futile, I agreed. After I found out how to get to Zabraty I headed for the station, boarded a local train and went to Sabunchi, the last stop on the line. From there I set out on foot to Zabraty, arriving around four or five in the afternoon. Sure enough, when I got near the school I saw a crowd of workers.

One of the workers, who was thin and taller than average, kept looking down the road, evidently waiting for someone. I realized that he must be my "organizer," so I headed straight for him.

He was Mukhtadir, all right, and within minutes he opened the meeting and then let me have the floor. I spoke for about fifteen or twenty minutes, following roughly the format outlined by Alyosha, ending my speech with an appeal to the workers to join the union. Mukhtadir then signed up those who were interested, collected the entrance fees, and handed out receipts. As I recall, quite a few workers signed up for the union, and Mukhtadir was very pleased with me. And that's how, with Dzhaparidze's blessing, I began my political work in Baku and my "debut" as a union activist.

Late in the evening on my return to the city I stopped off again at the Baku Party Committee headquarters and this time I found Shaumian there. I told him my reasons for coming to Baku and presented the envelope from Shaverdian.

This is what Shaverdian had written to Shaumian on the back of his visiting card:*

> Dear Stepan! The bearer of this note, Anastas Mikoyan, is a newly baptized Social Democrat who has had good training. I send him to you to help in the struggle against the Dashnaks. He's a very able young fellow. I ask you to give him your special attention. He'll tell you about the current state of affairs.
>
> Yours, Danush

I had never seen Shaumian before, but I had heard a lot about him from Shaverdian and other oldtime Bolsheviks. I knew that Stepan was one of the most prominent of our Soviet leaders, and that he was personally associated with Lenin and enjoyed his unlimited confidence. I also knew that he had embarked on the revolutionary path when he was still very young, organizing in 1899, when he was a student at the Tiflis Technical High School, the first Marxist circle in Armenia in the small town of Dzhalalogla (now the city of Stepanavan), and that after two years at the Riga Polytechnical Institute he was expelled because of his revolutionary activities among the students.

In 1902, together with Knuniants, Shaumian headed the first Armenian Social Democrat organization, "The Union of Armenian Social Democrats," which immediately became part of the Russian Social Democratic Workers' party.[6] That fall Shaumian entered the philosophy department of the University of Berlin where he participated in the work of his revolutionary minded Russian compatriots (living abroad), attended the party meetings of the German Social Democrats, and became acquainted with their distinguished spokesmen.

In 1903 he realized a cherished dream—he met Lenin. He worked under Lenin's leadership in Geneva for some time, proofreading a Marxist publication in Armenian and Georgian, thus beginning a last-

---

*I found out about the content of Shaverdian's note many years later when Shaumian's youngest son Sergei sent me this visiting card—which had been carefully preserved by his mother—as a gift.
[6]See Appendix.

ing relationship with Vladimir Ilyich which continued throughout Shaumian's life.

During the stormy years of the first Russian Revolution Shaumian was one of the leaders of the Transcaucasian Bolsheviks. During the years of reaction and the new revolutionary upsurge he fought actively against the liquidators, the Trotskyites, the recallists and the conciliators in staunch support of Lenin's policies. "We are firmly convinced of the correctness of [Vladimir] Ilyich's positions . . ." he wrote on July 27, 1908, in a letter to Tskhakaya.

In June 1911 at Lenin's suggestion Shaumian was made a member of the Russian Organizational Commission on the Convocation of the Sixth All-Russian (Prague) Party Conference. The meeting of this commission was organized in Baku by Sergo Ordzhonikidze, who arrived from abroad specially for the occasion. In the course of the meeting, Shaumian was arrested, along with a group of Baku Communists, but at Lenin's suggestion he was co-opted, in his absence, as a candidate member of the Bolshevik Central Committee.

Vladimir Ilyich valued Shaumian's opinion very highly and asked him to write to him abroad more frequently. Shaumian complied by providing Lenin with essential information on what was happening in the country and in the party. Vladimir Ilyich was especially interested in Shaumian's opinion on theoretical and practical problems relating to the resolution of the nationality issue.[7] Shaumian had long been concerned with the nationality issue. In 1906 he had put out a pamphlet entitled *The Nationality Question and Social Democracy*. In 1914 in an article "On National Cultural Autonomy," he exposed those nationalists among the Social Democrats who advocated the separation of the workers into different nationalities. From Poronin, Lenin asked Shaumian to write a summary of this article for the Bolshevik journal *Enlightenment*.

Shaumian returned from exile in 1914 and became head of the Baku party organization. He led a famous general strike of oil industry workers and provided Lenin with detailed information about its progress. A Leninist perspective always illuminated his speeches, reports, and resolutions on such subjects as the attitude to the war, nationality policy, and the building up of the party. Early in 1916 Shaumian was arrested again and later exiled to Saratov where he was living when the February Revolution broke out.

He had vast authority among the workers of Baku, and everyone who had any dealings with him had a warm and sincere affection for him. Stepan's devoted companion throughout his life, Ekaterina Sergeevna Shaumian, told how in the most difficult days, when Stepan was in Tsarist prisons, the workers would donate a few kopecks apiece and collect a few rubles which they sent, via their wives, to "Stepan's

[7]See Appendix.

old lady," as they called Ekaterina Sergeevna, of whom they had grown extremely fond.

Soon after our first meeting Shaumian took part in the first All-Russian Congress of Soviets in Petrograd, where he was elected a member of the All-Russian Central Executive Committee. A little while after that at the Sixth Party Congress he was elected a member of the Central Committee of the party (although he wasn't present), and later brought into the so-called inner circle of the Central Committee. Shortly afterwards he took part in the historic meeting of the Central Committee of the party in Petrograd where they discussed Lenin's famous letters on armed revolt.

Shaumian made an indelible impression on me on the first day we met. He was a little taller than average, well-built, and very handsome, with a memorable, wise, or "intelligent," as we used to say then, face, and a kind, gentle smile. His rather pale face with its light blue eyes—rather rare among Caucasians—was shown off to advantage by his dark moustache and his small, carefully clipped beard. (When I was young and tried to imitate Shaumian, whom I loved and respected, I cut my hair "à la Shaumian" for a long time.)

In contrast to Alyosha Dzhaparidze, who was loud and somewhat effusive, Shaumian was quiet and even-tempered. He wasn't talkative; you felt that he literally pondered every word. Everything he said was well thought-out, logical, and convincing. Afterwards, when I heard him speak at mass workers' meetings and rallies, I became convinced that he was also a passionate orator and a genuine tribune of the people. He and Dzhaparidze were similar in this regard: they both spoke simply and intelligibly and at the same time with such authority that it wasn't easy to enter into a polemic with them. But most important of all, the masses believed and followed them, and for my part, they provided me not only with an excellent education but with real happiness as well.

Let me return, however, to my first meeting with Shaumian.

When Shaumian had read Shaverdian's note, he said, "Well, it's good that you have come here. We need good party propagandists, and Shaverdian recommends you highly! We'll try to fix you up with some kind of job, even if it's as a telephone operator. That's an easy job, and you don't need any special knowledge or experience for it."

He then wrote a letter to a friend of his who was a manager at the Mantashev oil fields, asking him to take me on as a telephone operator.

I spent the night, with Tsintsadze's permission, at the party committee headquarters and early the next morning I set out for the address which Shaumian had given me. I reached Balakhani on foot, got hold of the manager and gave him Shaumian's note. He read it, shook his head, and said that, unfortunately, he couldn't fulfill Shaumian's request because there were no vacancies whatsoever.

There was nothing to do but to return to Baku.

I didn't have any money for a hotel so once again I spent the night on the newspaper-strewn table of the Baku Party Committee.

In the morning I went to see Shaumian at the Soviet, told him about my unsuccessful trip, and asked if there wasn't some other job possibility.

"You'll have to go to Bibi-Èibat (a section of Bailov)," Shaumian replied. "At the Pitayev oil fields there's an engineer named Pirumov. He's a Social Democrat and a very good comrade. Tell him I sent you and gave him this note." He wrote it out quickly on a scrap of paper. "Maybe he can find you some sort of technical work."

With difficulty, I managed to arrive at the right place and found an office, but they told me that Pirumov was out somewhere at the oil wells. I went to look for him, asking one person after another until I finally found him. He was embarrassed after reading Shaumian's note, and spent a long time apologizing because he couldn't do anything for me. I had to go back to Shaumian. Another "stroll," this time to Belyi Gorod, and again no success.

By now I felt very awkward about going back to see Shaumian, so I decided to talk to Tsintsadze about my situation since my financial reserves had dried up completely. Tsintsadze provided me with a small food allowance from the committee's funds which sufficed for ten days or so. And in the meanwhile I began to carry out various assignments for the party committee—I traveled throughout the districts, went to meetings, chatted with workers, and delivered speeches.

Shortly after, when the comrades on the committee had become convinced I could be a useful party worker, they hired me as a paid worker, and I became a propagandist for the Baku Party Committee.

* * *

I spent every day from early morning until late at night at the oil fields of Sabunchi, Balakhani, Zabrati, and Bibi-Èibat.

In Balakhani, Sabunchi, and Zabrati, the man who usually organized the workers' meetings was a middle-aged Dagestanian named Kazy Mamed. He was thin, of medium height, with a thin face, long, sharp nose, and bright "southern-looking" eyes. Unfortunately, he was illiterate and spoke Russian very poorly. He was, however, committed to the Revolution and to the cause of the working class to the point of fanaticism. He knew the city of Baku inside out and took me around from one site to another, introducing me to various people and informing me about the workers' jobs and their way of life. We went everywhere on foot since there was no other means of transportation at the time.

At first, meetings were held chiefly for the Armenian workers. There were no workers' clubs at the oil fields then, so the meetings were usually held in the workers' cafeterias, both before dinner and after. About fifty or sixty people would attend. The main speaker would talk for forty or fifty minutes, and then the workers themselves would speak, bombarding us with questions.

By this time I was totally immersed in propaganda work. The workers who made up my audience were, for the most part, politically backward and their heads had been filled with all sorts of prejudices, mainly of a nationalist character. In my speeches I hammered away at the following points: the unification of the workers around our party, the need to work for the cessation of the war and for a just peace, for the transfer of all gentry lands into the hands of the peasants, for worker control over production, and for a transfer of power to the Soviet of Workers' Deputies.

The basic content of my speeches was roughly the same at all the meetings because the audience was always different, although their level of development remained constant. I worked only occasionally at improving my speeches, therefore, and it was quite obvious when I surveyed my audiences that my arguments, my methods of reasoning, and the facts I cited were becoming clearer and more intelligible to the workers. I was happy that the size of the audience kept increasing each time.

However, as before, I had barely enough money for food and I couldn't even think about an apartment so I asked for permission from Tsintsadze and another worker on the committee—Sultanov—to spend the night in the Baku committee headquarters. Late in the evening when the work of the committee was over, I would spread newspapers out on the table, fashion makeshift pillows for myself out of stacks of different papers, and go to sleep. I didn't need a blanket because the weather was hot and, in addition, I slept in my clothes. It was agreed that I would get up early, around six o'clock in the morning, and put the room in order, because worker-activists came to the committee in the mornings on their way to work to pick up copies of the newspaper *The Baku Worker*,* for distribution among the workers in the oil fields.

I confess that things were rather hard for me at that time. Even though I was still very young and had a lot of strength, I almost never had a good night's sleep. Usually I would return "home" rather late, exhausted after my day's work, fall off into an uncomfortable sleep and then, bang, be up on my feet again at six o'clock.

But despite it all, I remember this period of my life with special warmth.

* * *

---

*The Baku Worker is a Bolshevik newspaper which began publication in 1906.

I was in the very thick of the political life of Baku and I knew what made our Bolshevik organization tick.

Once in May 1917 Shaumian invited me home. He lived on the outskirts of the city in a small house which stood right on the mountain slope. His lodgings were on the second floor and consisted of three small rooms which were very modestly, even meagerly, furnished. I entered his apartment from the yard, on the mountain slope. I didn't even have to go up a staircase; on the contrary, I walked down a few stairs and found myself right on the veranda. This was my first meeting with Stepan's wife and family. He had three sons and one daughter.

Shaumian and his family greeted me warmly and affectionately. I think that the comrades on the committee had given him a fairly good account of my work because he was favorably inclined to me and seemed to trust me immediately. I remember how once he proposed that I join the editorial staff of the Armenian weekly newspaper *The Social Democrat*. Stepan had been made editor of this paper but couldn't devote sufficient time or attention to it, and he wanted me, who knew Armenian very well, to help him edit the paper.

I agreed, after telling him that I would have to rely heavily on his help because I had had no experience whatsoever in newspaper work. And so I began to work on the newspaper, and subsequently became its editor, never ceasing, however, to carry out organizational and propaganda work for the Baku Party Committee.

At that time the moving spirit among the soldiers, their guiding light, was the Military Commandant of Baku—Ensign Avakian. I would like to say a few words about him. He was a courageous, self-sacrificing man. He had his eccentricities, to be sure; for example, he stubbornly insisted that he was a Social Democrat, but abstained from joining either the Bolsheviks or the Mensheviks. He had become commandant of Baku after the February Revolution, and he was totally preoccupied with soldiers' meetings. These were his passion.

On June evenings, when the heat had died down, the endless meetings on Freedom Square began. The soldiers had erected a wooden platform on the square, and on it a podium from which Avakian would also speak, sometimes two or three times a night. During meetings he could usually be found on the platform, not far from the podium, sitting on his sturdy portable stool which he always carried with him. Avakian knew that the soldiers loved him. He spoke very well, always using words and arguments that were both simple and convincing. His outward appearance was unusual: he wore a black cloak and strange headgear—it was not an officer's cap, but neither was it a regular soldier's cap. He was tall and very thin. I thought he looked like Mephistopheles.

The most diverse speakers talked at these meetings, not only official representatives of the political parties. In their speeches, which were awkward and confused at times but which came from the heart, the speakers expressed the secret thoughts and aspirations of the people. For this reason, the soldiers listened to them willingly, evincing a lively interest in the many speeches.

At that time the overwhelming majority of the soldiers were still prey to the slogans of "revolutionary defensism." I was anxious, therefore, to set forth the Bolshevik position on this issue, even though this was rather risky at the time because an open speech against the war could provoke the soldiers and turn them against me. I had already seen how they had shouted and unceremoniously interrupted those speakers who called for an end to the war! Nonetheless, I managed to get up the courage: the die was cast!

And so, at one of the meetings I spoke up on the Bolshevik position. I had thought over what I would say and decided that I would describe in an intelligible way the general principles of our policy. Then if I succeeded in capturing the audience's attention, I would bring up the issue of the war at the end of my speech. I told them that there had been enough speeches on the transfer of land to the peasants: for how many months had we talked about that! The peasants should seize the land themselves, I said, and for that you didn't need to wait for the Constituent Assembly.

Seeing that my words met with their approval, I continued, "While our brothers in the trenches are suffering incredible hardships and are perishing, ill-clad, hungry, and dirty, in the war, and their families are experiencing hunger and poverty, the war is reaping huge profits for the capitalists, who wallow in luxury, feast at restaurants, and meanwhile try to keep their sons a little farther from the front, at the rear." Since these words struck an especially sympathetic chord, I moved on to my main point: "It's clear who benefits from the war—it's only the capitalists. Therefore we have to put an end to it. But Kerensky's provisional government has neither the will nor the means to do this; *we* must overthrow it and transfer power to the workers and peasants."

My last words were listened to with rapt attention. Then shouts of displeasure were heard and a group of hostile soldiers began to approach the podium. A racket broke out. But by now I had finished my speech, and stepping down from the podium, I left without looking back. The comrade who accompanied me said later that I had been right to leave when I did, because they had wanted to settle accounts with me.

During my speech Avakian had been stubbornly silent. Shortly after this, however, he, together with the soldiers, began to "move quickly to the Left," and during the October Revolution he adopted a completely Bolshevik position. He led the Baku garrison in support of Soviet power and became one of the rightist parties' most hated enemies. Avakian died the death of the brave, as a Communist, together with the Baku commissars.

* * *

During that time we still hadn't split off organizationally from the Mensheviks, remaining joined with them in a single, unified group. Within the Baku Party Committee, however, the Bolsheviks were stronger than the Mensheviks, both in numbers (seven of the nine members were Bolsheviks), and in terms of influence. Such "unification" proved to be a special hindrance when it came to strengthening the Bolsheviks' influence among the working masses, and it also tended to undermine the policy position of our newspaper *The Baku Worker* in its efforts to elucidate the most important issues of current political life.

I had the occasion once or twice as a party activist to attend the meetings of the joint Baku Party Committee, and I was a living witness to the heated debates which took place between Shaumian and Dzhaparidze, on one side, and Aiyollo and Sadovsky, the Mensheviks' leaders, on the other. Naturally, their positions on the most important political issues were diametrically opposed to each other, and, what's more, mutually antagonistic.

This made it seem as if there were two separate parties inside a single organization. The Bolshevik members of the Baku Party Committee met separately every week to discuss their own matters, and to meet with fellow comrades from the many city districts where the Bolsheviks were gradually taking over command, thus preparing for an inevitable organizational split with the Mensheviks.

The evils of this situation became especially clear to us after Lenin's first speeches of April on his return from his exile abroad, which made us see that the task of escalating the bourgeois-democratic revolution into a socialist revolution called for an immediate and final split with the Mensheviks.

After the split had been agreed upon in principle at a meeting of the Bolshevik leadership of Baku, our comrades wished to put it into practice in such a way that most of the party apparatus in the different districts would be preserved intact, as would the allegiance of those workers who were vacillating between the Mensheviks and the Bolsheviks. In addition, we also wanted to keep in our camp the so-called Menshevik-Internationalists who were then assuming a Bolshevik position in regard to the war.

On account of these considerations, the matter of a split with the Mensheviks tended to drag out.

In the beginning of May Mikha Tskhakaya and Philip Makharadze arrived from Petrograd fresh from the Seventh All-Russian Party Conference of Bolsheviks which had taken place in April under Lenin's leadership. I attended a meeting with them, along with ten or twelve others, at the apartment of Victor Naneishvili, a member of the Baku Party Committee. Mikha Tskhakaya described in detail how Lenin's departure from Switzerland, along with that of a group of Bolsheviks which included Tskhakaya himself, had been organized.

He said that as soon as Lenin heard of the overthrow of the tsar he was most anxious to return to Russia. You could leave Switzerland at that time only through France and England, who were Russia's allies. However, only those Russian revolutionary exiles who were adherents of the policy of "defensism," that is, who supported the continuation of the war, were allowed passage by these allies into Russia.

Efforts were undertaken to secure visas, but they didn't lead to anything. With the help of Fritz Platten, a Swiss socialist-internationalist, who was later to become a leading figure in the Communist movement, Lenin arranged for negotiations with the German authorities to permit passage through their country into Russia for a group of Bolsheviks led by himself. Lenin didn't take part himself in the negotiations with the Germans and didn't meet with them. A stipulation was agreed upon whereby the railway car carrying Lenin and the group of Bolsheviks would be sealed so that they would have no contact with the Germans. All negotiations with the Germans concerning these matters were carried out by Platten directly, so that Lenin and his comrades could not be accused of having made a "deal" with the Germans. Tskhakaya told how the German Social Democrats, who were supporters of the war with Russia, wanted at one point to meet with Lenin, but he categorically refused their request.

Tskhakaya described in detail Lenin's reception at the Finland Station in Petrograd; he spoke of his first speeches and of the significance of his April theses which had already appeared in print by that time.

Philip Makharadze gave a report on the April Conference, which he had attended as a delegate of the Tiflis organization.

Shaumian was the next speaker and he informed his newly arrived comrades of the Bolsheviks' plans to split off from the Mensheviks in the near future and of our intention to bring about the split in such a way as to isolate the Mensheviks completely and retain for the party the allegiance of the majority of the Social Democrats among the workers.

When Tskhakaya found out that in Baku the Bolsheviks were working with the Mensheviks in one joint organization his indignation knew no bounds. He noisily got up from his seat and, after announcing that he could not attend a meeting with "the united," made a dramatic exit from the room.

Alyosha Dzhaparidze, Victor Naneishvili, and several other comrades went out after him and gave him the details of the situation. They told him the resolution to break with the Mensheviks had already been passed and that the Bolsheviks of Baku supported Lenin's April Theses and were in complete agreement with the resolutions of the April Conference of the Bolsheviks.

Tskhakaya returned and the conversation lasted until late at night.

Shortly afterwards, at a joint meeting of the Baku Party Committee, a resolution was passed to convene an All-Baku Party Conference. Meanwhile the Bolshevik leadership of the committee carried on extensive preparatory work in the district party cells in order to guarantee a majority of delegates from the Bolsheviks, and, where that was impossible, to work for the election of Menshevik-Internationalist delegates. After the delegates were elected the Bolshevik leadership set the day for the opening of the conference—June 25. The conference was to be held, moreover, not in Baku but Balakhani, and only the Bolshevik delegates and the Menshevik-internationalists were notified of the opening date. The Menshevik delegates did not receive any notification so that, when the conference opened, the Mensheviks ended up "overboard."

On the second day of the conference, when news of the split and of our unwillingness to maintain an alliance with the Mensheviks had become known, a delegation of Mensheviks, consisting of Isidor Ramishvili and Bogaturov, came to the conference. Isidor Ramishvili was a short, gaunt old man with a white beard, who had a good speaking voice and looked like a [biblical] prophet. He made this appeal: "Comrades, don't leave us, let's remain together in the Marxist ranks. If you leave, then you'll veer more to the Left, closer to the anarchists, and the Mensheviks will turn more to the Right and be closer to the parties on the Right. If we split up now, then we'll never come together again. I appeal to you, Comrades, to restore the unity of our ranks!"

Ramishvili's speech, even though it was inspired and eloquent, was not supported by anyone, not even the Menshevik-Internationalists. So the delegation of Mensheviks left the conference without accomplishing anything. The split was now official.

Mikha Tskhakaya had brought back from abroad some interesting Bolshevik publications of the year 1916, and two or three issues of the journal *Communist*, printed on thin paper in Switzerland. I remember the December issue. It contained an obituary notice from the party's Central Committee concerning the death of Central Committee member Timofei (this was the pseudonym of Suren Spandarian). He died in exile in Turukhan territory.

Tskhakaya left one copy each of all of the journals he brought back for the Baku Party Committee.

Lenin's article "On the Tasks of the Proletariat in the Current Revolution," which contained his April Theses, made a strong impression on me. Lenin illumined our future path like a flash of lightning, summoning everyone to begin immediate preparations for the Socialist revolution, clearly defining the driving forces of this revolution, planning party tactics in the conditions of "dual power" which then existed, and thoroughly substantiating his slogan "All Power to the Soviets!"

Lenin's April Theses were fated to play, on the whole, an extremely important role. They were the theoretical foundations of our party's new political policy at the time, and became the call to arms of its practical plan of action as well.

I remember, however, Lenin's proposal concerning the renaming of our party as the Communist party was not well received by many of the old party members in Baku: they had become accustomed, it seems, to the old name—"Social Democrat"—and didn't want to part with it. As is well known, it was almost a year later before our party came to be called the Communist party.

This proposal had immediately appealed to me, however. I couldn't understand how we could call ourselves "Social Democrats" after Marx and Engels had called their historic Manifesto the *Communist Manifesto*. I liked the name "Communist" so much at that time that I even chose it as my literary pseudonym and began to sign my articles more and more frequently with the pseudonym: "Communist," or, in abbreviated form, "C-st."

* * *

Toward the end of July 1917 my health got much worse. Too much work, lack of food and sleep had taken their toll. For several months I had spent every day from dawn to dusk visiting the industrial districts of Baku, going on foot from one plant to another, speaking two or three times a day at workers' and party meetings, but it was the Baku heat, to which I was unaccustomed, that finally undermined my health. I became terribly thin, and began to have dizzy spells.

One day at Shaumian's invitation, I stopped by at his apartment. He questioned me in detail about my work and impressions and was concerned about my health. When I explained the way I was living and how I had to work, he suggested that I go straight home to my family in the village, recover my strength, and return to work only after my health had improved. "The heat here in Baku can get even worse," he said, "so why ruin your health completely!"

Then Shaumian told me that he intended to send his second son, Leva, home to the country, so that he too would be able to rest and recover his strength. "Our family's village," Shaumian said, "is in the same district not far from yours. Why don't you and Leva go together?"

I agreed, feeling that I really needed to rest and gather my strength for the future.

By the beginning of August, Leva and I had left for Tiflis where we boarded a train for Erevan. My village was about 100 kilometers away, and Leva's destination—a small, summer-resort town (today Stepanavan)—was about 125 kilometers from Tiflis.

As our train entered the narrow Lori Gorge of the Debet River, we kept marveling at the beauty of the wild scenery and at the gigantic cliffs that stretched along both sides of the gorge. By some kind of miracle, large as well as small trees grew on these cliffs. The Debet River—small, but swift—was completely covered with foam at its steep rapids. As the air became fresher and fresher, it seemed to me that there was no more beautiful place in the world. Although at that time I still hadn't seen the world and couldn't imagine that there existed such beautiful places as the shore of the Black Sea, Lake Ritsa and the road leading to it, the resort of Dzhermuk in Armenia, Teberda on the northern slope of the Caucasus, the beauty of the Kolsky Peninsula, the astounding beauty of Cuba, Bali in Indonesia, the Grand Canyon in the United States, and other remarkable corners of the globe.

I got off at Alaverdy Station, and Leva continued on.

Upon meeting me, my mother was as usual beside herself with joy. She embraced and kissed me, exclaiming over how thin I had gotten. She immediately started worrying about what tasty foods she could feed me. My father, of course, was equally delighted to see me, but he appeared reserved, as always.

My younger sister and brother were particularly glad to see me. Naturally I loved them very much, but, like my father, I didn't give vent to my feelings—I didn't even kiss them—in a word, I tried to act like a "real man," imitating my father in everything.

My little brother and sister had really grown up. My sister had become a real lady: she was seventeen, while my younger brother was just about to turn twelve. He had gotten taller, and was thin and underdeveloped the way I had been when I was his age. He was in elementary school and when I asked him how school was going, Artem replied that he was doing well. My mother confirmed this and praised him.

When news of my arrival reached my older sister, who was married by then and living in her husband's house on the outskirts of the village, she came over for a visit.

I devoted the first part of my visit to regaining my strength, letting myself enjoy the pure mountain air, and the warm, sunny days. I slept a lot, ate fairly well, and did a little reading.

When I began to feel better, I started talking more and more often with my fellow villagers. They had changed. Formerly, they had never thought about politics, but now everything interested them: what was happening where, what would the future bring? The overthrow of the tsar seemed like something incredible to them, so firmly had they believed in his omnipotence. They had questions about literally everything, since they knew that I had just returned from Tiflis and Baku, otherwise they would hardly have been interested in my opinion on current events, since from their point of view I was still too young to have one. . . .

That was precisely why I felt that, in spite of the outward respect and attention they gave to my comments on the prospects for a socialist revolution, they could not hide their skepticism. They simply could not believe, for example, that ordinary workers, for the most part illiterate (as was the case in our Alaverdi Copper Smelting Factory), could chase out the bosses (who were, after all, French!) and run the business themselves. Nor could they believe that they would be able to join forces with illiterate peasants, drive out the princes and noblemen, seize their lands, and take power into their own hands. The Bolsheviks' demand to end the war greatly excited and moved them, however, since they had all been waiting for a long time for their sons to come home from the front.

Naturally, I got in touch with the party organization at our factory, and attended several meetings of the local party organization, where I made speeches about the political state of affairs in the country, the situation in the party and its tactical approach to the struggle. I also spoke at two general political rallies for factory workers, and I spent time in other, neighboring villages, including Akhpat, the most revolutionary village, where the majority of peasants were Communists. The Akhpat peasants were struggling actively to seize the gentry's lands in the entire surrounding area.

Antilandlord sentiment had existed in the towns of our district for a long time and it made the regional and provincial authorities very nervous. The majority of the peasants and factory workers in our district supported our party. At that time the Dashnaks were our main enemy; the SRs and Mensheviks were less important.

Characteristically, in contrast to their performance in other districts of Armenia and Georgia, the Communists of our district managed to win five times as many seats as the Dashnaks in the elections for local organs of self-government, and thus were able to ensure their decisive influence in these organs. At that time our party organization did not include a single member of the intelligentsia: leading workers and peasants comprised the entire active force. Some of them had had prerevolutionary service in the party and a lot of experience in the revolutionary struggle. Kakhoyan, a Bolshevik and native of Akhpat, was their leader and their inspiration.

After I made my first speeches, the whole village found out that I was a Bolshevik. The news finally reached my mother. One day she sat down beside me and began this conversation:

"You're such a learned, intelligent young man, yet everyone is saying that you're a Bolshevik. I've heard that there are many good parties: the Dashnaks, SRs, Mensheviks. . . . The most esteemed and respected people in our village have decided to support those parties. But people say you have joined the worst one and have become a

Bolshevik. You're such an intelligent man, why don't you quit the Bolsheviks and join another party?"

She spoke in such a pleading tone that I had to think how best to answer in order to make her change her mind.

I said that the Bolsheviks were the best party, that it was fighting for truth, for the poor and against the rich, for justice for working people.

"What you've heard about the Bolsheviks," I told her, "is a lie spread by the worst people. And those parties which they've told you are good are, in fact, bad."

My mother listened to me, but I could tell from her eyes that my arguments had little effect on her. Then, a new way of convincing her suddenly occurred to me.

"*Mairik* (Momma)," I said, "could you renounce Christianity and become a Moslem?"

My mother shivered, crossed herself, and said excitedly:

"Why, what are you saying, my son, how can this be! I'd rather die than do anything like that."

Then I said to her:

"I understand how you feel. So you should try to understand me, too. The Bolsheviks are my faith, the same as Christianity is for you. I can't renounce them. So, please, stop trying to persuade me—it's useless! In time, you yourself will realize that the Bolsheviks and I are in the right, but for the time being, you'll just have to take my word for it."

This had an effect on her, although it didn't make her very happy. Since she trusted me, however, she resigned herself to it, and never brought the matter up again.

From 1923 on, after my father had been dead for several years, she lived with me: first in Rostov, then in Moscow in the Kremlin, and was very pleased that her son commanded such great respect in the country.

It is typical of her that she never prayed or went to church in my presence. Perhaps she didn't know that there was an Armenian church in Moscow. For my part, I never spoke with her about God or religion in our family. I even thought that she had stopped believing in God.

In 1959, when I was returning from a trip to America on a Scandinavian Airlines plane, two of the four motors conked out while we were flying over the ocean, so disaster seemed imminent. Somehow news of this incident reached my mother, who said to me when I finally reached home safely, "I was worried about you and prayed constantly that you would come home from that country alive!"

I looked at her in amazement and said:

"*Mairik*, can it be that you still believe in God?"

"And how can one live without God?" she answered simply.

My mother died in 1960, in Moscow, at the age of ninety-three.

I never talked politics with my father. He never asked any questions about it, while I, for my part, felt embarrassed about getting involved in such conversations with him, especially since they might mean my having to teach him something.

Since my father was a carpenter at the factory, he heard the workers talking and knew what was going on politically. Somehow he found out that socialists had appeared at the factory and that they wanted to overthrow the bosses.

My father was working at the time for a certain Greek contractor, who valued him not only as a qualified worker, but in general as a deserving man. Sometimes my father would tell my mother in my presence that his boss had invited him home after work "for a cup of tea" on his terrace. These invitations obviously flattered my father: in any case, he always spoke of his boss with respect.

In all his sixty years my father responded to talk about revolution with skepticism. Once, in an obvious attempt to catch me off guard, he said to me completely unexpectedly:

"You know, some of those 'dodo socialists' have turned up at the factory. They say you're one of them. Give it some thought! Why, you're still wet behind the ears, yet you want to overthrow the powerful, respectable bosses. Nothing will come of it!"

I told him he was very mistaken. "The bosses aren't as respectable as you think," I retorted, "they all live off the sweat of the workers. At the moment, of course, there aren't many of us socialists, and we're still weak. But you will live to see us win and have our way!"

After that I made an effort not to argue with my father anymore, knowing that nothing would come of it except dissension.

My father was still alive during the October Revolution and came to understand a great deal by the time of his death. He died from pneumonia, without medical assistance, at the age of sixty-two, in 1918, before the victory of Soviet power in Armenia.

Chapter V

Preparing for Armed Revolt

Creation of the Union of Young People—I am elected Secretary of the Tiflis Party Committee—The All-Caucasian Party Congress—On the eve of revolution.

At the end of August 1917, completely restored to health, I left the village and returned to Tiflis where I spent the first few days seeing my comrades and visiting the Tiflis Party Committee to catch up on the latest political news.

At the time several of my former classmates had decided to go back to school, and they tried to convince me to do the same, arguing that the proletarian revolution would soon be victorious and that highly educated people would be needed to build socialism. Their arguments made sense, and for a while I was persuaded, particularly since the Tiflis municipal government had offered me a rather substantial scholarship, and our diploma now entitled us, after the February Revolution, to enter any institution of higher learning in Russia. In the end, however, I made up my mind not to go on to college, but rather to continue my education at "the university of the Revolution" instead. I have never regretted this decision.

My comrades from the Marxist circles, particularly Boris Dzneladze and Gevorg Alikhanian, encouraged me to start a Bolshevik Union of Young People in the Caucasus, and we set about drawing up the proposed statutes for this organization. I was given the job of writing a manifesto which was to replace the program of the union.

A constituent assembly for the Union of Young People was held in the party club in Avlabar, and this meeting generated great enthusiasm. After general discussion, the two documents proposed by our planning group were accepted, and a provisional committee for the Union was elected.

The manifesto of "Spartak" (as the newly created Union of Young Socialists and Internationalists was called) was directed at all working youth in the Caucasus. It contained, in particular, the following:

Comrades! Spartak, an organization of young socialists and internationalists, turns to you, to all who hold dear the interests and future of the proletariat, who are interested in liberating man from the yoke of slavery and oppression, whose heads have not been turned by chauvinist passions, who have not besmirched the lofty principles of militant socialism with patriotic trash; Spartak turns to you, the young fighters for the future, and calls upon you to create your own local and general organizations . . . and always and everywhere to give your ardent support to the revolutionary movement of the international proletariat . . . particularly in the Caucasus, where the air has been poisoned by chauvinistic passions and by nationalistic dissension, which provide fertile soil for opportunism of all kinds: here the organizations of internationalist youth will have to overcome even greater obstacles.

Comrades! Spartak, an organization of young socialists and internationalists, has dedicated itself to the difficult task of unifying and organizing internationalist youth in the Caucasus, and is guided by the basic principles stated in this manifesto.

The Spartak organization is beginning its work full of high hopes and aspirations, and it summons all young socialists and internationalists in the Caucasus to offer their total support and to assist in the internationalist organization of youth in the Caucasus.

The name "Organization of Young Socialists and Internationalists—Spartak" was borrowed from the revolutionary union "Spartak," created in Germany by Karl Liebknecht and Rosa Luxemburg. It was no accident that we chose the names "Socialists and Internationalists" instead of "Social Democrats and Bolsheviks." We wanted to encourage the participation of those young leftists who had still not defined themselves as Bolsheviks, but were sympathetic to our view of revolution.

It should be noted that around this time the Georgian Mensheviks and the bourgeois parties had begun to organize an interparty, or supranational, Union of Georgian Youth. The Armenian nationalists, headed by the Dashnaks, had done the same thing.

Because of this, in late September 1917 Spartak issued an appeal entitled "To All Working People in the Caucasus," in which it criticized the Georgian and Armenian youth groups, exposing the traitorous, antipopulist character of their plans. We called upon the youth to reject bourgeois national unity and a "class truce" in favor of an international union of workers, and an uncompromising class struggle against the counterrevolutionary propertied classes.

The appeal ended with the following summons:

Comrades, join us in the merciless war against the nationalism and chauvinism implicit in the creation of the Interparty Council and the Armenian All-National Conference.

Down with the counterrevolutionary capitalist and landowning classes!

Long live international solidarity among the workers of all countries!

Long live the All-Russian Revolution!

The Provisional Committee for the
Organization of Young Socialists and
Internationalists—Spartak.

During that time the Tiflis committee kept me busy each day with numerous organizational and propaganda assignments. I undertook them willingly since I had no other employment at the time. I was living at the apartment of my relatives, Lazar and Verginia Tumanian, and planned to return to Baku in the near future, as had been arranged with Shaumian.

My comrades in Tiflis asked me to postpone my trip to Baku since the All-Caucasian Party Congress was soon to be held in Tiflis, as a result of a joint decision by the Tiflis and Baku party organizations, and they wanted me to help prepare for the congress and to take part in it. The date of the congress, meanwhile, was rescheduled from September 15 to October 1 because the two most important Bolshevik leaders in Transcaucasia, Shaumian and Tskhakaya, had gone to Petrograd.

* * *

The work of the Tiflis Party Committee was now growing by leaps and bounds, but no one on the committee could devote full time to party work. The committee's assignments were carried out by its members and by individual communists in the evenings after their regular work. Representatives of lower party organizations would arrive daily from their respective locales only to find no one at the committee during the daytime. Because of this bizarre circumstance, the Tiflis Party Committee passed the following resolution in the middle of September 1917:

"All members of the committee agree on the necessity of having a full-time secretary in the office who will direct the business of the Tiflis committee, and who will, up until the convocation of the Regional Congress, maintain relations with the provincial organizations. Comrade Mikoyan has been elected secretary."

Thus I became the secretary of the Tiflis Party Committee and immediately plunged headlong into my duties, which had to do chiefly with handling various tactical and organizational matters.

To jump ahead a little, let me say that after the All-Caucasian Party Congress, some of the important members of the Bureau of the Tiflis Party Committee left to work on the staff of the Regional Party Committee. Because of this, a new staff was elected to the Bureau of the Tiflis Party Committee, and I was again elected secretary. The bureau provided the leadership for the Tiflis committee's work. Philip Makharadze usually presided at its meetings although officially our committee did not have a president.

At that time many people came to the committee both on foot and by vehicle not only from Tiflis, but also from Batumi, Sukhumi, Kutaisi, Kars, Aleksandropol, and Sarykamysh (where there were heavy troop concentrations), as well as from Erevan and Elizavetpol (now Kirovabad). They brought the latest news about the state of affairs in their locales and the rate at which Bolshevik influence was growing among the laboring masses. I remember that dramatic shifts of allegiance were occurring at that time in military installations, with whole regiments and garrisons coming over to our side.

The local representatives complained that the SRs and the Mensheviks (not to mention the officers) in charge of the local military organizations did everything they could to hinder the work of the Bolsheviks in the military sector, depriving them of the freedoms of speech and of the press, and forbidding them to distribute copies of *Petrograd Pravda* and *The Caucasian Worker*.

The Bureau of the Tiflis Party Committee met every evening to discuss the information we received and to make appropriate decisions about these matters. Our primary task, however, was the preparations for the All-Caucasian Party Congress, and we constantly dealt with specific issues related to the congress, while in the provinces party meetings were held to discuss the current situation and the tasks of the party, and to elect delegates to the Regional Congress of the party. Around this time I revisited my native area, giving a speech at the Alaverdi Factory, where the Bolsheviks elected me their delegate to the congress, and I also spoke at both general and party meetings in nearby Akhpat and Manes.

On October first, prior to the convening of the All-Caucasian Party Congress, a preliminary meeting was held by the members of the Tiflis Party Committee and the delegates to the congress, to find out how many delegates there were and where they had come from. We also set up a working agenda for the congress, along with a list of speakers. These would include Shaverdian, who would report on the preparations for the elections to the Constituent Assembly, and Korganov, who had just arrived from the front and was up to date on the political situation there and on the contravention of freedom of speech and freedom of the press, and who would speak on the question of preparations in the military sector for elections to the Constituent Assembly.

The congress opened on October 2, 1917, with a speech by Philip

Makharadze. He noted the great significance of the first All-Caucasian Party Congress which, at this unique moment in revolutionary history, had the task of unifying around one center the party organizations which were scattered over our large and multinational region.

To the unanimous applause of all the delegates, Lenin, the leader of the proletariat, who had at that time gone underground, was elected Honorary President of the congress.

Twenty-six voting delegates attended, including delegates from some districts of the northern Caucasus. I participated in the work of the congress as a delegate from the party organizations of Alaverdi, Manes, and Akhpat.*

The Tiflis delegate Kavtaradze reported on the work of the Sixth All-Russian Party Congress which had taken place that July and August.

The congress had been forced to meet illegally, and was prey to the wild outbursts of repression, terror, lies, and calumny perpetrated against the Bolsheviks by the bourgeois and so-called socialist parties. It seemed to us that the main significance of the congress lay in the fact that its resolutions were informed by the Leninist idea of preparing for an armed seizure of power. This had enormous importance for us in the Caucasus in reshaping the Communists' mentality; in providing us with the tactical reorientation needed in the struggle ahead; and in mobilizing our forces under local conditions toward solving the common task of the party.

Reports from the separate localities occupied an important place in the work of the congress. There were many of them, but they were extremely interesting since they gave the latest information on the political changes in these places, on the conditions under which the Bolsheviks were waging the struggle, and on their setbacks and their successes.

Georgi Sturua was the speaker from Baku. His report on the success of the Baku oil workers' general strike, which had aroused an impassioned response in workers all over Russia, was received with enthusiasm. When he reported, during the course of his speech, that the strike had been victorious, the Congress sent a special salutation to the courageous workers of Baku.

Dalakishvili, the delegate from Kutaisi, said that although an independent Bolshevik organization had been in existence in Kutaisi a mere two weeks, it had already managed to rally 239 people to its cause.

Arveladze was the speaker from Aleksandropol. He recommended paying particular attention to the way in which elections to the Constituent Assembly were carried out in the army. Our task, he said,

---

*The first Communist cell in Armenia had been created in the village of Akhpat in July of 1903. This organization's delegates to the Fifth Party Congress were Stepan Shaumian, Mikha Tskhakaya, and Asatur Kakhoyan.

"is to organize the masses around the banner of the revolutionary proletariat and to wage a merciless struggle against the petit bourgeois illusions of the Mensheviks and the SRs." Later, after the congress had concluded its work, Arveladze accompanied Shaumian to Aleksandropol, where the latter worked hard at strengthening the local party organization and spoke out effectively at broad-based soldiers' meetings.

The delegate from Elizavetpol, Mravian, said that because of the predominance of Bolsheviks in the united party organization there, the Mensheviks had withdrawn from the organization at the beginning of June. After collecting 1,200 votes under difficult conditions, we [the Bolsheviks] had sent three comrades to the Duma. The [military] garrison, he said, was sympathetic to the Bolsheviks.

I spoke on the situation in Alaverdi, Manes, and Akhpat, saying in particular that "this is the only province where the landless and petty landholding Armenian peasantry has been organized under the banner of revolutionary social democracy. The reason for this lies in the conditions of local economic life, in which the class struggle between the semiproletarianized peasantry and the landowning classes has been intensified to the extreme."

Korganov reported on the situation among the soldiers at the front. He also spoke about the preparations for elections to the Constituent Assembly that were underway in the army. Many delegates were struck by the fact that freedom of speech and of the press had been revoked in certain districts.

Danush Shaverdian gave a detailed speech at the congress that evoked a great deal of discussion and debate.

Shaumian, who had arrived a little late for our congress, having just returned from Petrograd where he had been a member of the Central Committee's inner circle and had participated in its preparations for armed revolt, responded to Shaverdian's report and focused the discussion in the right direction, stating:

"The Constituent Assembly is not an ordinary parliament, and it will have the greatest possible significance. We must draw people's attention to the fact that if a new revolution does not take place before the Constituent Assembly meets, then it can always happen after it has been convened if the assembly proves incapable of resolving these pressing problems. We should not allow the assembly to force us into foolhardy retreat. We must mobilize all our forces and exert pressure within the Constituent Assembly (from above) *and* outside of it (from below)."

Before Shaumian made his comments, I was also one of the speakers who responded to Shaverdian's report. Polemicizing in particular with Bagrat Borian—a highly esteemed Bolshevik and a political prisoner—I said that in resorting to revolutionary mass action, we wouldn't and couldn't avoid using the parliament to our own advantage. Therefore, I said, we can hardly be indifferent to whether the Constituent Assembly turns out to be revolutionary or reactionary. In preparing and organizing the masses for revolutionary action, we should participate in the elections for the Constituent Assembly and at the same time try to send as many of our own delegates to it as possible. I maintained that Borian had underestimated the importance of our struggle during the period of the election campaign for the Constituent Assembly, and had clearly exaggerated reactionary influences felt among the peasants.

Toroshelidze made a speech at the congress on the nationality questions. On the issue of what forms national self-determination would take, he adhered to the old party policy position on so-called local self-government. Everything he said suggested that he did not agree with the resolution of the April 1917 Party Conference, which had demanded regional autonomy as an integral part of the party's new platform on the nationality question, and in the draft resolution he proposed to the congress he put forward the demand for regional, all-Caucasian self-government (autonomy). His was a confused formulation which presented a clear issue in a "topsy-turvy" fashion. There is a fundamental difference between self-government and autonomy: they do not mean the same thing, and are mutually exclusive. Toroshelidze's position on this question had obviously been influenced by Makharadze, who had spoken out against regional autonomy at the April conference.

Shaumian then made a speech criticizing these confused and outmoded positions and putting forth his own proposal, which history later proved to be correct. It was summarized afterward in the paper as follows:

> Comrade Shaumian pointed out two weaknesses in Comrade Toroshelidze's report. He said that we must approach the nationality question more from a practical than a theoretical point of view. Comrade Toroshelidze lost sight of the fact that our [original] platform on the nationality question had been composed many years ago, whereas now we are on the verge of a social revolution, and the nationality issue must be resolved within this new context and not within the framework of a bourgeois system. What forms of self-determination do we propose, then? Finland didn't wait to see what Russia would say about the self-determination of nations, neither did the Ukraine, and we can't keep putting everything off till the Constituent Assembly. We must pose the question in concrete terms. The Georgian and Armenian nationalist parties have already worked out their platforms.
>
> This is what was said in our program: regional self-governments are created by taking into account economic and social factors. In addition, our program contained a plank concerning the self-determination of nations right up to the creation of an independent

state, the general democratization of the country, and the right to give instruction in the native language. At the 1913 Conference of Bolsheviks, however, our platform on the nationality question was expanded. Following the suggestion of Comrade Lenin, we replaced [the idea of] regional self-government with that of regional autonomy. What is the difference between regional self-government, autonomy, and a federation?

A regional self-government has no legislative powers. It fulfills cultural and economic-administrative functions. It has the right to levy taxes.

Autonomy differs from self-government in that it has a representative assembly with broad legislative powers. Autonomy means that a well-defined complex of issues is relegated to the local autonomous regions by the central government.

A federation represents a union of equal units. It has its own Constituent Assembly.

In accord with our present program, we must recognize the autonomy of the regions. Since we are on the verge of a social revolution, we should have less fear of decentralization. In regard to Transcaucasia, I consider the restructuring of certain administrative units to be necessary.

Shaumian suggested creating three autonomous national regions in Transcaucasia and proposed a rough plan for determining their boundaries. He favored maintaining the federal character of these autonomous regions' relations with Russia.

In his speech at the congress Shaumian pointed out that Toroshelidze had failed to take into consideration those changes in the Bolsheviks' views on the nationality question which had occurred as early as the immediate pre-World War I period. Those changes were reflected in the resolutions of the April Party Conference, which recognized that all nations which had become a part of Russia had the right to be free and independent, and to form independent states. Finally, Shaumian pointed out that Lenin, in speaking at the First Congress of Soviets and exposing the chauvinistic, monolithic policies of the SRs and the Mensheviks, particularly in regard to Finland and the Ukraine, had created the slogan: "Let Russia be a union of free republics!"

Unfortunately, the majority of the delegates, in dogmatically adhering to the positions of the old party program, neither understood nor supported Shaumian's proposals. Nevertheless, as a result of his speech, corrections were made in the draft resolution proposed by Toroshelidze. In the resolution which the congress passed, the phrase "regional, all-Caucasian self-government [autonomy]" was replaced by the wording "autonomy of the Caucasus, with the creation of a representative assembly . . . broad, local self-government, including the establishment of new administrative and economic units within the Caucasus."

The resolution that was passed contains the following: "The congress declares that it recommends neither separation, nor the formation of federal states for the Caucasian nationalities." As a result, Shaumian's proposal on this issue was rejected.

It must be admitted that the Caucasian Bolsheviks' position (on this issue) constituted an untimely and serious political mistake. A year and a half later, in Baku, we corrected this mistake. Other comrades, however, led by Makharadze, did not change their position on this issue until three years later when independent Soviet republics had arisen in Georgia and Armenia.

Shaumian also gave a very interesting and substantive report on the current situation. He summarized the progress of the revolution to date and noted the appropriate changes that had been made in party tactics. In particular, he said, "Even after the fifth of July, one could speak only of the *struggle for* power, but not of its transfer. . . . Our task is to assume leadership of the revolution and to take power into our own hands."

The congress was satisfied with this report and accepted it without discussion.

The congress' resolution, which was proposed by Shaumian and entitled "Concerning the Current Situation," contains the following statement:

> In complete agreement with the Central Committee of the party, we consider the next task of the Revolution to be the convocation of an All-Russian Congress of Soviets of Workers', Soldiers' and Peasants' Deputies which, after declaring itself a revolutionary convention and beginning the immediate implementation of the revolution's fundamental demands, must take all state power into its own hands.
>
> The congress is well aware of the fact that the bourgeoisie and its servants, the Kerenskys, Tseretellis, and past and present so-called socialist ministers, will not give up their positions without a fight.
>
> But the workers, soldiers, and revolutionary representatives of the peasantry should not be cowed by these difficulties and sacrifices; rather, they should go boldly forth into the battle for power and save the country and the revolution from destruction. . . .

Our regional All-Caucasian Party Congress' resolution on the financing of the election campaign for the Constituent Assembly might perhaps be of interest. The election campaign required money, and quite a lot of money at that. Money was needed for the publication of special literature, posters, and leaflets, and for dispatching our propagandists and agitators into the cities and towns of the region. Where to find this money was the problem, for, except for membership dues and the rather meager receipts from paid lectures, we had no other sources of revenue.

The precarious state of our party's finances reminds me of an incident involving Mikha Tskhakaya. Upon his return from Switzerland, Tskhakaya came to Tiflis and arranged to live in the Zubalov Home for the Aged: they gave him free room and board. At the end of September the Tiflis committee discussed the possibility of the party's giving Tskhakaya some financial assistance so that he could leave the Home for the Aged.

Tskhakaya's innate modesty and uncommon scrupulousness in financial matters kept us from doing this, however. Knowing how pressed we were for cash, Tskhakaya was adamant about not becoming a burden on our party budget, and refused any kind of financial assistance, declaring that he was "quite satisfied with [his] material circumstances."

We could not persuade him to change his mind. The Tiflis party organization was really in very straitened circumstances at that time, and Tskhakaya knew it. Thus, he remained in the Home for the Aged until mid-1919, when the Mensheviks, who had come into power, arrested and imprisoned him in the Kutaisi Prison.

To solve the matter of financing the election campaign for the Constituent Assembly, the Regional Congress passed a special resolution on ways of supplementing our party funds. The resolution obligating all party members to donate one day's pay to the party budget, and in addition, to take up a special collection of voluntary contributions and to increase the number of paid lectures and concerts, all the proceeds from which would be used to augment our party funds. The congress also passed a resolution mandating the transfer of half of all our revenues to a special Central Committee party fund that would be used for setting up a country-wide election campaign.

On October 8, 1917, our Regional All-Caucasian Party Congress concluded its work. In a great surge of enthusiasm, the congress sent Lenin a telegram of greetings and salutations:

> In concluding its work, the First Congress of the Caucasian Organizations of Social Democrats and Bolsheviks sends its warmest greetings to Comrade Lenin, the leader of the Russian Revolution, the best representative of the Third International, and an energetic fighter for worldwide proletarian revolution. . . .
>
> The congress expresses its firm belief that in the very near future the working class of Russia and the revolutionary peasantry (by means of its own new revolution) will throw off the yoke of the Miliukovs and their lackeys, the Kerenskys, and make it possible for Comrade Lenin to emerge from exile and openly assume his position as leader of our party and of the great Russian Revolution.

* * *

At that time I had many commitments to speak at workers' and soldiers' meetings as well as before audiences of young workers and students in connection with organizing the youth union, "Spartak."

The soldiers' enthusiasm for the war had now begun to wane. It grew noticeably weaker after Kerensky ordered an offensive that ended quickly in shameful defeat. The provisional government's decision to institute capital punishment at the front only made matters worse by provoking widespread unrest throughout the country, especially among the soldiers.

At almost all the soldiers' political meetings in September and October of 1917 the Bolsheviks prevailed in debates with the Mensheviks and right-wing SRs on the issues of war and peace. I remember the ovations that greeted the pronouncement of the Bolshevik slogans: "Down with the Imperialist War that is enriching the capitalists and impoverishing the country!"; "Stop the war now and conclude a just peace without annexations and reprisals!"; "Down with the Kerensky government that is pursuing the policies of the Russian capitalists and their foreign bosses!" The resolutions which the Bolsheviks proposed at these meetings were almost always accepted unanimously, leaving the SRs and the Mensheviks to crawl away like beaten dogs.

At one of the soldiers' meetings, I failed to win a majority vote on an issue of great topical interest for soldiers and peasants alike— the land issue. In spite of my efforts to swing the soldiers over to the Bolshevik platform on the nationalization of the land, the SR agitator managed to put his resolution through.

The SRs turned out to be quite well-prepared on this question, and agitated skillfully by practicing their usual demogogy. They demanded "socialization of the land," and its transfer to the rural communes, declaring that no matter what government was in power, it was alien to the peasants, and therefore they should reject the Bolshevik proposal for nationalization of the land, that is, for its transfer to the state, and strive for the transfer of the land to the peasant communes. This sort of demogogy made a strong impression on the soldiers, and I was defeated at that meeting.

Depressed by my failure, I decided to turn to Mikha Tskhakaya for help. We all loved and respected him as a learned, celebrated revolutionary, and Lenin's comrade-in-arms. It seemed to me that while I myself was convinced of the rightness of nationalization, I wasn't able to find the arguments that would convince these peasants dressed in soldiers' greatcoats that my position was correct.

After explaining this to Tskhakaya, I asked if he would speak on the agrarian question at the next soldiers' meeting and give us younger agitators and propagandists a lesson from an old and experienced party propagandist on how to defend our agrarian program.

Tskhakaya readily agreed to speak.

We gathered the soldiers for a meeting in the club building which was located next to the barracks, packing the hall. Tskhakaya was, as they say, in top form, and managed to rivet everyone's attention. He spoke fervently about the Entente, against the war, even touching on ancient Greece. He got so carried away, however, that to our chagrin he neglected even to mention the agrarian issue.

The meeting ended with an ovation, and Tskhakaya was so pleased that I couldn't bring myself to ask him why he hadn't brought up the one issue for which we had arranged the meeting.

I must say that we were all very fond of Mikha. To the end of his life he commanded universal respect as a man of crystalline purity and revolutionary conscience. He participated in all the Comintern congresses (except the first), was a member of the Comintern Executive Committee, and then a member of the presidium of the International Control Commission. He was elected a deputy to the Supreme Soviet of the USSR and died in 1950 at the age of eighty-five.

But, to return to the events of those days.

I could not reconcile myself to my failure at the meeting and kept trying to find a way out of the situation. Then I decided to change my tactics by shifting the emphasis of my speeches to the demand for immediate seizure of the lands by the peasant committee. It is known that the SRs had postponed a decision on the land issue for review by the Constituent Assembly and had come out against immediate seizure of the land.

At the next soldiers' meeting I tried out this new approach. "Many of you," I said to the soldiers, "intend to vote for the SRs policy on land. This means that you support them on this issue. But why don't you ask them this: how many months have the SRs and the Mensheviks been in control of the provisional government, and made fools of you by promising that sometime in the future, they'd pass a law at the Constituent Assembly, that would transfer the land to the peasants?! Why should the peasants wait, and go on laboring under the landowners' yoke?! Let's not argue today about how the peasants should use the land: today that issue is not of primary concern.

"The Bolsheviks are struggling to realize Lenin's slogan: 'Immediate seizure of gentry lands and their transfer to the peasant committees for allocation!' We Bolsheviks are against putting off the resolution of the land issue until the Constituent Assembly. The SRs do not agree with us.

"This is the main point of disagreement on the agrarian question between the Bolsheviks on the one hand, and the SRs and the Mensheviks, on the other.

"The issue at hand is this: should the peasants seize the land now, right away, or should this matter be postponed until that far-off moment when the 'establishment,' after wasting no one knows how much time discussing the question, passes no one knows what kind of law in regard to it.

"Therefore, let's not argue today about the nationalization or the socialization of the land—we'll be able to find a better solution to that question when the land is actually at the disposal of the peasant committees."

As I was saying all this, I could tell from the soldiers' eyes and from their shouts that I had achieved the necessary breakthrough. The SR representatives had not expected my maneuver and were thrown completely off-guard.

The Bolshevik resolution was passed by majority vote. The soldiers had repudiated the SRs.

After winning a majority of the soldiers on the issues of war, peace, and land, we had little difficulty turning them against Kerensky's government in favor of the Soviets and a government of workers and peasants.

By the beginning of October 1917 the majority of soldiers, not only in the Tiflis garrison, but in units and garrisons all over the Caucasian front, had come over to the Bolsheviks' side. The Mensheviks still had the support of the Junker and Officer Training Schools, of staff members in the military units at the front, of isolated military units, and of almost all the officers, but the masses of soldiers were ready, weapons in hand, to fight for the establishment of Soviet power in the Caucasus. As regards the local population, however, the situation was somewhat different, and not at all favorable to us. Aside from the center of worker strength in Baku, where power had passed to the Soviets in the early days of October, the proletarian stratum in the other districts was still very weak.

The peasants' struggle for land against the gentry in Azerbaijan and in the western regions of Georgia had intensified.

If everything were taken into account—the position and influence of the Baku proletariat, plus an active minority that supported the Bolsheviks everywhere in the towns and in the country (particularly after the victory of Soviet power in Russia)—then it was completely possible, with the enormous armed strength of the revolutionary soldiers of the Russian army, to establish Soviet power in Transcaucasia as well, and to oust the Mensheviks, Dashnaks, and Muscavatists. After that, we could then win the support of the indecisive masses of the cities and towns, and thus strengthen Soviet power in the area to such extent that even after the Russian soldiers had returned to their homeland, it would remain secure, with the support of the central Soviet government.

In other words, during and immediately after the October Revolution, conditions were propitious in Transcaucasia for the spread of Soviet power to all regions.

Two days after the shot on the "Aurora," the majority of the Caucasian Regional Committee, excluding Shaumian (who was then in Baku), made a special appeal to the party organizations of the region not to take power by force, but "to effect its painless and peaceful transfer" to the Soviets, even though the necessary conditions for such a transfer of power did not exist in the region.

The conference of delegates, who had been elected by the majority of the 80,000-member Tiflis garrison and who were under the leadership of the Bolshevik Kuznetsov, supported the platform of the Petrograd Soviet and demanded that the local organs of power recognize the Soviet government. It won the support of more and more military units, controlled the issuance of weapons from the Tiflis arsenal, and made preparations for an armed demonstration in order to achieve its revolutionary goals. The leaders of the Tiflis Menshevik Soviet immediately assumed an adversary position in regard to the garrison's revolutionary Delegates' Conference and resolved to disband it. Led by Kuznetsov, the Bolshevik faction of the Tiflis Soviet issued a sharp protest against this resolution, announced the withdrawal of the Bolsheviks from the Executive Committee of the Tiflis Soviet, and warned the Soviet's leadership that if the Executive Committee initiated any disbandment proceedings whatsover against the Delegates' Conference, the latter would offer substantial resistance.

I remember the Regional Committee's discussion of this issue, for although I was not, in fact, a member of the Regional Committee, I was present at the meeting as secretary of the Tiflis committee.

Proponents of a peaceful transfer of power, headed by Makharadze, spoke against the disbandment of the Delegates' Conference and for the countermanding of an armed demonstration. Tskhakaya wavered. At first he favored revolutionary tactics, but later, after acrimonious debate, he abstained from the vote taken on the resolution. His behavior really disillusioned us, for as young Communists it was on Tskhakaya whom we had placed our hopes.

The Regional Party Committee's incorrect resolution on the peaceful transfer [of power] underestimated the significance of the Delegates' Conference and the role of the Russian revolutionary troops in the fate of the revolution in Transcaucasia. The Mensheviks took advantage of this mistake and went into action. They tried to force several Bolshevik-oriented regiments out of Tiflis and they kept new revolutionary military units, which were returning from the front, from entering the city simply by sending them to a destination farther away. They created nationalist units, and in uniting themselves with the landowners and capitalists of the region, they strengthened the forces of counterrevolution.

At that decisive moment in the fate of the revolution in Transcaucasia, the majority of the [Party] Regional Committee was unable to agree on the correct tactical position for resolving the conflict between the Bolshevik-oriented Delegates' Conference, which contained representatives from seventy-six military units of the Tiflis garrison, and the Tiflis Menshevik Soviet. It was against this complex background, and after examining the situation in the region and satisfying itself that the Menshevik had made certain verbal "concessions" to the Bolsheviks (e.g., the reelections of the Tiflis Soviet, the convocation of the Second Army Congress, etc.), that the Regional [Party] Committee declared on November 11, 1917, "the function of the Delegates' Conference had been fulfilled," and recalled from it all Bolsheviks "in view of the peaceful resolution" of the conflict. By so doing, the Regional Committee sealed the fate of the one revolutionary organ that could have played a decisive role in the struggle for Soviet power in the region. By so doing, the [Party] Regional Committee deprived itself of the only powerful armed force in the struggle or the victory of the proletarian revolution.

The very next day, the Mensheviks declared a state of war in Tiflis, and instigated an armed attack on the arsenal which had been taken by Bolshevik-oriented soldiers. They used the weapons they seized for arming their own military units.

Lenin's reaction to all this is interesting. Tsintsadze, who was sent to Petrograd by the Regional Committee in December 1917 to report to Lenin on the situation in Transcaucasia, makes reference to it in his memoirs:

> When I mentioned that the Mensheviks had captured the arsenal, he [Lenin] stopped me and asked, "Do you mean they conceded the arsenal to the Mensheviks?" I wanted to explain how it had happened, so I continued. But he interrupted me a second time and asked, "Still, you're telling me, aren't you, that they *gave the arsenal* to the Mensheviks?"

Clearly, Lenin had realized the seriousness of the Regional Committee's mistake.

On his arrival in Tiflis during those critical days Shaumian immediately began to proselytize for revolutionary tactics. Based on his experience with the Baku Bolsheviks, he suggested preparing the workers for open armed combat against the counterrevolutionary, bourgeois-nationalistic power of the Transcaucasian Commissariat, and called for reliance on the Russian revolutionary troops of the Tiflis garrison and of the entire Caucasus.

The lack of agreement on his proposals in the Regional Party Committee forced Shaumian to telegraph Lenin on November 23:

> We have declared war on the Transcaucasian Commissariat as

counterrevolutionary. The majority of the garrison is on our side. With the help of the army we can force the Commissariat to recognize the power of the Soviet of Peoples' Commissars. We request immediate orders on how to proceed.

No answer was forthcoming.

On November 25 Shaumian sent Lenin a second, similar telegram. It too remained unanswered.

At that moment it was still possible to set matters right due to the deployment of powerful Russian troops, both in Tiflis and in Transcaucasia, who were supporters of Soviet power.

By sending his telegrams and relying on the Central Committee and Lenin, Shaumian wanted to force the Regional Committee to revoke its mistaken plan of action and secure the victory of Soviet power in the Caucasus. However, the telegrams were intercepted by the Mensheviks, and never reached Lenin.

Finally, Shaumian sent Kamo to deliver a letter to Lenin in person, while he himself left for Baku.

After enormous difficulties, Kamo executed Shaumian's orders. A month later he arrived back in Tiflis with a mandate designating Shaumian as Special Commissar of the Caucasus.

Before he left for Baku at the end of November 1917 Shaumian advised me to return there too, since there was much work to be done to bolster the Soviet victory.

"Return to Baku!" Shaumian repeated to me several times in saying good-bye.

I followed his advice, and soon left for Baku.

Part II

THE BAKU COMMUNE

Chapter I

The Victory of Soviet Power in Baku

Transfer of Power to the Baku Soviet—The March 1918 counterrevolutionary uprising and its suppression—Formation of the Baku Council of People's Commissars (Sovnarkom)—The struggle for the peasantry.

The Bolsheviks' influence on the Baku proletariat increased enormously after the successful outcome of the general strike in September of 1917 by the oil workers, led by Alyosha Dzhaparidze and Vanya Fioletov. A great victory had been won there by the workers in a collective agreement with the capitalists, which established, in particular, an eight-hour working day.

On October 15, 1917, an expanded session of the Baku Soviet was held which included representatives of the industrial-plant commissions and the regimental, naval, and company-commanders' committees. The majority of those at the meeting were already on the side of the Bolsheviks and their allies, the left-wing SRs.

Shaumian suggested that the meeting declare itself to be the provisional, expanded Baku Soviet of Workers' and Soldiers' Deputies, since the previously elected Baku Soviet was, for all intents and purposes, already defunct. Sadovsky, the leader of the Mensheviks, disagreed, claiming that this meeting had nothing to do with the Soviet. Sukhartsev, the head of the left-wing SRs, supported Shaumian's proposal, however. He said that the meeting *could* consider itself to be fully representative, since the Soviet would no longer convene in its former state and the present assembly fully and authoritatively expressed the will of the Baku proletariat and the military garrison. When the vote was taken, Shaumian's proposal was adopted by an overwhelming majority.

After electing delegates to the second All-Russian Congress of Soviets, the Soviet passed a resolution proposed by Shaumian to "seize power from the enemies of the people and transfer it to the people as embodied in the Soviets of Workers', Soldiers', and Peasants' Deputies."

On October 26, 1917, the news reached Baku that the revolutionary forces of Petrograd had overthrown the provisional government and proclaimed the power of the Soviets in Russia.

The Bolsheviks greeted this news with general rejoicing. A meeting of the Baku Soviet was called for the following day, but on this occasion the Mensheviks, SRs, and Dashnaks, who refused to recognize Soviet Russia, succeeded in blocking the resolution to transfer local power to the Baku Soviet. Thereupon, the Baku Party Committee appealed directly to the workers of the city to rise *en masse* in favor of "all power to the Soviets." Mass meetings took place in plants and factories, as well as in military units, and resolutions in support of Soviet power were passed.

This fierce struggle with the right-wing parties resulted in the following resolution, passed by the Baku Soviet on October 31, 1917: "In an effort to counterbalance the bourgeoisie and the Kaledinites, the Baku Soviet supports the new government that has come into being on an All-Russian scale, and deems it necessary to delegate to itself the task of expanding the power of the Soviet in the Baku region until all power is transferred into its hands."

When the next meeting of the Soviet began—on November 2, 1917—the Menshevik Bogdanov, as a sign of protest, called upon his allies—right-wing SRs, Mensheviks, and Dashnaks—to walk out of the meeting. Of the 468 people present, 124 walked out, and they subsequently organized the so-called Committee of Public Safety which they set up in opposition to the Baku Soviet and tried to pass off as the only democratic organ of local power. Only Bolsheviks, left-wing SRs (who were aligned with them at the time), and independents remained in the Baku Soviet.

At Shaumian's suggestion, the Soviet elected a new Executive Committee capable of pursuing a policy of Soviet power. The "Executive Committee of Social Organizations," a local organ of the provisional government was abolished, and the decision made to put the garrison's armed forces under the control of the Baku Soviet and not allow them to be moved without its knowledge and consent. At the same meeting a resolution was passed that established the "Executive Committee of the Soviet of Workers' and Soldiers' Deputies" as the highest authority in the city of Baku. Thus the transfer of power to the Baku Soviet of Workers' and Soldiers' Deputies was accomplished without resorting to armed struggle.

After October, the Baku Bolshevik organization intensified its agitational and propaganda work among the masses and exposed the enemies of Soviet power. Primary attention was paid to strengthening armed support. The Baku Party Committee created its own militia, and detachments of Red Guards were organized in the industrial districts.

The Soviets had the committed support of the sailors of the Caspian Naval Fleet; one of their leaders was Arkady Kuzminsky, who is still alive and thriving. The Baku garrison, composed of separate units of the old army, was, for the most part, also on the side of the Soviets.

After the announcement of demobilization and the enactment of the decree transferring the land to the peasants, the soldiers of the Baku garrison, who were mainly peasants from Central Russia, began to return to their homeland. Thus the remnants of the old army which supported Soviet power dwindled away day by day until finally, as in other localities, they disappeared entirely, making it mandatory to create a new, socialist army.

In February of 1918 units of the Red Army were already being formed in Petrograd and Moscow. The Baku party organization was faced with the same task. After a slow start, the situation began to improve drastically after a group of Bolshevik leaders of the Caucasian Region Military Soviet were transferred to Baku from Tiflis. (In Tiflis they had not had enough support since the Mensheviks, who were in the majority there, had created their own military leadership.) The following comrades were included in this group: Korganov, Sheboldaev, Malygin, Solntsev, Gabyshev, Koganov, Ganin, and others who had agreed to remain in Transcaucasia for the sake of creating a Red Army.

The work of organizing an army really picked up speed by the end of February and the beginning of March 1918. The old Bolshevik, Grigorii Korganov, who commanded great authority among the soldiers at the Caucasian front, was in charge of this operation. Boris Sheboldaev, who was also a Bolshevik with a prerevolutionary service record, became his assistant. All their activities were carried out under Shaumian's leadership.

International battalions, regiments, and reserve military units were created, and three armored trains were formed. The service personnel consisted mainly of soldiers who had formerly been workers in Baku and young people who had not yet seen military service. Shortly after, Sheboldaev informed Moscow that by June 1918 as many as 13,000 men had been conscripted into the Red Army, formed into battalions, and organized into four brigades, and that the corps staff had essentially been formed. An army instructors' school was also founded in Baku. It was headed by Solntsev, a very capable military organizer and a member of the Military-Revolutionary Committee, who had come in from the Turkish front.

Until May of 1918 the military units of the Red Army were international in composition. Only when the threat of a German-Turkish attack hung over Baku did the Armenian National Council propose transferring their own Armenian national units into the Red Army. Let me say a few words about these national military units, however. They, too, have a history.

The bourgeois nationalists, and the Dashnaks in particular, verbally concurred (in order to pass themselves off as "socialists") that workers and peasants of all nations share the same class interests, but they nevertheless maintained that all Armenians, regardless of class origin, and in order to protect their "common" interests, should be organized into a special national council and have their own special military units, etc. And, indeed, such national Armenian and Azerbaijanian councils did exist in Baku at that time. Combining forces with the other reactionary bourgeois parties, they armed the Muscavatists and the Dashnaks, relying on the enormous financial support of the Baku oil manufacturers. These councils used demobilized soldiers and officers to create their own national military units in Baku.

We had to expend a lot of energy to convince the toiling masses of the great harmfulness of antisocialist "ideas" inimical to the proletariat which constituted a particular danger at the time within the multinational context of the Caucasus. Rather than leading to unification, the bourgeois nationalists' policy sowed deliberate dissension between workers and peasants of different nationalities and weakened the class ties between them, and we made a point of speaking out decisively against this policy.

However, when Baku was threatened by a German-Turkish invasion in May of 1918 we were forced to accept the Armenian National Council's suggestion that we include the Armenian national units in our army. It was a compromise forced upon us by circumstances, and it obviously contained hidden dangers. The overwhelming majority of soldiers in the Armenian national units fought honestly and well against the Turks, and even many of the officers fulfilled their duties conscientiously. But there were Dashnaks among them, especially in the higher command positions, who were taking orders from their party. They were the group that played the most treacherous role during our army's retreat later that summer, at the end of July.

During this time the publication of the newspaper *News of the Soviets* in Azerbaijanian played a very important role in strengthening the Soviets' influence on the Azerbaijanian masses. Our choice of Tukhula Akhundov as editor of the paper was felicitous. He was a member of the left-wing SRs. During our brief collaboration I came to know him very well and we became fast friends. Young, well-read, politically astute, a firm supporter of Soviet power, and a good writer, Akhundov joined the Communist party during the British occupation under the Muscavat government in mid-1919, and played an active role in the plans to overthrow the Muscavat government. After the establishment of Soviet power in Baku in 1920 he became one of the outstanding leaders of the Azerbaijanian Communist party. (I shall speak about him in more detail later on.)

My articles appeared in the years 1917 to 1919. They were published in *News of the Baku Soviet* (in Russian and Armenian), and in the newspapers *The Social Democrat* and *The Baku Worker*, and later on in *The Workers' Path*, *Communist*, and others. Almost all of these articles have been preserved. While glancing through them a short while ago I seemed to relive that distant time and to feel the pulse-beat of life during those hot, complicated, and at times very anxious days.[8]

Of course, the development of social thought has made such giant leaps forward since then that some of the things I wrote may now seem slightly antiquated or dubious, or just simply "immature." This is understandable; we were moving, literally feeling our way, through virgin territory. Much of what we were forced to contend with was completely unfamiliar to us.

Nevertheless, why is it that all this brings me so much satisfaction? When one contemplates these early articles from the perspective of one's later experience, one can see that, by and large, we were historically in the right, even then; life subsequently confirmed the majority of our theories, prognoses, and conclusions.

And for this we are indebted, first of all, to the great teachings of Marx, Engels, and Lenin which we culled in those days not only from their books, and from Lenin's articles and speeches, but also from the day-to-day struggle, in the battles and fierce confrontations with the forces of the Old World.

To return to the national military units in Baku—which were widely subsidized through the national councils by the local bourgeoisie—the Azerbaijanian national armed forces were at that time organized into the so-called Savage Division. On the strength of that division, the Azerbaijanian bourgeois-gentry circles launched an uprising against the Baku Soviet of Workers' Deputies in March of 1918. After three days of street fighting, however, the uprising was suppressed—due in large part to the mammoth efforts of Narimanov, Azizbekov, and Dzhaparidze to settle the dispute peacefully. The soldiers and officers of the "Savage Division" scattered and joined up with counterrevolutionary units from districts of Azerbaijan where Soviet power still didn't exist. Their center was Elizavetpol.

It is worth noting that, prior to these events, an understanding had been reached between the Azerbaijanian and Armenian national councils that the latter would provide armed support for the uprising planned by the Muscavatists' party. However, when the uprising began, the Armenian National Council declared its neutrality; and when the victory of the Red Army was imminent, certain Armenian national units even fought on its side.

At the same time as the uprising in Baku, reactionary bands led by the Imam Gotsinsky from Dagestan mounted a campaign against the town of Khachmas, and advanced on Baku. Their forces included a regiment of the "Savage Division" which had gone through the

[8]See Appendix.

training school of the World War. However, they were still fifteen kilometers from the city on the day when the uprising in Baku was suppressed, and were unable to hitch up with the rebels. Units of the Red Army, numbering about 2,000 men, attacked them and drove them far from Baku. Then the districts of Azerbaijan that had been seized by Gotsinsky, as well as the Dagestani cities of Derbent and Petrovsk (today Makhachkala), were liberated. The detachment of the Red Army that had been dispatched from Baku to assist their Dagestani comrades carried out this operation under the command of Victor Naneishvili, who had been named Special Commissar of the Dagestani region.

This victory led to the opening of land communications between the northern Caucasus and Baku, which were crucial for getting foodstuffs to the starving inhabitants of the city. Shaumian sent a special communication to Lenin about the Red Army's successful advance to Dagestan.

Later, reactionary outbreaks also took place to the south of Baku—in the important grain-producing regions of Lenkoran and Salian. The Red Army units which were sent there were soon victorious.

I was wounded in the March street fighting and ended up in a military hospital. When Shaumian found out that I would soon be discharged but that I still didn't have a place to live (I was still spending my nights in the office of the Baku Party Committee) he literally demanded that I come to live in his new apartment. I lived there for a long time, practically a member of the family, with the opportunity to socialize in the evenings with Shaumian and the people who visited him. As a result, I had a unique opportunity to become privy to party business and affairs of state, a privilege that greatly enhanced my political growth.

The suppression of the March uprising did much to strengthen the power of the Baku Soviet. The city Duma, which aspired to power, was dethroned once and for all and the influence of the national councils noticeably decreased, with the Baku Soviet assuming control over their activities. Repressive measures were forthwith instigated against the oil manufacturers. In particular, the Baku Committee on Revolutionary Defense imposed an indemnity of fifty million rubles on the oil manufacturers: a great deal of money then; money that was crucial in bolstering our armed forces and strengthening the organs of Soviet power. When the oil manufacturers refused to pay, the committee decided to arrest certain well-known Baku oil magnates. I was one of those directed by the committee to arrest three prominent oil manufacturers: Lessner, Tagianosov, and Gukasov, all of whom had played a leading role in the Union of Oil Manufacturers. Shaumian's oldest son, Suren, took part in this operation. Later, he was one of the organizers and commanders of the Soviet Tank Corps.

I remember that we took a small truck, two or three Red Guards, and drove up to the two-storied house where Lessner lived. His private residence was located in one of the especially dusty and dirty working-class districts of Baku. Around us poor workers' huts huddled together in disarray; not one tiny tree could be seen, but Lessner's residence, which was situated in the middle of a beautiful, shady park filled with lush, decorative trees, shrubbery, and flowers, made the extreme contrasts typical of those days even more striking.

It was evening. The front doors to the residence were already locked, so we went in the back way and found Lessner, a fat, sleek man, lying on his bed. He had obviously been expecting his arrest for some time and showed no signs of surprise at our arrival. We put him inside our small truck which had no seats. To our great satisfaction, he got bounced around quite thoroughly as we drove over the broken cobblestone streets of Baku.

The other Baku oil magnates were arrested in approximately the same way. The arrests achieved their purpose: that same night they ordered their firms to pay the indemnity that had been imposed on them.

After the events of March all the newspapers of the bourgeois parties and the Mensheviks were closed down. The publications of the other petit bourgeois parties continued to come out, however, and the parties themselves continued their activities. As long as things were going well for us, these parties didn't do us any particular harm, but as soon as the food situation became critical, they exploited the population's dissatisfaction for the purposes of their own struggle with us.

Soon after the nationalist parties had succeeded in severing Transcaucasia from Soviet Russia, the situation became more complicated. Part of this was due to the fact that sultan-ruled Turkey had violated its end of the Brest-Litovsk Peace Treaty, and had moved its troops into Transcaucasia. Under attack from the Turkish forces, the Red Army retreated to Baku, causing the right-wing parties to bring up the question of inviting the British to Baku.

The conflict-torn situation had now reached a breaking point. We made no concessions, but, as always, we allowed representatives of the opposing parties to argue their sides of the issue. However, very complicated tasks confronted the Baku party organization, and whatever the cost, Soviet power had to be maintained in Baku and extended to all of Transcaucasia, or, at the very least, to Azerbaijan.

Lenin's advice to Shaumian regarding the policy we should pursue at that time is particularly interesting in this respect. Lenin wrote as follows:

> Dear Comrade Shaumian! Many thanks for your letter. We are very enthusiastic about your firm and decisive policy. Just manage to combine it with the kind of extremely cautious diplomacy that is

obviously warranted by the present very difficult situation—and victory shall be ours.

The difficulties are immense. *For the time being* our *only* salvation lies in the contradictions, conflicts, and struggles within the imperialists' camp. Know how to exploit these conflicts: *For the time being* [we] must learn to play the diplomacy game.

Best regards, wishes and greetings to all [our] friends.*

Back in November when power was first transferred to the Baku Soviet, the Mensheviks and right-wing SRs led by the Menshevik Bogdanov had demonstrated by walking out of the Soviet's Executive Committee, but now, on realizing that they had "run aground," they returned to the Soviet and began to participate in its work.

This complicated things. A situation arose in which the Executive Committee, representing as it did a huge variety of views, was unable to pass radical resolutions in the area of socialist changeovers, and could not offer effective resistance to the reactionary elements. Therefore, in April of 1918 the Baku Soviet of Workers' and Soldiers' Deputies formed the Baku Soviet of People's Commissars (Sovnarkom), made up exclusively of Bolsheviks and a few left-wing SRs. Representatives of the rightist parties were not allowed into the Sovnarkom although they tried very hard to get in.

Shaumian was confirmed as President of the Sovnarkom; Dzhaparidze, Narimanov, Korganov, Fioletov, Kolesnikov, Vezirov, Zevin, Karinian and others were made Peoples' Commissars. Azizbekov, the commissar of the Baku Province, was given the most difficult task of creating rural Soviets in the Baku Province—in other words, the job of consolidating the Azerbaijanian countryside into the Soviet system.

With the aim of attracting the villages to the side of Soviet power, we began to organize national friendship societies for the various ethnic groups that comprised the Baku work force—namely, Azerbaijanians, Armenians, Russians, Dagestanis, and others. As early as January of 1918 the Provisional Committee of the Union of International Workers' Societies had been created. I was assigned to participate in several organizational meetings of that committee and to make speeches at the meetings of the workers' international friendship societies.

The main job of this organization was to establish regular contacts with and to exert an influence on fellow-villagers and fellow-countrymen, with the aim of mobilizing them for the struggle against the landowners; the seizure of the lands, and the organization of peasant Soviets.

The need to attract the great mass of Azerbaijanian peasants to the side of Soviet power was made even more critical by the threat of the German-Turkish invasion of Baku. The peasants were still largely

*V. I. Lenin, *Complete Collected Works*, Vol. 50, pp. 73-74.

under the influence of the beks, the khans, the Muscavatists, and the reactionary clergy. In spite of their general political backwardness, the peasants' economic conflicts with the landowners were so acute that the class struggle for seizure of power and for the destruction of the landowners' oppression grew more and more intense. Particularly severe class skirmishes took place in the Kazakh district and in Karabakh.

Lenin's decree on the granting of the land to the peasants also stirred up the Azerbaijanian peasant masses.

In April of 1918 the Baku Soviet of Peoples' Commissars (Sovnarkom) published its decree granting all gentry lands to the peasants. Measures were taken to ensure the dissemination of this decree in the Azerbaijanian language to the districts and villages. A special department was organized in the Baku Soviet for working with the Azerbaijanian and Armenian peasantry.

The Bolshevik organization called Gummet did a great deal of work with the Azerbaijanian peasantry at this time. Its organizers and outstanding activists included Narimanov, Azizbekov, Efendiev, Sultanov, Buniat-zade, Israfilbekov, and others. Nariman Narimanov was the director of all Gummet's activities. His poor health, however, forced us to send him to Russia for treatment in June of 1918.

With Azizbekov's active participation a Representative Congress of Soviets of Peasants' Deputies of the Baku district, chaired by Narimanov, was convened at the end of May 1918. Dzhaparidze, Shaumian, Narimanov, and Azizbekov made speeches at the congress, and a joint meeting was arranged between the Baku Soviet and delegates to the congress, resulting in a decision to include the congress' delegates in the Baku Soviet, and thus transform it into a Soviet of Workers' and Peasants' Deputies. It was also decided to set up congresses of Soviets in other districts of the province, and then to convene a provincial congress of Soviets of Peasants' Deputies. Work toward this goal proceeded smoothly. It went particularly well in the Shemakha district, where Azizbekov was in charge, and also in the Kuba, Salian, and Lenkoran districts, where Soviet power was soon established. The worsening of the military situation, however, made it impossible to convene a provincial congress of Soviets of Peasants' Deputies.

While traveling from Tiflis to Baku, Shaumian came across incidences of spontaneous, mass peasant uprisings against the gentry which we had known nothing about. He told us about them later in a newspaper article. The peasants were seizing the land and setting fire to the gentry's estates, but the age-old peasant hatred of the gentry-oppressors was so intense that this struggle often acquired elements of savagery. In the Elizavetpol province where there were not enough Bolshevik cadres to organize and lead the peasant movement, the gentry succeeded in suppressing the peasant rebellion with the aid of the Turkish regular troops who had arrived there.

In recollecting this period, I would like to say something about our fiery revolutionary, Meshadi Azizbekov. He was an experienced, well-trained party activist. In appearance he bore a marked resemblance to Alyosha Dzhaparidze, with a personality that was just as hot-tempered and irascible as Alyosha's. Meshadi was a revolutionary enthusiast—I would even say, a revolutionary fanatic, in the best sense of the word. He moved from village to village, constantly mingling with the people and stirring up and organizing the peasants. He was neither disturbed nor cowed by the constant danger of being killed by the Muscavatists or by the gentry. Even in the most difficult moments of the struggle he remained optimistic and convinced of the rightness of our cause. Meshadi won the deserved love and respect of the masses. We were particularly impressed by the close relationship, comradely intimacy, and mutual understanding and trust which always existed between him, Shaumian, Dzhaparidze, and the other comrades.

I remember how thoroughly the Baku party organization discussed the most pressing issues of general and local significance. Usually at meetings and conference sessions different opinions on the issues under debate were expressed straightforwardly and openly. We had our share of sharp disagreements and heated debates but I don't recall a single case of anyone disputing a resolution that had been passed. On the contrary, the person who had objected most during the discussion usually would try to carry out the resolution with painstaking care so that his point of view would not interfere with putting the resolution into effect. The spirit of Bolshevik discipline reigned supreme in the organization.

The following episode is typical: in February of 1918, at the demand of the SRs and the Mensheviks, the issue of the Brest-Litovsk Peace Treaty was scheduled to be discussed at a meeting of the Baku Soviet. The SRs and the Mensheviks had campaigned furiously against the treaty. The night before the meeting the Baku Party Committee met to discuss the treaty. Shaumian, who supported Lenin's policy on concluding peace, was away in Tiflis at the time. Dzhaparidze, who was second in command in the party organization and President of the Baku Soviet at the time, spoke out against the treaty. It was warmly defended by Victor Naneishvili and myself, among others. The majority voted in favor of a resolution supporting the party's peace policy, leaving Dzhaparidze in the minority. Therefore, he said that it would be both difficult and awkward for him to speak in defense of the treaty at the meeting of the Baku Soviet and asked that a comrade who had supported the treaty speak in his place. We unanimously rejected this idea, however, because we knew that if Dzhaparidze didn't speak, the Mensheviks and SRs would immediately realize that we lacked unanimity on this question and we would run the risk of a general defeat. We therefore assigned Dzhaparidze the task of speaking at the Baku

Soviet in defense of the Soviet government's policy favoring the Brest-Litovsk Peace Treaty.

One would have had to have been present at the meeting of the Soviet to have seen and heard how magnificently, and with what weighty arguments, Dzhaparidze defended the Leninist peace policy! One could not help feeling that in thinking out his speech, he had reexamined his own views in many respects and had become convinced by the irrefutable logic of the Leninist position. Dzhaparidze's speech made an enormous impression on all the deputies of the Soviet, convincing many of those who had been undecided, so that the resolution of the Bolshevik faction passed by an overwhelming majority of votes.

In addition to the debates on the Leninist peace policy, another extremely controversial issue confronted us at that time: whether or not to include in the Baku Soviet of Peoples' Commissars (Sovnarkom)—along with the Bolsheviks and the left-wing SRs—the members of the right-wing socialist parties represented in the Baku Soviet. The majority of the Baku Party Committee were against including them.

After the Baku Party Committee had made its decision, I wrote an article for *Izvestia* (News) entitled "Why We Are Depriving the Dashnaks and SRs of Representation in the Baku Soviet of People's Commissars." In that article I wrote that the Sovnarkom should be restricted to representatives of those parties who "are sincere and consistent supporters of Soviet power," and who "go beyond token verbal support and actually pursue a Soviet policy in life."

"The Dashnaks and the right-wing SRs," I wrote,

> who seek representation in the Sovnarkom, cannot be consistent and determined supporters of Soviet power because up until now they have waged "a furious, 'no-holds-barred' struggle against Soviet power." Many things testify to this: their walking out of the Soviet's Executive Committee in repudiation of the power of the Baku Soviet; the two-faced policy they pursued in the municipal Duma in regard to the Soviet; their relentless struggle against Soviet power and a worker-peasant government in connection with the breaking up of the Constituent Assembly . . . their participation—despite their declarations to the contrary—in the Transcaucasian commissariat and representative assembly, which pursued the following policies: severed Transcaucasia from Soviet Russia; gave every possible assistance to her [Transcaucasia's] class enemies (General Kaledin and others); nationalized the troops; disarmed the Russian soldiers returning [home] and instead of thanking them for liberating Transcaucasia from its many years of suffering, plunged them into rivers of blood; and, finally, organized the forces of the beks' counterrevolution against the Baku Soviet.

This article vociferously, but with justification, disputed the claims of the right-wing, openly bourgeois parties to representation in the Baku Sovnarkom. At the same time the article contained no hint of the divergence of opinion that existed among the Bolsheviks, although certain well-known and highly respected comrades had spoken out against the party's position in this regard.

Discussions at the leadership level of the party in Baku did not breach party discipline or disrupt the atmosphere of comradeliness. Neither our official or political relationships, nor our personal ties suffered as a result of these disputes, even the bitterest of them. Therefore, disagreement over controversial issues did not give rise to inter-party cliques or factions, and while such issues sometimes required lengthy debate before being resolved; these discussions did not spill over onto the pages of the newspapers. The Communists considered it their duty to carry out resolutions passed by the majority whether they had voted for or against them.

I remember that when the question of the nationalization of the petroleum industry was being decided Dzhaparidze expressed doubts on the matter. As a result, Shaumian suggested that he hold a meeting in the center of the industrial district of Sabunchi and give a speech summarizing both sides, pro and con, of the nationalization issue to find out what the Communists and oil workers themselves thought about it. This is precisely what was done.

Gogoberidze and I (he was also a young Communist at that time) decided to go to the meeting, and we found the speeches calm, businesslike, well-argued, and hewing to the basic principles at issue. The general impression was that the people had come to consult with the Communists; nobody tried to foist his own point of view on the meeting.

Almost all the Communists who spoke out on the issue expressed themselves in favor of nationalization and gave their own supporting arguments. The meeting did not pass any resolutions; they were, in fact, unnecessary. At the Baku Party Conference held later Dzhaparidze voted for nationalization of the petroleum industry along with practically everyone else; it seems likely that he had left all his doubts and waverings behind him at the Sabunchi meeting.

I have mentioned this incident in order to illustrate and reemphasize the first-rate, genuinely Bolshevik, Leninist style of Shaumian and Dzhaparidze's work. The entire Baku party organization was nurtured on precisely this spirit.

The Baku Communists owed most of their solidarity, unity, and faithfulness to Leninism—to the crystalline, party-spirited honesty, genuinely Marxist upbringing, and unshakeable Leninist integrity that characterized the entire life and work-style of Shaumian, Dzhaparidze, Azizbekov, Narimanov, Fioletov, and the other leading activists of the

Baku Bolsheviks. Many other party workers of Baku, as well as simple, rank-and-file party members, followed their example.

For me personally, as a young Communist (I had not yet turned twenty-three), this was an invaluable Bolshevik training-school.

Chapter II

The Commune's First Steps

*The nationalization of the banks and the oil industry—Bicherakhov's regiment—The beginning of foreign armed intervention in Transcaucasia
I leave for the front.*

The question of the nationalization of the banks came up at the Baku Sovnarkom (Soviet of People's Commissars) about three weeks after the events of March 1918. It fell to my lot, by mandate of the Sovnarkom, to carry out the nationalization of the Russian Bank for Foreign Trade. Shaumian's middle son, Leva, who was then a fourteen-year-old adolescent (and who is now Assistant Editor-in-Chief of the *Great Soviet Encyclopedia*), was a member of the detachment that took part in this operation.

To be honest, I had only the vaguest notion of banks and finance. Although I had read the book, *Finance*, written by Hilferding, the famous German economist, I had never set foot in a bank—there had been no necessity for me to do so—although, of course, I knew that a vast array of valuables were kept there in special safes.

In any case, at a precisely appointed time a small detachment of men and I, armed with revolvers, entered the bank, occupied all the entrances and exits, and mounted to the second floor. There we saw a lot of people sitting in a big room, diligently writing at desks and clicking away at their abaci. We walked in and shouted: "On your feet, hands up!" The manager was summoned and ordered to open the safes. "I can't," he replied. "There are two keys: I have one, but the owner has the other, and you need both to open the safe."

At first I was indignant and thought that the manager simply didn't want to submit to us, but then I believed him. On looking around and seeing that everything seemed in order, I allowed the employees to sit down and continue their work, and after posting a commissar in the bank with several men from the detachment, I left. The bank safes were first sealed, and then later opened up, and after an inventory had been taken of all the bank's valuables they passed into the hands of the Soviet authorities.

The other ten or so banks in Baku were nationalized in the same way. It couldn't be said, of course, that the nationalization of the banks took place in an atmosphere of calm. There was stubborn opposition, angry protests, and threatening telegrams to the Baku Sovnarkom, but in the end all these obstacles were overcome.

How was the nationalization of the oil industry in Baku to be carried out? From February on, this question was the subject of numerous discussions at Baku party conferences and at sessions of the Baku Party Committee. Shaumian was usually the one to raise the question, setting off endless debates among the Communists on this issue. Several of the leading workers expressed their fear that the nationalization of the oil industry would curtail the flow of oil so necessary for Soviet Russia: they felt that the engineers and technicians and administrators of the oil fields who firmly supported the private owners would oppose nationalization and engage in sabotage. At that time we had almost no engineering and technical cadres of our own.

The question of nationalization of the Baku industry was of exceptional importance for the entire country. The power of the Soviets in Russia depended in many respects on the quantity of oil being shipped from Baku. Shaumian consulted with Lenin several times on this issue and received his support on a decision to proceed with immediate nationalization. A corresponding resolution was then passed by the Council of People's Commissars. However, the representatives of the oil companies were then in Moscow engaged in lobbying "activities" among the specialists of VSNKh (All-Russian Council on the Economy), who were, in turn, also apprehensive about a shutdown of the oil fields in the event of nationalization. The oil manufacturers took advantage of the stalemate by sabotaging the delivery of oil in every way possible, while VSNKh hesitated, taking no practical measures for a long time and giving Baku no directives on the matter. What's more, a telegram was sent from VSNKh which spoke of a temporary postponement of the plans for nationalization, and there was a telegram from Glavneft (the State Oil Trust) that actually countermanded nationalization.

Shaumian, who was always calm and even-tempered, was beside himself on this occasion. He was forced to appeal to Lenin with the following telegram: "such a policy is inimical to us and extremely harmful, as I've insisted again and again. I am totally opposed to it. After all that has been accomplished so successfully, there can be no return. These telegrams only create confusion. I ask you to interfere personally in order to forestall serious consequences for industry."

In May 1918 a resolution was passed at the Baku Party Conference to proceed with the immediate nationalization of the oil industry, even though a directive concerning nationalization had not been received as yet from Moscow.

Following upon the party conference, the Baku Soviet of People's Commissars also passed a resolution to nationalize the oil industry. To this end a Baku Economic Council was formed and the entire oil industry was made subordinate to it. Simultaneously with this, the Sovnarkom nationalized the Caspian Merchant Marine, which was largely responsible for transporting the oil to Astrakhan.

Vanya Fioletov, a Bolshevik who had been a leader of the oil industry workers' union for many years, was put in charge of the Baku Sovnarkom. He was, by the way, one of those who had been against the nationalization of the oil industry, but when nationalization was finally decided upon he gave himself up to it with total enthusiasm—a fact which only served to increase our affection and esteem for him.

All the fears about the nationalization of the oil industry leading to a breakdown in the extraction and exportation of oil proved to be unwarranted. On the contrary, the drilling of oil was not curtailed, and the exportation of oil—which was of special importance at this time—was sharply increased.

* * *

In the spring of 1918 it was learned that there was a well-trained body of Cossack troops in Iran which was from the northern Caucasus and was headed by Colonel Lazar Bicherakhov, who had fought with the Turks as part of the Russian Expeditionary Corps. This regiment was financed by the British, insofar as they, as a rule, financed the entire Russian Expeditionary Corps, with the stipulation that they would receive compensation from the Russian government at a later date.

Bicherakhov had more than once made requests through different channels that his regiment, which numbered more than 1,500 men and was completely armed, be evacuated in an orderly fashion from Iran to Baku. He made it known that although he didn't understand anything about the Soviets [i.e., our form of government], and didn't want to meddle in our internal politics, he was ready to take up arms against the Turkish-German troops and support the central Soviet authorities, because, aside from them, there was no other power in the country that could save Russia.

During May and June of 1918, negotiations were set in motion concerning the disposition of this regiment. There were various reasons for being apprehensive: was Bicherakhov a British agent and would the entry into Baku of such a large and well-organized regiment have unforeseen consequences?

Shaumian wrote to Lenin several times concerning this issue, expressing his doubts and setting forth all the available information he had on it. On the other hand, he spoke of the advantages of having this strong military arm. We began to receive intimations, moreover, of a growing sentiment in favor of Soviet Russia within the regiment. After long hesitations, meditations, and negotiations, those who favored admitting Bicherakhov's troops gained ascendancy. This decision was eased by the fact that Bicherakhov would bring with him a rather large reserve of arms and military supplies, then so crucial to Baku.

After Shaumian had reached an agreement with Lenin concerning Bicherakhov's entry into the city, he gave a report on the entire matter at a session of the Baku Soviet in June 1918.

It was then that Shaumian made public the last telegram which he had received from Bicherakhov:

> I accept all the conditions which have been put forward by Commissar Korganov, not because I wish to secure some responsible position, but because I see the salvation of Russia in Soviet power. I don't believe that Russia's salvation lies in the Constituent Assembly or in the Assembly of the Land, because no one can implement their decrees until Soviet power is strong. I deem any speech against Soviet power to be an act of criminal treason. I consider the sabotage of Soviet power to be a crime. I do not aspire to power or to a position of responsibility. I don't understand anything about politics or socialism, I'm not prepared for the building of a new life, I'm a Cossack, I know a little something about combat and military matters, and that's all.

Shaumian reported that the Baku Council of People's Commissars had concluded the following agreement with Alkhavi, an emissary of Colonel Bicherakhov: "Bicherakhov recognizes Soviet power—both in Russia at large and in Baku. He is appointed commanding officer of one of the units of the Caucasian Red Army and is under the authority of Korganov, the Commissar of Army and Navy Affairs. In matters relating to strategy and tactics, he has an independent voice, but all his orders are to be countersigned by the commissar. The temporary suspension or cessation of hostilities is a decision that rests in the Baku Council of People's Commissars. Combat missions are authorized by headquarters and are executed by the commanding officer independently. Bicherakhov's regiment receives support from the Baku Council of People's Commissars, which assumes responsibility for the support of all the military units which might subsequently be formed by Bicherakhov in the northern Caucasus and which would also form part of the Red Army."

Bicherakhov and his regiment arrived in the Baku area early in July. To be on the safe side, it was decided not to have him enter Baku, but to have him halt at the Aliat Station, which was in any case a strategic embarkation point for the front. Bicherakhov was to command the troops on the front's right flank.

At first, the troops under Bicherakhov's command acquitted themselves quite well in combat with the Turks. However, the subsequent course of events—namely, the retreat of the Red Army to Baku, and also the successful ploy of the right-wing parties to invite the British to Baku—made Bicherakhov change his behavior. Without warning he withdrew his troops toward Dagestan and thus exposed the front. Bicherakhov's treachery played a fatal role in the defense of Baku.

* * *

During the summer of 1918 a serious military threat hung over the Baku Commune when the German-Turkish command began a campaign in Transcaucasia to seize Baku—the great oil reservoir.

You have to keep in mind that in Transcaucasia at that time, Soviet power existed solely in Baku and its suburbs, and in several districts of Baku Province. In the rest of Transcaucasia the Counterrevolutionary Transcaucasian Commissariat was in power, headed by Georgian Mensheviks, Azerbaijanian Muscavatists, and Armenian Dashnaks.

The Counterrevolutionary Commissariat was in mortal fear of the growing revolutionary upsurge in Transcaucasia, and hence went to extreme measures by waging a campaign of terror against the workers and peasants. Realizing soon enough, however, that this treacherous policy was only provoking the indignation of the masses, the parties supporting the Commissariat decided to resort to a new strategem. In February 1918 they set up a "broadly representative" Transcaucasian Assembly which, however, consisted as before, of Mensheviks, Muscavatists, Dashnaks, SRs, and Cadets. The assembly created its own government, headed by the Menshevik Tskheidze, and declared the Transcaucasian Republic to be independent, thus legalizing its separation from the Russian Soviet Republic (although in actuality the Transcaucasian Commissariat had not previously recognized the central government of Soviet Russia because they anticipated the downfall of the Bolsheviks and the accession to power of an SR-Menshevik government).

Soon after, the Transcaucasian Assembly openly entered into negotiations and eventually into an agreement with Germany and Turkey. The formation of a Georgian republic headed by Jordania, the leader of the Georgian Mensheviks, was announced, and was formally recognized by Germany, which thereby violated the conditions imposed by the Brest-Litovsk Treaty. The new Republic of Georgia gave permission for the German troops to enter the Republic and march on Baku, and soon afterward the Azerbaijanian and Armenian landlords and capitalists did the same thing and announced the formation of their own governments.

The Transcaucasian Bolsheviks unleashed a broad campaign against the Transcaucasian Assembly and laid bare the real motives of its organizers and its guiding lights. One should add that in general during this period the Communist groups in Transcaucasia were increasing in number and had become stronger in organizational terms—a fact that contributed greatly to the revolutionary upsurge of the popular masses.

It should be noted that up till now neither the socialists of Transcaucasia nor the bourgeois parties had ever included a demand for separation from Russia in their party programs. The only exception to this was a small, reactionary Pan-Turk group in Azerbaijan and Dagestan, whose determined goal was union with Turkey. The Georgian party of Federalists and the Armenian party of Dashnaks demanded federal status within the Russian state. Before the October Revolution the Mensheviks and the Bolsheviks had come out in favor of the unity of the Russian State—and for regional self-government. This was stipulated in the program of the Russian Social Democratic Workers' party as well.

The nationality question was, as I have noted before, discussed even at the First All-Caucasian Regional Congress of Bolsheviks at the beginning of October 1917. The speaker—Malakiya Toroshelidze—had tried to justify the old position of regional self-government without adding anything new to the nationality policy of the party.

Shaumian had put forward a new proposal at that congress: separate the Transcaucasus into three autonomous national territorial divisions, which would take into account the composition, by nationality, of the population of these territories. Unfortunately, the congress had not understood the new demands imposed by the changing political situation and had rejected Shaumian's proposal. As a delegate to this congress (and as someone who thought he understood the nationality policies of our party) I had been among those who had neither understood nor supported Shaumian's proposal. That was a political error on our part.

The Communists of Baku recognized their mistake only in 1919, when they advanced the correct proposal for a Soviet Azerbaijan closely tied to Soviet Russia. After trying in vain to bring the leading Tiflis Bolshevik comrades over to this position, we reached a compromise with them: if they didn't come out against our proposal for a Soviet Azerbaijan, we wouldn't raise the issue of forming Soviet republics in Georgia and Armenia. After the victory of Soviet power in Azerbaijan in April of 1920, the old position of the Tiflis comrades became simply intolerable.

* * *

During the spring of 1918 the ruling circles of sultanate Turkey had been trying in every possible way to make Azerbaijan part of the Turkish State. Eventually, in May, Turkish troops appeared and marched on Baku; they were aided and abetted by the Azerbaijanian landowners, whose government was at Elizavetpol from where it ruled over those areas of Azerbaijan where there was no Soviet power.

What had, in fact, begun was armed intervention in Transcaucasia by the German-Turkish imperialists, in league with the forces of internal counterrevolution. At that time the position of Soviet power in Central Russia was extremely grave, and although aid did come to us from Russia, it was not on the scale that was needed at such a critical time.

Shaumian had sent Laak Ter-Gabrielian* to Moscow to see Lenin and bring back military and material aid. He was received by Lenin, who promised to provide aid, and who, in turn, gave him the responsibility for stepping up sevenfold the flow of oil from Baku to Russia, issuing him an authorization signed in his own name. At the same time, Lenin wrote the following note to Trotsky on May 24, 1918: "The bearer, Comrade Ter-Gabrielian, is traveling to Baku with a military detachment, with money, etc. Please give him first priority and use the detachment train to carry out the *emergency measures* of providing the people of Baku with the needed military aid."** These lines clearly illustrate the importance which Lenin attached to the defense of Baku.

Lenin's instructions were carried out. Ter-Gabrielian was accorded top priority and given everything that was necessary, but difficulties arose in transporting the detachment put at his disposal. On June 5, while Ter-Gabrielian was at a session of the Sovnarkom, he wrote Lenin a note: "Dear Vladimir Ilych, I beg you, order Tsentrobon to dispatch the Baku detachment tomorrow without fail, otherwise I shall be powerless to do anything to implement its speedy departure."

Lenin wrote back to him on the bottom of the same note, "Why didn't you contact me before? What's your telephone number? What exactly is *Tsentrobron*? What is its telephone number?"

Ter-Gabrielian answered: "Tsentrobron is an organization which has command over all armored vehicles. It has no telephone. Its address is: 5 Znamenka, Dolgorukov's residence. My telephone is: 5-29-00, room 436."

As a result of Lenin's spirited intervention, Ter-Gabrielian was

---

*An old-time Bolshevik who was President of the Extraordinary Commission on the Struggle with Counterrevolution, Ter-Gabrielian later became President of the Council of People's Commissars of Armenia and filled his post very successfully.
**The Lenin Anthology, XXXVII, p. 83

able to depart with the armed detachment and the money, but unfortunately this aid didn't arrive in time.*

The decisive battle was approaching. Where should we meet the enemy? Should we wait for him at the walls of Baku and give battle to him there, or should we assume the offensive ourselves and go to Elizavetpol and smash him there? The latter plan of action was decided upon as the most feasible.

At the beginning of June 1918 when the resolution was passed regarding the Red Army offensive, I was allowed to leave for the front. I was named Commissar of the Third Brigade, which was commanded by the well-known Armenian Dashnak, Amazasp. The location of my command—the right flank of the front, north of the main railroad line to Baku—was Elizavetpol, in the direction of the Shemakha and Geokchai districts.

I was not in the easiest situation. I, who had no military experience, would be meeting both the command cadres and the soldiers for the first time during the course of the offensive. I had to win the confidence of my subordinates, and organize instructive political work among them, in addition to participating in decisions regarding military operations and putting them into effect immediately.

During the first month we made considerable strides and things went quite well, although we had our share of difficulties. The Muscavatists, for example, did everything they could, while retreating, to carry off the local Azerbaijanian population with them at the time of the grain-harvest. We carried out a mobilization drive among the local population for help with the harvest, but even so, there was a shortage of hands. It was impossible for the military units to delay any longer. I sent a telegram to Dzhaparidze and Shaumian, as follows:

> . . . In the river valleys of the Shemakha and Geokchai districts, which we have occupied, the grain has already ripened, and there is no one to harvest it. . . . I have issued orders to the commissars of Geokchai and Shemakha districts to organize forthwith the harvesting of the grain in the designated locales, and I have proposed declaring a mandatory labor conscription and training men to carry out the job. . . . Regarding the essential tools of labor we appeal to you. . . . In a week's time we shall have new foodstuffs for the army. In the meantime I await your instructions and your aid.
> Commissar of the Third Brigade Mikoyan
> (*The Caucasian Red Army*, No. 29, June 1918)

About two weeks after the beginning of our offensive against Adzhikabul Station, the Turkish headquarters, Shaumian arrived. He

---

*Ter-Gabrielian's correspondence with Lenin was first published in the middle of 1970 in the Lenin Anthology, XXXVII, pp. 86–87.

called at once for a general military conference which I was to have attended as Brigade Commissar, but I was unable to go because events developed so swiftly that my leaving the brigade for even a day was an impossibility. I regretted very much having to miss this conference.

Our advance was successful and we soon reached the district center of Geokchai, but then the Turks poured in reinforcements and attacked our flank. In order to prevent our vanguard from being surrounded we evacuated them from the city in a retreat that began amidst heavy fighting. My participation in the fighting gave me a good military education. Up until then I hadn't had any run-in with either the brigade commander or the chief of staff, but by now the men in the ranks had come to know me and, judging by everything, they gave me my due. What influenced the troops was not only my political work—my speeches and my conversations with them, etc.—but most likely what appealed to them was the fact that I, in contrast to Brigade Commander Amazasp, did not sit around headquarters all the time but instead spent time in the trenches with them—first in one battalion, then in another.

From the beginning of our retreat I marched with the rear guard in the rearmost ranks in order to prevent panic among the soldiers. After crossing the pass we then consolidated our forces on the other side, but the mountain peak was in the hands of the Turks, giving them the advantage over us. Moreover, we had no reserves. Amazasp, our brigade commander, and Kazarov, the head of the new regiment (which had been formed in the Shemakha sector under the command of Colonel Kazarov, formerly of the Tsarist army, as part of our brigade and the other adjoining units) were constantly pointing out that the Turks, with cavalry support, could attack our left flank, break through it, cut it off from the rear, and smash it completely. In any case, there was nothing to do but await reinforcements. We had received word that part of Petrov's regiment from Russia was expected in Shemakha in a couple of days.

A militia unit of several hundred Molokani*—Russian peasants from the Shemakha district—also occupied a position at the front, and while this was a help we still had nowhere near enough of the reserves we needed to back us up. We had only a company of cavalry at our disposal, and even this was incomplete.

In the midst of this difficult situation Amazasp, the brigade commander, began to complain that he had a severe stomachache and hadn't the strength to stay there any longer. He took a horse and several bodyguards and rode off. It was days later before I realized that Amazasp's illness had been a fabrication.

The following morning the Turks stepped up their fire. Reports

---

*The Molokani is a religious sect that fled to Transcaucasia from the repressions of the Tsarist authorities and the Holy Synod in the eighteenth and nineteenth centuries.

began to flow into headquarters from the battalion leaders concerning the seriousness of the situation, along with urgent requests for reinforcements, making us aware of the constant danger that one of the units might abandon its position and flee—an eventuality which would have disastrous consequences for the entire regiment. Kazarov, the commander of the regiment, said that the only way out was to retreat in orderly fashion to Shemakha, and then to Marazy; but this should only be done at night because a retreat during the day over open land would give the Turks the opportunity to destroy all the units.

Suddenly, before dinner, Kazarov informed me that he, too, didn't feel well and had to leave for the hospital in Shemakha. That disturbed me: yesterday it was the brigade commander who had taken ill, and today it was the regimental commander. Kazarov was an experienced military commander and his presence in the unit was extremely necessary, especially at such a crucial time, but since he persistently maintained that he had taken seriously ill, I couldn't do anything with him, and he left. The troops were now without any command. I, who was the commissar and had almost no experience in the military, turned out to be the senior person.

As I understood it, my main task under these circumstances was to hold out until dark, and then at night to begin an orderly retreat to the hills in front of Shemakha.

I contacted the battalion commanders via the field telephone, informed them of the situation, and gave them strict orders to hold their positions until night, when at a given signal they were to retreat in orderly fashion to the designated lines. Although they complained of difficulties, the commanders promised to carry out my orders.

I went to check the situation at the battery. There were only some fifteen shells left but the commander of the battery and the entire gun crew turned out to be calm, experienced soldiers who knew enough to save the shells for the decisive battle.

When I returned to the command tent to see if there were any new communications, I found out from the battalion commander that the armed militia of Molokani wanted to go back to their homes. Their departure would leave a gap in the front lines and we had nothing to cover it. I immediately set out for the positions occupied by the Molokani, adjacent to our brigade. The volunteer militia of Molokani consisted, as I have said, of Russians who had settled in that district in villages that were near our forward positions. Up until that day the Molokani had behaved well, without any vacillation.

When I reached their positions I asked them what the trouble was. They replied that their present position was uncomfortably close to their native villages, and that if they were forced to retreat any farther their villages would fall into the hands of the Turks. Since they were convinced that we would be unable to hold our positions they intended to return to their homes. I began to admonish them, but they were adamant. "We are peasants and we came to your aid voluntarily," they said, "but now we cannot abandon our families to the mercy of fate."

After lengthy negotiations I managed finally to persuade them to hold their positions until our brigade could take them over; if we were forced to retreat they could withdraw in orderly fashion along with us, and then disperse to their homes without causing a break in the front lines. This was the course finally agreed upon.

When I returned to headquarters I found Safarov there. He was the commander of a cavalry company in Tatevos Amirov's partisan detachment whom I had met several times before and who had made a very good impression on me. He was strong-willed, straightforward, and had a favorable attitude toward Soviet power. He and his superior Amirov had especially good relations with Shaumian. (I'll give more details about Amirov and his peculiar and contradictory personality later on. He was to perish as one of the twenty-six Baku commissars.)

Just at this time we received word that the Turkish troops had begun some suspicious movements on the left flank. We had Safarov's cavalry company standing in reserve in the ravine, and we decided to have him lead his company out of there in order to confuse the enemy and make it appear as if we were moving our cavalry out. He accomplished this successfully, moving his company onto the mountain slope as if he were maneuvering to surround the Turkish positions. This move took place in the late afternoon and I was sure that we'd be able to hold our positions until nightfall. However, the situation was precarious, and was further aggravated by the fact that in the meantime the battalion commanders had been calling headquarters and had learned that neither the commander of the brigade nor the regimental commander were at their posts. Naturally, this had a disturbing effect on them.

I set off to the battalion that occupied our central position, that was settled into two lines of trenches. On chatting with the soldiers I didn't notice any particular anxiety on their part, but their commander was in a rather depressed mood, although he managed to appear composed on the surface. On the other hand, he did not display any initiative.

We were in the second line of trenches; fighting was going on in the first line, which was about three hundred meters away. The path to it led across the ravine, and though part of this path was exposed to enemy fire, the ravine itself was relatively safe. I asked the battalion commander if he had visited the forward ranks that day.

"No, not today," he said. "But I was there last night."

"Let's go there together, talk with the soldiers, and find out what the situation is," I proposed. His answer was that it was dangerous to go there now with the Turks firing all the time.

I became furious. "Yes, but it's always dangerous at the front!" I said. "Why are you so afraid? We'll go together."

"No, I won't go until it's dark. I don't want to risk my life for nothing!"

It seemed to me that it was not so much caution as cowardice that was speaking. His cowardice not only disturbed me, it disgusted me. But I restrained myself and merely said, in a sharp tone of voice, that if that were the case I would go alone. And then I headed down into the ravine.

I had gone only a short distance when bullets began to whistle by on all sides. I fell to the ground as if I were dead while bullets continued to whistle all around me. Instinctively I began to move the nearby stones over to me so that I could protect my head, otherwise I might perish, if only from a stray bullet. Later, when I had rested for a while, I took advantage of the lull to get up and run quickly forward. Again the bullets began to whine past me and again I fell to the ground. It occurred to me that by jumping into the ravine I had shown my zeal and also had, on the whole, acted rather stupidly: I could easily be killed or, at the very least, wounded, at a time when there was no brigade commander and no regimental commander. Then I became furious at the battalion commander, this time not because he hadn't wanted to accompany me, but because he hadn't dissuaded me from going myself.

The Turks evidently thought that I had been killed and stopped firing. It wasn't more than twenty steps from the safety zone; so again I dashed forward and tumbled into the ravine. A feeling of relief swept over me on reaching safety, and I calmly made my way up the mountain-slope to the forward trenches where the soldiers stared at me in surprise, marveling at how I could appear there.

It turned out that the soldiers' mood was good: they had both ammunition and bread, and they didn't have any particular complaints as they questioned me about the general situation at the front.

"The situation, as you know yourselves, is not very good," I replied. I complimented them on their performance, saying that they should continue as before and not retreat so much as a step without an order. Then I made the rounds of the entire line, stopping to talk with the commanders of two companies. I told them straight out that we would begin a retreat that night, and that their companies would be required to screen our retreat and would be the last to fall back.

I liked the company commanders very much, feeling that one could rely on them. Also, judging by everything, my appearance among them had made a good impression on them. But they chided me in comradely fashion not only for exposing myself to danger but also for taking the wrong route. It turned out that one should cross the ravine a little farther on where a significantly smaller stretch of land was exposed to fire. One of the Red Army men, who was familiar with this less dangerous route, was to show me the way. The company commander took leave of me and told me that when I had reached the danger zone they would open fire with their rifles to divert the Turks' attention away from me. And that's just what happened.

In a word, I made it back safely. I found out at headquarters that everything had been peaceful during my absence, and that the troops were still holding their positions.

With the onset of darkness our units began an organized retreat in the direction of the city of Shemakha. The Turks took no notice of our retreat and did not pursue us while our troops took up defensive positions on the hills in front of the city.

In Shemakha I came across Kazarov, our regimental commander. There was nothing about his appearance that suggested that he was seriously ill. When I asked him what we should do next, he said that it would take a couple of days or so to get our troops in order and that we should not engage the Turks in combat near Shemakha but should withdraw to the area around the village of Marazy, which was one day's march from Shemakha, on the way to Baku. He responded to my objections to his plan in this fashion:

"The troops are tired, and meanwhile the Turks have received new reinforcements. If we engage them in combat, then we could incur additional losses which would impair our defense of Baku. Moreover, we do not hold advantageous battle positions. If we carry out an orderly retreat to the village of Marazy, we can receive provisions and consolidate our forces. The Turks, on the other hand, will be cut off from their supply transports and will find themselves in a more difficult situation. In addition," he said, "Red Army units are retreating on our left flank."

I was not in a position to contradict him since I myself wasn't sure whether I was right or not. Besides, Kazarov had been in the World War and was a truly experienced soldier. Unaware that Kazarov had long ago come to a general understanding with Amazasp behind my back, I had no grounds at the time for suspecting them of treachery.

The next morning I checked to see how the evacuation of the wounded was going. I found that they had succeeded in removing all of them, and that they had also managed to load our last, small reserves of ammunition onto horse-drawn lorries and to send those off, too. In the middle of the day, however, one of the company commanders ran up to me and reported that his soldiers had abandoned their positions without permission and were heading down the adjoining street in the direction of Baku. The two of us ran off after them and when we caught sight of them I grabbed my pistol and shouted: "Stop, or I'll shoot!"

I was not confident of the wisdom of such a step, since there were a lot of them and they were armed, while there was only me and their commander. Also, I was threatening them with a gun; what did they care whether they killed us? Nonetheless I shouted at them once again. To my relief, they obeyed and went back, but about fifteen minutes later I was informed that these same soldiers were heading off once again, this time down a different street.

It was precisely at this moment that a unit of Petrov's detachment arrived in Shemakha. A reliable group of sailors were put at my disposal, along, moreover, with a truck and a machine gun. I went off with them in pursuit of the fleeing Red Army men. We caught up with them and aimed our machine gun at them, finally managing to stop them and return them to their positions.

When it was dark we roused the infantry, and keeping the cavalry in the rear guard on the outskirts to the city, we set off for the village of Marazy in the second stage of our retreat toward Baku. The officers accompanied the infantry on horseback. We had not gone more than five or ten kilometers when a pouring rain began which lasted throughout the night; everyone got soaked to the skin and walked knee-deep in mud. We marched until dawn over the pulpy black earth, along with all the combat equipment. It was hard for the horses to move through the mud, and the soldiers were completely worn out.

We rode our horses at a walking pace, staying close to the infantry. The commander of the regiment rode beside me. We were silent, each of us absorbed in his own thoughts. I was troubled by the situation at the different sectors of the front, and wondered what kind of instructions we would receive from Baku when we arrived in Marazy. It was then that I began to ponder the strange attitude of the regimental commander and the brigade commander, both of whom had "taken sick." The idea suddenly occurred to me of possible collusion between them; after all, they had both "got sick" at the same time, and their opinions were somehow suspiciously similar.

But despite their outward politness toward me, I began to become suspicious. I realized, however, that I couldn't deal with them by myself: I had neither the strength, nor the expertise. I decided that when we arrived in Marazy, I would request that Sheboldaev or someone else from Baku help us to reconsider our defensive position and organize it along correct lines.

Drenched to the skin, and completely worn out, I slept that night on horseback for the first time in my life. I would doze off and suddenly feel that I was falling, and would immediately wake up, a routine that repeated itself all night. Till then I couldn't have imagined it was possible to sleep on horseback. It turns out that the human organism is capable of adapting to even that expediency.

In Marazy we summoned all the officers and ordered them to put their troops in battle formation, take up their positions, post guards, and send mounted forward observers in the direction of Shemakha. The senior commander of Petrov's regiment reported to me that during the retreat their armored car—the only one in our sector of the front—had got stuck in the mud, and we sent the cavalry to pull it out.

By that time the forward observers had reported back that the Turks had entered Shemakha, but were not advancing further. Evidently, the Turks had put their troops in order, and we realized we were now cut off from the enemy by a large distance. I informed Baku of the situation, and asked that measures be taken to ensure that the troops receive supplies, and I also asked that Sheboldaev be sent to aid us in forming a plan for our defense.

In Marazy I stayed aloof from the regimental commander, lodging myself in a peasant cottage. On the strength of the promise received from Baku that Sheboldaev would soon arrive, I slept peacefully.

When I got up the next morning I beheld an amazing sight: the troops were all lined up in marching order and at the front, on horseback, were the regimental commander Kazarov, along with our brigade commander, Amazasp, who had also unexpectedly reappeared. It turns out that the command had already been given to advance toward Baku to the district of Vodokachka, which was several kilometers from the Sumgait Station.

I was dumbstruck: how could such a decision be made without me, the commissar? What had provoked it? No Turks were in sight, why beat such a hasty retreat? I directed these questions at Amazasp, and he replied, "I command the brigade, and I am the one who is responsible for the decisions that are made." With that, he urged his horse forward, leaving me numb with astonishment.

Soon after, on meeting Safarov, the commander of a cavalry company, I proposed that he and his outfit remain under my command, to which he readily agreed. We set off at once for the telegraph office to report to Baku on what had happened.

On the way there I asked Safarov whether he thought the retreat was justified. The enemy was nowhere in sight and our rations had not yet been received, which meant that the regiment had left without provisions or water over a route—from Marazy to Vodokachka—that was waterless. Safarov agreed that nothing had warranted such a hasty retreat. It was then that I came to grips with the strange behavior of both the regimental commander and the brigade commander during the past few days, realizing that a series of actions had taken place which had obviously been plotted out beforehand. They were traitors! After reaching this conclusion I sent the following telegram to Shaumian's address in Baku: "Despite my efforts to the contrary, the transport units have fallen back by Amazasp's order, and moving out gradually behind them is the infantry. The culprits responsible for this should be brought to trial."

After I had sent the telegram, I instructed Safarov to send a mounted reconnaissance unit to Shemakha to secure new information on the whereabouts of the advance column of Turkish troops. Meanwhile we remained in Marazy.

Later in the day the scouts returned, and reported that the Turks were still in Shemakha and had not advanced in our direction. On receipt of this news, I left for Vodokachka, together with Safarov's company of cavalry.

I spent the night billeted with Petrov's regiment (Petrov was already in position in the Vodokachka district), and without going to see either Kazarov or Amazasp I visited the various battalions to talk with the soldiers and ascertain the morale of the troops and their fighting efficiency. The soldiers complained of fatigue because of the daily skirmishes and of the hardships of their recent retreat without water and rations. When I asked them if they were ready to take up their positions, they answered that they would be if they could have three days' time to rest, get washed, and have a good sleep. And in the meantime they asked that they be moved to the rear and temporarily replaced by fresh troops.

This, of course, would have been fine, but we didn't know whether Baku would consider it feasible. At the same time, we realized that the Turks would need a few days more before they could make contact with our troops. This time could certainly be used for rest.

The battalion commanders were given the appropriate orders. They asked only that they be provided with the necessary rations, and I promised to fulfill their request.

That same morning a motor car from Baku was unexpectedly put at my disposal. It arrived most opportunely, and I left immediately for the Sumgait railroad station where there was a telegraph link with the front. After receiving a guarantee that our units would be supplied with rations without fail, I returned to the brigade.

As I drove up to the building in Vodokachka where the brigade headquarters was located, I saw several hundred Red Army men resting on the ground near the road, about a hundred meters away. The car was without the roof and I was sitting in the rear seat. Suddenly I saw one of the Red Army men raise himself lazily from the ground, and leaning on his rifle, ask my chauffeur to stop the car. Why would he address such a question to the chauffeur and not to me, the commissar? The chauffeur ignored him, and the car continued on. Then I ordered the chauffeur to stop the car, got out, and said sharply to the Red Army man:

"What's going on here, what happened?"

Several other soldiers came over. The first one asked in an embarrassed and nervous way:

"Is it true, Comrade Commissar, that you want our commander Amazasp to be courtmartialed?"

This question took me completely by surprise. How could they have known of my telegram to Shaumian? Immediately the thought flashed through my mind that some plot had been rigged up against me by Amazasp, since the soldiers could not have known of the telegram except through him.

I didn't answer right away, but instead asked this question by way of rejoinder: "Did you see any Turks when you were pulling out of Marazy?"

"No," they replied, "we didn't."

"Then why is it that you were moved to new positions? You had no bread and water, did you?"

"No, we didn't."

Then I continued, "If the military situation necessitated a withdrawal, we should have waited until the provisions arrived. The Turks are far away, and there was no immediate danger of any confrontation with them. All these questions demand answers, and that's why I asked the proper military authorities to look into who was to blame for this."

By this time large numbers of Red Army men were clustered around me, and the situation had become defused, settling down into an ordinary peaceful conversation between the soldiers and their commissar.

During the conversation I noticed that Amazasp was standing about a hundred meters away with several of his retinue, staring fixedly in our direction and only too aware that the lynching he had wanted to stage for me had not taken place. Then, obviously on Amazasp's order, two of his bodyguard-cavalrymen strode over to us and pushed the soldiers surrounding me out of the way. One of his bodyguards then swung his arm and struck me on the head and neck with his whip. Instinctively, I grabbed for my pistol and he pulled out his Mauser, but then the Red Army men intervened and separated us, thereby averting inevitable bloodshed. Without a word I got into the car and drove off. All the while I kept thinking: "How did my telegram to Shaumian become known to Amazasp?" But at the time I simply couldn't find the answer to the question.

My first order of business in Sumgait was to try and get hold of a copy of *The Baku Worker* to find out the news. There, in the issue dated July 22, I saw the verbatim text of my telegram to Shaumian. How could this happen? I thought indignantly. Amazasp had not been arrested or brought to trial, and yet the telegram was published in the paper. Who made it possible, and why? To whose advantage was it?

Later, in Baku, it turned out that my telegram had been handed over to the newspaper because of the inexperience of Olga Shatunovskaya, Shaumian's secretary: she had wanted to publish proof of the Dashnaks' treachery at the front in order to undermine their influence on the Armenian workers.

For myself, I decided then and there to go to Baku and do everything in my means to bring about the arrest of Amazasp, and after that to return to the front.

Chapter III

The Armed Struggle Continues

The Mensheviks, SRs and Dashnaks favor inviting in the British Shaumian brands the traitors—Revolutionary Baku struggles.

I read something in the July 25 newspaper that disturbed me even more. Apparently, the evening before, mass political meetings had been held in all districts of Baku to discuss the question of inviting British troops into the city. The Bolsheviks opposed it, whereas the Mensheviks, the SRs, and the Dashnaks, who worked directly for the secret service of the British Military Command, supported the step.

The close alliance at that time between the Baku SRs and the British has been described quite eloquently in the memoirs of General Densterville, who was in command of the British occupation forces in Baku. "My communications with Baku," writes the general, "were arranged by means of couriers who were despatched almost daily. . . . Our friends, the Social Revolutionaries were . . . in a position . . . to overthrow the Bolsheviks in the near future, establish a new form of government in Baku, and invite the aid of the British.

"I frequently," Densterville admits later on, "carried on negotiations with leaders of the SR party, whose programs conformed much more closely to our goals. . . . They desired our help, particularly in the financial area. I maintained friendly relations with the SRs, and they knew that they would be able to rely on us in many ways if they seized power."

The workers of Baku, deceived and misled by the SRs, Dashnaks, and Mensheviks, tormented by hunger, frightened by the Turkish invasion, and apparently not clearly understanding the dangers of British intervention, considered the British to be a lesser evil than the Germans and Turks, and consequently during the meetings favored inviting in the British troops.

In essence, this was an expression of a lack of faith in Bolshevik policies on the part of some workers. I was amazed that they would organize such dangerous discussions on such a critical issue in the very shadow of imminent military attack, when the British forces had already seized Murmansk and Arkhangelsk, and when a general intervention against Soviet power had begun. An invitation to the British armies to enter Baku would, in fact, open the door to their occupying the city. In my opinion, it was an irreparable mistake.

True, I saw other news in the paper as well—about meetings of workers and Red Army men where Bolshevik resolutions were passed opposing British aid and demanding the mobilization of workers to the front. This was the voice of the politically conscious vanguard of the Baku proletariat.

Extremely disturbed by all that was happening I left at once for Baku, arriving there on July 25. On that same day there was a meeting of the Baku Soviet attended by members of the district Soviets, and of naval committees, and by representatives of the Red Army.

* * *

The meeting of the Soviet was already nearing its end when I arrived. There was a recess, after which Shaumian spoke. He reported on the political and military situation and roundly condemned the appeals made by the Mensheviks, the SRs, and the Dashnaks for inviting the British into Baku. As Special Commissar on Caucasian Affairs, he issued a statement on behalf of the Central Soviet authorities which declared that the Fifth All-Russian Congress of Soviets had come out in favor of Soviet Russia's pursuing an independent policy—independent of the Germans and the British. Shaumian urgently demanded that the question of inviting in the British troops not be included on the agenda. He introduced data which testified to the aid we had received from Soviet Russia, citing figures on the number of weapons which had already arrived or were en route from Astrakhan to Baku. While we were waiting for this new aid to arrive, Shaumian proposed that we mobilize and strengthen the Army and Navy and thereby create a powerful defense for Baku.

Then, one after another, the right-wing party representatives spoke out against Shaumian's proposal and viciously attacked the Baku Sovnarkom and the Red Army. Azizbekov, Dzhaparidze, Zevin, and other Bolsheviks in their turn gave impassioned speeches against the right-wingers.

After fierce debate, and in spite of all the Communists' efforts, the right-wingers' resolution to invite the British troops into Baku was passed by a narrow margin. The resolution also called for the formation of a new coalition government in which all the parties represented in the Soviet would participate. (The jointly sponsored resolution of the Mensheviks, Dashnaks, and SRs received 258 votes, and that of the

Bolsheviks, left-wing SRs, and leftist Dashnaks, which had been proposed by Shaumian, received 236.)

A fateful role was played by defection to the right-wing parties of a group of sailors from the Caspian fleet, who had been partly misled and partly bribed by the British secret service. Shaumian made a direct reference to this affair in his speech. He also said that the sailors would come to regret their fatal mistake, but by then it would be too late.

After the vote Shaumian declared that, as a representative of Central Soviet authority, he protested this treason.

A recess was called, during which the Bolshevik faction of the Baku Soviet convened privately.

I had now arrived and was in the hall when the Soviet reconvened in time to hear Shaumian make a statement to the assembly on behalf of the Bolshevik faction. The invitation to the British, he said, was nothing more than an act of rank ingratitude and high treason directed at revolutionary Russia. He branded the SRs, the Mensheviks, and the Dashnaks as enemies of Soviet power, and declared that the Bolsheviks renounced responsibility for this criminal policy and refused to serve as People's Commissars any longer.

I was stunned by the unexpectedness of all this, since I was ignorant of the events that had led up to it. Later on, when I was able to reconsider the whole complicated scenario of events of those days in a more calm and meticulous fashion, this is what I wrote:

> The Mensheviks, SRs, and Dashnaks indeed succeeded in deluding a significant segment of the Baku workers with promises of aid and support from the British in order to fight off the Turks laying siege to Baku. The workers faltered. There was no food: Kazakh bands had cut off Baku from the northern Caucasus. Famine stalked the workers' quarters. Baku had been cut off from its supply of potable water. Larger and larger counterrevolutionary mobs were gathering at the walls of the city. The roar of artillery fire drowned out the muted factory whistles. Fear of possible counterrevolutionary reprisals undermined the workers' strength. By exploiting legal sanctions, the SR-Menshevik scoundrels were able with impunity to poison the workers' consciousness with the idea that they had only to say the word and countless numbers of "cultured" British troops would come to the aid of their "allies and brothers" against the Turks, and would save the city. At district meetings and at sessions of the Soviet of Workers' Deputies, the social-traitors' proposal to invite in the British received the majority of votes.

That same night, immediately after the meeting of the Soviet, an emergency, expanded meeting of the Baku Party Committee was held. Many of us spoke out against the decision of the Bolshevik faction to have the Peoples' Commissars leave their posts. A resolution was passed not to surrender any [existing] authority nor, in addition, to establish any new authority, and it was decreed that all previously existing organs of power—the Soviet's Executive Committee, the Soviet of Peoples' Commissars—should continue their work. The decision was made to launch a mobilization drive among the Baku workers to fill the ranks of the Red Army, to proclaim a state of emergency in the city, and to take measures against those who were engaged in counterrevolutionary agitation. In view of the extreme complexity of the situation it was decided to convene an All-Baku Party Conference. This took place on July 27, 1918, and confirmed all the resolutions passed at the meeting of the Baku Party Committee.

The next day an emergency session of the Baku Soviet was convened and chaired by Dzhaparidze.

On July 28 the Baku Sovnarkom ordered the conscription of additional men into the Red Army. That same day the Executive Committee of the Baku Soviet made an appeal to the population in which it made known its resolution "to inform the population of the city of Baku that the full complement of the Executive Committee continues as before to guard the Soviet Government of Workers and Peasants, and that any attempt to undermine the [Baku] Soviet will be ruthlessly suppressed by same. . . . The Executive Committee appeals to all army units, sailors, and workers, and to everyone who holds Soviet authority in sincere esteem, to guard the Soviet and close ranks around it, encircling it in an iron ring, so that those enemies of the Soviet who dared to raise their heads at such a critical moment will be repulsed, and an unwelcome reception provided for the insolent enemy who marches against us."

That same morning a mass meeting and a major demonstration were held on Freedom Square. The following signs could be seen everywhere: "Long live the new mobilization!"; "Shame on the cowards who avoid being mobilized!"; "Death to the traitors and betrayers!"; "To arms! Join the ranks of the worker-peasant socialist army!"; "Long live the RSFSR!"; "Baku favors only a Soviet republic!"; "Down with the German, Turkish, and British imperialists!"; "The British have seized all of Persia and now they want to seize eastern Transcaucasia. Down with the aggressors!"; "The Germans and Turks have seized all of Transcaucasia and now they want to seize Baku too. Down with the aggressors and the invaders!"; "Down with the traitors who want to surrender Baku to the British!"; "Agitation for the British sows dissension in our fraternal ranks and weakens the front. Down with this agitation!"

On behalf of the Baku Soviet, Dzhaparidze welcomed those taking part in the rally and demonstration, declaring that the Baku proletariat was ready, weapons in hand, to defend Soviet power. Impassioned speeches were made by Korganov, Zevin, Petrov, Vezirov, and representatives of the Red Army men and the sailors. Vezirov said, in particular, that "the Moslem peasantry was awakening from its century-long sleep and casting in its lot with the fate of socialism in Russia. By enlisting into the army in scores, the Moslem peasants are defending and will defend mighty Russia."

In the evening a meeting of the soldiers and commanders of the Baku garrison was held in the packed hall of the opera house. Shaumian spoke at the meeting and said that the "revolutionary front would be squeezed from two sides, from within and without." To the question of where the people of Baku could get help, Shaumian replied: ". . . only from Russia! Only from our revolutionary comrades at the center can we get aid. According to the principle of revolution, the proletariat cannot appeal to any oppressor and kiss his hand just because that hand, which holds a whip, will strike less savagely than that of another oppressor. That's a slave's logic, a slave's conception of the question, a slave's psychology." He called upon the Baku proletariat to fight "not only for their city and their homes, but also for all of Russia. This historic mission," said Shaumian, "should fill our Baku comrades and workers with pride and summon them to fight this glorious battle to the end—to win, or to die with honor."

The commander of the regiment, Petrov, made a speech at the meeting and also called upon the soldiers to continue their heroic defense of proletarian Baku. On behalf of the men of the Red Army he thanked Shaumian for his unflagging efforts to strengthen Soviet power in Baku.

* * *

Shaumian informed Lenin of the situation and of the measures being taken to deal with it, and asked that new military units be sent as soon as possible.

Lenin had been kept in constant contact with Shaumian either directly from Moscow, or through Stalin, who was in Tsaritsyn. ". . . You know that I have complete confidence in Shaumian,"* wrote Lenin in August of 1918 to the President of the Astrakhan Sovdep (Council of Deputies).

Shaumian's urgent request for additional troops was answered by Lenin's telegram of July 29: "Regarding the dispatching of troops, shall try, but can't promise anything definite."**

This rather vague reply was fully understandable and explainable. 1918 was an extremely difficult year for Soviet Russia. The Germans had occupied the Baltic region, Byelorussia, and the Ukraine. British troops were moving from Murmansk and Arkhangelsk toward the

*V.I. Lenin, Complete Works, Vol. 50, p. 142.
**Ibid., p. 129.

center of Russia. Japanese and American troops had landed at Vladivostok. The Czechoslovakian expeditionary corps had started a rebellion and with the support of the White Guards had occupied several strategic cities in the Urals and on the Middle Volga. A Cossack counterrevolution was looming large in the Northern Caucasus. Krasnov's bands were attacking Tsaritsyn. The Orenburg Cossacks had pushed their way through to Astrakhan. Left-wing SR mutinies were flaring up in the center of the country, in Moscow, and then in Yaroslavl and Astrakhan. The British offensive on Turkestan had begun. Armed forces were needed everywhere at a time when the Red Army was still in the process of being formed. It was difficult to transfer troops from one front to another, and shortages were felt everywhere.

Soviet power was fighting for its life against both counterrevolution and the foreign intervention of fourteen nations.

Our best propagandists in those days went to the masses, to organize meetings of workers and soldiers in support of Soviet power and against the traitors. The Gummet organization conducted meetings of Azerbaijanian workers in different parts of the city. Party activists made appeals to defend Soviet power and proletarian Baku against the invasion of the Turkish-Muscavat hordes. Those who participated in these rallies firmly resolved to join the Baku proletariat in defending the cause of revolution.

Our opponents were not sitting on their hands, however. Secret meetings were taking place between the conspiratorial leaders of the Mensheviks, the right-wing SRs, the Dashnaks, the bosses of Tsentrokaspii (the Central Committee of the Caspian Naval Fleet), and a representative of Colonel Bicherakhov, who acted as though it was his function to transmit to them the wishes of the British military command.

The Military-Revolutionary Committee of the Red Army did everything possible to strengthen the front.

As a member of the Baku Party Committee, I took part in those days in all kinds of party, political, and war-related work. And I still continued to try to return to the front as soon as possible. While taking our overall difficulties very much to heart, I simply couldn't forget (why not admit it!) the outrage of Amazasp's betrayal, and my own sense of personal grievance. I demanded from Shaumian, as categorically as possible, that Amazasp be arrested immediately and replaced by one of the battalion commanders.

"Once that's done," I said to Shaumian, "I'll return to the front right away, and I'm convinced we'll manage to get the brigade— which is on the whole an experienced outfit—into top fighting shape, and to organize the defense of Baku in our sector of the front."

Shaumian understood my feelings and shared them, but at the same time, using great patience, he urged me not to be hasty:

"You must realize," he said, "that Amazasp is in the brigade, he knows you're demanding that he be brought to trial, and he has probably already taken steps in his own defense. The Armenian National Council is backing him up, and we don't have enough power to do what you're asking us to. If we have to send an armed detachment there, an armed confrontation will be unavoidable, and that would only further exacerbate the situation in that sector of the front."

Under these circumstances it was impossible for me to return to my brigade.

* * *

Once the resolution had been passed not to relinquish power in Baku but to stay and fight to the end, we decided to evacuate the families of party and Soviet officials to Astrakhan. As before, I was living in Shaumian's apartment and saw that Ekaterina Sergeevna tried every way she could to postpone her departure, not wanting to leave Shaumian and her two adolescent sons, Suren and Leva, who were members of the Bolsheviks' armed militia. Shaumian was busy at work day and night, attending various meetings, and he rarely went home. I tried to persuade Ekaterina Sergeevna to leave as soon as possible because her continued presence in Baku, particularly with the younger children, upset Stepan. Dzhaparidze's wife—Varvara Mikhailovna, who had two small daughters, Elena and Liutsia—didn't want to leave either. Finally, we managed to get both families to collect their most essential belongings.

On the evening of July 29 Shaumian telephoned me.

"The news from the front is very bad," he said. "The Turks have broken through our front lines, and our troops have retreated to Baladzhary, which is just one railroad stop from Baku. I want you to go there at once and look over the situation, see what's happening with your own eyes, do whatever you can, and then report to us."

I called the railroad station and ordered them to have an engine ready to take me to the front. On arriving at the station, where the train was already at the track, I saw Olga Shatunovskaya standing by the train with a rifle in her hands. When I asked her what she was doing there, she said that Shaumian had instructed her to accompany me. I protested that there was absolutely no reason why she should go with me, and told her she'd be more useful doing her job as Shaumian's secretary, but she insisted on having her way, and there was nothing I could do about it. She came along with me.

When I arrived in Baladzhary, I located the frontline headquarters, in an official clubcar, and presented myself to the Headquarters Commander. Avetisov was an experienced commander, a former officer in the Tsarist army, a colonel, and much older than I. He didn't make a very good impression, however. Greeting him politely, I asked:

"Please tell me the present state of affairs at the front. What are the positions of our troops, and where are the Turks?"

On his desk I saw a map of the front with the positions marked out. Somewhat overwrought, Avetisov replied that the situation was very bad and that he really didn't know what was happening. He was obviously confused. "A man in your position," I said, "is obliged to know how things are going at the front. Besides, I can see that the disposition of our forces and those of the enemy is marked on the map. It probably represents the latest information. Please don't be nervous, and report calmly."

In the same agitated manner, Avetisov replied that none of the markings on the map meant anything since troops were retreating without asking permission of their commanders, and the entire front was in a state of chaos. I began to inquire exactly which military units he was positioning at the approaches to Baku. He named a few battalions, two armored trains, and Petrov's regiment. Then I asked:

"And where is Bicherakhov's regiment?"

"Today," he replied, "Bicherakhov announced his refusal to continue fighting the Turks as part of the Red Army, and, taking his units with him, he exposed the right flank of our defense."

This was a new and serious blow. What could be done in the face of such conditions? I suggested that Avetisov set about reestablishing contact with the frontline units and decide which ones to transfer to the sector of the front exposed by Bicherakhov. He began babbling something to the effect that "it's not in my power to do thus and so." I realized that it was useless to go on talking with him and went off to find the car where Ganin, the Staff Commissar, had his quarters. I found him with Gabyshev, the Brigade Commissar. I told them about my conversation with Avetisov and began inquiring about the real state of affairs. They confirmed that the situation at the front was serious, and that it had gotten much worse because of the withdrawal of Bicherakhov's regiment. In answer to my question about which units could be moved to the positions vacated by Bicherakhov's Cossacks, they said that the safest thing was to send some units from Petrov's regiment, which was stationed there (in Baladzhary), and to bolster them with reinforcements from Baku.

There was no military discipline in Baladzhary. As my comrades had said, groups of Red Army men had left their positions and were hanging around the railroad station, thereby adding to the considerable disorganization that everywhere prevailed. We decided to give a small group of sailors from Petrov's regiment the responsibility of establishing order at the station, and we summoned the commanders of the subunits of Petrov's regiment, which were stationed in Baladzhary, and ordered them to occupy the positions which Bicherakhov had given up. The order was carried out.

Then we got in touch with Sheboldaev and asked him to send

reinforcements from Baku for Petrov's regiment without delay. In addition to that, we asked him to send two or three companies from among the newly mobilized workers who were undergoing training. He promised to fulfill our request insofar as possible. He gave us the news that Petrov had replaced Bicherakhov as commander of that sector of the front, and told us that Bicherakhov had fled because he disagreed with the policy being pursued against the British, and at the wish of his Cossacks was returning with them to the Northern Caucasus.

After a few hours' sleep, we got up and went to the station. We were gratified to see for ourselves that the group of revolutionary sailors had succeeded in establishing order there. It soon became known that Petrov's regiment was now occupying Bicherakhov's designated positions.

When I stopped by at Avetisov's headquarters that morning to learn whatever news I could, I again found him in a state of panic, complaining that we didn't have enough forces to resist, that the Turks could wipe us out in a flash, and that the help that had been promised from Baku had not arrived and probably never would. Nonetheless, I did find out from him that certain Armenian national units, which had been trained to go to the front, were still in Baku. I went to the station and put a call through to Shaumian, but was told he was busy at a meeting of the Soviet. I went back to Ganin and Gabyshev and together we began to take measures to strengthen the front.

News came through almost at once that Petrov's regiment had valiantly withstood a Turkish attack in its new sector of the front and, although it had sustained heavy losses, had held on to its positions in a bloody skirmish. This made the immediate reinforcement of our units even more crucial. Such reinforcements were expected at any moment.

For the third time I went back to Avetisov, to find him in the same state as before. The fact that Petrov's regiment had repulsed a Turkish attack made absolutely no impression upon him. I told him that I had no way of getting in touch with Shaumian, but while we were talking, a man from the railroad station came in and said that it was now possible to put a call through to Shaumian. Avetisov wanted to come with me. At first I was against it, but then it occurred to me that he might prove useful in providing some information that might interest Shaumian. And so we went together.

There was no one besides Avetisov and myself in the room where the telephone was. I immediately asked Shaumian what was happening and what we should do. He replied that the political situation had turned out to be more complicated than the military one, and that round-the-clock meetings and endless conferences of the representatives of the right-wing parties, the Tsentrokaspii (the Central Committee of the Caspian Naval Fleet), and the Armenian National Council were taking place. The Mensheviks and the SRs had managed to get the leaders of Tsentrokaspii on their side: they had agreed to send ships to the Persian port of Enzeli in order to transport the British to Baku. The Armenian National Council had not only refused to send several well-organized units to the front to fight the Turks, but was demanding the initiation of peace negotiations with the Turks, and had already got in touch with the Swedish consulate to act as an intermediary in this regard. There was a suspicion that the National Council was preparing to send their own representatives to the Turks independently. This was being done under a false pretext: that the front would not, in any case, hold up, and that peace negotiations could save the Armenian population from being slaughtered, which was a possibility in the event of a Turkish seizure of Baku. At the same time the National Council had got in touch with the SRs from the Tsentrokaspii and was playing a double game: on the one hand, it was supporting the German-Turkish policy and on the other, the British.

It goes without saying that we could not agree with the Mensheviks, the SRs, or the Dashnaks. In effect, they were destroying the rear and weakening the front. We intended to continue the struggle. Then, Shaumian said something that really startled me: it turned out that Colonel Avetisov had informed the Armenian National Council as early as the night before that the Turks would occupy Baku in three or four hours and for that reason he suggested raising the white flag. In connection with this, the Armenian National Council was asking the Council of Peoples' Commissars to give the order to the front to raise the white flag. This made me so indignant that I shouted into the phone:

"What white flag?! We don't intend to and shall not raise any white flag here!"

"The Sovnarkom is also against raising the white flag," Shaumian said.

At the end of the phone conversation, Avetisov, nearly beside himself, said to me:

"No, Mr. Commissar, we'll *have* to surrender, and we'll force you, as commissar, to raise the white flag personally!"

Unable to contain my rage any longer, I grabbed my revolver and, enunciating every word very clearly, said: "Mr. Colonel! This scheme of yours to surrender won't work! You should keep in mind whom you're dealing with, and know that there's a bullet in this gun for you!"

Avetisov blanched, afraid that I would shoot him there on the spot. But I was only warning him. He realized that and walked out without saying a word.

I went to Ganin and Gabyshev and told them everything that had happened. We composed a telegram to send to Baku which all three of us signed there and then. In it we said that Avetisov was threatening to force us, the commissars, to raise the white flag and surrender to the Turks. Not even the threat of execution could make us do such a shameful thing.

Then we began to discuss how the struggle for power in Baku might end, whether our comrades would manage to hold on to power or whether the Mensheviks, SRs, and Dashnaks would seize it. What kind of orders could be expected from Baku? All of us still had hopes that Soviet power would manage to endure in Baku.

News began to arrive from the front that the Turks had begun an advance and were occupying a small elevation not far from Baladzhary. Later, information reached us that Korganov, and then Sheboldaev, were going to come to the front. This raised our spirits a little. But time passed, and still they didn't come.

At nightfall the Turks mounted their artillery on the hill they had taken and began firing on Baladzhary. It became clear that our headquarters couldn't remain in Baladzhary any longer. We summoned Arveladze, the Chief of Military Communications in the Caucasian army, whom I knew quite well from our work in Baku, and after consulting with him, we decided to begin sending the transport trains with the troops to Baku in relays.

It was about eleven o'clock in the evening when our train pulled into the station in Baku. I took a rifle with me and went out onto the platform. It was quiet in the station, without any hustle and bustle, as if everything were all right. On the platform I ran into Ashot Ter-Saakian, a left-wing SR and commissar of an armored train, who had been a student in Moscow and whom I had known before as a good revolutionary. He said to me point-blank: "You know, don't you, that there's been a coup!"

I became very agitated.

"What coup? How do you know?" I asked him.

"Everyone talks about it."

"I don't believe it," I replied. "I'll go to the *Revkom* [Revolutionary Committee]."

"Be careful, you might be arrested!"

But I went anyway. I walked down the street to the Revkom, which was located in the Hotel Astoria on Freedom Square. No changes were visible in the streets, and around the Revkom everything was the same as before, with the same sentries at the entrance. With an emphatically self-confident look on my face I entered the building and went up to the second floor, but instead of heading toward the offices of the upper administration, I went to the rooms at the end of the corridor where the rank-and-file clerical workers were usually stationed. I opened the door to one of the rooms and saw Polukhin sitting there. He was a member of the Naval Board who had come to Baku as a plenipotentiary from the center. He was tall, about thirty-five, and a sailor whom everyone respected. Solntsev, the Director of the Baku

School for Cadre Commanders, was sitting with him. They were conversing quietly.

"What are you doing here?" I asked them.

"We don't know ourselves."

"How is that?"

"Well, we just got here, too, and found out that our comrades had been evacuated to Astrakhan."

"Is that really true?"

"Unfortunately," they replied, "it's a fact."

Indignant, I said: "How can it be that after deciding to evacuate, they couldn't have found a way to let us know about it in Baladzhary?"

"Maybe they tried to let you know, but couldn't get through," my comrades replied. "Nobody informed *them* either—since it was an emergency and everything happened so unexpectedly." Then they told me that on August 1, 1918, the Mensheviks, SRs, and Dashnaks had taken power into their own hands and had formed, on behalf of the Tsentrokaspii, which had actually ceased to exist, the so-called Dictatorship of the Tsentrokaspii and the Provisional Presidium of the Executive Committee of the Soviet. The Caspian fleet had already sent ships to Enzeli for the British. In a word, the counterrevolution had won a victory.

"Do you happen to know who is the head of the counterrevolutionary government?" I asked them.

It turned out that as far as they knew, the Menshevik Sadovsky had been appointed head of the government, and Bicharakhov had been put in command of the army.

"How can you sit here and chat so peacefully?" I asked. "What do you intend to do?"

"There's only one thing we can do," they replied, "and that is, whatever happens, to try to get to Soviet Russia, where we'll still be needed."

"And how will you get there?"

"We still haven't decided, that's why we're sitting here and thinking. The road to the northern Caucasus and to Transcaucasia is closed. In the Transcaspian area there's a counterrevolutionary government bolstered by British bayonets. There's no way of getting to Astrakhan by ship, but boats *are* running between Baku and Enzeli. We think we'll try to get to Enzeli and from there make our way through Persia and Afghanistan to Tashkent, which is under Soviet rule. But what about you? What are you going to do?"

"I'll stay in Baku, go underground, and continue doing party work, though at the moment I don't know where I can hide. I'll try to go to the barracks of the party militia. Maybe there's still someone there who can help me find a hiding place at some worker's house."

I made haste to leave, advising them to find some place of refuge with friends. Unfortunately, I couldn't give them any address, since

I didn't know where I would hide myself. I told them that I would try to convey the news to Ganin and Gabyshev, who still didn't know what had happened.

Although I was a nonsmoker, I asked them to give me a cigarette so that, by lighting up, I could pass by the sentries "with an air of importance." In saying good-bye we agreed that we would try to keep in touch, via the local inhabitants, until they left.

After that I walked down the corridor, descended the stairs, and passed the sentries without any trouble. I felt good about having got out of the building and into the open air. I walked down Telefornaya Street to the large, multistoried building we had requisitioned as a barracks for the party militia. I went up to the second floor, where I saw people sleeping on the parquet floor in a large hall. It was almost midnight. Among the sleepers I recognized a close friend from my schooldays, Artak Stamboltsian, who was now commissar of the regiment. Annoyed that at a time like this, Communists—among them, Artak—were sleeping peacefully, I gave him a poke in the ribs with my foot. He jumped up, and stared at me, not understanding what the trouble was.

"Where are Shaumian, Dzhaparidze, and Azizbekov?" I asked him.

"I don't know."

And he told how he had stayed with his regiment to guard the State Bank, until he had received orders a short while ago to return to the barracks. And here he was, resting with his men and awaiting instructions. We immediately roused the militia and ordered them to return to their homes without being seen by the counterrevolutionaries and to wait there for further instructions.

I asked Artak if he knew some place where I could spend the night. Where did he intend to go himself? He gave the address of one of his comrades in the militia, adding that he knew him well and thought he could stay with him. He suggested that I spend the night at Tatevos Amirov's apartment, if, of course, I trusted him. I replied that although Amirov was not a Communist, he was a truly decent fellow and besides, he had been nice to Shaumian: when he was at the front with his cavalry division, we had met several times and became friends. I didn't think that he would refuse me refuge.

Artak gave me his address.

It was midnight. I went to his house and knocked on the door. Tatevos Amirov himself opened it, having obviously been roused from his bed. I told him that our comrades had left, but that I was staying on, and asked if he could let me stay the night.

"Please, come in," Tatevos replied cordially.

I spent the night at his house. When I awoke the next morning, Amirov was already up and about. It turned out that he had been to town and brought back the news: all the ships on which our comrades,

including Petrov's regiment, had tried to leave had been returned to Baku, and were now anchored at Petrovsky pier.
 I immediately set off for there.

Chapter IV

The Fall of the Baku Commune

Will the Soviet stay in power?—The arrest of the Baku commissars and our struggle for their release—We do battle with the Mensheviks in the Soviet—The Turks enter Baku.

When I arrived at the pier I learned that the ships with Petrov's regiment aboard had not even managed to leave the harbor before they were apprehended and returned to their moorings. At sea a cutter with members of the Tsentrokaspii aboard had pulled alongside Shaumian's ship to demand his extradition and arrest, along with that of Dzhaparidze and Sheboldaev. Dzhaparidze was not on the ship, and Sheboldaev had managed to hide in the hold by mingling with the sailors on the ship, but Shaumian was arrested and transferred to the battleship *Astrabad*. After putting back to shore the comrades on Shaumian's ship told Petrov and Amirov what had happened. Petrov and Amirov set out at once for the Tsentrokaspii to demand Shaumian's immediate release, threatening to call out their troops if this was not done. Shaumian was released on the spot.

I went on board the *Kolesnikov*, where Shaumian was now ensconced, to find out what had happened. He told me that the situation in the city had become critical by the evening of July 31. Colonel Avetisov had reported that the Turks would be in Baku in three to four hours, and he and the Armenian National Council had begun insisting on surrender. To be honest, Shaumian said, we even thought that the nationalist units had already surrendered, or were about to do so without our consent. Under the Tsentrokaspian leadership, the SRs, Mensheviks, and Dashnaks had, in fact, on British orders, already set up their own counterrevolutionary government. The chief of staff of Petrov's regiment had reported to Shaumian that they had sustained heavy losses in the fighting and were retreating toward the city under enemy fire. "All is decisively lost at the front," he had declared.

With a Turkish invasion of Baku imminent, Shaumian continued, we did not feel it was the right time to provoke a civil war. Therefore, the Soviet of Peoples' Commissars decided to relinquish its authority and to evacuate its military units by ship to Astrakhan, along with the state property of Soviet Russia. It was later revealed that under pressure from the British, the Tsentrokaspii had threatened the Armenian Council and the latter, expecting British aid, had not surrendered, and had even agreed to send new military units to the front.

On the first of August, the Turks began to step up their major offensive in an all-out attempt to take Baku, finally breaking through the Wolf Gates into the Bibi-Èibat district, whereupon the Tsentrokaspii's troops fled. After consulting with the other leaders, Shaumian proposed that Petrov unload his artillery, open fire from the pier on Bibi-Èibat where the Turks were positioned, and send in reinforcements of Red Army soldiers and sailors, who were stationed a few miles away from the pier, to this sector of the front.

Panic changed to joy among the populace when intensive artillery fire began to bombard the Turkish positions. Shells whistled over the heads of the citizens of Baku and fell into the midst of the Turkish troops. The shelling caught the Turks off guard, and after sustaining heavy losses they retreated. When it became known in the environs of Baku that the commissars and Petrov's regiment were in Petrov Square, workers and Red Army men—singly and in groups, some armed and some weaponless—began to show up to enlist in Petrov's regiment, which swelled from 400 to 2,000 men.

On the second or third of August, I don't remember which, a party conference was held to discuss the situation and to decide what to do next. After long debate the conference reached a decision: instead of evacuating our troops to Astrakhan, the Bolsheviks would take advantage of the populace's change of mood to seize power once again. This was possible in practical terms, since the enemy had fewer forces at his disposal in the city than we did, and would be unable to move troops up from the front. By consolidating all our forces in Baku and getting the assistance of troops from Astrakhan, we hoped to defend our city.

Since our party newspaper, *The Baku Worker*,* had been shut down, an appeal to the workers of Baku was published in one of the presses immediately after the conference. It was written by Shaumian and signed by the Baku Committee of the Workers' Communist party (of Bolsheviks). The appeal summarized in detail the events of the past several days, exposed the calumny that had been spread against us by the counterrevolution; pointed out the treason of the Dashnak, SR, and Menshevik parties; and explained the circumstances that had forced us to issue an evacuation order for our armed forces. The appeal

*After Shaumian, Arsen Amirian—a stalwart and high-principled Bolshevik—had become the editor of *The Baku Worker*. His brief, pointed, and substantive editorials appeared in the paper almost every day. Amirian was one of the twenty-six Baku commissars who perished.

said that during this critical period Petrov's regiment had fought heroically, and, aided by the Baku proletariat, had defended the city from the Turks.

The appeal ended with an expression of confidence that troops would soon arrive from Russia, and that Baku would be saved ". . . not by the white flags of the cowardly Avetisovs and Amazasps, and . . . not by the mythical aid of the British and their lackeys, but by the armies of worker-peasant Russia who are coming to lend us a hand, by Soviet troops.

"The Baku proletariat, which has heroically come to the defense of its city, has true friends and protectors only in the Soviet troops from Russia. Only they, joining hands with the Baku proletariat, will save Baku from the German, Turkish, and British imperialists. The regiment of the military commissar from the central government, Comrade Petrov, is already waging this struggle.

"Comrade workers! March under the banners of Comrade Petrov's regiment, under the banners of our dear worker-peasant Russia who sent him to us!

"Long live the struggling Baku proletariat!

"Long live the Great Russian Socialist Republic of Soviets!

"Long live the revolutionary city of Baku, free from the imperialists and indivisible from Russia!"

* * *

Outposts were set up in Petrov Square over an area of several blocks, near the buildings occupied by our troops. Party organizers were dispatched to various districts of the city in order to enlist Communists and other supporters in Petrov's regiment. A training program for the troops was successfully set up.

On the fourth of August we watched from our pier as the ships filled with British troops, invited to Baku by the "Tsentrokaspian Dictatorship," began to pull into the harbor. After disembarking, the British soldiers were marched through the main streets of Baku in a triumphal welcome organized by the Tsentrokaspian. In all, 200 men arrived on the first day, but in order to make an impression on the local population they were marched down the same streets twice in succession. British troops continued to arrive over the next few days, but there still were very few of them; only about 1000 soldiers came to Baku instead of the 16,000 that the SRs had promised. Moreover, the British did not send all their troops to the front, but left a significant number of them in the city. Clearly they were pursuing two goals: first, to maintain control over the government by using their troops to carry out purely occupational functions; and second, to avoid heavy losses at the front.

After Petrov's artillery had successfully beaten back the Turks, the morale of the populace improved notably. I walked around the city and saw for myself how willingly large groups of workers were undergoing military training, and how columns of other workers were marching off to the front singing military songs. At that time there were about three thousand combat-ready Red Army men both on the ships and in Petrov Square itself, with Baku Communists making up the overwhelming majority of them.

The general situation still remained very vague, however. It was not by chance that Lenin, who was concerned about the fate of Soviet rule in Azerbaijan, sent a telegram to the President of the Astrakhan Soviet on August 9 in which he inquired about the situation in Baku, asking who was in power there, and trying to determine the whereabouts of Shaumian. He did this because he "could not grasp the situation there from Moscow, and saw no possibility of sending immediate aid." Lenin therefore instructed us to act in response to local conditions as they arose.

The second party conference after the fall of Soviet power in Baku took place on August 10, 1918. In his report at this conference Shaumian surveyed the external situation and the conditions prevailing at the fronts of Soviet Russia. All reports forecast a worsening of the situation, he said, especially in the northern Caucasus. The Cossack counterrevolution there, led by Denikin and Krasnov, had won several victories and now occupied Krasnodar. Having cut off the northern Caucasus, Krasnov's troops were now marching on Tsaritsyn. At that time Serge Ordzhonikidze directed the struggle with the attacking forces of the counterrevolution in the northern Caucasus.

Shaumian reported that, according to available intelligence, the Turks were stepping up their efforts to take Baku by transferring new units from the Mesopotamian front for a decisive blow. At the moment, he said, there was no information as to whether Soviet troops from central Russia would come to our aid; in view of the extremely critical situation on the Volga we could not count on such aid. We still had enough strength to seize power, Shaumian continued, but there was some doubt as to whether we could manage to hold onto it while fighting against the Turks on the one hand and the British on the other. Therefore, it made more sense, he suggested, to pull our troops out of Baku and get them to Russia via Astrakhan. By doing this we would save them for the fighting in the Volga region and for a subsequent return to Baku in six months' to a year's time to reestablish Soviet rule. The majority supported Shaumian's proposal.

A small group of comrades spoke out against this proposal. They insisted that the resolutions passed by the previous party conference be carried out, that is, that we overthrow the "Tsentrokaspian Dictatorship," take power into our own hands, and with the support of the Baku proletariat, use our forces to chase out the interventionists. We must not forget, these comrades said, that the local population had already become disillusioned with the effectiveness of British aid, that it would soon recognize the treasonous character of the right-wingers' activities, and would begin to have more and more confidence in Soviet power.

Twenty-two votes were cast in favor of evacuating our troops to Astrakhan, and eight (including mine) against.

Even before the conference, we had considered renewing publication of *The Baku Worker*, which had been suppressed by the "Tsentrokaspian Dictatorship." The day before the party conference I spoke with Shaumian about this, and he thought it would be a good thing to do, but pointed out that we didn't have a press.

"At the moment we don't," I replied, "but not far from here, beyond the zone occupied by our troops, there's a fairly good press in the 'Cooperation' Society. What would stop us from seizing it, particularly since the Mensheviks would hardly have the guts to offer us armed resistance?"

After consulting with the other members of the Baku committee, Shaumian approved my idea. He said that he and Dzhaparidze would prepare material for the first issue, and he gave me the job of seizing the press and running off the copies.

I put together a group of Red Army men that included Shaumian's two sons, Museib Dadashev, two typesetters, and a ready-made proofreader—Olga Shatunovskaya—and that evening after dark we drove in an armored car to the area where the press was located. There we surrounded the building, posted sentries, entered the press room, and set to work. After setting up the first run we corrected the proofs and started to print the paper. With the whole operation going so well we were in excellent spirits, but we were soon informed that Menshevik patrols were approaching our sentries, although there had been no skirmishes as yet. We had only managed to run off about a hundred copies of the paper when the lights went out. The press stopped and we found ourselves in utter darkness, but, nothing daunted, we lit the candles and kerosene lamps that we collected from the residents of the building and continued to print the paper, turning the press by hand.

Late that night, after we had printed several hundred more copies, we gathered them all together, took a box of typeface and a handpress (which might come in handy for printing proclamations) and took our leave. The next morning, Red Army men, workers, and Komsomol members helped us to organize a network for distributing our paper all over Baku; the issue contained two articles by Shaumian which dealt with the resolutions of the last party conference.

Shortly after the party conference of August 10, which had voted to evacuate our troops to Astrakhan, a meeting of the Baku Party Committee was held. Discussion centered on two questions: how to arrange party work in Baku after our departure, and which members of the Baku committee should be left behind in charge of underground operations. Everyone unanimously rejected the idea of leaving either Shaumian, Azizbekov, Dzhaparidze, Fioletov, or Korganov in charge —they were all too well known. We decided against even discussing them as candidates for the job. Shaumian and Dzhaparidze were asked to give some thought to the matter of who should be left in charge.

The next morning I went to Shaumian and said that I would like to stay behind in order to do illegal work in Baku as a member of the Baku Party Committee. He replied: "It's good, of course, that you want to stay, but it could be doubly dangerous for you: you know, of course, that beside everything else, Amazasp is extremely embittered against you and has ordered his cutthroats to annihilate you. Therefore, you have to be very, very careful."

I replied that I was well aware of this: Safarov, the commander of the cavalry company, had already warned me. "And besides, as the saying goes, the devil is not so black as he is painted," I said to Shaumian, but of course I promised him to be careful.

"Still, you can't stay in Baku for long," Shaumian continued, "and so, as soon as we get to Astrakhan, we'll send a replacement for you." Then he suggested devising a code to use in maintaining contact with Astrakhan, and advised me to use an extremely simple code in my correspondence. A pamphlet published in Moscow and entitled *The International Situation and Our Tasks* happened to be at hand on the ship. Shaumian opened it at random and said, "The code will begin at the tenth line from the top of this page." Beginning with the first letter in that line, each successive letter (not counting repeats) was to be designated by a number—from one to thirty-three. These numerical symbols would be used to encipher the text. (As it turned out, we never did use this code.)

Our people spent another whole day at the pier at Shaumian's quarters on the *Kolesnikov*. Those of us who were to stay behind made arrangements for an illegal apartment and for operating an illegal press, for which we intended to use the typeface and the handpress we had taken from the printing office of the "Cooperation" Society.

That same day we decided to transfer the typeface and the hand press to the illegal apartment and set up the press there. Unfortunately, we entrusted this critical job to completely inexperienced people. A certain Azerbaijanian student had offered us his apartment as a press room, and we gave "Little Gevork" (the underground nickname of one of our comrades, the Bolshevik Avetisian) the job of accompanying this student and taking the typeface to his apartment. It was located in the Moslem section of the city, which was very advantageous under the circumstances, particularly if the Turks seized the city. Instead of transporting the typeface with due caution, however, Gevorg, in his naivete, simply wrapped the heavy box containing the typeface in a blanket and loaded it onto an *ambal* (a Persian porter who had a special contraption on his back for carrying heavy loads). "Little Gevork" walked alongside the porter, but while the blanket-covered load was not very large it was so heavy that it caused suspicion, and an armed secret police agent on the street stopped the porter and asked him what he was carrying that was bending him nearly in two. He ordered the load removed so that he could examine it. Seeing that it was a lost cause, "Little Gevork" fled, leaving the typeface in the hands of the police.

Before our elder comrades left for Astrakhan, we held the first organized meeting of our underground contingent. First it was made clear who exactly was staying behind, then duties were assigned, arrangements were made about how to get in touch with each other, and secret meeting places were established, etc. Also, since we had lost our underground press, we had to find a way to publish proclamations and leaflets. Artak promised to find accommodations for the new printing office, and since my stint as an editor had made me familiar with the printing business, I took upon myself the job of trying to get hold of some typeface.

* * *

As soon as the Baku Party Committee had passed the resolution to evacuate the Soviet regiment from Baku, delegates had begun to negotiate with the "Tsentrokaspian Dictatorship" for the regiment's unimpeded departure for Astrakhan. The Mensheviks, glad to be rid of such an important Bolshevist force, immediately consented to the evacuation, but they demanded that we leave them all our weapons and ammunition. Shaumian flatly refused to do this. He told our delegates to say that the arms belonged to Soviet Russia and were needed on other fronts, so they could not be left behind. After lengthy arguments the "Dictatorship" grudgingly consented to the evacuation of the Soviet regiment along with its weapons. At the same time Polukhin, the Commissar of the Caspian Naval Fleet from the Central Sovnarkom, got the crew of the ship *Ardagan* to agree to assume responsibility for the safety of the entire evacuation.

On the afternoon of August 14, 1918, the seventeen ships carrying our military units and materiel began, one by one, to depart from the Baku pier. The first to leave was the *Kolesnikov*, with the People's Commissars and other high-ranking workers aboard. We stood on the pier and waved good-bye to our friends, thinking that we were sending them off on a safe journey. We left before the last ships departed lest a detachment from the "Tsentrokaspian Dictatorship" descend upon the square at any moment and arrest us.

We felt depressed as we walked away from the pier. We had just said good-bye to the people who were our close friends, as well as our more mature, experienced leaders. But at the same time another feeling welled up within us—a feeling of great party responsibility at the realization that we were staying behind in their place to fulfill our duty to the party. And—I won't hide it—this consciousness filled our hearts with revolutionary pride.

Three days later we were shocked to learn that the seventeen ships had been surrounded near Zhiloy Island by the "Tsentrokaspian Dictatorship's" naval fleet and been forcibly returned to the military port. The troops were then disarmed and sent on to Astrakhan on these same ships. But the Baku commissars and the other leading workers, with Shaumian at their head (thirty-five people in all), were arrested and imprisoned.

In their reports on this event the "Tsentrokaspian Dictatorship" went to great lengths to compromise us in the eyes of the populace and to justify their treachery by libelously accusing our commissars of fleeing the front and attempting to make off with stolen valuables.

Our main task now was to save our comrades from execution. At a hastily called meeting we came to the following decision: as an emergency measure we would join forces with representatives of the left-wing SR and the left Dashnak parties and present the "Tsentrokaspian Dictatorship" with a joint ultimatum.*

Insofar as our comrades had been detained illegally, for no valid reason and in contravention of the authorities' guarantee of their safe evacuation, the ultimatum asked for the immediate release of all our comrades under arrest, and demanded that they be allowed to depart for Astrakhan. The Baku committee of the Bolshevik party, as well as the left SR and left Dashnak parties, warned the members of the "Dictatorship" that if any of our comrades came to any harm whatsoever, a personal campaign of terror would be instigated against all the members of the "Dictatorship."

Representatives of the left-wing SRs and the Dashnaks and myself signed the appeal and took it to the "Tsentrokaspian Dictatorship." There we demanded a meeting with the president, Sadovsky, with whom we were already acquainted, and handed him our appeal. When he had finished reading it, I waited a moment and then said:

"You're well acquainted with the parties that wrote this appeal. You cannot doubt our ability to follow through on the threat we have

---

*At that time the three parties—the Bolsheviks, left-wing SRs, and the left Dashnaks—had formed a Bureau of Left-Wing Socialist Parties in order to more effectively coordinate their activities and their joint appearances before the "Tsentrokaspian Dictatorship."

The left-wing Dashnak party was made up of a small group of workers who had broken from their party and supported Soviet rule. It was led by Avis Nuridzhanian, who soon joined the Bolshevik party during the second British occupation of Baku. In 1920 he took part in the struggle to overthrow the Dashnak government and became a member of the government of Soviet Armenia.

made. If you arrest us before we leave this building, nothing will change. Each of our parties has already formed groups of terrorists which will take concerted action as soon as it becomes necessary."

All this made a great impression on Sadovsky. He began to justify himself and to assure us that no one was threatening the lives of the arrested men or intended them any harm. There would be an investigation, he said, after which a decision would be made on how to deal with them. Rather than try to argue with him, however, we just gave him the appeal and left. We knew very well that traitors cannot be trusted and that our ultimatum could only hold off the execution temporarily. In order to save our imprisoned comrades we had to wage a full-scale political war.

Yet, we had achieved something. Within a few days a rather large number of Communists had rallied around us, especially when they found out about the arrest of our comrades.

It was at this time that Georgi Sturua, a well-known Baku party activist, appeared in the city. He had spent the last few months in the northern Caucasus where he had been authorized by the Baku Soviet of People's Commissars to procure and deliver grain to Baku. Since there were no rail communications between Baku and the northern Caucasus, he had sent the grain on ships that were then unloading at one of the piers in the Kizlyar district. Although he had heard of the events in Baku he was uncertain as to exactly what had happened, and he was surprised to find himself arrested as soon as his ship docked. He was a stern man, a strongwilled, unyielding, and experienced conspirator who had received excellent schooling in underground activities during the Tsarist period, when he had been working at an illegal party press in Moscow as early as the turn of the century.

It didn't take him long to prove the illegality of his arrest, since the only thing that he had done during the past few months was to try to get grain delivered to the hungry inhabitants of Baku. We were overjoyed at the arrival of this Communist experienced in underground work, who was, moreover, a member of the Baku Bolshevik Party Committee.

We set about reorganizing the staff of the Baku Bolshevik Party Committee, deciding in view of the current situation to broaden the scope of our legal activities, to speak out regularly at various conferences, meetings, and workers' organizations, and at the same time to strengthen the strictly secret party organization, both at the center and in the outlying districts. We singled out certain members of the Baku committee who would then devote themselves entirely to conspiratorial organizational work and therefore would not speak at open meetings or expose themselves in any way. This assignment was given to Sarkis and Pleshakov. They were joined by Mir-Bashir Kasumov, who did work among the Azerbaijanian workers. Kasumov was an active worker-Communist in charge of the party organization at the Lieutenant Schmidt Factory; he later became President of the Central Executive Committee of Azerbaijan.*

We could not count on publishing our newspapers, but we considered organizing an underground press for printing proclamations and leaflets. Artak had found a location for such a press on Stanislavsky Street at the home of the old-time (woman) Bolshevik, Maro Tumanian, who is still alive and well today. Artak and Marusia Kramarenko, who were living there, took upon themselves the job of "managing" the press. They got hold of a Communist typesetter who had worked at the press that had formerly published *Izvestia*. Once again I had to find some typeface, so I went to the head of the *Izvestia* press (we had been good friends before) and told him that if he happened to notice any typeface missing from the press room he should ignore the loss—it would not be a case of theft. I said that I was counting on his discretion, and was confident that he wouldn't put any of the workers on the spot if he should catch them taking typeface. He gave me an understanding look and silently nodded his approval.

Over the course of the next several days the Communist typesetter at *Izvestia* and his comrades hid typeface in their pockets and brought it to us. Through them we also managed to get a small handpress, which made it possible for us to print proclamations. The first of these was an appeal from the Baku Bolshevik Party Committee to the workers of Baku, calling on them to demand at their meetings the immediate release of the illegally arrested leaders of the Baku proletariat. Then we distributed handbills at any and all meetings of workers, factory and plant committees, and other workers' organizations, exposing the treasonous policies of the "Tsentrokaspian Dictatorship" and demanding the immediate release of the Baku commissars.

The Mensheviks and the SRs, sensing how the arrest of the Baku commissars had stirred up the people, hoped to enhance their standing with the working class by convening a special conference of factory and plant committees and trade union boards that August of 1918. As might be expected, very few workers attended this conference. The best of the workers (and the Communists above all) were then in the Red Army troops that were exiled to Astrakhan, and for the most part, the organizers of the conference tried to get their own supporters to attend, people who in no way represented the interests of the Baku workers.

*Other active party workers included the young printer Dzhafar Babaev, one of the founders of the Baku Union of Young People, and the three Agayev brothers, Persian emigrants and active Communists who headed the party organization of Azerbaijanian workers. Rukhula Akhundov led the left-wing SRs operations among the Azerbaijanian workers and coordinated his party's activities with ours. We assigned Gubanov, the president of the Water Transport Workers' Union, and the still very young Levon Mirzoyan to do trade union work. Martikian, Shatunovskaya, Agamirov, Stamboltsian, and Kramarenko were also very active in Baku Party Committee operations at that time.

The Bolsheviks gave me the job of making an unscheduled announcement at the very beginning of this conference, a short summary of which was published in the Menshevist paper *(The Spark)* on August 26, 1918.

"The floor was given to Comrade Mikoyan," the paper reported, "who made an announcement on behalf of the Bolshevik faction. He protested that the conference had not been well publicized in the districts, with an official notice of the conference appearing only the day before it was to take place. Comrade Mikoyan said that his faction viewed this as deliberate obstructionism."

I remember how difficult it was for me to get the floor a second time in order to reply to the speech of the Menshevik Sadovsky, who was the ringleader of the "Dictatorship."

I must say that Sadovsky was an artful demagogue and a past master at juggling arguments. First, he tried to maintain that the existing dictatorship was a provisional authority pending the convocation of a new Soviet of Workers' Deputies which would, in fact, determine what kind of rule Baku would have; then he called for the replacement of the "Bolshevik regime" by a government of genuine representatives of the people elected on the principle of "universal suffrage." By covering up the criminal alliances between the leaders of the "Tsentrokaspian Dictatorship" and the British troops, he tried "to prove" that these troops took their orders from the local Russian command, that they had no particular political goals in Baku, and were not interfering in the internal affairs of Baku in any way. In spreading a false version of the evacuation of the Red Army units and Baku commissars, and in justifying the treasonous arrest of Shaumian and the other Bolshevik leaders, he falsely claimed that the Mensheviks wanted to "safeguard Baku for Russia, because without Baku oil, Russia is nothing, and without Russia, Baku is a still bigger nothing."

In the second speech I made at the conference I cited facts which proved how totally dependent the "Tsentrokaspian Dictatorship" was on the British high command. I characterized the "Dictatorship" as a toy in the hands of the invaders; pointed out the great harm brought upon Baku by the treason of the SRs, the Mensheviks, and the Dashnaks, and described the white terror that the "Dictatorship" was using against the Bolsheviks. They have deprived us of our leaders, I said, hidden them away behind prison bars, shut down our newspapers, forbidden us to speak at meetings among soldiers and workers, and they have done all these things under the false slogans of "democracy." In concluding, I demanded the immediate release from prison of the leaders of the Baku proletariat.

I think that my readers will be interested to see how the editors of the *Tsentrokaspian Bulletin* summarized my speech in the August 24, 1918, issue. I quote this "summary" word for word.

Mikoyan (from the Bolshevik faction): "Many provocative attacks against our party have been made here. It is impossible to answer all of them in twenty minutes' time.

"Petrov's regiment and the other troops were at our positions all the time, while they, the Dashnak sympathizers, wanted to surrender the city [a great uproar]. Why did we give up power? Because at the critical moment when the enemy was attacking, the Dashnak sympathizers refused to take up the defense. There was no one at the front, but there were many forces at work in the city. We saw evidence of the right-wing parties' sabotage everywhere. On July 25, after the Soviet had decided to invite in the British, we, who were against the move, submitted our resignation, so that the parties which desired to take power into their own hands could do so.

"How shameful it was to tell Sadovsky that Petrov had 200 men. He had not 200 men, but 1,300, and 600 of them died at their battle positions, up until July 31 they were the only ones who held their positions on the left flank and in Baladzhary [noise].

"I could cite many reasons for the defeat, but I must answer the other attacks that have been made against us. I'm sure you recall the rumor that a representative of the German General Staff had brought Shaumian a letter from Lenin. A commission was appointed to look into the matter, and it was revealed to be completely untrue.

"There has also been talk that the Bolsheviks made off with certain amounts of money. Ten days ago, the local authorities, the so-called Dictatorship, suggested to the Soviet of People's Commissars that they leave the city. If the 'Dictatorship' knew that they had money, then by giving them permission to leave, they'd be criminals [applause from the left]. Everyone knew that Shaumian, Korganov, and Dzhaparidze could not be accused of taking money.

"They write in the *Bulletins* that all people who are detained will be brought to trial and put in prison.

"They are trying to shut us up by forbidding us to agitate within the army and not allowing us to express our opinion. When we were in power, all parties, except the Mensheviks, had their own organs, and even the Mensheviks were allowed to express their thoughts freely at meetings.

"As early as last July you wanted to poison the masses against us, and at that time the proletariat was not well-disposed to us, but later it joined our cause. And it will follow us again now.

"When they shout about the necessity of having *one* Russia, with a *single* central government, they forget that they themselves destroyed this principle when they decided to call in the British. At the Fifth Congress, where all the Soviets were represented, it was resolved not to invite in any allies except the central government.

"Comrade Shaumian, Comrade Petrov, and myself, as representatives of the central government, could not become military allies of

England at a time when the British were arming Czechoslovaks, and the German and Austrian prisoners of war [laughter].

"When the mercenaries of British imperialism began to crush Russia and occupy Russian cities, when they tore to pieces those who called themselves Bolsheviks [great uproar], when German imperialism laid its paw on Russia in the west [great uproar that drowns out the speaker's words].

"Your savior Bicherakhov fled to the northern Caucasus and demanded your help.

"Under these circumstances the representatives of the central government, Petrov and Shaumian, had to leave the city. Petrov's regiment did not run away. Only shameless liars could say that. They spent two full days making preparations to depart; everything was done openly for everyone to see.

"They have also said that the British will not interfere in our internal affairs. Thus we are to conclude that the British will sacrifice their lives simply because they're smitten with us. No, on the contrary, they have already gotten their hands on the provisions by saying that the front cannot be strengthened if the current poor rationing system is allowed to continue. All the right-wing SRs say that the food rationing issue is a political one. This indicates that the British are already interfering in political affairs.

"The example of Georgia and the Ukraine has amply demonstrated [huge uproar].

"They say that a government will be formed here on the basis of universal suffrage. Thereby they deny the power of the workers. However one-sided the previous government was, it was nevertheless a workers' government, while universal suffrage will mean that fat cats and speculators will participate in goverment . . . [noise. The speaker descends from the stage.]"

Nuridzhanian, from the left-wing Dashnak party, spoke out at the conference in support of our policies. He was followed by the Bolshevik speaker, Comrade Bliumen, who exposed many of the false statements made by Sadovsky and the two-faced position held by the right-wing Dashnaks.

In order to nullify the effect made by our speeches, Sadovsky made a second wide-ranging and libelous speech, and after him came the Menshevik Bagaturov. They were supported by their stooges and yes-men, who constituted the majority of those present at the conference.

After debate at the conference was concluded, the SRs' streamlined, demogogic resolution was proclaimed, which contained, however, a direct expression of confidence in and support for the "Tsentrokaspian Dictatorship." The Mensheviks went still farther: they proposed writing into the resolution that the Baku commissars were traitors and enemies of the people. The resolution was passed 117 to 20.

Then, Comrade Baker made a declaration on behalf of the Bolshevik faction. I shall quote it as it was published in that very same *Bulletin:* "We consider the conference to have been improperly convened and therefore lacking in authority [loud noise]. There are no workers present [loud noise, the speaker tries to outshout the audience]. Therefore, we do not intend to carry out any of the resolutions!"

After the conference was over, we reached an agreement with the left-wing SRs and the left-wing Dashnaks, and prepared a written appeal in the name of our three left-wing parties, in which we demanded the immediate release of the Baku commissars, and the creation of an interparty committee of inquiry, composed of representatives from all the socialist parties in Baku, that would be empowered to examine the accusations leveled against the Baku commissars. We took this document to Sadovsky, who, obviously wishing to hide his intentions from us, refrained from arguing with us at the time, and simply said that he would bring our appeal to the attention of the "Dictatorship" and report their decision to us.

The next day it became clear from the newspapers that the Tsentrokaspian Government had rejected our demands and had transferred the case of the arrested commissars to the jurisdiction of its extraordinary commission, which was headed by Vasin and Dalin, both working in the "Tsentrokaspian Dictatorship."

Our first concern was to establish contact with our imprisoned comrades, find out how they were, and supplement their rations. We started collecting money and buying food. We put Agamirov—one of our younger active workers—and Stepanova, Shaumian's secretary-typist, in charge of organizing this effort. Stepanova (she is still alive and living in Moscow) had maintained contact with Korganov, whom the prisoners had chosen to talk with the prison administration.

Dzhaparidze's wife, Varvara Mikhailovna, soon became involved in this operation. After having been evacuated to Astrakhan with Shaumian's family, she had left her two young daughters in the care of Shaumian's wife and returned to Baku on one of the steamships.

Stepanova has kept a letter which Korganov sent her from the Bailov Prison. In this letter Korganov described in detail his ideas on how best to organize the delivery of food to the men in the prison. He recommended that Stepanova talk it over with me and pass on to me his practical opinions in this regard. It is interesting to note Korganov's request that everything sent to the prison be wrapped in fresh newspapers and that some news about life outside be written on the inside folds of these "packets." "But," he warned, "crucial messages must be sent via specially designated persons."

We succeeded in organizing several meetings between the imprisoned men and their relatives. In particular, Varvara Mikhailovna obtained a meeting with her husband, Dzhaparidze. After persistent requests, I managed to get permission for one meeting with Shaumian. He stood on one side of the bars and I on the other (during the Tsarist period Shaumian had had similar meetings with comrades and members of his family in that same prison). He looked tired and pale, listened attentively to what I told him about our activities, and I even thought I detected a shade of surprise in his eyes when I told him that the young people who had been left to their own devices were still "green," but were working so hard and so well. He was truly happy about that.

I remember how outraged I was that the Mensheviks, who had at one time been considered our comrades and had even been in the same party with us, were now so shamelessly slandering genuine, bona fide revolutionaries; I said that not only were the workers indignant at this, but the simple people of the city as well.

Then Shaumian said to me:

"There's no need to be upset. A lie won't stand up for long, it will be exposed by life itself. Truth is on our side, and it will prevail."

I asked him to indicate what tactics we should use in the future.

"The plans that the Baku Party Committee has outlined are correct," Stepan replied, "and they must be carried out."

Soon after the commissars were arrested, we secured the release of Shaumian's son, Leva, since he was a minor (he had just turned fourteen). A short time later we got his older son, Suren (who was eighteen), out on bail.

From Suren we learned the details of how the ships had been detained at Zhiloy Island, which is not far from Baku. They had waited there for the other ships to arrive. (They had to wait a particularly long time for the last two ships, which had been caught in a storm that was especially hazardous for the oil tankers, the lower decks of which were filled with Red Army men and weapons.) They had been surrounded by ships from the Caspian Naval Fleet which had ordered them at gunpoint to return to Baku. Shaumian and the others had refused to heed this demand because it was illegal. In reply the naval fleet had opened fire and there were casualties.

At that point many comrades had demanded that Shaumian, Dzhaparidze, and Azizbekov disembark on Zhiloy Island, and make their way to Astrakhan on barges or fishing boats. But all three flatly refused to leave their comrades.

Soon military vessels from the "Tsentrokaspian Dictatorship" had approached, herded our ships back to Baku under convoy guard, and disarmed them. The Baku commissars, as well as some other leading Soviet and party workers, were then arrested and jailed.

* * *

The political situation in Baku had become very hot. At meetings and at rallies the vanguard of the workers demanded the release of the commissars. The right-wingers, however, relying on British bayonets, were preparing to execute them rather than release them.

In order to establish a public democratic power base for the then existing government of Baku, and simultaneously to rid themselves completely of the Bolsheviks in the Soviet, the Mensheviks and the SRs conspired to hold elections for a newly constituted Baku Soviet of Workers' Deputies. The elections were to take place on August 28, 1918.

We decided on our part to participate in these elections by turning the election campaign into a broad workers' movement for the release of our comrades. However, we didn't have a chance to prepare for this very effectively.

The Mensheviks conducted the election campaign in such a way as to destroy the basic democratic rights of our party. The Bolsheviks were not allowed to publish any legal printed material, or to participate in military meetings. In an all-out effort to gain a few extra days to extend our influence over several important enterprises, we (the Bureau of Left Socialist Parties) issued a declaration to the central election commission requesting that the elections be postponed for a week. We based our request on the serious infringements of the rules that had resulted from the conduct of the preelection campaign.

The elections were to take place on August 28, but on the twenty-seventh, the day we submitted our declaration, the lists of candidates (for deputies) had still not been published in the papers. We announced that we would view the rejection of our request as another example of their obstructionist tactics.

They rejected our request but continued to play the democracy game by publishing the text of our declaration in an SR newspaper two days after the elections had taken place.

Nevertheless, on our illegal press we managed to print a preelection leaflet that contained an appeal for support of the Bolshevik deputies. In various districts of Baku we had nominated as candidates the imprisoned Baku commissars and also the most active workers. Despite all obstacles, the workers of Baku managed to elect twenty-eight Bolshevik deputies to the Soviet, including nine of the imprisoned Baku commissars—Shaumian, Dzhaparidze, Zevin, Fioletov, Azizbekov, Basin, Korganov, Malygin, and Bogdanov. Sturua, other Communists, and myself were also elected. Seven deputies, including the People's Commissar Vezirov, were elected from among the left-wing SRs and the left Dashnaks.

The Mensheviks' and the SRs' gamble on isolating the Bolsheviks completely had failed. The vanguard of the Baku workers had also supported our demand for the immediate release of the imprisoned comrades.

All of this coincided with a new Turkish offensive on Baku. The attack was repulsed, with the Turks, as well as the British units, sustaining heavy losses. The Turks soon occupied one of the elevations [around the city], however, set up their artillery on it, and began to subject the city to systematic bombardment. At the first meeting of the new Soviet, a shell hit the building, and the meeting was reconvened on Morskaya Street at a restaurant with windows overlooking the sea that gave relative security.

After the results of the elections had become known, we submitted to the "Tsentrokaspian Dictatorship" a written demand for the release of the imprisoned commissars who had been elected deputies to the Baku Soviet, and had thereby received a vote of confidence from the working class.

At the meeting of the Baku Soviet on September 5, Georgi Sturua, acting as spokesman for the Bolshevik faction, made a speech in which he demanded outright that the comrades be released from prison and that they be given the opportunity to take part in the work of the Soviet in order to fulfill their function as elected representatives of the working class.

Three days later the *Tsentrokaspian Bulletin* reported that "the Bolshevik representative Sturua, followed by Mikoyan, Beker, the left-wing SR internationalists Ter-Saakian, and others, had raised the question at the meeting of the immediate release of the former People's Commissars, who were languishing in prison and had been elected to the Soviet on the party slates."

A few representatives from the left bloc spoke out on this issue because several Menshevik and SR leaders came out against releasing the imprisoned commissars. The main speaker Sadovsky announced that the Communists' demand for the release of the commissars, as made known to the "Tsentrokaspian Dictatorship," had been examined and taken under advisement, i.e., rejected. He and the other Menshevik and SR speakers tried to prove that the imprisoned commissars could not be released because criminal charges had been brought against them which would ultimately be proved in court.

The Bolsheviks and the left-wing SRs tried to insist that the articles of accusation against the Baku commissars be withdrawn from the investigatory organs of the "Tsentrokaspian Dictatorship"—which were composed of legal underlings who had served under tsarism—and transferred for examination to a special interparty commission made up of members from all the political parties represented in the Soviet. In this regard we cited a letter from the imprisoned Baku commissars in which they announced their refusal to testify before the investigatory authorities of Tsentrokaspii, and demanded that they be allowed, after their release from prison, to give a full account of their actions before the Baku Soviet.

There were lengthy debates. Finally, the right-wing parties succeeded in shutting off the discussion altogether, since continued debate had revealed the falseness of the accusations leveled against the Baku commissars.

The chairman of the meeting called for a vote on the Communist faction's motion that the Soviet immediately release all the imprisoned deputies without waiting for an investigation and trial. The results of the vote were: twenty-seven—yes; one hundred twenty-two—no; nineteen—abstentions.

Then a vote was taken on our second motion—to name a special investigatory commission, consisting of members from all the political parties represented in the Soviet. Forty-two voted in favor of this motion, three abstained, and the rest voted against it.

The defeat of both the Bolshevik factions' proposals made us extremely indignant. Despite the shouts of the chairman and his attempts to bring the meeting to order, our deputies leapt up from their seats, protesting loudly. Loud noise broke out from the right-wing parties' benches as well, resulting in a skirmish between the two opposing camps.

With a great deal of difficulty the chairman succeeded in restoring order and, taking advantage of the overwhelming majority of his supporters at the meeting, he managed to vote through a resolution proposed by the right-wing SR faction: "Having heard the Communist faction's declaration and also the explanation of the provisional 'Tsentrokaspian Dictatorship' and the Executive Committee, the Soviet turns to the other issues on the agenda."

The Menshevik leader Ayollo—an enemy of the Bolsheviks' to the end—made a speech. He moved that the Soviet make an appeal to the populace and to the army.

Then I introduced a motion on behalf of the Bolshevik faction, moving that the basic issue of authority be decided before the question of an appeal was considered.

"The 'Tsentrokaspian Dictatorship,' which carries no authorization and was never elected by anyone, is now ruling Baku," I said. "Now that a new Soviet has been elected it must dissolve this illegal body and take the full measure of power into its own hands (tactical considerations made this proposal expedient at the time). Only then can the question of an appeal and its contents be discussed."

The Soviet ignored this motion of ours and did not even discuss it.

Ayollo read out the "Theses of Appeal," which consisted of a few planks composed in the spirit of Menshevik politics.

I said the following in regard to these "Theses":

"Without entering into any discussion of the proposed appeals, the Communist faction goes on record as saying that our party rejects with indignation all attempts to use the name and the flag of the Soviet to cover up the activities of the bourgeois rulers and to exploit the Soviet for the struggle against the Communists. It will not take part, therefore, in the voting."

After the appeal was passed without our participation [in the vote], the chairman returned to the question I had raised concerning the ruling power in Baku and said:

"As soon as the Soviet convened, the authority of the dictatorship was legally curtailed; it is in our power to reinstate it."

Sturua spoke out in protest against this motion. The Soviet, he declared, could not transfer its authority to another body. It should itself implement the full measure of its power and abolish the provisional "Tsentrokaspian Dictatorship."

The representative of the right-wing SRs, Velunts, moved that until that time when the Soviet elects a new government, full ruling power should be left in the hands of the "Tsentrokaspian Dictatorship." By using their majority in the Soviet, the Mensheviks and SRs succeeded in pushing this motion through.

The Soviet decided at the next meeting to hear the declarations of the political parties and to discuss the question of forming a government.

* * *

On September 7 the extraordinary commission of the "Tsentrokaspian Dictatorship" issued a communiqué in which Dzhaparidze, Petrov, and certain other commissars were charged with committing "criminal offenses." This was so senseless and absurd that the very next day, under pressure of the protest which we had made during talks with representatives of the SR and Menshevik parties, the extraordinary commission was forced to publish a letter that made mention of "incorrect formulations" that had been given in their communique.

At the second and third meetings of the Soviet, which took place on the tenth and eleventh of September, the clash between the two camps in the Soviet intensified to the extreme. The reading of the right-wing parties' declarations, which lauded the policies of these parties and contained malicious attacks against the Communists, was accompanied by loud protests from the left-wing deputies' side of the room.

I must say that the Communist faction's declaration had been well-prepared in advance; it had been discussed in the Baku Party Committee and by the Bolshevik faction in the Soviet, and had been secretly transmitted to the prison where it received the approval of Shaumian and our other comrades.

Having taken the floor, I began to read out the declaration as the spokesman for the Bolshevik faction. Unfortunately, the text of the declaration has not been preserved. Therefore, I summarize it here from memory.

The declaration began with a brief assessment of the international situation, the World War, and the perfidious role played by the SRs, Mensheviks, Dashnaks, Muscavatists, and other forces of internal counterrevolution that had unleashed a bitter civil war in the country. It described how the internal counterrevolution had closed ranks with the foreign imperialist powers of the Entente, and it exposed the treason of the right-wing parties in Baku, which declared verbally that they supported Russia and that they had invited the British into Baku to try to save the city for Russia, but which had, in fact, refused to recognize the central Russian government, thereby attempting to tear Baku away from Russia. The declaration maintained that the British had come to Baku not to preserve it for Russia and not to fight against the Turks, but to seize the city. Thus, their behavior here, our declaration continued, was exactly the same as it had been in all of Transcaucasia, Central Asia, and Russia. The British had set themselves the task of suppressing the socialist revolution in Russia, dismembering Russia into small pieces, seizing her rich border regions, and transforming them into colonies. The other imperialist powers of the Entente were pursuing the same goals. In this mortal confrontation between revolutionary Russia and the imperialist aggressors, the parties which called themselves "socialist," as well as the government which they had created in Baku, were playing the role of the interventionists' puppets and were entrenched in the counterrevolutionary camp.

The declaration exposed the perfidy of the Dashnaks, and the treason of the Caspian Naval Fleet, the personnel of which had ultimately been partly deceived and partly bought off by the British.

As a result, the declaration said, Baku had been cut off from Russia. The British had become the real masters here, and the "Tsentrokaspian Dictatorship" their fawning maidservants, while the leaders of the Baku proletariat had been put behind bars. We rejected with indignation the absolutely baseless, libelous accusations that had been leveled against the imprisoned Baku commissars, and demanded their immediate release.

The declaration concluded by saying that in spite of all the hardships that Soviet Russia had endured, the civil war would end in the complete victory of Soviet power, and world revolution would rise victorious from the ruins of capitalism.

The declaration was read in an extremely tense atmosphere. I was frequently interrupted by the chairman's bell and by the hysterical shouts of the more frenzied Mensheviks, SRs, and Dashnaks, which impeded the reading. In spite of all this I persevered, raising my voice more and more in order to make myself heard over the noise.

The chairman of the meeting was the old Menshevik Mikhailchenko, a member of the first State Duma. He did not play an important role in Baku at that time and had been made chairman in order to lend an aura of "respectability" to the Menshevik Soviet. Although he interrupted me with his bell, Mikhailchenko could not cope with the situation, with the result that Osintsev, who was sitting next to him, grabbed the bell.

When I came to the part of the declaration that stated the Caspian Naval Fleet had been partly deceived and partly bought off by the British, the more hotheaded of my opponents rose from their seats and started to hurl themselves upon me with shouts and threats, while Osintsev, shaking the chairman's bell, almost hit me in the face with it. Then, unable to restrain myself, I grabbed for the revolver which I almost always carried with me, tucked behind my military waist-belt. Fortunately, it wasn't there; I had accidentally left it home that day. When they saw what I intended to do, several people aimed pistols at me. Then I thrust my hand into my trousers pocket. Thinking that I was going after my revolver, they threw themselves upon me and grabbed my arm. When they pulled my hand out of my pocket, they saw that it contained a handkerchief. Meanwhile, all the left-wing deputies, chairs raised above their heads, moved toward the president's table in my defense. A general melee ensued.

Nevertheless, Osintsev managed somehow to restore a modicum of order. Our opponents demanded that I be deprived of the floor and forbidden to read the declaration to the end. The chairman, however, in an obvious attempt to prove his "democratic attitude" began persuading the others to allow me to finish. I did so amidst incredible noise and shouting from all sides. Voices began to be heard in the hall demanding that I be stripped of my deputy's rights and brought to trial for defaming the sailors. The sailors' representative declared that they would drag the entire Bolshevik party to court for libel.

After my speech a representative of the SRs made a motion that condemned "the Communists' desire to defame our glorious fleet" and "offered the fleet an expression of the Soviet's gratitude for its wholehearted and unselfish service in defending the city." Noise and shouting broke out again. The leftist deputies heaped shouts of shame and abuse on the traitors, while an incredible uproar resounded from the right-wingers' side of the room. It was no longer possible to restore order, and the meeting had to be stopped.

The next day, on the morning of September 11, we read a report in the *Bulletin of the Dictatorship* which infuriated us: the military-investigatory commission had concluded its work, and the imprisoned Baku commissars had been transferred to a military court. A military field court had been granted provisional status especially for this case. All signs clearly indicated that an execution was in the works.

As a result, we concentrated all our efforts on obtaining the immediate release of our comrades, intensifying our campaign against the military field court in the districts. We decided to use the third meeting of the Soviet, which was set for that day, entirely for a discussion of the plight of the imprisoned commissars. I demanded the floor on behalf of the Bolshevik faction as soon as the meeting was convened in order to make an unscheduled announcement concerning the "Dictatorship's" latest act of illegality—the transfer of the leaders of the Baku proletariat to the jurisdiction of a military field court.

In spite of a warning from the chair, I began in the sharpest and most extreme terms to expose the Mensheviks, the SRs, and the Dashnaks, and demanded that the Soviet countermand its shameful order to bring the commissars up for court martial, and immediately release them. I declared that the leaders of the ruling parties in the Soviet would answer with their heads for the fate and life of the Baku commissars.

Speaking on behalf of the Mensheviks, Ayollo once again cast aspersions on the Baku commissars and our party.

Then Georgi Sturua took the floor. Using sharp words, he persuasively unmasked the perfidious policies of the right-wing parties, accusing them of becoming the executioners of revolutionaries, and declaring that they would not escape the judgment of the working class and of history. A general uproar ensued, drowning out the speeches of Sadovsky and Beker. The chairman would not surrender the floor to our other comrades, so they were forced to seize it, so to speak, in any way that they could. The resulting situation made the continued work of the Soviet impossible, and the chair declared the meeting adjourned.

In order to illustrate the tendentiousness of the Menshevik press's treatment of the last two Baku Soviet meetings, I shall quote some excerpts from their newspaper, *The Spark:*

". . . The atmosphere in the hall intensified to a fever pitch the moment the Communists proclaimed their extensive declaration. Like a water-soaked sponge it was saturated with the usual verbal abuse, narrow-mindedness and insinuations . . ."

> The Communists made a very long declaration that contained their usual lamentations in the hackneyed "style" of *The Baku Worker* newspaper. This document had such a profusion of curses, libelous attacks, and lies that the cries of indignation coming from all quarters made it impossible for Mikoyan to continue his reading of it. The Communists' insinuation that the majority of the fleet had been bought off by British gold evoked a particular sense of outrage. The members of the Soviet lost their patience. An overwhelming majority of them demanded that Mikoyan be deprived of the floor and the rest of the declaration not be read. Calm was restored with difficulty, so that Mikoyan would be able to go on with his reading.

. . . If one were to exclude from this—and we use the word advisedly—"declaration" the abuse heaped upon all non-Bolsheviks and the insinuations and libel directed at individual naval and army units, then what did this declaration offer the laboring masses? Absolutely nothing.

Yet you will learn from it the following very important, great, but mainly "very new" truths:

1) that there was a world war;
2) that a world socialist revolution followed in its wake;
3) that Bolshevik Soviet power is the only torch lighting up the path of this world revolution and leading it forward, like the fiery column in the Bible which led the Jewish nation out of Egyptian captivity; and
4) that as a consequence of the sum total of all these circumstances, the Communist party will strive for continuous civil war, the annihilation of the bourgeoisie, the seizure of power, and the establishment of a Communist order on the ruins of the "old regime."

. . . With the definite aim of breaking up the meeting, the Communists began to introduce endless questions and to make announcements. The form these questions took was so improper that cries resounded from all sides of the Soviet demanding that the Communists be deprived of the floor. Comrade Mikhailchenko wore himself out in an effort to restore order. Finally he gave the chair to Comrade Osintsev, whose indubitable gifts as a presiding officer enabled him to restore calm and establish order. The mood was so tense, however, that it was considered impossible to continue the meeting.

Early on the morning of September 14, I was still asleep on Martikian's* balcony when Varvara Mikhailovna Dzhaparidze ran in and said that the Mensheviks and SRs were fleeing Baku, that the Turks would soon enter the city, and that we had to rescue the imprisoned men right away. Just the day before we had met and worked out plans for rescuing our comrades from prison. But with the approach of the Turks the situation had changed, time was running out on us.

As a member of the Baku Soviet of Workers' Deputies I was assigned to go to the "Tsentrokaspian Dictatorship" authorities and demand the release of our comrades from prison. In case of a refusal, I was to demand that they be evacuated in order not to leave them to be killed by the Turks. Furthermore, we organized a fighting detachment of six or seven men headed by Shaumian's oldest son. This outfit was armed with revolvers and grenades. It was agreed that if I did not succeed in obtaining the release of our comrades, this detachment, without awaiting further instructions, was to attack the jail as soon as

*Sergo Martikian—a Bolshevik and close friend of Shaumian. Later President of the Central Executive Committee of Soviet Armenia.

the Turks appeared in the city and free the prisoners. In order to do this the detachment would be stationed at all times in the district of the Bailov Prison. We made arrangements with the commanding officers of the Soviet ship *Sevan* to take the freed prisoners to Astrakhan.

The *Sevan* had arrived from Astrakhan several days before with a group of delegates who had come officially to initiate trade negotiations between Astrakhan and Baku. We had learned, however, from the local Communist sailors that the real purpose of the *Sevan*'s arrival in Baku was to ascertain the general situation in the city and the fate of the imprisoned Baku leaders. Therefore, it was easy for us to reach an agreement with the ship's crew. We arranged that by evening the *Sevan* would be standing at anchor in the Bailov district, not far from the prison.

I went to Tsentrokaspii where I spent several hours trying to get a hearing. None of the bosses were there; they had all fled in panic. I found out that during the night the remnants of the British army had hurriedly left Baku by ship.

Finally, in the evening, Velunts, a member of the "Tsentrokaspian Dictatorship," appeared. I literally pounced on him and declared that they were cowards and scoundrels if they left our comrades to be torn to bits by the Turks, and that this whole nefarious business was, in fact, their filthy handiwork. I said that it was baseness of the worst kind and they would not get away with it.

"You, Velunts," I said to him, "will answer for this with your head."

He said that he couldn't release the prisoners, that he had no right to do so, but that he also didn't want to leave them to the Turks. Then I asked: if he had no right to free them, then why couldn't he have them evacuated?

Velunts replied that he had no way of doing that, and that he himself had no idea of anyone who could. Then I said that I could do it since I was a member of the Baku Soviet of Workers' Deputies.

Unexpectedly Velunts agreed with me and said that he would give the appropriate instructions to Dalin, the assistant chief of counterintelligence. I asked him to provide a written order to that effect. He did so, and I took the document to Dalin. Fortunately, I found him at his post.

"Here is an order for you," I said to Dalin, "I request you to undertake the evacuation of the imprisoned Baku Bolsheviks."

"I can't do that," Dalin said to me. "I have no ships, and there is no one to accompany them; I don't have an escort-guard for them."

I replied that we had a ship in reserve which could be used for this purpose. He became interested as to where this ship was and where it had come from. Naturally I refused to answer and said only that we had already organized everything for the evacuation. Again he declared that he had no convoy guard and that he couldn't release the prisoners without one, since that would amount to setting them free. To this I replied: "As a member of the Baku Soviet of Deputies elected under your rule, I have the authority to escort and evacuate the arrested men. I'll go to the prison and do everything myself."

But he kept on insisting that without convoy soldiers he could not carry out the evacuation order. Just before, I had noticed that there were a few old worker-soldiers in the Cheka building. I went over to them and began trying to persuade them to come with me as an escort guard to release our comrades from prison. At first they refused, saying that they were in a hurry to get back to their families since the Turks were already breaking into the city. But in the end I succeeded in persuading them.

"All you have to do is get to the prison," I told them, "and from there you can go where you like."

Dalin still hemmed and hawed for a long time, but at last he signed the order which entrusted A. Mikoyan, a member of the Baku Soviet, and the convoy guard attached to him, with the evacuation from Baku of the imprisoned Baku commissars.

I set out on foot at once with the "convoy guard" from the center of the city to the prison building in Bailov. It was between eight and nine o'clock in the evening by now. First of all, I wanted to find our detachment, headed by Suren Shaumian, which we had agreed was to be stationed near the prison, but I couldn't find it. (I learned later that before my arrival the detachment had been detained by some SR sailors.)

The warden of the prison was standing at the front door when I arrived. He was in a very agitated state, not knowing what to do. I introduced myself, showed him the document, and asked him to hand over the prisoners to me, which he was very happy to do. I literally ran into the prison where all my comrades were standing expectantly at the doors of their cells, having heard the artillery and rifle fire of the approaching Turks. I announced loudly that they were about to be released, taken out of the prison, and put aboard the *Sevan*, which had arrived from Astrakhan.

Once we were outside the jail with our released comrades we saw that the detachment was still nowhere to be found. Worse yet, the ship was not in the agreed-upon place. Passersby told me that troops from the "Tsentrokaspian Dictatorship" had seized the *Sevan* and sent it off somewhere. There was only one way out: to go back into the city and try to find a place to hide.

By this time the Turks were already firing on Bailov Prison with rifles and machine guns. The bullets whistled over our heads as we made our way behind the walls of buildings. When we reached the area of the embankment we saw people rushing aboard the ships standing at anchor in the bay. In the midst of the crowds and confusion we ran into Tatevos Amirov, who was on horseback. He was very glad to see us and immediately said to Shaumian: "You can be evacuated from Baku on the *Turkmen.*"

He led us to the pier where the *Turkmen* was docked. The ship was packed with refugees and armed soldiers, but Amirov ordered the upper deck and the wardroom cleared for the Baku commissars and those in their party. Amirov arranged a place for himself there, too, just as the ship was about to weigh anchor.

Thus ended the first period of the Baku underground after the fall of the Baku commune. The Turks had entered the city.

* * *

I shall continue my story about what happened to us aboard the *Turkmen* and about the tragic deaths of the twenty-six Baku commissars. But at this point I want to quote a story told me by one of the Baku Communist activists who remained in Baku after our departure, Olga Shatunovskaya:

> The Turks broke into Baku on September 15, 1918, and remained in the city until November when, according to the Versailles Treaty, they were to leave and the British take their place.
>
> The Turkish interventionists plundered, killed, and committed atrocities. They slaughtered many inhabitants—Armenians, Russians, and Azerbaijanians. Those were terrible days! People who had stayed in Baku sat them out in cellars, in underground pipes, and anywhere that they could hide.
>
> Two Bolsheviks—Suren Agamirov and Alexander Baranov—and I hid in Sergo Martikian's apartment. Later we were joined by Levan Gogoberidze and Kostia Rumiantsev. At that time the Baku underground still numbered among its members the Agayev brothers (Bakhram, Magerram and Imran), Levon Mirzoyan, Julia and Vanya Tevosian,* Amalia Toniants, and many others.
>
> When it became known that rail communications with Georgia had been renewed, Levan said that we should take the first available train to Tiflis since it was dangerous to remain in Baku any longer.
>
> But the question was how to get past the guards at the station. Gogoberidze decided to obtain a Georgian officer's uniform and ID, and to make Kostya Rumiantsev his orderly. This plan was successful: the two of them left safely on the first train.
>
> Agamirov, Baranov and I also managed to get ourselves tickets and false documents as students but we were recognized at the station by two former Military-Revolutionary Committee workers who had

---

*Ivan Fedorovich Tevosian (1902–1958)—an outstanding Soviet statesman and party worker. One of the organizers and leaders of heavy industry, Vice President of the Soviet government.

turned traitor after the Turkish seizure of Baku and had begun to work for the Turkish secret police. We were immediately surrounded by a small group of gendarmes. Besides us, the gendarmes had accidentally pushed aside two other young people, whom these same provocateurs identified as Beker and Rumiantsev. They tried to protest, and so the gendarmes began beating them up. Then they took all of us off to the secret police, which was located in the home of the former governor. The chief of the secret police was Bekhaèddin-bay, nicknamed "the Red-haired Turk." The two young people who had been detained along with us by mistake were released when their relatives came to get them and proved who they were.

But the secret police kept us in custody. They beat up Suren and Shura. They also beat me on the head with their fists: knowing that I was Shaumian's personal secretary, "the Red-haired Turk" tried to get me to tell him where Shaumian was. Then they began to beat my comrades with *nagaikas* [prerevolutionary leather truncheons]. But we stubbornly kept on saying that we knew nothing and that we didn't even know each other. "Where is Shaumian?"—This was the question they wanted an answer to. I said that they had all left by ship. They didn't believe me, screamed that I was lying, that they had accurate information that Shaumian was here in Baku.

They held us for several weeks and then announced that the commander-in-chief of the Turkish army, Nuri-pasha, had signed and issued the order for our execution by hanging: the next morning at six o'clock we would be hanged on Parapet Square.

Several hours passed. An escort guard entered my cell, and gave the command to exit. Where—I did not know. They led me through the former governor's garden, up along Nikolaev, and then Gubernia Streets. The courthouse was located on Gubernia Street. I thought that they were taking me to court. Opposite the courthouse was the home of the oil industrialist Rothschild. For some reason they took me into that house, leading me into a huge office where, to my surprise, I saw Beybut Dzhevanshir. He was a childhood friend of Stepan Shaumian: they had grown up together and gone to Germany as students. Stepan had finished his studies in the Philosophy department and become a professional revolutionary, while Dzhevanshir became an engineer at a factory and a capitalist. To be truthful, I must add that in deference to his old friendship with Stepan, Dzhevanshir had, even before the Revolution, gotten Stepan out of prison more than once, given him financial assistance, and hidden underground materials (for him), etc.

I had met Dzhevanshir for the first time under the following circumstances: during the March 1918 Muscavatists' revolt, our regiment was firing at the house where Dzhevanshir lived, returning the machine gun fire which was coming from the roof. Dzhevanshir informed Shaumian about it and asked for help. Shaumian sent Suren

Agamirov and his son Suren to bring Dzhevanshir and his wife to his apartment where I also happened to be at the time. They lived there in safety for about two weeks. That is how Dzhevanshir came to know both me and Suren Agamirov, who had once led him out of his house under fire.

When the Turks formed the Azerbaijanian government, Dzhevanshir was named Minister of the Interior. In this capacity he had been informed that Suren Agamirov, Shura Baranov, and I would be hanged on Parapet Square the next morning.

When I saw Dzhevanshir standing in front of me, I was thunderstruck. We were alone. "Come closer, Olga," Dzhevanshir said to me. "I'm the Minister of the Interior. I've been informed that Stepan is in the city and that you don't want to give his address. I am Stepan's friend. He saved my life, now I want to save him. If *they* find Stepan, they could kill him on the spot. Give me his address!"

I replied that Shaumian wasn't in Baku, that if he thought that, he was mistaken.

But Dzhevanshir didn't believe me. He kept begging me to give him the address, swearing that he was duty-bound to save Stepan. I tried to prove to him that I was telling the absolute truth, and swore oaths to that effect. Then he flew into a rage and shouted: "You damned fanatics, you're doing Stepan in!" In the end he got really nasty, then called the guards, and ordered, "Take her back!" I was hardly in a condition to remind him that tomorrow Nuri-pasha had condemned us to be hanged. They took me back to the prison.

Toward evening a senior prison guard (a Turkish prisoner of war who spoke a little Russian) came to my cell and, looking at me, said sympathetically: "Poor girl, tomorrow morning . . . you know what's coming, don't you?"—and he made a noose-like motion around his neck. I replied that I knew. Then he left me a bunch of grapes and left.

He came by again in an hour: "Oy, poor thing, tomorrow you'll (again the same gesture around his neck). Have yourself some wine, drink up!"

And he came a third time, bringing a pillow: "It's your last night. Sleep on a pillow!"

Then I made up my mind and said: "My brothers are downstairs. I want to say good-bye to them, take me to them." He refused, obviously very frightened. Then I threw the pillow at him and the grapes he had brought me, and shouted: "Get out! I want to see my brothers!"

He "melted." "Wait," he said—"in the evening when the warden leaves, I'll take you."

Later in the evening he took me across the courtyard and from there into a deep cellar where I found Suren and Shura. They rushed to embrace me. They, too, knew that tomorrow we would all be hanged.

The guard was afraid to leave me with my friends for long and dragged me back, but we had managed to agree that when they led us to the gallows, we would sing the "Internationale."

I sat in my cell and waited for the first rays of dawn to be visible through the glass strip in the doorway: my last day. Suddenly I heard some noise, the scraping of the gendarmes' swords, and some footsteps near my cell. Could it be they were coming for me? But it wasn't daylight yet! "The Red-haired Turk," accompanied by a translator and a guard, entered my cell. Bekhaèddin-bay turned to me and said in French (knowing that I spoke some French): "By order of Minister Dzhevanshir you are freed. Your sentence has been changed from death to exile from Azerbaijan."

At first I couldn't believe it. I thought: "Perhaps he wants me to go to the gallows compliantly, without offering any resistance?" I said: "Why are you deceiving me? Take me to the gallows—I'm ready." But he repeated again in French: "You're free." And the translator repeated it in Russian.

At that point I began to understand what had happened. Apparently Dzhevanshir had interceded after all! But a horrible thought occurred to me at that moment: what if this was only for me? I asked, "And what about my friends?" The chief of the secret police chuckled: "So now they've become your friends?! Why, didn't you deny it just a little while ago, and claim that you didn't even know each other?" Finally, when he had stopped laughing, he said that they too would be released.

They took me to the office where I signed a statement saying that I would appear at the police station in three days' time in order to be exiled from Azerbaijan.

Then they pushed me out through the gates. Suddenly I saw my father was standing there. It turned out that he had been told that the gendarmes had picked me up at the station, and he had been spending all this time going around pleading for me. Yesterday he had been informed that I would be led out to the square tomorrow to be hanged, and he had stood by the gates at the (secret) police station all night, waiting for me to be brought out.

He stared at me, unable to grasp anything. And I, for my part, could not believe my own eyes: what was my father doing here? My father could not restrain himself; his legs buckled from under him, he fell on his knees, grabbed my hands and began to sob very loudly . . .

Ten to fifteen minutes later Suren Agamirov came out through the gates, followed in a few moments by Shura Baranov. They, too, had signed a statement that they would show up at the police station in three days to be exiled.

For two days and nights my father begged and pleaded with me to give it all up and go and become a student somewhere. "The Bolsheviks deserted you," he said, "and then flew the coop themselves, so leave them, enough is enough . . ." He didn't give me a moment's peace with his pleas and demands. On the third night I couldn't stand it any more. We had a quarrel and at one in the morning I left home.

I walked along the streets in a city that was in utter darkness. Somewhere shots and shouting could be heard. I decided to go to my friend Zina's—a seamstress. She took me in.

The next day I saw Suren and Shura, and we decided not to go to the police. After all, they could exile us abroad, to Petrovsk, where the Whites were and Bicherakhov and his band, where they, too, were hanging Bolsheviks. We decided to make our way to Georgia on our own.

I lived at Amalia Toniants' house for a short time and soon we managed to leave for Tiflis. There we had a secret rendezvous. And there we found Gogoberidze and Rumiantsev.

Chapter V

We Leave To Return Victorious

We decide to go to Astrakhan—The treachery of the ship's committee—Our arrest in Krasnovodsk—In the house of detention.

It was two hours before the ship, the *Turkmen*, cast off from the pier of the besieged city with its cargo of refugees and our comrades from prison. Before we sailed we were joined by Suren Shaumian and members of his outfit, who had found out that his father and the other imprisoned commissars were on board.

Lev Shaumian was also on the ship with us. Following his release from prison after the committee of the left-wing SRs had posted his bail, he had lived in Tatevos Amirov's apartment, and then Dzhaparidze had sent him to Sergo Martikian's apartment where Dzhaparidze's wife and the wives of our other comrades were staying. Shortly after we had boarded the *Turkmen* Leva brought all of them to see us on the ship.

While we were all sitting in the passenger's lounge Dzhaparidze's wife suggested to her husband that he leave the ship and find a safe place in Baku to hide from the Turks. She then added that she knew a man who could hide him at his apartment. This astonished many of us because, practically speaking, it was virtually impossible for anyone as well known in Baku as Dzhaparidze to find a safe hiding place there. Moreover, no preliminary security measures had been taken. What kind of person would undertake such a thing, and who would vouch for the safety of his apartment? Dzhaparidze's comrades felt that such a plan was even more dangerous than staying on the ship, and Dzhaparidze himself decided to stay on board with his comrades.

When at last the *Turkmen* weighed anchor and began to move out of Baku harbor, I went to Stepan Shaumian and told him that the captains of all the vessels in the convoy had instructions to go to Petrovsk where the Tsentrokaspian government had resettled and where Colonel Bicherakhov was in charge. I suggested to Shaumian that, for our own safety, he persuade the captain of the *Turkmen* to pull away gradually from the convoy under the cover of night and set course for Astrakhan. Given the situation, the naval flotilla could hardly pursue us.

Shaumian put up the idea for discussion and everyone responded favorably to it. Then Shaumian had a private conversation with the captain, who turned out to be a compliant fellow: he was a Latvian whose family lived in Riga, and since he was anxious to go to Astrakhan himself in hopes of making his way from there to his family, he consented to Shaumian's plan. He asked only that Shaumian see to it that he be allowed to proceed from Astrakhan to Latvia. As a representative of the central authorities, Shaumian gave the captain a full guarantee.

We were overcome with joy, and dreamed of reaching Astrakhan and Soviet Russia that much sooner—a dream that seemed so close to being a reality. The weather was good and the sea calm, and when we went out on deck we saw our ship was gradually moving away from the lights of the other ships in the convoy. Soon we were on the open sea where the lights from the other ships could not be seen.

Then our short-lived dream was shattered. The ship's captain appeared in the passengers' lounge and, addressing Shaumian, showed him the position of the *Turkmen* on the map, as well as that of the convoy which headed for Petrovsk. It was clear that we had broken away from the convoy and were on a direct course to Astrakhan, but trouble had arisen from an unexpected quarter. It seemed that the ship's seamen's committee, which consisted of SRs, had found out we had changed course for Astrakhan and had held a meeting during which they had decided not to proceed to Astrakhan but to change course for Krasnovodsk.

"The sailors say," reported the captain, "that there is famine in Astrakhan, whereas according to their information, there are sufficient food supplies in Krasnovodsk, so they refuse to go to Astrakhan."

This news took us completely by surprise.

Shaumian then proposed that Polukhin, a representative of the central naval authorities, go with the captain and try to convince the ship's committee to revoke their decision and continue on to Astrakhan. Polukhin returned after a while, looking gloomy. Nothing had come of his efforts, he said, the sailors insisted on having their own way and didn't want to bend to anyone's will. It turned out that their leader, the ship's mechanic, was an inveterate counterrevolutionary.

It also turned out that the refugees and soldiers on board were also against going to Astrakhan, and supported the decision of the ship's committee. Thus, not only was the ship's crew against us, but also the armed soldiers and the refugees, who occupied virtually the entire ship with the exception of the upper deck. Later we learned that a fierce propaganda campaign had been unleashed against us among the soldiers and refugees by the Dashnak commanding officers, who had joined forces with two British officers of General Densterville's regiment, who also happened to be on our ship.

It's not hard to imagine how depressed we all were as we tried to figure out what steps to take. Till then I hadn't taken part in the discussion because I had assumed that my more experienced comrades would solve it without me, but when they began to discuss how to avoid docking at Krasnovodsk I couldn't restrain myself, and said that in view of the danger of ending up in British-occupied Krasnovodsk, we should use force to make the crew continue on course to Astrakhan.

"But where can we find such force?" I was asked.

"The chief of Amirov's outfit," I replied, "can find about twenty armed men on board who are loyal to him and whom he could summon inconspicuously, one at a time, to the upper deck. With their help he can disarm all the soldiers who don't want to go to Astrakhan, and distribute their weapons amongst us. If we're armed, we'll be able to get the crew to follow our orders by threatening to throw overboard those who offer resistance."

Alyosha Dzhaparidze, who was the most emotional among us, shouted at me: "What are you, some kind of animal? What are you saying, throw them overboard . . . ?"

I should say that I loved—no, that I idolized—Alyosha, but I found his words very offensive. I didn't answer him, and from then on kept my thoughts to myself. Naturally, I realized that it was no simple matter to disarm the soldiers; there might be a skirmish, and there was no guarantee of success. Nevertheless, there was a definite possibility that we might gain the upper hand and by force of arms break the resistance of those hostile to us. No one said anything more about my plan, one way or the other, however, and an oppressive silence fell. Then I thought: why get excited? After all, they're more experienced than I am, they know what they're doing. Feeling tired and depressed, I lay down under a table in the passengers' lounge and fell asleep.

I slept for a long time, and when I woke up it was already day. There was no one in the lounge. Outside the sun was shining brightly, and several of our comrades were sitting on deck in groups of twos or threes, talking with each other, while others strolled quietly around the deck. This outward calm—both of men and Nature—stood in sharp contrast to the inner anxiety about our fate which still plagued me and, I think, many of my friends.

I recall how Petrov was lying on his back on a low projection which overhung the deck, his hands folded behind his head as he gazed pensively into the cloudless blue sky. Petrov was one of the left SRs, but he hadn't supported his party's leaders in their betrayal of Soviet power and had served conscientiously in the Red Army. Although I did not know him personally, I had got a close look at him on Petrov Square on the day when his artillery had smashed the Turkish positions and staved off the Turks' invasion of the city. Petrov was a calm, cool, spirited commander much respected by his subordinates. I had been told that at times his behavior was uncontrolled, but on the whole his conduct toward the Baku Soviet of People's Commissars and the Special Commissar of the Caucasus Shaumian had been both disciplined and loyal. Looking at Petrov I wondered why he hadn't offered one suggestion yesterday on how to get to Astrakhan instead of sitting quietly without saying a word. Seeing him lying there so pensively I wanted to know what he was thinking and what ideas he had, but I didn't know how to ask him.

I couldn't for the life of me understand why such experienced military leaders as Korganov and others had not supported my proposal yesterday of taking over the ship by force and proceeding to Astrakhan. However, I felt it was inappropriate to raise this question again.

Although my fears over my comrades' fate didn't cease, their calm and steadfast behavior did allay them somewhat. Also, my long sleep had calmed my nerves, and since I was still very young and it was my first time on a ship in the open sea, I was soon luxuriating in the magnificent seascape, the blue sky, and the sparkling Caspian Sea. Looking at the peaceful faces of my comrades around me I wondered if their seeming serenity was due simply to their personal courage, or whether they still cherished the vague hope that the British authorities would turn out to be civilized European rulers, so to speak, and would be guided by the rules of international law and obviously, they still didn't realize their terrible personal danger.

Before nightfall on the evening of September 16, we entered the approach route to Krasnovodsk Harbor, where we were stopped by the harbor launch *Bugas,* with armed men aboard. They ordered our captain to drop anchor for the necessary quarantine procedures. Only the two British officers (the reasons for their appearance on board were unknown) were allowed to leave the ship, along with an Armenian wearing a St. George's medal, who said that he had "important information to impart to the local authorities."

This was our first alarm signal that something was in the works, since normally a ship would enter port peacefully to be unloaded there. As on the preceding night our comrades slept, some sitting at the table in the passengers' lounge, others settling down in corners on the floor.

We awoke in the morning to another bright and sunny day. The *Bugas* pulled up to our ship once again, and by its order the *Turkmen* proceeded slowly to the Urf harbor, which was located several kilometers from Krasnovodsk. At the pier an unusual picture confronted us: lined up on both sides of the moorage were two columns of soldiers wearing Turkmenian *papakhas* and carrying rifles, and standing in front of them were three or four officers, a detachment of militia, and an armed squad of local SRs. Over to the side stood a battery of British artillery, and pacing up and down the pier were several British officers, including the two who had disembarked from the *Turkmen* the day before; they were not, however, openly involved in what happened. In addition, there were also officials from the government apparatus, headed by Kondakov and the aforementioned Armenian with the St. George's medal. It was obvious that we were about to be arrested.

We gathered together in the passengers' lounge to try to make some plan. Shaumian said that since they evidently intended to arrest us we should immediately go below, mingle with the passenger-refugees, and attempt to slip through the inspection. Once we'd made our way into the city, we would hide from the authorities with the aim of eventually reaching Astrakhan or Tashkent, where the Soviets were in control. No one had any other suggestion, and under the circumstances, there was little else to suggest.

I then informed everyone (I had told only Shaumian till then) that I had some money from party funds, and at Shaumian's suggestion this was distributed equally among us. As I recall, it came out to about five hundred rubles apiece, which was roughly equivalent to the average white-collar worker's monthly wage.

Our comrades had no belongings except for coats and some trifles, and not everyone even had that. Suren Ovsepian, who had been the editor of the newspaper *Izvestia* during the last few months, was the exception to this rule. He had a rather heavy wooden chest. When we were leaving Bailov prison under a barrage of Turkish bullets and were forced to run across dangerously open places, he had carried this chest on his shoulders. In view of the situation now I asked him to please get rid of it.

I remember how offended he was by my disdainful attitude to his baggage. He said that it wasn't so difficult to carry the chest, and that he wouldn't get rid of it since it contained things that were precious to him—a very understandable attitude. He had been an active party worker and Leninist, who had worked at *Pravda* and *The Enlightenment* before World War I.

The lower deck was crowded with people; there was no room, as they say, for a needle to fall. We mingled with everyone else. It was very cramped but I didn't see any of our comrades near me.

The closer we got to the pier, the more obvious the situation became.

When the *Turkmen* had docked and the gangplank had been lowered, they began to disembark passengers. Outwardly it seemed as if a routine check were being carried out as people went through slowly, one by one. When my turn came the search was over quickly because I didn't have anything with me. I was wearing a rough soldier's blouse, belted at the waist, riding breeches, and (high) boots; on my head I wore a regulation military cap without a cockade. I wasn't carrying anything in my hands either, but behind my belt in the folds of my blouse was my pistol. I was very thin at that time and I hoped that they wouldn't find my gun during the search. During the first check I raised my hands over my head, some soldier quickly frisked me and, not finding anything, let me go.

After I had walked twenty or thirty steps down the pier, I noticed a second checkpoint. I was again frisked in cursory fashion and again let go, since nothing had been discovered. They didn't ask us for any documents since the passengers on the ship were refugees who had escaped the Turks, and, naturally, most of them had no documents. I had gone another ten steps when I saw that there was yet a third checkpoint. However, having already passed two "tests" with flying colors, I felt I'd do fine on the third one as well.

I didn't notice anything suspicious as I approached the third checkpoint, but nonetheless I looked around cautiously so as not to attract attention. Suddenly, someone grabbed me by the shoulder. I turned around and saw a uniformed port official dressed in a white, high-collared naval jacket and cap. He gestured at me and said:

"Follow me."

We walked in silence. He took me to the edge of the pier where a small harbor launch, the *Vyatka,* was moored.

To my surprise I saw Shaumian and Dzhaparidze on the deck of the *Vyatka* sitting at a table and conversing quietly with each other. Fioletov and his wife sat at another small table. Dzhaparidze's wife was also there. I saw Azizbekov go down into the lower quarters of the boat, carrying a teapot in his hands. Soon he came back up on deck and, holding the teapot and some glasses, said quite cheerfully:

"Friends, I've made you some good tea. Let's have some."

You could tell from his behavior that he wanted to cheer up his comrades and put them in a good mood.

Everyone turned around when they caught sight of me, but there was no look of surprise on their faces: many of our comrades were already on the boat, and the others were being brought in gradually. They began to tell me who had been detained and under what circumstances. It seems that agents-provocateurs among the refugees (first and foremost, the fellow with the St. George's medal) knew many of our comrades by sight, and the police had picked us out of the crowd and taken us to the *Vyatka.*

The number of those arrested reached thirty-five. Everyone sat calmly, chatting with each other, not knowing what to expect, until the authorities came with armed men and a search began. Each of us was searched individually, and very thoroughly this time as they made us take off our outer clothing and frisked us from head to toe. It was obvious that they were looking for money. They were especially strict in their search of those few paltry things which some of our comrades carried with them, and they looked to see if any documents or valuables had been sewn up in our clothing.

Shaumian, Dzhaparidze, Azizbekov, Petrov, Korganov, Vezirov

and Fioletov were subjected to an especially thorough search. It turned out that the only money my comrades had was what I had given to them; only Petrov was discovered to have an additional small sum.

As my turn neared I kept wondering what to do about the "Colt" I had hidden under my shirt. After all, it might come in handy in prison, and I didn't want to part with it, so while the search was going on, I decided to hide it somewhere. I asked if I might go below to the toilet, hoping to find a hiding place for it, but a guard was stationed there and I had to go back.

Vexed at the idea of parting with my Colt, I decided on the spur of the moment to give the policeman a scare.

"Your things?"

I thrust my hand behind my belt and with an abrupt movement pulled out the Colt.

Evidently, the policemen thought I was planning to shoot them. In terror they recoiled from me and one of them cried:

"What are you doing!" and grabbed me by the arm.

"What do you mean, what am I doing?" I cried indignantly. "I'm giving you the one thing I have. I don't have anything else!"

Deep inside I was gratified that the policemen had revealed themselves to be such cowards, and I could see, judging by the smiles of my comrades who had observed this scene, that they were pleased, too.

There were no unpleasant incidents during the search, nor were there any interrogations. Those who searched us were obviously disappointed we didn't have the valuables which they had hoped to find. The women who were with us were also searched, but policewomen carried out the search below in the cabins.

Finally, our boat pulled away from the side of the pier and dropped anchor. It seems that the Krasnovodsk authorities feared a negative reaction from the townsfolk if they witnessed the imprisonment of the commissars, and so we were to be detained until nightfall on the *Vyatka* and brought to the prison after dark.

We were now officially informed that we were under arrest, and were under the jurisdiction of the Krasnovodsk Strike Committee (the counterrevolutionary government of the city, led by the SRs). Shaumian made a protest against their use of force, saying that we were not guilty of anything, and that, furthermore, the local government had nothing to do with us whatsoever and certainly didn't have the right to arrest us insofar as we still hadn't set foot on Transcaspian territory.

Kondakov, the senior official among the arresting officers, said that he had no intention of entering into any discussion of the issue, since he was merely carrying out his government's order.

We then began to discuss our predicament among ourselves, and came to the conclusion that we were in a more dangerous situation now, in the clutches of the Transcaspian right wing SRs and the British authorities, than we would have been had we stayed under the jurisdiction of the "Tsentrokaspian Dictatorship." We had felt that the Baku Mensheviks and SRs in the Tsentrokaspii were directly dependent on the British military command, but since they were also anxious about their accountability to the Baku working class, they might have taken fright and not have resorted to violence against us. Now we began to feel that it might be in our best interests to ask to be sent to Petrovsk. True, Bicherakhov was in Petrovsk with his regiment, but we felt that his treatment by the Soviet authorities had given him no cause for violence against us.

I expressed the opinion that I should make use of my position as a member of the Menshevik-SR Baku Soviet, and also of the official authorization for the evacuation of the commissars which I had in my possession, to send a telegram to the "Tsentrokaspian Dictatorship" in Petrovsk, so that they might demand that we be sent free from Krasnovodsk to their jurisdiction in Petrovsk.

Shaumian agreed, but said that that wasn't enough. It was essential that all official representatives of Soviet power arrested by Tsentrokaspii communicate by radio directly to the "Tsentrokaspian Dictatorship." Everyone agreed with this, and Shaumian accordingly summoned the representative of the local government and demanded that he take down a radio-telegram for transmission to Petrovsk. Consent was received, and Shaumian and Dzhaparidze composed the text of the radiogram, which has been preserved.[9]

The local authorities took our telegrams and said they would send them, but we were not confident they would. When time passed and no answer was received, we began to think that we had been deceived and that our telegrams had never been sent. It was only several years later that the telegram from Shaumian and the other comrades was discovered in the archives. Mine was never found.

In the evening our boat moved toward the city pier. We were informed that some of us would be put in the house of detention, since there wasn't enough room in the prison. Alaniya, the head of the Krasnovodsk police, read out the names of those who were to be sent to the house of detention. Included in the list were Shaumian, eleven other men, and five of the women who had been arrested with them. My name was not on that list. When I realized they were sending me to the prison I went over to Shaumian and said that I wanted to stay with him, as I could be very useful in the event of an escape. Shaumian told me to try and see if I could persuade the police to put me in with his group. I then appealed to Alaniya and said that I wanted very much to be put in the house of detention and not the prison. He

[9]See Appendix.

looked at me very attentively for a moment and then, without asking me any questions, said, "All right." This made me very happy.

* * *

Krasnovodsk is a small city. The house of detention turned out to be fairly close to the pier, and we were taken there on foot, under escort and the men put in one cell, and the women in another.

It was a one-story building with, insofar as I can remember, six cells—three along each side of the corridor. The cell where we were put was relatively small, with several rows of wooden benches along the wall. The floor was cement and there were no mattresses, no pillows, no blankets, and no furniture, except for the customary latrine bucket in the corner. The older comrades settled down on the benches and went to sleep, while the younger ones—about six in all—lay down right on the floor.

Three tedious days went by. The weather in Krasnovodsk was stiflingly hot. It was terribly cramped in the cell, they didn't let us out for walks, and the food was poor, but none of this really disturbed us very much, as each man was preoccupied with his own thoughts about the uncertainty of our fate. On the surface, Dzhaparidze, Azizbekov, Fioletov, and Amirov seemed relaxed, even cheerful: they had made themselves comfortable on their benches and had begun a noisy game of preference with some cards Amirov had produced.

Arsen Amirian, however, the editor of *The Baku Worker*, was in a depressed frame of mind. His own brother—Tatevos—was in the cell with us. A week or so later, two more of their brothers who were not members of the Bolshevik party and had no interest in politics, were also put into our cell. The older of the two—Alexander—was sick with tuberculosis and had a terrible cough. I met the younger one—Armenak—for the first time in the cell. From his stories I learned that he was a great lover of billiards and spent all his time in the billiard rooms of Baku while living off his brother, Tatevos. At that time I had absolutely no conception of the game of billiards and was amazed at his passion for it. Neither of these Amirian brothers played any political role whatsoever.

Arsen Amirian paced up and down the cell, traversing it diagonally in five or six steps, with a somber, self-absorbed expression on his face. One had the impression that he was feeling extremely pessimistic and, possibly he was the only one of us who felt our situation was hopeless and that it would end badly. He didn't strike up a conversation with anyone, and no one interrupted his thoughts. He never reacted to the conversations of other comrades, remaining indifferent even to the witty stories "from my life" of Ensign Avakian. I referred earlier to some of Avakian's idiosyncrasies; even in the cell he didn't part with his portable stool. While some were sitting on the benches and others on the floor, Avakian would hoist his stool up on the benches and sit on it there. He had also kept with him a glass with a holder. I was surprised that they hadn't taken these things away from him during the search, but he said he had implored them not to and they had let him keep them. But he complained that as far back as his first arrest, in Baku, they had taken away his bird cage and canary. He would recall his beloved canary with great sorrow and bitterness as if he were speaking of a profound loss.

Avakian hoisted his stool up next to Shaumian, who usually sat on the benches with his back leaning against the wall, and entertained him with his stories, to which, it seemed, there was no end. Avakian was a master of the word, loved to tell stories, and evidently got a lot of enjoyment out of it himself. We listened to him with great interest. One episode from Avakian's life has stuck in my mind particularly, and I frequently recalled it later and related it to my comrades. Avakian related with humor how for many years he had been an "eternal student" at Moscow University. As a Social Democrat, he had taken part in the student movement, and had been an active and even somewhat wild youth. Once, on the eve of World War I, he attended the All-Russian Pirogov Congress of Doctors, sitting with a group of students in the gallery of the hall where the congress was held. One day after several of the participants' papers and speeches had already been heard, Avakian got up from his seat and cried out imperiously with his thunderous voice:

"Gentlemen, may I have the floor!"

The entire hall heard this exclamation, but the presidium ignored him. Avakian repeated again, even louder:

"Gentlemen, may I have the floor!"

The president, wishing to avoid an unpleasant incident, was planning at that point to give the floor to the next speaker, but Avakian asked for the floor yet a third time, even more loudly and insistently, and then went down into the hall and walked up onto the rostrum. Turning to the gathering he addressed them as follows:

"My dear gentlemen! A lot has been said here about the protection of the people's health, about the difficult conditions, the poverty and the backwardness in which the simple people live both in the villages and in many of the cities; we've heard how pathetically few hospitals, doctors, and medicines there are, and how many villages and settlements are without any medical assistance whatsoever. The esteemed speakers who have addressed you here have put forward all sorts of proposals on how to extricate ourselves from this intolerable situation, and how to set up-to-date standards of health care for our people. I feel that everything that has been said and proposed here is correct. But in order to be true to our people and to our consciences, it should be added that all these proposals are merely palliative measures. In order to cure the severe diseases that plague our people, we have to accomplish the primary goal. And this primary goal is to overthrow the autocracy, the source of all the misfortunes of our life. Down with the autocracy!"

The assemblage listened to Avakian in a tension-filled atmosphere, but with great attention. When he had come to the end of his speech, something extraordinary occurred. Some shouted and others swung their arms. Realizing how all this could end, Avakian descended hurriedly from the platform, and taking advantage of the confusion slipped backstage and left the building by the back door, thus escaping certain arrest. He told us all this with great animation and with his characteristic narrative skill.

Shaumian didn't have much to say. He only answered questions, but didn't raise any serious issues himself.

The majority of those in the cell had been in prison before, and the situation was not a new one for them. They acted calmly, with dignity, and with what seemed to me to be a kind of unconscious sense of optimism.

Toward evening on the first day of our stay in the house of detention, the door of the cell opened and in came a solidly built man in a quasi-military uniform, who was close to fifty, and was accompanied by two or three other men. This was Kun, as we learned later, head of the counterrevolutionary government of Krasnovodsk, and his retinue.

They wanted to ascertain who was Shaumian, Dzhaparidze, Azizbekov . . . They looked at each of us as if they were studying us . . . They didn't ask any particular questions. Dzhaparidze asked the reasons for our arrest and demanded an explanation of the charges against us. They replied that they didn't know. They had informed the government of the Transcaspian region of our arrest and were awaiting a reply as to what to do with us.

On the second evening Alaniya, the chief of the Krasnovodsk police, arrived with some high-ranking officials to whom he displayed our leading comrades. Then Alaniya informed us that our comrades in the prison had drawn up a written request, signed by Korganov and the others, to provide all those arrested with mattresses and pillows, standard rations, and everything that is given to a prisoner by law. (The next day we learned from this same Alaniya that Kun had appended instructions to this appeal of our comrades: "Prison is not a place for comfort!")

Then Alaniya said to us in a tone of spite and mockery:

"As far as the pillows are concerned, we'll give you those!"

He was obviously imparting a sinister meaning to his words. To Dzhaparidze's question whether an answer had been received to our radio-telegram from the "Tsentrokaspian Dictatorship," Alaniya answered shortly, "No."

On the fateful night of the nineteenth to the twentieth of September we were awakened after midnight. Again it was Alaniya with a group of some high-ranking officials, some of them obviously in their cups. The group included Funtikov, the president of the SR Transcaspian government in Ashkhabad, Kun, and several other members of their governments.

Our nocturnal visitors entered the cell and stood by the half-open door surveying us, while Alaniya announced that, according to the decision of the Transcaspian government, a number of those arrested were to be transferred that coming day to the Ashkhabad regional prison where they would be tried, and that the remaining prisoners would be released. He began to read out the names of those comrades who were to be transferred to Ashkhabad prison. We all stood up.

When the list of names had been read and mine hadn't appeared on it, I knew that I was among those who were to be released. I went over to Shaumian and told him that I wanted to ask to be put in their group: I still cherished hopes of organizing an escape. Shaumian looked at me and said, "Give it a try."

Addressing Alaniya, I said that I wanted to be with my comrades who were being transferred to Ashkhabad. He replied that he had no right to make any changes in the list of names.

Then Shaumian took me aside and said:

"It doesn't matter that your request was denied. You're being set free—you, and Surik, and Leva (his sons). Try to get to Astrakhan, and from there to Moscow to meet with Lenin, and tell him everything that has happened to us. Make the suggestion in my name that several prominent right-wing SRs and Mensheviks be arrested (if none have already), proclaim them hostages and offer them to the Transcaspian government in exchange for us."

I replied that naturally I would do everything he asked. Then Shaumian went over to Surik and Leva, placed his hands on their shoulders, and said:

"You must get to Astrakhan with Anastas, and then to Moscow—to Lenin."

He instructed them to send his greetings to their mother, look after her, and also to send his regards to Manya and Serezha (his other children).

"Tell Mama," he said, "not to worry. Nothing terrible will happen to us. Lenin will see to it that we are exchanged for the SR and Menshevik prisoners. Soon we'll be together."

We began, like brothers, to say an affectionate goodbye to each other. It should be said that everyone was in a fairly cheerful mood. We naively thought that if some of us had been promised release and others a trial, then our situation wasn't so bad after all, because those comrades who were going to trial hadn't committed any crimes. We still hadn't realized the full extent of the tragedy that was brewing.

We never doubted in those minutes that we would be released, and the thought never occurred to us that by the following day our comrades would no longer be alive. It was only in Shaumian's gaze, when he said good-bye to us, that I sensed some hidden anxiety.

* * *

What guided the Transcaspian government and the representatives of the British command in drawing up the list of twenty-six from among the thirty-five who had been arrested is evident in this written testimony of Suren Shaumian in June 1925, which was given when he was examined as a witness at the trial of Funtikov:

> ... In the middle of August 1918 we were arrested in Baku by the Anglo-SR-Menshevik government. In addition to the twenty-five comrades who subsequently perished, there were also: Mudryi, Meskhi, myself, Samson Kandelaki, Klevtsov—thirty men in all.
>
> The one we chose to be our prison elder was Pavel Zevin* (from among the twenty-six), who kept a list of all those arrested which he used in distributing the provisions which were brought to us by our comrades on the outside.
>
> Several days before the Turks occupied Baku and we were "released" from prison, Kandelaki fell ill with dysentery and was put in the prison hospital. For that reason, his name was crossed off the list of those who were to receive provisions.
>
> I was released on bail two days before the evacuation from Baku. My name was also crossed off the list.
>
> Meskhi, Mudryi, and Klevtsov did not end up with us on our ship to Krasnovodsk but, rather, found themselves on some other ship with refugees on board, which went to Petrovsk (to Bicherakhov territory), and from there to Soviet Russia.
>
> When we were arrested and searched in Krasnovodsk, the list which I spoke of earlier was found accidentally on our elder, Comrade Zevin. After this, they began to use this list to arrest people and to draw them out of the general mass of refugees (600 people).
>
> In addition to those men on the list, several other comrades were also arrested, namely; (1) Anastas Mikoyan, (2) Samson Kandelaki, (3) Varvara Dzhaparidze, (4) myself, (5) my younger brother, Leon, (6) Olga Fioletov, (7) Tatevos Amirov, (8) Mariya Amirov, (9) Satenik Martikian, and (10) Maro Tumanian. All those enumerated were not known by the Krasnovodsk authorities and were arrested only on the instructions of the agents-provacateurs included among the refugees. Tatevos Amirov was the only one familiar to them as a well-known

*This is a mistake: the prison elder was not Zevin, but Korganov.—A. M.

Soviet partisan, and for that reason his name was subsequently added to the twenty-five and thus the number "twenty-six" was obtained.

This explains why such prominent Bolsheviks as Anastas Mikoyan and Samson Kandelaki survived, and correspondingly, why included among the twenty-six there were some party workers of lesser magnitude (Nikolaishvili, Metaksa, the younger Bogdanov) and even some chance comrades (Mishne), who had been arrested in Baku because of a misunderstanding. Arrested accidentally in Baku, they ended up on the elder's list which subsequently became the black list.

If Comrade Kandelaki hadn't gotten dysentery he would have ended up the same way that I would have ended up, if I hadn't been released on bail on the eve of the evacuation.

The Krasnovodsk SRs reasoned along the following lines: if the people whose names were on the list were arrested in Baku, that meant that they were precisely the ones they wanted, and that they should be exterminated.

In the event that this list of Comrade Zevin had not been found, then one of two things could have happened: (1) they would have shot all of the thirty-five arrested, or (2) they would have shot the most important of the [party] workers, whose names were familiar to them . . .

A member of the Military-Revolutionary Committee of the Caucasian Red Army, Emmanuel Gigoyan, who is now well and thriving, found himself in exactly the same situation as Kandelaki. He was arrested in Baku together with all his comrades, but he too didn't appear on the list found on Korganov because of the fact that he had taken sick and had ended up in the prison hospital.

I wish to add to this that all the arrests of the Baku comrades were carried out on board ships en route from Baku to Astrakhan, which is another reason why my name was not on the list since I, as a rule, never left Baku to go anywhere, and as a member of the Baku Party Committee had stayed behind to do illegal party work during the days of the counterrevolutionary regime and after the victory of the Turks.

Being free, I went about trying to rescue my imprisoned comrades.

For the very reason that I had not been arrested in Baku at that time, my name had not appeared on the list of the prison elder referred to above, and according to which the Baku commissars were later arrested in Krasnovodsk (the names of the commissars' wives—the Bolsheviks Varvara Dzhaparidze and Olga Fioletova who also had not been arrested in Baku—were not on the list either).

As a result of these circumstances, three of us—Kandelaki, Gigoyan, and myself—responsible workers in the Baku Commune, and also Varvara Dzhaparidze and Olga Fioletova, escaped the tragic fate of the twenty-six Baku commissars.

Chapter VI

The Twenty-Six

The savage execution of the Baku commissars—In memory of our fallen comrades—Who is responsible for the crime?—Malleson, Jr., intercedes—Some reflections.

After our comrades had left for Askhabad prison, we slept peacefully through the rest of the night, awakening refreshed and full of hope that we would soon be released. We knew that some of our comrades on the *Turkmen* who had managed to escape arrest were still in the city, the old Communist Sergo Martikian among them. His wife had been arrested with us, but he had managed to slip through all the checkpoints. We began to think about how, once we were set free and had made contact with our local comrades, we would try to find a way of getting to Astrakhan.

When the guard finally appeared in the afternoon we asked him when we would be released. He replied that he had no orders on that score. This perplexed us, but we continued to deceive ourselves, supposing that the paper ordering our release had still not been drawn up.

Two or three days later the Ashkhabad newspapers for the twentieth of September found their way to our cell. They contained reports that the Baku Bolsheviks, headed by the "Caucasian Lenin"—Stepan Shaumian—had been arrested in Krasnovodsk, and that an enormous amount of money and weapons had been found on them. We found such libel both astonishing and extremely disturbing: after all, this vile report was being circulated after meticulous searches of the arrested men had turned up nothing of the kind.

The newspaper contained the threat that the Transcaspian government would punish the commissars, take vengeance upon the Bolsheviks "for their savage acts in Russia," and would go so far as to consider quartering them. This made us extremely anxious about the fate of our comrades.

About a week later a new prisoner who had been brought to our cell reported rumors that the arrested Baku commissars had been handed over to the British command, which was sending them to India via Persia. We racked our brains trying to decide what was better: a trial in Ashkhabad, which could end with their being condemned to death; or their being sent to India. At times the latter even seemed preferable to us since it would win us time for bargaining for their release.

About a month later the horrible news reached our cell. A railroad conductor told his Krasnovodsk comrades that he had been on the train which was transporting the Baku commissars, and had seen with his own eyes how at dawn on September 20, between the stations of Akhcha-Kuyma and Pereval, not far from Krasnovodsk, the commissars had been taken out of their car and led onto the desert where some had been shot and others stabbed to death.

We made every effort to hide this terrible news from Leva Shaumian, and for a while we were successful until he accidentally heard it from one of the prisoners. In spite of his age, Leva withstood this blow as stalwartly as the rest of us.

The head of the Transcaspian government, Funtikov, had had plenty of experience in executing Bolsheviks in the Transcaspian region. A peasant emigrant from Saratov Province, he had worked as a machinist on the Turkestan railroad and had been considered a member of the SR party since 1905.

By using politically backward workers and relying on Menshevik support, Funtikov had organized a campaign against Soviet rule in Kizyl-Arvat in June and July of 1918, then in Ashkhabad, and finally in Kazandzhik and Krasnovodsk. So-called strike committees of workers were organized there, which became the de facto organs of counter—revolutionary power. These strike committees instigated mass arrests of Bolsheviks.

Because of this the Soviet government of the Turkestan Republic had sent Poltoratsky, the People's Commissar of Labor and well-known revolutionary-Bolshevik worker, from Tashkent to the Transcaspian region in order to clarify the local situation and establish order. When Poltoratsky arrived in the city of Merv, which was still under Soviet rule, Funtikov used deceit to invite him to come to Ashkhabad for talks. On the way Poltoratsky was arrested by a group of Funtikov's SR cutthroats and was shot without investigation or trial.

Several days later the same Funtikov organized the killing—also without investigation or trial—of nine arrested Ashkhabad commissars: Molibozhko, Rozanov, Batminov, Zhitnikov, Telliya, Petrosov, Kolostov, Smeliansky, and Khrenov. During the night of July 23 they were transported on a special train to the stage between Annau and Gyaure stations and shot under Funtikov's personal supervision.

On that tragic day of September 20, 1918, according to the villainous plan hatched by SR hirelings who were in league with the

British high command, and at the instructions of the latter, the following brave men were put to death by the executioners Funtikov, Sedykh, and their assistants:

Stepan Shaumian—Extraordinary Commissar of the Caucasus, president of the Baku Sovnarkom, member of the Central Committee of the Communist party, People's Commissar of Foreign Affairs, member of the Military-Revolutionary Committee of the Caucasian Army, member of the Constituent Assembly;

Prokofii Dzhaparidze (underground nickname: Alyosha)—chairman of the Baku Soviet of Workers', Peasant, Soldiers' and Sailors' Deputies, People's Commissar of the Interior and People's Commissar of the Food Industry, candidate for membership in the Central Committee of the Communist party;

Meshadi Azizbekov—Baku Provincial Commissar, chairman of the Executive Committee of the Peasant Council, Assistant People's Commissar of the Interior, Communist;

Ivan Fioletov—chairman of the Council on National Economy, Communist;

Mir-Gasan Vezirov—People's Commissar of Agriculture, Left SR;

Grigorii Korganov—chairman of the Military-Revolutionary Committee of the Caucasian Army, Commissar of Naval Affairs of the Baku Sovnarkom, Communist;

Yakov Zevin—People's Commissar of Labor, Communist;

Grigorii Petrov—chairman of the Central Military Authority in Baku, commander of a regiment, Left SR;

Vladimir Polukhin—Commissar of Naval Affairs from the Center; Communist;

Arsen Amirian—editor of *The Baku Worker*, Communist;

Suren Ovsepian—editor of *The Baku Soviet News*, member of the Military-Revolutionary Committee, Communist;

Ivan Malygin—Assistant Chairman of the Military-Revolutionary Committee of the Caucasian Army, Communist;

Bagdasar Avakian—Town Major of Baku, Communist;

Meer Basin—member of the Military-Revolutionary Committee of the Caucasian Army, Communist;

Mark Koganov—member of the Military-Revolutionary Committee, Communist;

Fyodor Solntsev—military worker, Communist;

Aram Kostandian—Assistant People's Commisar of the Food Industry, Balakhan District Produce Commissar, Communist;

Solomon Bogdanov—member of the Military-Revolutionary Committee, Commissar of State Security, Communist;

Anatolii Bogdanov—Soviet functionary, Communist;

Armeniak Borian—Soviet worker, journalist, Communist;

Eyzhen Berg—sailor, Commissar of Communications in the Caucasian army, Communist;

Ivan Gabyshev—brigade commissar, Communist;

Tatevos Amirov—commander of a cavalry detachment that joined the First Caucasian Red Corps;

Iraklii Metaksa and Ivan Nikolaishvili—Communists, assigned by the party as personal escort guards for Shaumian and Dzhaparidze;

Isay Mishne—chief clerk of the Military-Revolutionary Committee, no party affiliation.

Flowers of the people, in the full bloom of their lives, young, with everything still ahead of them, the Baku commissars were unselfishly devoted to the cause of the proletariat's struggle. Each of them had his own, vivid revolutionary biography, years of heroic struggle against tsarism, and strong ties with the working class. Spontaneous organizers of and participants in many revolutionary events, they were representatives of eight different nationalities who set out on their revolutionary paths in various parts of Russia. Fate led them all to the land of Azerbaijan, within the walls of ancient Baku, where they created the Baku commune—the bulwark of Soviet power in Transcaucasia—and spilled their blood fighting to defend it.

They fell at their battle stations. But they did not die for the Revolution. Their names and their deeds became the banner of the Baku workers and of all laboring people in the Caucasus in their struggle for the subsequent unification of revolutionary forces against the bourgeois, counterrevolutionary, nationalistic governments in Transcaucasia; for the ultimate victory of Soviet power in the Caucasus.

Upon leaving Baku, Shaumian declared on behalf of the Bolsheviks: *"We are leaving, but we shall soon return."*

And these words turned out to be truly prophetic. Twenty-one months later, on the night of April 28, 1920, Soviet rule was proclaimed in Azerbaijan—Soviet Azerbaijan was created. On November 29, 1920, Soviet power emerged victorious in Armenia. At the end of February 1921 the red flag of revolution began to wave above Tbilisi (Tiflis). The Caucasus had become Soviet.

* * *

More than half a century has passed since the day of the tragic deaths of the Baku commissars. But time has not erased, nor could it erase, the bright images of those fearless knights of the revolution in the grateful memory of a nation. It is impossible to forget the deeds and the very lives of these people—heroic lives, filled with revolutionary fervor, permeated with struggle, and so selflessly given at the dawn of our victory, in the name of the future triumph of the great ideas of communism.

I cannot help but recall Stepan Shaumian once again. How much unrealized human potential vanished with him! When he died he was only forty years old; for a major statesman and political figure that is only the beginning of an active life. And Stepan was richly endowed for such a life with the most various talents: a highly educated Marxist, an outstanding organizer, a brilliant orator, a man of unshakable ideological convictions, decisive, bold, politically highly principled and at the same time flexible in matters of the strategy and tactics of the class struggle. It was not by chance that the great Lenin who, as is known, was not given to excessive praise, wrote Shaumian in 1918: "We are delighted at your firm and decisive policies . . ." In several newspapers of that time Shaumian was referred to as none other than the "Lenin of the Caucasus." And there was a great deal of truth in that epithet.

Shaumian had the good fortune to combine in himself the lofty virtues of a professional revolutionary with a soft, somehow uniquely charismatic personality. All decent people liked him; one felt at ease in his presence, under his leadership you could find an outlet for your strengths. In the newspaper supplement which the Baku Party Committee published in 1919 in memory of the fallen commissars, I wrote about him as follows:

> Shaumian was a great revolutionary and an outstanding personality not only of Russian, but of international stature. He possessed inestimable gifts of mind and heart, he knew how to penetrate to the essence of things, to define the causes of every social phenomenon, to orient himself quickly and easily in numerous political issues, to reveal the underlying class causes of all phenomena, and formulate the correct approach to them from the standpoint of the interests of the international proletariat. In victory as in defeat he exhibited a cool-headed confidence rarely encountered in most people, which he combined with fervent enthusiasm and unshakable devotion to the cause . . .
>
> . . . In spite of the great bitterness caused him by the filthy streams of lies and calumny fomented by such newspapers as *The Bulletin of the Dictatorship, Forward,* and provocative or irresponsible speeches made by rascals like the late-lamented Sadovsky and others; in spite of their open calls for pogroms and lynch law against the Bolsheviks, he always refrained from stormy explosions of indignation and hatred against the impudent boors, whom the waves of the human sea had cast amidst the filth and litter on the shore of Baku life, and whose moment in power was brief. He once told me in a meeting we had at the Bailov Prison: "You shouldn't give in to your feelings of rage and should try to control yourself. Don't forget that all this will pass, that one can't lead the masses with baldfaced lies and calumny for more than a day or two. The real truth will soon be revealed, and then the victory attained in such bastardly fashion will be destroyed

by the earth-shattering crash and downfall of all the scoundrels' and liars' bull-market speculations . . ."

Another Baku commissar—Meshadi Azizbekov—was a living example of incorruptible ideological integrity and boundless devotion to the ideals of communism.

Although he came from a wealthy family and had received a university education in Tsarist Petersburg, he found the inner strength to break all ties with his native milieu and to bind his life forever to the laboring mases. A participant in the revolutionary movement since 1890, he sustained his faith in the working class and the laboring peasantry in spite of numerous arrests and imprisonment. At the orders of the RSDRP he created a militant organization of Moslem workers in Baku (the *Beydagi nusret,* which means "Victory Banner").

"Moslem proletarian democracy," I wrote in the same newspaper supplement of 1919,

> was so well acquainted with the activities of its leader, a revolutionary, that it was particularly bent on sending Azizbekov to the Soviet of Workers' Deputies, and Azizbekov proved to be a most active member of the workers' parliament and socialist government. But Azizbekov was not only a leader and defender of the interests of the urban proletariat: heedless of all dangers and the threats of the counterrevolutionary "Muscavat," he traveled through the towns and villages waging an energetic campaign of socialist propaganda on behalf of the Soviet in order to build ties between the Moslem village proletariat, which had been terrorized by the khans and beks, and the revolutionary ranks of the international proletariat.
>
> Alyosha Dzhaparidze will also always remain a brilliant star in the magnificent constellation of Baku Communists.
>
> He was all fire, all noble ardor. He was literally aflame in revolutionary-constructive work. A genuine leader and stalwart Communist, Alyosha was everywhere. To the trade unions and institutions of the workers' government he brought order, businesslike enthusiasm, and an honest attitude to his duties; he inspired the workers with his indefatigable revolutionary energy . . . He was a remarkable organizer and practical man of action who knew to perfection both Transcaucasian life and the soul of the Caucasian proletariat, which burned with the unassailable desire to destroy the rule of the feudal lords and bourgeoisie and to spread the power of the workers and peasants of socialist Russia . . .
>
> Who among the Caucasian proletarians did not know our Alyosha?
>
> The old workers speak with particular enthusiasm about the energetic and tireless work he did among them, about how he organized

trade unions for the workers and popularized the ideas and fundamental precepts of scientific socialism.

The young workers—those who knew Alyosha in the second half of his life—are well aware of the fact that he managed to hold three responsible positions simultaneously: he was chairman of the Soviet's Executive Committee, Commissar of the Interior, and Commissar of Food Production . . . "

With great love and affection I remember Vanya Fioletov, the son of a poor Russian peasant, who from youth cast his lot in with the fate of the Baku proletariat. A participant in the famous Baku strike of 1903, he spent a lot of time in Tsarist prisons and in exile. Along with Dzhaparidze he was the soul and real organizer of the Union of Oil Workers.

Everyone loved our Vanechka for his youthful, bubbling energy. He didn't miss a single important event in the sociopolitical life of Baku. He was everywhere . . .

In spite of poor health that was greatly undermined by his ceaseless travels, Vanya Fioletov amazed [everyone] with his tireless capacity for work. He threw himself so wholeheartedly into his work that he hardly noticed the base, perfidious web in which his enemies intended to catch him and those who shared his beliefs . . .

While still an adolescent, during the Tsarist period, Grigorii Korganov joined our Bolshevik party. Finding himself in the ranks of the old army, he conquered the hearts of the revolutionary soldiers at the Caucasian front; after the Revolution he headed the Military-Revolutionary Committee of the Caucasian Army, and then became People's Commissar for Naval Affairs of the Baku Commune and commander of the Caucasian Red Army.

This firm and principled revolutionary, an outstanding military leader who possessed a charismatic personality, was the heart and soul of the defense of Soviet Baku.

Grigorii Petrov. Selflessly courageous, boundlessly devoted to the Revolution, he came to us with his regiment from the far north—as a symbol of mutual support and of Soviet Russia's assistance to the Baku proletariat which was exhausted by the unequal struggle. Side by side with us he defended the positions of the Soviet; we fought together . . . And how he fought—the Baku workers still have not forgotten . . .

Arsen Amirian. Not possessed of the gift of eloquence, he was never seen at committee or Soviet meetings, but his rebellious, subtle thinking always found its way into party life . . . He was intransigent; his uncompromising spirit and intuitive grasp of the complicated real-life situation in Baku amazed many of us who knew him to be like a naive child in run-of-the-mill matters of everyday life . . .

Suren Ovsepian. Lured into politics from his years at the gymnasium, comrade Suren was very energetic in his work both as an organizer and as a propagandist . . . In the prisons of Baku and the Transcaspian region as well as on the place of execution—everywhere he could be found with a pile of manuscripts and issues of *Izvestia* . . .

I wrote the following in the newspaper supplement:

> The memory of the Baku commissars will remain alive in our hearts forever . . . And perhaps like them we too shall fall at the halfway point of our journey and shall lie down next to our comrades, but [we shall do so] as they did, with the same bright hope in our hearts that new fighting men will replace us and lead our cause to its victorious end; we shall fall with the same cry on our lips which emerged from theirs at the last moment, but was drowned out by the volley of rifle-fire and the whistle of bullets:
> "We are dying for the Communist revolution!"
> "Long live communism!"
> The life of each one of the twenty-six Baku commissars is a bright page in the history of our revolution, of our state, and of the great multinational friendship between laboring people.
> Yes, the lives of people like Shaumian, Azizbekov, Dzhaparidze, Fioletov, Korganov, Petrov, Polukhin, Vezirov, Zevin, Amirian, Berg, and the other Baku commissars, people who represented eight different nationalities, but whom the October Revolution had unified into a single international brotherhood, should become an educational model for the new, maturing generations of builders of communism. To continue their glorious revolutionary deeds and traditions, to nurture in themselves the qualities which were so magnificently embodied in those remarkable revolutionaries—this is one of the first tasks confronting our Soviet youth . . .

* * *

At the end of August 1920 a special commission was sent from Baku to the Transcaspian region. It included among its members Suren Shaumian and the old Baku revolutionary, Gandiurin, among other comrades. They were sent to bring the remains of our fallen comrades to Baku for burial with honor.

The funeral services took place on September 8, 1920. On the seventh the Baku Soviet and the Baku Committee of the Communist Party of Azerbaijan published an announcement declaring the day of the funeral a day of mourning for all Baku workers.

On the day before the funeral the newspaper *Azerbaijanian Poor* contained the following:

> Tomorrow is a day or mourning. Tomorrow the Baku proletariat will commit to the earth the ashes of our twenty-six comrades who were savagely shot by British executioners and White Guards. The twenty-six self-sacrificing fighters for communism were treacherously apprehended, carried away, and ignobly and stealthily shot by the agents of imperialism.
>
> What a blow for the proletariat . . . How much revolutionary energy and love, how many possibilities perished along with them!
>
> Can there be even one worker in the city of Baku who cannot recall the figures of comrades Shaumian, Dzhaparidze, Fioletov, Azizbekov, and the others.
>
> They are no longer with us, but their bright spirits hover invisibly above us. They will be with us always, in every place where the victorious banners of the Third Internationale, born amidst the thunder and lightning of civil war, will wave.
>
> But tomorrow, tomorrow, on a day of grief and mourning, on the day of our fallen brothers' funeral, we shall quietly make obeisance before their freshly dug graves.

On September 8, at the last meeting of the Congress of Eastern Peoples, after a short speech in praise of the twenty-six Baku commissars, the delegates rose in honor of their memory. A funeral march was played.

The eighth of September was a clear, sunny day in Baku. The streets were filled with countless columns of workers who marched carrying flags of mourning and portraits of the Baku commissars.

The families of the deceased attended the funeral. Delegates to the Congress of Eastern Peoples and members of the Executive Committee of the Comintern who had attended the congress marched in the funeral procession.

The ship with the twenty-six coffins containing the remains of the deceased slowly approached the pier.

Party and state figures from Azerbaijan, comrades-in-arms of the fallen commissars, delegates to the Congress of Eastern Peoples, and vanguard workers took turns as pallbearers.

The procession wound its way through the main streets of Baku to the sounds of revolutionary and funeral marches toward its destination in Freedom Square, the place of burial.

There was a mourning rally. Local activists and representatives of the Comintern delivered speeches.

On behalf of the government of Soviet Azerbaijan and the Central Committee of the Communist party of Azerbaijan, Nariman Narimanov said:

Now we commit to the earth of Soviet Azerbaijan our best and dearest comrades, who so gloriously stood at their revolutionary posts until the last. These stalwart, honest heroes fell at the hands of the British, of the Britain which always touts its humanitarian attitudes. Here lies the result of that humanitarianism—twenty-six coffins. Today fate was pleased to give a demonstration of the meaning of that humanitarianism to the East. May the East come to know it. But know also that the hour of retribution draws nigh. Yesterday the East swore to join forces in order to overthrow these world robbers. And it won't be long before the gray-haired East will awaken, and easily shake this yoke from its shoulders. Sleep, dear comrades! The ideas for which you fought will cast their bright light all over the earth!

When we heard and read Narimanov's moving words about our fallen comrades, many of us recalled how closely connected Narimanov himself had been with the long revolutionary struggle of the Baku proletariat and with these, its best sons, whose remains were buried with such honor on that day in Baku, in Freedom Square.

When he headed the Bolshevik organization Gummet, Nariman Narimanov had always had the unwavering support of Shaumian and Dzhaparidze and had been close friends with them. Their mutual struggle in the ranks of the Baku proletariat had made them as close as brothers. Even before the February Revolution, during the Tsarist period, when Shaumian and Narimanov were both in exile in Astrakhan, a feeling of mutual concern and personal friendship had grown up between them. Narimanov often came to the assistance of Shaumian and his family when they needed his help.

On the first day of the March Muscavatist rebellion in 1918 in Baku, when Shaumian had learned that the house in which Narimanov and his family were living was located in the fire zone, he had immediately sent a group of Red Guards there, headed by his son Suren, to bring Narimanov's family to Shaumian's apartment, which was located in a safe area. Narimanov's family joined Azizbekov's in spending several anxious days at Shaumian's.

In the summer of 1918 when Narimanov fell ill, Shaumian had managed to have him sent to Russia for treatment. This helped Narimanov not only to recover his health, but also to escape the tragic fate of the Baku commissars. After the formation of the USSR, Narimanov became one of the chairmen of the Central Executive Committee of the USSR. He died at his post in 1925 in Moscow and was buried in Red Square.

It turned out that Shaumian had actually saved the lives of two other Baku People's Commissars. He sent Nadezhda Nikolaevna Kolesnikova, the People's Commissar of Education, to Astrakhan, where she was elected chairperson of the Astrakhan Provincial Committee and worked with Kirov during a critical period for Astrakhan. In 1919 Nadezhda Nikolaevna became Krupskaya's assistant in the division of adult education. She was frequently present in Lenin's Kremlin apartment and took part in talks with him.

In the spring of 1920, as soon as Soviet power was victorious in Azerbaijan, Kolesnikova returned to Baku and resumed her old position as People's Commissar of Education; from 1923 to 1929 she was in charge of the Provincial Party Committee's Propaganda and Agitation Division, first in Moscow and then in Yaroslavl. In 1929 she was recalled to Moscow and became the rector of the Krupskaya Academy of Communist Education.

In 1954, in commemoration of fifty years of sociopolitical activity, Kolesnikova was awarded the Order of Lenin, and a year later, on the fiftieth anniversary of the first Russian Revolution (Nadezhda Nikolaevna took part in the December armed revolt in 1905), she was deemed worthy to receive this high honor a second time. Her husband, People's Commissar of Labor Jakov Zevin (party conspiratorial name, Pavel), was one of the twenty-six Baku commissars who perished.

The Baku Commune's People's Commissar of Justice, Artashes Karinian, also left Baku on Shaumian's orders. For a time Karinian worked in Moscow, then he moved to Armenia where he is living and working at the present time. He is a full member of the Armenian SSR Academy of Sciences.

A majestic monument bearing the name of the twenty-six Baku commissars has been erected on the spot in the Baku square where they were buried. It is a group sculpture depicting their execution and was created by the outstanding Soviet sculptor, Merkurov.

The children of the Baku commissars have proved worthy of the immortal names of their heroic fathers.

I have already mentioned Stepan Shaumian's children.

I also know Alyosha Dzhaparidze's daughters—Elena and Liutsia. They are Communists and live in Moscow. Elena Prokofievna Dzhaparidze, a specialist in power engineering, took part in the construction of Magnitka, and was for many years the Assistant People's Commissar to Tevosian in the field of ferrous metallurgy.

The sons of Meshadi Azizbekov, the Communists Aziz and Aslan, held responsible posts in Azerbaijan.

Jakov Zevin's son, Vladimir, is engaged in party work in Moscow.

* * *

Less than a month later, the rumors started by the British and the SRs, that the Baku commissars had been taken to India, were dispelled. It became clear that savage reprisals had been carried out against them and everyone was disturbed by the question: who was the real villain in this business? Who bore the legal responsibility for this crime?

The arrest of the commissars in Baku, their transfer to the jurisdiction of a military field court created for this purpose by a special military tribunal, and the confirmation of the military court's right to hand down the death sentence—all of these things had been done by the "Tsentrokaspian Dictatorship" as a special punishment for the Baku commissars. The "Dictatorship," as I've already said, consisted of leaders of three political parties—the right-wing SRs, the Mensheviks, and the Dashnaks—which acted on the approval and direct instructions of the high command of the British Occupational Forces, headed by General Densterville and Consul MacDonnell.

The plan to do away with the commissars under the banner of "legality" (by setting up the military field-court procedure) was not carried out. To their own surprise, the Turks had broken through the front, and the members of the "Dictatorship" along with the British high command had fled Baku in panic. As was later revealed, after the Baku commissars were arrested in Krasnovodsk, the leader of the local counterrevolutionary government, Kun, sent a radiogram to the "Tsentrokaspian Dictatorship" and its military commander, Bicherakhov, in Petrovsk. Kun requested their consent to transfer the arrested men to the jurisdiction of a military field court, and immediately received an answer from Bicherakhov stating that he, Bicherakhov, and the "Tsentrokaspian Dictatorship" agreed to this. But just three days later, the Transcaspian government, headed by the right-wing SR Funtikov, who was following the instructions of the British high command, furtively, secretly, and savagely annihilated the twenty-six Baku commissars without trial or investigation. The facts of the shooting were not revealed anywhere.

How could this have happened? After all, for the sake of "justifying" their action, the SRs needed to preserve at least the appearance of legality. Moreover, since they were in the habit of "playing at democracy," they would have tried to "formalize" the execution of the Bolshevik commissars with written documents, even crudely fabricated ones. They had, after all, tried to interrogate Stepan Shaumian in Baku (in order somehow "to legalize" the procedure being followed in this case). In a letter to his wife from the Bailov Prison, dated September 2, 1918, Shaumian wrote the following: "Yesterday, one of the old Tsarist investigators came to interrogate us. We chased him out and declared that we do not recognize the authority of any investigators or courts." Thus, the attempt to interrogate Shaumian and our other commissars had failed at that time.

Counter to their verbal agreement with Bicherakhov, the authorities in Krasnovodsk did not even try to interrogate the commissars, or any of the arrested men, for that matter. They merely took down our listing of the first and last names of the prisoners and decided to shoot all those whose names appeared on the prison elder's list, plus the regimental commander Amirov, who was known to them personally.

We were at a loss. Among the local authorities there had been a certain Yakovlev—a lawyer and notary. When he had come to our cell we had asked him: why haven't they interrogated us or read us the charges? The lawyer replied with calm impudence:

"For thirty years I served the tsar faithfully and truthfully, then I endured the rule of the Bolsheviks. Now it's your turn to endure. . . ."

The chief of police, Alaniya, evaded responsibility with a few meaningless phrases. All this gave us a basis for concluding that the local counterrevolution was acting under pressure from their British bosses. And so far as it concerned them, it was not necessary to interrogate us. Therefore—the absence of any interrogation of the Baku commissars; the haste with which their savage punishment was carried out; the vain attempt to cover up the murder itself; and the fact that after the execution not one of us who remained in prison was interrogated—all these things are links in a single chain. The British high command did not want to leave any traces or documents which could testify to its involvement in this crime. As Funtikov testified in court in 1926, Tighe-Jones, in his haste to execute the commissars, tried in every way possible to "buoy up" the SRs and promised to help them hide the evidence of the crime: on behalf of his commanding officers he assured them that an official communiqué would be issued saying that the commissars had been taken to India. This promise, however, as one might have expected, turned out to be another deceit.

In March of 1919, immediately after we had returned from prison, Vadim Chaikin appeared in Baku. He was a well-known lawyer, a member of the SR party's Central Committee, who had been elected on their slate of candidates for the Constituent Assembly from Turkestan. He arranged a meeting with me, and we met in a secret apartment.

Chaikin reported that he had spent more than a month in the Transcaspian area trying to ascertain the facts surrounding the execution of the Baku commissars. He had done this not only as a professional lawyer, but also as a member of the SR party who wanted to convince himself personally whether any members of his party in the government had taken part in the murder. He wanted first to expose and then to condemn them in order to erase the shameful blot on the SR party, in which he believed, while not, however, approving of many of its actions.

Chaikin was firmly convinced that if the British high command had not been working behind the back of the Transcaspian SR government, that government would not by itself have dared to commit a crime.

In conclusion Chaikin told me that he considered it his moral duty to mount a campaign on an international scale in order to expose and bring to justice all the parties actually responsible for this crime—both Russians and British.

I must admit that these were not just empty words. Chaikin began to act on them immediately in Baku and Tiflis, and we gave him our full support. I think it would be interesting to quote an excerpt from a letter Chaikin sent, shortly after our conversation, on April 3, 1919, to the chairman of the House of Commons in London:

> A month ago, as a member of the All-Russian Constituent Assembly from Turkestan and secretary of the committee on convening the Turkestan Constituent Assembly, I visited the Transcaspian region where I uncovered a serious capital crime that had been organized in September of last year with the participation of the head of the British legation in Ashkhabad, Captain Reginald Tighe-Jones, and the chief of the Transcaspian Bureau of Investigation, Semyon Druzhkin, whom the British legation later named chairman of its Transcaspian directory. After I had obtained proof of the complicity of Tighe-Jones, Druzhkin, and others in this ignominious crime—the organization of the murder of twenty-six captured Baku commissars on Turkestan territory—I considered it my civic duty to come out in the local press with an open accusation of the criminals and at the same time to appeal to General Malleson, the highest-ranking diplomatic and military representative of Great Britain in Turkestan and Persia, with a declaration of the necessity of initiating a preliminary investigation into this affair. Instead of promoting the cause of justice, General Malleson took upon himself the initiative of hastily sending one of the chief culprits—Semyon Druzhkin—from Ashkhabad to England, providing him with the Turkestan British Legation's safe-conduct out of the Transcaspian region to Constantinople.
>
> Taking into account the actions of General Malleson, who was Captain Tighe-Jone's immediate superior at the time of the crime, I appealed to the current highest authority in the British command in Transcaucasia, General Thompson, with a new declaration of the necessity of immediately forming an investigatory commission to look into this matter, made up of British, Turkestani, and Baku representatives, a commission to which I was prepared to report the proofs in my possession of the complicity of Tighe-Jones and Druzhkin. General Thompson categorically refused to form such a commission and tried, on his own, in complete disregard of existing laws, to carry out an interrogation of Semyon Druzhkin* inside the British Legation on Georgian territory. In spite of the charges that had been made, and the protest of the sociopolitical organizations in Transcaucasia,

---

*Here Chaikin has in mind the fact that General Thompson personally "interrogated" Druzhkin with the aim of obtaining evidence he needed to exonerate the British high command.—A. M.

he once again offered Druzhkin the possibility of safe-conduct out of the country under the escort of British arms.

As a chosen representative of the population of Turkestan, and as secretary of civil rights and justice in the area, I consider it my public duty to make the above-mentioned facts known to the House of Commons and to request that it take exceptional measures toward investigating the nightmarish and perfidious murder of twenty-six unarmed captives.

Natually, neither the British Parliament, nor the British judiciary, nor the British military leadership in particular had any desire to investigate this affair.

Our investigation of the murder of the twenty-six Baku commissars began in 1925. In the spring of 1926 a trial session of the Supreme Court of the USSR convened in Baku to hear the Funtikov case. The judges included Chief Justice Kameron (from Moscow), members of the court Mir-Bashir Kasumov and A. I. Anashkin, well-known Baku workers, and Bolshevik activists who had known the Baku commissars well in conjunction with work they had done together. An old Bolshevik, Sergey Kavtaradze, was the Prosecutor for the State.

Many of the people who had been involved in the murder of the Baku commissars had managed to escape prosecution by fleeing abroad before the victory of Soviet rule in the Transcaspian region and Transcaucasia. But one of the chief culprits, the instigator and person directly responsible for executing the Baku commissars, Fyodor Funtikov, was apprehended in 1925 and brought to trial.

The Supreme Court of the USSR established that Funtikov's counterrevolutionary aims led him to organize armed rebellion against Soviet rule in several cities in the former Transcaspian region during 1918. With the support of the so-called Union of Front-line Soldiers and fighting cadres of the SR party in the Transcaspian region, he had overthrown Soviet power there, and together with other members of the SR party had taken power into his own hands. In the same year (1918) he had entered into criminal alliance with the British high command and set himself the following goals: seizure of the Transcaspian region with the aid of imperialist British troops; armed struggle with the forces of Soviet Turkestan; the overthrow of Soviet power there, and seizure of Turkestan's rich cotton-producing regions. On July 24, 1918, he organized and carried out the murder of nine commissars from Ashkhabad, and joined other members of the SR party and representatives of the British high command, to wit: General Malleson, and chief of staff of the British Expeditionary Forces in the Transcaspian region, Reginald Tighe-Jones, in planning and executing the savage killing of Stepan Shaumian and the other twenty-five Baku commissars.

For his crimes the traitor Funtikov was sentenced to death by the Supreme Court of the USSR and shot.

In connection with my recollections of the distant past, I think it would be useful to relate certain later incidents which took place not so long ago.

In August 1966 I received a letter from the son of the English general, Malleson.

Here is what the younger Malleson, a retired lieutenant-colonel in the British Royal Navy, wrote to me:

> . . . On more than one occasion you have blamed the British military commander in the south for planning that mass murder. That man was my father—Major General Sir Wilfred Malleson (who died in 1946). I want to assure you that everything he ever told me points to his complete innocence in that affair. As far as he knew, the whole thing was thought up and carried out by the Whites themselves. I am writing you because I do not want you to think for the rest of your life that my father planned the murder of your comrades. . . .

Malleson's son did not cite any objective evidence that would confirm his father's innocence; he had not made a study of the issue, and wrote on the basis of what had been told him by his father, who, naturally enough, tried to vindicate himself in the eyes of his son and of public opinion.

Malleson's son was not satisfied with this and sent another letter, saying approximately the same thing, to the editors of the journal, *Soviet Union,* after that journal had published an interview with Shaumian's son concerning the death of the twenty-six Baku commissars.

Lev Shaumian's answer to Mr. Malleson, Jr., was published in the third 1967 issue of *Soviet Union* and contained, among other things, the following:

> . . . In Transcaucasia and in the Transcaspian region, and not only there, the British used all possible means for supporting any anti-Soviet forces. Moreover, in Baku, Ashkhabad, and Krasnovodsk the counter-revolutionaries were directly dependent on the will and decisions of the British officers. It was not without some self-satisfaction that General Malleson wrote that at that time he and his officers had "in the course of about eight months, come to control a territory to the east of the Caspian Sea equal to half the land area of Europe." The general had obtained official sanctions for "supporting the provisional government of the Transcaspian region against the Bolsheviks." General Malleson wrote all of this in a 1933 article (in the *Fortnightly Review*), and in 1922, in the journal of the Central Asian Society he openly asserted: "I had no fear whatsoever of being called

to account and was convinced on the basis of the reports I had received that our first-rate troops would encounter no difficulties and would settle accounts with the Bolshevik riff-raff in the manner it deserved. . . ."

Further on [in his reply] Lev Shaumian cited an interesting document—the testimony of the former president of the Provisional Executive Committee in the Transcaspian region, Funtikov:

> . . . The representative and head of the British Legation in Ashkhabad, Tighe-Jones, spoke to me in person before the shooting of the commissars about the necessity of their being shot, and after the shooting, he expressed his satisfaction that the shooting had been carried out in accordance with the views of the British Legation.
> Ashkhabad. On this, the second day of March, of the year 1919, 4:35 p.m.

Funtikov's notation of the year, day, and hour of his declaration is a very important detail: it shows that he wrote it in his own hand while the British troops were still masters of Ashkhabad.

It must be emphasized that Tighe-Jones, who was chief of staff of the British Expeditionary Forces in the Transcaspian region, in speaking of the fate of the Baku commissars was expressing not only his own opinion, but that of his direct superior—General Malleson.

In his book *The Transcaspian Incident*, which appeared in London in 1967, the British officer Ellis tries to exonerate Malleson's legation. The author himself was both an officer on the legation's staff and was stationed at the time in Ashkhabad. His efforts at exoneration do not succeed, however. On the contrary, the facts he cites, in spite of all the things he leaves out and the "corrections" he makes, convince one of the opposite.

Ellis writes that Dokhov, the representative of the Ashkhabad government attached to Malleson, requested an immediate meeting with the general on the morning of September 18 and reported "in extreme excitement" the contents of a telegram which he had just received from Ashkhabad. The SR government announced the arrest of the commissars, with Shaumian at their head, and requested instructions from Malleson on how to proceed in regard to the arrested men. "General Malleson replied," the book says, "that in his opinion the commissars should not be allowed to complete their railroad journey to Ashkhabad under any circumstances." The general left it up to the Directory "to decide exactly how this could be prevented."

After such a straightforward answer, Ellis introduces Malleson's alleged, albeit far from firm suggestion, of wondering whether the commissars could not be delivered to him as hostages. And this is viewed as an exoneration of [Malleson's actions]!

According to Ellis's testimony, Captain Tighe-Jones was also completely knowledgeable about this conversation and about everything that happened after our arrest in Krasnovodsk. He was immediately telegrammed instructions in this regard—apparently, two days before the shooting. During those two days he was in constant communication with Funtikov, Zimin, and other Ashkhabad leaders. And we knew without Ellis' help what sort of contacts Funtikov's government had with the British. The latter assumed direct leadership of the Directory's activities.

Finally, *The Transcaspian Incident* relates that on the morning of September 18 Malleson reported to his superiors in Simla—at the staff headquarters of the viceroy of India—both the arrest of the Baku commissars and the measures he had taken.

All this leads one to suppose that the archives of the British authorities in India at that time could shed additional light on the actual course of events. It is interesting that relevant files of documents are mentioned in the inventory of the present-day Indian National Archives (to which the archives of the Viceroy's government were added on August 15, 1947). My son Sergo happened upon this fact by accident when he was on a scientific mission in Delhi in 1970. When he tried to familiarize himself with these documents, it turned out that the relevant files were missing from the archives—they had either been removed, or destroyed. If the role played by Malleson's legation was so innocent, as certain people have tried to maintain until now, then why, on the eve of granting India independence, did the British authorities find it necessary to destroy or remove archival documents concerning the legation's activities?

No dodges will help the British imperialists. They will never be able to wash away the shame of their complicity in the deaths of the twenty-six Baku commissars!

* * *

Certain questions could arise in the minds of the readers of these memoirs: which decisions or deeds of the Baku commissars were correct, and which were not? How would things have turned out if the commissars had taken other decisions or, in general, if they had acted differently? How could all this have influenced the course of events and the overall fate of the Baku Commune?

In anticipating my answers, the reader will, of course, start from his *own* conceptions of these events and will attach his *own* interpretations to them, interpretations suggested by the circumstances and positions of the *present* day and that take into consideration everything that has happened since the fall of the Baku Commune.

I would like to emphasize once again in this regard that I have tried to describe these events as I have remembered them; I have tried to assess them as I did *then*, at the time they actually occurred. *That* is how it was. *That* is the way I interpreted it. *That* is how I remember it.

In concluding my reminiscences of the days of the Baku Commune, I want to find the answer to one question: did any possibility at all exist of avoiding the tragic fate of the Baku commissars?

As I examine the situation I think such a possibility might have existed in two cases. The first possibility arose when the commissars and Petrov's regiment were evacuated from Baku to Astrakhan. Their fate could have been avoided if they had not stopped at Zhiloy Island to wait for the other two vessels with the armed Red Guards aboard. Or if they had left a military command on the island—the head of Petrov's regiment and perhaps Korganov, the People's Commissar of Military Affairs—while the commissars and other political workers had continued on to Astrakhan without stopping. After all, it took the Tsentrokaspian Naval Fleet twenty-four hours to encircle them, and in that time they could have been far away. The stormy weather that prevailed at the time would have helped them. But they did not do this out of a political leader's sense of responsibility for his followers, out of a feeling of comradely solidarity.

The second possibility arose when the ship *Turkmen* managed to break away from the convoy of Tsentrokaspian vessels and set course for Astrakhan. The Baku commissars could have used their weapons to try to force the commander of the ship to go to Astrakhan.

Now let me say a few words in regard to the chances of maintaining Soviet power in Baku.

It seems to me now that in 1918 we had a good chance of holding out for two or three months longer, of not allowing the British into Baku and of using our own forces to defend ourselves from the Turks while awaiting assistance from Astrakhan. (Documents later revealed that shortly after we had relinquished authority, a Red Army regiment assigned for Baku on Lenin's orders did arrive in Astrakhan.) At the same time we were continuing to get aid (true, not as much as we needed) and food supplies from Astrakhan through the Shendrikov Pier in the Kizlyar district. In addition, the Soviets were still in power at that time in certain regions of the northern Caucasus, including some of its eastern portion. Therefore, although the Caspian Sea was closed to ships, it was still possible for us to maintain contact with Astrakhan via the northern Caucasus.

Soon after, however, the White Guards' counterrevolution in the northern Caucasus defeated Soviet rule and the Red Army. Toward the end of 1918 and the beginning of 1919, the Eleventh Army, which had sustained great losses in the fighting with the counterrevolutionary forces pressing in on the northern Caucasus, began its retreat from the Pyatigorsk region. During its subsequent retreat in the direction of

Astrakhan, it suffered further substantial losses. A second group of Red Army units under the command of Ordzhonikidze, after fighting to the last bullet, retreated to the Caucasian mountains, and, after organizing the mountaineers into partisan detachments, it began to cross over into Transcaucasia.

Thus, by the end of 1918, Baku had been deprived in essence of any aid from the northern Caucasus. The beginning of winter would have meant an end to navigation at the Volga estuary and in the northern portion of the Caspian Sea. As a result, the procurement of military supplies as well as any other form of assistance from Russia would have ceased entirely. All these things would have meant a complete blockade of Baku, the defeat of Soviet power in Baku, and the still greater losses that would entail.

It is true that Turkey, followed by Germany, soon suffered defeat in the war and signed an armistice with the Entente, and that at the beginning of November 1918 the Turkish troops left Baku. But the British occupational forces returned to the city literally on the heels of the Turks. When they first arrived in Baku, the British had had no more than about a thousand men, but the second time they had a significant force at their disposal and set up their garrisons in various cities in the Transcaucasian region.

After the Entente's victory over Germany, Soviet power in Baku, even if it had held out to the end of the year, would inevitably have had to contend with the numerically far superior British forces, and would have had little chance of surviving under the blockade. With the British no longer involved on the Turkish front, they had more forces at their disposal, while we would have had no contact with Soviet Russia until the spring of 1919 and could not have received the aid we needed. In the fall and winter of 1918–1919 Soviet Russia's military and political situation was critical, particularly in the southern part of the country. During that period all land and sea communications between Baku and Soviet Russia were completely curtailed.

Therefore, when one examines the situation *now*—and takes into consideration what happened after the fall of Soviet power in Baku—one comes to the conclusion that the temporary defeat of the proletarian revolution in Baku was the inevitable consequence of the fact that Baku represented the arena in which two belligerent coalitions of world powers clashed, both of which were attempting to gain control of Baku oil as well as an important strategic point from which to rule the Near East.

At that time Soviet Russia was weaker than its opponents in the southern part of the country and suffered a defeat there.

The genius of Lenin—who, with only a small number of facts at his disposal, was able to draw the correct political generalizations and interpret them properly—was also revealed in his assessment of what happened in Baku at that time.

On July 29, 1918, at the time of the fall of Soviet power in Baku, Lenin spoke at a joint session of the All-Russian Central Executive Committee, the Moscow Soviet, and Moscow factory-plant and trade union committees. Since contact between Baku and Moscow had already been broken at that time, we did not know about this speech of Lenin's.

Lenin spoke about the critical military and economic situation of Soviet Russia and about the danger threatening her. Drawing upon all the facts he had at his disposal, Lenin analyzed in particular detail the tactics and methods of the struggle between Anglo-French imperialism and Soviet Russia.

Lenin said that it was obvious from the reports he had received that the British were attempting to attack Soviet Russia from their bases in India and Afghanistan; the British had clearly participated in the counterrevolutionary revolt that had gained control of some Central Asian cities. "And now," Lenin said, "when the separate links [in the chain] become clear to us, the current military and general strategic situation of our republic is now quite determined. Murman in the north, the Czechoslovakian front in the east, Turkestan, Baku, and Astrakhan in the southeast—we see that almost all the links in the chain forged by Anglo-French imperialism are interconnected with each other."

He read an extensive report he had just received concerning the desperate military situation that had arisen in Baku, and the fact that the left-wing SRs and Bolsheviks had been in the minority in the Soviet during the vote on the question of inviting British assistance in the struggle against the Turks.

Lenin approved of the Baku Bolsheviks' refusal to collaborate with the British interventionists against the Turks. "A decisive rejection of whatever kind of agreement with the Anglo-French imperialists," Lenin said, "was a correct step on the part of our Baku comrades since one could not invite them in without transforming an independent socialist government, albeit territorially isolated, into the slave of an imperialist war."

It is significant that Lenin called attention to the fact that the Baku left-wing SRs had been on the Bolshevik side at that critical moment in the history of Soviet power. Lenin said of these left-wing SRs, "unfortunately, very few, who did not support the vile adventurism and perfidious treason of the Moscow left SRs, but allied themselves with Soviet power against imperialism and war."

Further on Lenin said, "We know that the position of our comrades and Communists in the Caucasus was especially difficult because they were surrounded by Mensheviks who betrayed them by entering into direct alliance with the German imperialists, on the pretext, naturally, of defending the independence of Georgia."

In speaking subsequently about the international situation, Lenin said: ". . . we are confronted by a new success of Anglo-French, and consequently, world imperialism. If German imperialism continues to represent an aggressive, military, imperialistic force in the west, then in the northeast and south of Russia, Anglo-French imperialism has seized the opportunity to get stronger. . . . Anglo-French imperialism has won a major victory, and, after encircling us in its ring, has directed its efforts toward crushing Soviet Russia. We know very well that the success of Anglo-French imperialism is indissolubly connected with the class struggle. . . . Therefore, it is not in the least surprising that the current exacerbation of the Soviet Republic's international situation is connected with the intensification of the class struggle within the country."

Lenin said that when the Bolsheviks realized the obvious numerical superiority of the interventionists' forces in whatever segment of the front they were attacking, they refused, unlike the Mensheviks and other appeasement-oriented parties, to enter into any alliances with imperialists, or to invite them in as "saviors," but retreated for the time being in order to regroup and strengthen their forces. Although we had not known about these pronouncements of Lenin, Shaumian's decision to temporarily remove the Bolshevik forces from Baku seemed to echo Lenin's thoughts.

Everything that has been said points to a single fundamental conclusion—that the main reason for the temporary defeat of Soviet power in Baku lay in the unpropitious, from the standpoint of the proletarian revolution, alignment of forces in the Caucasus, and in the conjunction of the contradictions and clashes of two warring imperialist coalitions which had intervened militarily in Transcaucasia.

As I review the entire course of events, I come to the following conclusion: Shaumian's proposal to evacuate the Bolshevik forces in mid-August 1918, which was passed by the majority of the Baku Party Conference, was a wise one, and subsequent events proved its correctness.

Naturally, I cannot pretend to have made an exhaustive analysis of the events which took place at that time. That is the job of scholars and historians.

PART III

TRANSCASPIAN PRISONS

Chapter I

In Krasnovodsk

We learn the details of the commissars' execution—Assorted "visitors"—We're transferred to Kizyl-Arvat—The story of Artak and Marusia Kramarenko.

And so, on the night of September 20, 1918, as I related earlier, the Baku commissars were removed from prison in Krasnovodsk and a bestial punishment was administered to them at dawn.

Only four men were now left in our prison cell: Shaumian's two sons Suren and Leva, Samson Kandelaki, and myself. The four women, Varvara Dzhaparidze, Mariya Amirova, Olga Bannikova (Fioletov's wife), and Satenik Martikian, were still prisoners in the women's cell of the house of detention.

Up till then we had lain on the floor of the cell because there hadn't been enough room for everyone on the benches, but now the four of us made free use of them, sitting with our legs dangling over the side or with our backs against the wall. Soon afterward when two other prisoners were put into our cell—the Amirov brothers, Alexander and Armenak—there was still plenty of room on the benches.

For several days we vainly waited to be set free. Whenever we were visited by the prison guard or by Alaniya, the local police chief, we would ask when we were to be released, but they invariably replied that "they were awaiting the order," or that "they still didn't have the order," or that they, in general, had no idea when such an order would arrive. We were soon convinced that the promised release was a routine deception and we stopped asking such questions.

After the commissars had been taken away there were still some of the Baku Bolsheviks left in the prison—a fact which we were not aware of, sitting in our cell in the house of detention. These people had not been taken out with the commissars because their names hadn't been on the list of the "twenty-five" found on the elder, Korganov. Included among them was Gigoyan, a prominent figure in the military, and a member of the Military-Revolutionary Committee of the Caucasian Red Army, and also two or three other comrades whose names I can no longer remember. All of them were subsequently released from prison.

I wanted very much to have Leva Shaumian released from prison but the prison guard said that he couldn't do anything about it, despite all my appeals, although he did promise to pass on my request to the chief of police Alaniya.

Leva was silent during my negotiations on his behalf, but it was evident that he wasn't pleased by my efforts. Perhaps he didn't want to leave his comrades, or maybe he was motivated by a desire, very natural at his age, to share the hardships of imprisonment with us. He was a courageous boy who always wanted to appear older than his fourteen years. He had in prison with him a Circassian coat and hood from Dagestan which he would occasionally dress up in and which made him appear older and more mature. I forbade him to wear this outfit, especially when a visit from the authorities was anticipated. Leva understood my motives and sometimes he obeyed me and sometimes not. Once it happened, as if on purpose, that Alaniya, the chief of police, entered our cell unexpectedly when Leva was dressed up in his outfit. I turned to Alaniya and said: "Mr. Alaniya, you're not letting us, the adults, go—and to hell with you! But let this boy go. Keeping him in prison is in violation of every law." Alaniya turned toward Leva, looked him up and down, and said: "What are you crazy? What kind of a boy is this? It's obvious he's a real bandit from the Caucasian Mountains." My objections had no effect. Leva, however, was clearly pleased with this outcome. Suren, Samson, and I chided him for the childishness of his behavior, but Leva went on proudly flaunting his Circassian coat practically the entire time we were in prison, and we couldn't stop him.

I remember another episode from Leva's life which he told me about during one of our numerous prison chats. Once he had sent Lenin, via his father, a pin with Karl Marx's picture on it. This is how it happened: soon after the February Revolution an enterprising operator in Baku manufactured and put on sale an assortment of pins celebrating the Revolution. On one of them—a small metal square with a pin-clasp soldered onto the back—a tiny picture of Karl Marx had been mounted in a bright-red frame. When Stepan Shaumian went to Petrograd to the First Congress of Soviets, Leva gave him this pin with strict instructions to present it to Lenin personally. Leva told me with pride that, according to his father, Lenin wore the pin on his chest.

Many years later I recalled this story of Leva's when I was reading Krupskaya's reminiscences of Lenin. In the place where the events of June 1917 are described, these lines can be found:

> When Vladimir Ilyich returned home exhausted I didn't have the heart to question him about what had been happening. But we both

felt like talking, as we always did, when we took our walk together. Sometimes, but not very often, we would walk down the remoter streets on the Petrograd side. Once, as I recall, we took such a walk with Comrades Shaumian and Enukidze. It was then that Shaumian gave Ilyich some red pins which his sons had ordered him to give to Lenin. Lenin smiled.

Among Ekaterina Sergeevna Shaumian's papers, Leva's letter to his father in Petrograd has been preserved.

"Dear Papa," he wrote. "How do you feel? What date did you arrive in Peter? Did you give Lenin [the] Karl Marx [pin] and a greeting from me? If not, then do it. Bring me back some leaflets and collections of songs, if you see them anywhere."

Little did Leva know, that exactly a year later, in June 1918 he would be sent to Moscow as an emissary carrying Shaumian's letters to Lenin, making his way there, via Astrakhan and Kamyshin, with great difficulty. In Moscow he was warmly received and treated with affection by Bonch-Bruyevich, the manager of affairs of the Sovnarkom, but when the latter wanted to take him to see Vladimir Ilyich, Leva refused to go, mainly because he was embarrassed that his shoes were ripped, with one flapping sole tied on with a string. The following day, however, after he had managed to repair his shoes he agreed to have his visit with Lenin.

* * *

Our monotonous, unchanging life in prison continued. Devastated by the news of the tragic deaths of the Baku commissars, our own fate had ceased to concern us; from then on we scarcely thought about it. All our thoughts were tied up with the Revolution, which was repulsing the attacks of the counterrevolution that had arisen with the aid of the imperialist powers. We did not doubt that Soviet power would emerge victorious from this historic struggle, but we were frustrated by our inability to help bring about this victory.

None of us talked about what we had gone through, and we avoided discussing the death of our friends because we didn't want to add to the anguish of Shaumian's sons, and also because it was hard for us, too, to discuss this subject. I remember that at night we were wakened by nightmarish visions. The emotions I experienced at that time found public expression in an article I wrote four years later in Rostov, where I was then working. It was published in the newspaper *The Baku Worker* on September 20, 1922, under the title, "To a Bloody Anniversary." I shall quote an excerpt from it:

It was exactly four years ago, in a small dirty cell of the Krasnovodsk house of detention.

Three days had passed since we had found ourselves in this jail house. Shaumian, Dzhaparidze, Fioletov, and others—sixteen men in all—were in our cell.

The night of September 20. Some were lying on the benches, and others on the floor. Some were sleeping, others were just dozing. Through the window there shone a perfidious moon, bright and radiant.

Two o'clock in the morning. Noise is heard, the grate of iron. The door is opened. The band of "socialist executioners" bursts in. They begin a roll call. The inmates are divided up into three groups. The first two groups are led out of the cell . . .

To the question: "Where are you taking us?" they answer: "There's not enough room here. We're transferring you to Ashkhabad, to the regional prison."

The oppressive moment of parting with our best friends and comrades approaches, with the selfless fighters of the workers' cause, the revolutionary leaders of the Baku proletariat.

Then come tortuous days and weeks of not knowing . . .

Where they are, where they've been taken—we cannot ascertain. Some say that they're in Ashkhabad, others—in Persia. We don't know the whereabouts of our dearest friends, the Baku proletariat, of its leaders, and Soviet power, of its best fighters.

Doubts creep into our hearts, and torment them. It's painful to think that Stepan, Alyosha, Meshadi, and Vanechka are no longer alive . . .

Suddenly someone from the outside unexpectedly communicates to us in prison the terrible truth, and unveils to us the unheard-of crime of the SR executioners. A railroad worker from Krasnovodsk, who was unknown to us and had no knowledge of us, was, it seems, the transmitter of this information.

That night he was at the station. He saw suspicious-looking people who were armed and dressed up, wearing white turbans. In the train cars there were shovels.

Having a sense that something foul was about to take place, he decided to be a witness to the bloody execution being prepared. He crept between the cars, without being noticed, and found himself a place on the buffers. With his own eyes he saw how they stopped the train in the remote steppes, took the captives out of the cars in groups, and not far from the embankment, brutally cut them down.

Through the clatter of the machine guns he heard the last words of the victims of the white terror: "We're dying for communism. Long live communism!"

. . . Dawn was breaking. The murderers were in a hurry and were unable to bury all the corpses.

The leaders of the Baku proletariat stained the desert sands of the far-off Transcaspian steppes with their blood. They offered up their lives as a sacrifice to the cause of communism. With their heroic death they opened up the eyes of those Baku workers who had been deceived, and they set them on the correct path in the struggle for communism.

The darkest chapter of the workers' movement was brought to a close with their death, the chapter displaying the treachery and adventurism of the capitalist hirelings—the SRs and the Mensheviks.

May the Baku workers remember the last session of the Baku Soviet (July 25, 1918) when the majority of the Soviet under the leadership of the socialist-traitors took the side of the British against the Russian proletariat. With the impertinence of lackeys, the triumphant SRs and Mensheviks proposed that the Bolsheviks follow their lead in the service of imperialism, otherwise they were threatened with death and destruction. Comrade Shaumian, pale with disdain, pronounced these prophetic words: "The triumphant traitors are making every effort to steer us onto the path of treachery, too. But we prefer to die at our glorious posts as revolutionaries rather than save our lives through treachery and betrayal."

And they died the glorious death that befits true fighters of the proletariat. . . .

* * *

Frequently it would happen, especially at night, that the door of our cell would open and the prison guard would dump on the floor someone else who had been arrested.

We gradually became accustomed to this. Usually they were either daredevil youths who had been picked up during a drunken debauch on the street or in a restaurant, or they were persons who, for some reason or other, had fallen into the clutches of the night police patrols. As a rule, they turned out to be ordinary residents who willingly related their nocturnal adventures to us. They brought a certain animation into our monotonous life, isolated as we were from the outside world. After they told us how they had ended up in the house of detention and what they had done, we'd ask them what issues the people were concerned about, what was being written and said about the Civil War, and what new and interesting things were being reported in the newspapers, which rarely came our way. And so, after spending a couple of days with us, some of these "visitors" became attached to us, and when they were set free they'd promise to send us books, newspapers, and food. It has to be said that the majority of them kept their word.

There comes to mind one such nocturnal "visitor," who introduced himself next morning as the lawyer, Count Tarnovsky. Whether or not this Tarnovsky was really who he made himself out to be was of little interest to us at the time. He turned out, however, to be uncommonly stimulating company, and we listened to his stories with pleasure. I remember the count as a tall, middle-aged man, with a black moustache, who was handsome, plump, and well-dressed. He told us that he had come from Persia via the port of Enzeli (Pekhlevi). He had been arrested at the pier in Krasnovodsk while they were checking his papers, but he had managed, via someone at the pier, to get word of his predicament to his acquaintances in Krasnovodsk. Tarnovsky was confident that he would soon be released since he claimed to have important connections, and had moreover been arrested for no good reason whatsoever. He told us many interesting things about Persia, the British occupation, the lives of the Russians who had remained in Persia, the situation on the various fronts, and about certain events of international life. Evidently, he liked our company, for he readily shared entertaining stories of his own life with us, and furthermore, he could tell witty anecdotes so colorfully that he invariably made us burst out into loud guffaws. All this helped somewhat to relieve the gloomy atmosphere of prison life.

The count had been put into prison almost without any belongings, and without a coat. He had a large satchel with him, however—a saddlebag—and at night when it was cold he devised this scheme: he spread his saddlebag out on the bench, tucked his feet into one half of the bag and his head and shoulders into the other, and he covered his middle with a blue jacket. He looked silly, but on the other hand, he kept warm.

Seeing that we didn't have enough books and newspapers, Tarnovsky promised to send us some books the minute he was set free. After his release, within two or three days of his arrest, we received from him five or six volumes of Maupassant, a book of Amfiteatrov, and also a basket of superb grapes. The books were a real boon for us. And the grapes were exceedingly welcome, too.

Once, an Armenian from the merchant class was brought into our cell. When he realized that he had been put into a cell with Bolsheviks, he was afraid that he would have another serious accusation "pinned" on him and that he'd be jailed for a long time. He soon calmed down, however. He was totally astonished at how badly we were fed, and was sincerely amazed that we put up with it so calmly. He vowed that as soon as he was freed the first thing he would do was send us a good meal. And, in fact, the day after he was released we received a large crock containing a marvelously prepared Armenian *dolma* (stuffed grape leaves), and also a delicious fruit pilaf. We had many good things to say about our "benefactor" that day: we hadn't eaten so well or abundantly for a very long time.

Another Armenian, Zarafian, who was sharing our cell with us, took seriously ill one day and had to be hospitalized immediately. Contrary to all the prison rules, the authorities enjoined us to carry him on a stretcher (escorted by a guard). We were happy to have the chance to be on the outside, if only for a short time. As we walked along the streets of the city we saw for the first time what the city of Krasnovodsk looked like. Narrow streets, one- and two-story brick buildings. Not much greenery. We found out that Krasnovodsk had no fresh water supply; they desalinated sea water at a special plant, and some fresh water was brought from Baku by small tankers.

At that time Krasnovodsk was a small town with a population of something less than 10,000, and had no big industry, only cottage industries. There was a railroad station and a seaport with their workers and stevedores, that was all. The fairly large number of shops in the city were maintained not only by the population of Krasnovodsk, but also by the Turkmen population, which lived in the adjacent districts and was engaged also in cattle-raising and fishing. There was no agriculture under irrigation because of the absence of fresh water. On the whole, the population of the city consisted of Russians, Persians, Azerbaijanis, and Armenians. The Turkmen came to the city on short visits—to the marketplace or on other business.

We walked along, delighted to have an opportunity to breathe fresh air and see the city and the people. Our procession through the city under the guard of an armed soldier provoked the curiosity of many passersby. Thoughts of escape occurred to us but we didn't have any time, nor did we have a plan of escape. We could, of course, have scattered, but we wouldn't have been able to escape from the city anyway, since Krasnovodsk was bordered on one side by a waterless desert, and on the other by the sea. The only escape route was the railroad into the heart of the Transcaspian region, but the stations and the railway lines were under heavy guard, and no doubt in the event of our escape we would have been caught immediately.

The seaport was also under heavy guard, and since there were few ships there, it offered little hope of escape. At that time we had no contacts and no secret meeting places in the city, nor could we count on the help of our chance "visitors" in the prison cell, since they were, as a rule, ordinary citizens who were in fear of their lives. In addition, we didn't know where they lived.

Around the end of October or the beginning of November 1918 yet another prisoner was brought into our cell. He was a man of about forty, taller than average, thin, well-built, with curly black hair that set off his attractive, intelligent face. He turned out to be Ibragim Abilov from Baku. I had heard of him there, although I had never met him. Abilov had been sent to Krasnovodsk by the Soviet authorities in Astrakhan to ascertain the fate of the Baku commissars.

Besides Abilov, the Astrakhan Council of Deputies (Sovdep) had also sent two young Communists: Marusia Kramarenko and Artak Stamboltsian. They were provided with the necessary resources and put on board a small ship, the *Avetik*, which was sent under camouflage ostensibly to transport to Krasnovodsk those Baku refugees who wanted to escape the famine in Astrakhan. Abilov, Artak, and Marusia boarded this ship under the guise of refugees seeking shelter in the Transcaspian region, pretending during the journey to be strangers to each other. The voyage passed without incident, but upon their arrival in Krasnovodsk Abilov was unexpectedly arrested after the obligatory search, and ended up in our cell. He did not know what happened to his comrades.

Abilov was an educated man who turned out to be pleasant company. He brought us up to date on the general state of affairs in Soviet Russia, and in Astrakhan in particular. We also learned from him that Shaumian's wife and two small children had not gone to Moscow as we had supposed, but had remained in Astrakhan in the hope that the Baku commissars would soon be rescued: she still didn't know about their tragic deaths.

We learned that Abilov was himself the son of a craftsman and had worked in the cottage industries for several years. He had joined the Russian Social Democratic Workers' party (RSDRP) in 1905, and in the ensuing years he had taken an active part in party work in Baku until 1913, when he was arrested and exiled to Astrakhan. He knew Shaumian and Narimanov well from his work in Baku as well as from Astrakhan where he had met them when they were in exile there. Although he was not as yet a Bolshevik Abilov nonetheless supported Soviet power and enjoyed its confidence.

To jump ahead somewhat, I'd like to add that soon after we were removed from Krasnovodsk, Abilov was released and left for Baku. When we arrived in Baku early in March 1919 we found Abilov there as a member of the Azerbaijani parliament. At that time he sided with the socialist faction. In parliament he spoke out against the Muscavatist government and after the proclamation in 1920 of the Azerbaijani Soviet Republic he sided completely with the Bolsheviks.

At the conclusion of the Congress of the Peoples of the East, Abilov worked as secretary of the Council of Propaganda and Operations, which had been formed by the congress, and later in 1921 he became a plenipotentiary of the Azerbaijanian Soviet Socialist Republic in Turkey and in this high post demonstrated his real talent as a Soviet diplomat. The Soviet ambassador to Turkey, Aralov, has written very warmly about his activities there in his memoirs. The work of Abilov was held in high regard by Nariman Narimanov and Alexander Miasnikian in the articles which they dedicated to his memory. Abilov died in Turkey in 1923 after a serious illness.

Our life in prison proceeded along its monotonous course, in isolation from the outside world. The only people we were in contact with were the warden Altunin and his right hand, the prison guard. Plump, red-cheeked, with a long fluffy moustache, Altunin had the look of a provincial Russian merchant. His wife, or so it seemed to us, was not a bad woman; she was, on the contrary, even kind. Altunin himself was what he called a conscientious old veteran who carried out his duties meticulously. He didn't create any particular difficulties for us, but at the same time he rarely departed from the established rules. We were not let out into the prison yard, nor even allowed to mix with the prisoners in the adjoining cells.

Only Leva Shaumian could go out into the yard because it was his job to boil the huge samovar from which the entire prison drank its tea. This chore gave Leva a lot of pleasure since it gave him a chance, if only briefly, to be out in the fresh air and at the same time made him feel he was doing something useful and pleasant for us. The day after Abilov had been put into our cell, Leva informed us that while walking down our corridor he had caught sight of a new prisoner through the peephole of one of the cells—Marusia Kramarenko, whom he knew but whom, out of conspiratorial considerations, he gave no indication of recognizing.

After that, whenever we had occasion to pass Kramarenko's cell (and this happened every time we were taken to the toilet) we too tried "not to recognize" Marusia if we happened to look through her peephole. We kept this subterfuge up until the end; otherwise Marusia could have been linked to us and thereby exposed as a Bolshevik.

And thus we learned that of the three comrades who had been sent to Krasnovodsk from Astrakhan, two were together with us in jail, and fortunately, not discovered. Hence the third—Artak Stamboltsian—could have slipped through all the obstacles of searches and examinations and outwitted the Krasnovodsk police.

Artak was, of course, my old friend from schooldays when we were classmates together in the Tbilisi (Tiflis) Armenian Seminary. He was a professional revolutionary. Before the revolution we had been in the same underground Marxist circles; together we had joined the Bolshevik militia of the Baku Party Committee and had taken part in street fighting while defending Soviet power in March 1918. Our friendship and mutual confidence in each other had been put to the test on more than one occasion, and our relations had never been clouded by misunderstanding but had only grown stronger with time. Artak was distinguished by his selfless devotion to the Bolshevik party and to his comrades, and we all loved him very much.

He was bold and cunning to an audacious degree, and could always extricate himself from any situation, however hopeless. These qualities did not jibe with his unprepossessing appearance: he was small, puny, and had a childish expression on his face which he could, however, change at will, due to the exceptional mobility of his features. Artak was, moreover, an expert comic actor, with a vast repertoire of funny stories. He also had a sharp tongue and mimicked his comrades to a tee, poking good-natured fun at them by capturing their slightest peculiarity or weakness. But a certain characteristic, soulful gaze of his, and his ability to control himself in any situation made you know that standing before you was a strong-willed, brave-spirited man.

The harsh conditions of our life in jail, plus our meager, near-starvation rations, completely undermined our strength. Our remaining money was not sufficient for even a few days of substantial nourishment, but we resolved not to spend it anyway, feeling that it might still come in handy. Constant undernourishment showed up especially sharply in my physical condition, since I was emaciated even before my arrival in Krasnovodsk. My gums began to hurt so that I could barely eat the hard bread. Finally I appealed to the warden and asked him if I could be sent to a dentist. Altunin agreed to my request, and sent me, under escort guard, to a private dentist's office in the city, which was located on the first floor of a small house. The guard opened the door and let me in, and I waited there alone. The dentist's office was small, very poorly furnished, with hardly any equipment or furniture.

When the doctor came in and greeted me she turned out to be a young woman with a pleasant appearance. Since she had been forewarned about who I was, I didn't have to introduce myself nor did I ask her her name. She examined me carefully and said, "The only treatment you need is to eat fresh fruit and to spend as much time as possible in the open air." I was astonished, and wondered if she were mocking me. After all, she had been told about my circumstances. But she said it in complete seriousness, and I even felt that she regarded me with a certain respect, as evidenced by her eyes, her pleasant smile, and her gentle voice, so that while I went on thinking, "How can I buy fruit and breathe fresh air when I'm a prisoner?" pride forbade me from saying this to her. And besides, how could she help me? I thanked her and left.*

Our internment in prison continued. The days passed, one just

---

*Such strange events occur in life! Forty-one years later this woman doctor, who was living in America, happened to see me during my trip to the USA in 1959. I was in Chicago at the time attending a performance of the Soviet troupe of "Beryozka," under the direction of N. Nadezhdina. The well-known impressario of Soviet artistes in America, Mr. Hurok, had kindly sent me tickets to a private box. After the concert I went backstage to greet and congratulate our splendid young women.

Subsequently, I received a letter from the editor of a Wisconsin newspaper, Fred K. Steinecke, enclosing an interview with Judith Shuisky, who was the same woman dentist who had examined me in 1918 in Krasnovodsk. She had been at the performance of the "Beryozka" and had recognized me by my eyes as her former patient. The interview gave the details of our meeting fairly accurately. And although the tone of it was generally benevolent, Miss Shuisky made haste to dissociate herself from "the Bolshevik leader." She ended the interview with these words: "Don't give the impression that I'm the friend of a leader of Russia."

like another. We didn't know what was in store for us, but we had grown so accustomed to our situation that it didn't give us much concern. Then one day toward evening in the beginning of November we were told quite unexpectedly that within two hours we were be to transferred from Krasnovodsk to the city of Kizyl-Arvat. We were stunned and tried to figure out the reason for the extraordinary haste. What was the point of transferring us from one city-jail to another? There was only one detail that provided us with any clue: not all of us were being transferred—Abilov, who'd been in put in jail separately from us, was staying behind.

We immediately recalled that the Baku commissars had been taken out of our cell under the same pretext and had been shot en route, and we came to the dire conclusion that our fate was decided, and that the same thing that had happened to our comrades was going to happen to us: we'll be taken somewhere en route to Ashkhabad, we thought, and tonight somewhere in the desert we'll be shot.

Although at the given moment it was totally unexpected, we had become so psychologically prepared for such an outcome that, strange as it may seem, we didn't feel any particular emotion. But we had no doubt that we were being taken to be shot.

I remember one detail: we had one hundred twenty rubles left. By common consent I gave most of it to Abilov, keeping forty rubles on me ("Kerensky notes")—just in case. They might come in handy at the very last moment—say, for example, to give to some soldier so that he would give us water if we were thirsty. We also left Abilov our teapot, which we also wouldn't need anymore.

After sharing a warm farewell with Abilov, we were removed from the cell under guard. As we walked down the corridor we saw Marusia Kramarenko looking at us through the peephole of one of the cells. Our eyes met, and silently, unbeknownst to the guards escorting us, we bade each other farewell.

Just then the door of the women's cell opened, and the four women who had been arrested with us stepped out into the corridor. With them was a fifth woman—Maro Tumanian, who had been arrested somewhat later than we. This was the first time during these past three months or so that we had met. We were all taken out into the yard under guard. It was already dark, and we were taken on foot through the now quiet streets of the city to the railroad station. We went with dignity, showing no emotion. Only the elder Amirov brother, who was sick, was obviously in a nervous state, but he, too, made every effort to appear in control of himself.

At the station we were loaded onto a (heated) freight car which was set up for carrying the mail. A third of the car had been partitioned off, and was evidently intended for the postal employees. The women and men prisoners were put into the freight section of the car, and the escort guards settled themselves into the other. We were separated by a partition that had a small barred window through which several pairs of guards' eyes constantly watched over us.

Our part of the car was very small and there were no benches or planks so we sat and lay on the floor: the women on one side, the men on the other. The weather was cold and a strong wind blew in through the cracks in the thin walls, chilling us to the bone. I was especially cold, for I was dressed only in a soldier's blouse, trousers (riding breeches, as I recall), boots, and a military cap. My other comrades had coats. Suren Shaumian was dressed like I was, in a military uniform, but fortunately he had an overcoat as well, and his brother Leva was especially grateful for his Circassian coat, which was very useful to him on this occasion. I was shivering from the cold; so I lay down between Leva and Suren to warm myself.

There was only one thing that troubled me: if they took us out to be shot before dawn, before the sun had a chance to warm me up a bit, wouldn't our murderers think that I was trembling with fear rather than with the cold? That at a time like this such thoughts should have bothered me more than the prospect of my own death is obviously one of life's psychological mysteries.

The conviction that we were going to be led out to be shot grew ever stronger. What confirmed us further in this conviction was the fact that we had not been put into an ordinary convict car, but rather into a mail car and, most telling of all, we had been put together with the women, for at that time male and female prisoners were usually not put into the same cells or cars. In addition, there was no toilet or latrine-bucket in our section of the car, which fact filled us with the deepest indignation, embarrassment, and shame. Deprived of any possibility of using a toilet, the women, along with the elder Amirov, who had a stomach condition, suffered long torments and were finally forced to perform their natural functions right there on the floor. The floor of the car got wet and it became harder to breathe. Our awareness of the shame our unfortunate women were experiencing, coupled with our own, doubled our humiliation and pain.

The hours stretched out with agonizing slowness.

Everyone was wrapped in his own thoughts. As before, I lay between Leva and Suren, covered with part of Suren's overcoat and Leva's Circassian coat. We huddled close to one another until I got warm and the shivering from the cold passed. But we couldn't get to sleep. Thoughts such as they kept flashing through our minds: "What have we managed to accomplish? What did I do for the revolution? I had only just begun to be useful, begun to acquire strength and experience and self-assurance by being tested in revolutionary work, I had just begun to receive some recognition from my comrades when suddenly my life had to end so absurdly! How good it would be if I could serve the Revolution for another two or three years, and then, if I have to die, to fall in open combat after I had accomplished something really useful." I felt so sorry for Shaumian's sons, who had entered the Revolution so early and had been so stalwart. Such capable and devoted young men—and suddenly, death! All this seemed absurd and unfair to me.

I also thought about what was happening in Baku, wondering who, among the comrades who had remained behind, were still alive and who had perished, and wondering whether our party organization still existed or if it had been destroyed. All these things caused me concern, but it was impossible to find an answer to my questions. Conjectures could not be considered answers.

I thought of my family: I was particularly upset that my mother had most likely decided we had perished, since there had been no news of us for such a long time. She had probably heard that the Baku commissars had been shot, but she couldn't know that I was still alive. I felt terribly sorry for her. I thought of my father, too, although at that time he was no longer living; I thought of my sisters and my brothers, and, of course, of my second cousin Ashkhen. Even back in 1913, four years before the February Revolution, I had realized that my feelings for her were more than for a sister. But according to age-old Armenian custom, second cousins may not marry each other; we were one generation short of the "seven generations" between us decreed by folk custom. Popular traditions carried their own force, and I knew that our parents and relatives would be against our marriage and would censure me for it.

The more I had come to love Ashkhen the harder it was for me to conceal my feelings, and yet although I had tried on more than one occasion to declare my love to her, I had never been able to get the words out. It seemed to me that her friendly feelings toward me were those of a sister for a brother, and the possibility of her refusing me kept me mute; masculine pride did not allow me to risk my honor. For four years I acted cold and aloof, not daring to make my feelings known to her, but in the summer of 1917 I could no longer contain myself and I confessed my love to Ashkhen. She said that she had loved me for a long time, too, but had been put off by my coldness toward her, attributing it to my evident dislike of her.

This had happened shortly after the events of July 1917, when we were anticipating a long and difficult armed struggle before the proletariat would be able to seize power. In the face of these difficulties I told Ashkhen that we would have to wait a while before we could get married: I might perish in the impending revolutionary struggle and she would be left a widow, perhaps with a child. "When everything has calmed down and the Revolution is victorious, then we'll get married," I said. Ashkhen agreed with me. She understood everything only too well, having herself joined the Bolshevik party three months earlier. We resolved that no one should know of our decision since Ashkhen was still in school. A year later she finished her studies and left to teach in the country, in the Sukhum district. I had been overjoyed to find out that during this time she had been doing illegal party work. In particular, she had gone on an important mission for the party *kraikom* to the Bolshevik Party Committee in Erivan in provincial Armenia, giving them the *kraikom's* instructions, and receiving information in return concerning the situation in Armenia.

And so, as the train carried us along on what seemed to be our last journey, I thought of Ashkhen. Did she know about the death of the Baku commissars? Very likely she thought I had perished with them and was grieving for me. As hard as it had been for me to make us wait, I took comfort in the fact that I had done the right thing by not marrying her. This way she would find my death easier to bear.

Finally, after many long hours the train stopped at a station that was unknown to us. Through the cracks in the floor we heard someone give the command: "Unhitch the car!" And then a little later: "It's unhitched." We thought we had come to the end, but several minutes later our car unexpectedly began to move. A sigh of relief was heard in the corner, and Olga Fioletova's voice came out of the silence, "Whew, we're still alive!" We realized that during the night, unbeknownst to us, another car must have been added to our train, and that it had now been unhitched at this station. Pretty soon, however, it would be our turn to be unhitched. But by the time we had passed several more stations it was completely light outside and for the first time the thought occurred to us that perhaps they were not taking us to be shot after all, but were actually transporting us to another city.

Time passed inordinately slowly. Their inhuman treatment of us, especially of the women, and our frustration with our own helplessness was a constant torment, but finally, at long last, the train came to a stop at Kizyl-Arvat Station.

We no longer had any doubts now that we would be taken from the train and led through the streets of the city to the jail, but to our surprise we saw that we were being led away from the city streets, and soon found ourselves in a ravine, walking along the riverbed of a dried-up river in a direction leading away from the city, and once again the thought came to me that they were taking us to be shot. However, we had lived so long with this thought that we didn't even ask the guards where we were being taken. And then, all of a sudden, we were led out of the ravine and found we were on the other side of the city at the walls of the jail. The gates in the clay wall surrounding the jail were open and we saw before us a small one-story stone building, slightly smaller than the one is Krasnovodsk. This was our new jail, and once again were were separated, with the men and the women put in separate cells. When we asked why we had been taken from the station along such a strange route, through the ravine, our guards answered that it had been done so that the local population wouldn't see the prisoners.

Our hearts leaped when we saw that there was a stove inside our cell, one, moreover, that we could tend ourselves. We were delivered at last from the nighttime cold and could stoke the stove whenever we felt like it with the wood that was piled up in abundance beside it. How often we sat around that open stove and gazed at the tongues of fire, delighting in the warmth and even in the cozy quality of the atmosphere. We were further relieved, both for our sake as well as his, when the prison administration transferred the old and seriously ill Alexander Amirov to the hospital.

In this prison another boon lay in store for us, and for me in particular: they began to serve us tasty, juicy onions with our meals, since they were sold very cheaply in the local market there. We baked the onions in the cinders of our stove and consumed them in great quantities. I ate an especially large number of them as they were good for my bleeding gums. We were also given plump raisins for our tea, as a substitute for sugar, which was unattainable at that time. As a result of these additions to our diet my health began to improve somewhat.

It had been several months since we had had a bath or changed or washed our underwear, and as a consequence we had lice. After persistent demands we finally got them to take us to the railroad workers' bathhouse at the station. The bath was clean, with plenty of hot, steamy water. In short, it was a real Russian bathhouse, and it was the first time I had been in one in my entire life. It felt wonderful to wash. We steamed ourselves, but we had no soap—and what was a bath without soap! We also washed our underwear without soap. When it dried out it seemed almost as dirty as it had been before, but despite these drawbacks, the visit to the bath cheered us up, and we felt pretty good afterward. Later, on the same day, the women were taken to the bathhouse, too.

On our way from the prison to the bathhouse we had an unexpected and very cheering encounter. Not far from the prison gates we were astonished to see Artak. In order not to give himself away he, of course, pretended that he didn't know us. However, we could tell that he was as pleased with the meeting as we were. We realized that he must have walked past the jail every day in hopes of catching a glimpse of us, and we were deeply moved by Artak's regard for us: we hadn't even known whether he had managed to escape the clutches of the police in Krasnovodsk, and here he was in Kizyl-Arvat, trying to find some way to help us.

To jump ahead a little, I wish to say that later on, after we had been transferred to Ashkhabad, I managed to meet with Artak in the prisoners' wing of the regional hospital, and it was then that he told me how he had managed to evade arrest in Krasnovodsk. He wrote about this in 1926 in his memoirs, which I feel might be interesting to quote here, if only partially:

"Since word has reached us in Astrakhan that they searched you very thoroughly, nearly stripping you, before letting you enter Krasnovodsk, we decided to hand over our money to someone on the crew," Artak recalled.

> We arranged with the captain's mate, a certain Adolf Rogachik, that he would hand over the packet to any one of the three of us [Abilov, Marusia Kramarenko, or myself, who pretended to be strangers on shipboard] who gave him the password of "Red (street) light."
>
> Some time after the twentieth of October we set sail for Krasnovodsk. At first everything went well, but then one night a party-member, Leon Arustumov, who was also traveling on our ship, came over to me, and indicating the Armenian who had visited Shaumian's family in Astrakhan, said: "Beware of him."
>
> This Armenian was a British counterespionage agent who had somehow managed to ferret out everything about our plans. After drinking a glass of home brew with Arustumov on the ship, he had become overly sentimental and revealed too much to him: "I know why that man (a nod in my direction) is traveling on this ship. I'm going to set him up in such a way that he'll have his head chopped off . . ." And then he had sketched out for Arustumov the general outline of my mission. It was too late to change the plan, however, and I decided to wait and see what happened next. As before, I acted like a stranger with Marusia and Abilov.
>
> The first order of business in Krasnovodsk was that we were met by a launch carrying a British counterespionage team. Our Armenian got off the ship and walked over to them, and the British agents greeted him as if they were old friends. It was obvious to me that we had failed. All the passengers were subjected to a search, and we were singled out from among the general mass and searched in an especially meticulous fashion, right down to ripping open the linings of our clothes. When they didn't find anything they took away our papers, released us, and imposed on us the duty of appearing the following day at the headquarters of the British command.
>
> Why weren't we arrested right away? Evidently, they wanted to ferret out our secret meeting place. Mine was with Sergei Martikian.
>
> I stayed that night at a seaport hotel, while Marusia and Abilov stayed at a place near the station, all three of us still acting like strangers. In the meantime I decided that it was imperative to get the money from the captain's mate, Rogachik, so I met him and asked him about the packet. He acted ill at ease, saying that he was afraid, but would give me the money in half an hour. After an hour he finally reappeared. He was again in a state of confusion, and again said that he was afraid to hand over the packet, but promised to do so tomorrow at twelve noon on the square, by the pier.

I didn't like it, but there was nothing I could do. I decided then to go see Sergei Martikian and see if he could tell me anything about the fate of our Baku commissars. Convinced that I was being shadowed, I decided to throw them off the trail. After strolling around the city for a while I entered a hotel through one door and went out through another door onto another street. After taking another turn through the streets I finally arrived at the meeting place with Sergei. When he saw me Sergo became very agitated. "Shaumian and Alyosha are not in Krasnovodsk," he said. "Possibly, they have been sent to India, or maybe they've been shot. In any case, there's no hope, and you should skip out of here as soon as possible. Don't go back for the money—it's probably a trap."

Sergo infected me with his mood of panic and my spirits fell. After leaving him I tried to think of a way to get out of Krasnovodsk. The Armenian who had left the ship with the counterespionage agents came over to me on the street then and told me frankly that he knew about my mission. He demanded money in exchange for his silence, otherwise, he would turn me in. I decided to stall for time and promised to give him the money the following evening at the Armenian church. On this note we parted, and I returned to my hotel room.

In the morning I met Marusia, and after I told her all that had happened to me the day before, we decided to try to get out of Krasnovodsk. I arranged with Marusia to wait for me at the pier, and then left to go to the ship. On the way I met a Bolshevik sailor whom I knew. He was from the ship *Africa*, and agreed to give us refuge on his ship, but when I told him that I was going to get the packet, and asked him whether he was still willing to give me refuge, he cowered in fear and began to back down. Nevertheless, I decided to go and get the packet, feeling that if it was a trap, I'd be able to buy myself off with part of the money. It was a fairly large sum of money for that time—ten thousand rubles in Nicholas II bank notes. Without this money it would be difficult to help those who had been arrested.

I left the sailor and walked over to the square. It was almost twelve-thirty by then, and the square was empty, neither the captain's mate nor Marusia were anywhere in sight. I was heading off in the direction of the bazaar when I suddenly saw Marusia being conducted down the street under escort.

I learned subsequently that Marusia, while waiting for me, had decided to go and get the packet herself, so as not to expose me to any risk. At the agreed-upon place she saw the captain's mate pacing back and forth, and she went over to him and said the password, but as soon as he handed her the packet she was suddenly grabbed by several armed men. She managed to throw the packet onto the window sill of the nearest building, but when they didn't find the money on her the counterespionage agents returned to the scene of the arrest and found it.

The mate, Adolf Rogachik, most likely got scared when he saw that we had been given away and thinking there were important papers in the packet, and fearing for his own life, he had turned us in.

Depressed by this turn of events, I set off to walk aimlessly about the streets. Following me like a shadow were the agents. I had no luck in finding Abilov, either, since as I later found out, he had already been arrested. In the end I managed to shake off the agents who were trailing me and doubling back to the pier, I lay down under a stack of wood, staying there until evening. I decided to leave that night, no matter what.

My overcoat was still at the hotel, however, and since I didn't want to lose it I decided to get it. On arriving at the hotel and seeing armed men standing at the entrance, I made a wide circle and went around to another door, but there were armed men there, too, and with them was that same Armenian double agent. It was obvious that they were after me. I abandoned the idea of recovering my overcoat and went to the station, where I met a stream of people coming away from the depot with their suitcases. When I asked what was going on I was told that some Bolshevik spies were somewhere inside the station and therefore no one was being admitted to either the station or the trains without a pass from the British command.

I made a sharp detour along the tracks to the freight cars, and through the cracks I began to observe what was going on at the station. Walking along the platform accompanied by armed patrols was the Armenian agent, looking everyone up and down. He was giving everyone a pretty thorough screening. When the first departure bell sounded I crawled out of the freight car and walked over to the train with the idea of making an arrangement with the engineer. I approached him and asked him to take me on the locomotive with him, making up a mythical ailing mother in Ashkhabad, explaining that I hadn't had the time to get a pass. He refused me point-blank.

Then I decided to try my luck with the conductors, I went over to one of them, gave him my tale of a sick mother, and offered him thirty rubles. But he knew that a check was being made of every car, and so he was afraid and he refused me too.

When the second bell sounded the inspectors moved into the car. I found another conductor and offered him three hundred rubles. This evidently alarmed him and he, too, refused. Meanwhile, an armed detachment of British soldiers was marching toward the train from the platform. The conductor went into the car then, and in despair I climbed in after him and immediately caught sight of a small ladder leading up to the roof of the car. I clambered quickly up it and lay down on the dark side of the train roof.

The detachment had cordoned off the train. Time dragged on until finally the third bell sounded, but the train didn't move. Evidently the search was still going on. A half-hour passed, and at last the train started. Once it picked up speed it went very fast the cold wind pierced me through and through. At the next station there was another search before the train was allowed to proceed. The farther we went, the colder I became, until my hands were so numb that I could barely hold on. It was only a matter of time until my benumbed hands would refuse to obey and I would fall off.

Finally, at the fourth station I climbed down from the roof and taking advantage of the darkness, slipped into the crowded car that was crammed with refugees. I was chilled to the bone and shivered so badly from the cold that one of the refugees, an Armenian, took notice of me. I told him that I was having an attack of malaria and asked him to buy me a ticket to Ashkhabad, saying that I had lost mine, and to my vast relief, he agreed to do so.

When the train pulled into Kizyl-Arvat I recalled that my former student Navasartian lived there. He had lodged with us in our house in Tiflis and I had tutored him for entrance into the theological seminary. After thinking it over quite carefully I decided to get off at Kizyl-Arvat and stay with Navasartian. I didn't know he address but I knew his father had a bakery there.

I got off the train and after skirting the depot headed down the street. I was morning by now and when I saw a line of people outside a small house I knew I had reached my goal, since in 1918 most lines were breadlines. I walked over and asked for Navasartian. "That's me!" answered a thickset man, laughing. I asked if his son was at home, and on receiving an affirmative reply I told him that I was his son's former teacher. My friend Navasartian received me cordially, and in the end I confided in him and asked his protection. He has rendered me great assistance.

* * *

On the day following my meeting with Artak on the street, another pleasant surprise awaited us at the jail. Using go-betweens from among the local population, Artak contrived to send me, and the women prisoners as well, parcels of delicious dishes of Armenian cuisine. Along with the food came short notes from Artak (unsigned, naturally) written in Aesopian language, giving us information of interest to us, and in turn asking several questions. We would answer him in the same way, hiding short notes in among the empty dishes. We were able to get away with it because the prison guard, an old veteran, was in general a quite nice fellow; polite in manner, and not overly suspicious, he either didn't notice our communications or pretended not to.

Later on Artak told us how hard it had been for him to earn

enough money in Kizyl-Arvat to support himself and also help us. He worked at his comrade's father's bakery, and also worked at odd jobs involving the sale of grapes and fruit at the bazaar. He said that he wasn't very good at these roles, which were so highly unsuited to him.

Despite our privations, no one, aside from Amirov, who, as I mentioned earlier, had been taken to the hospital, came down with anything during our internment in Krasnovodsk and Kizyl-Arvat until one day when young Leva Shaumian began to have pains in his leg. The fact is that about a year before when he had entered the Red Guards a school-friend had accidentally shot him in the leg with a pistol. Doctors had taken care of his foot, and it had appeared to be healed, but in prison the pains came back again and the leg began to swell. We asked that a doctor be sent to the jail. They sent a medical assistant instead, who after examining Leva in the cell, gave instructions to massage his swollen knee in order to make the swelling go down and to relieve the pain.

Not having any experience or knowledge of medicine I followed his instructions, and began to massage Leva's ailing knee diligently. It was very painful for him, and he moaned. "Never mind," I told him, "it's good for you, have patience." He put up with it, but even so, he moaned. Then the pain became stronger and he was no longer able to stand my massage, but even after I stopped giving them the pain became worse. The next morning we raised a fuss and demanded that a real doctor be sent to us. When a doctor finally appeared and examined Leva, we told him about the massage. The doctor looked at me reproachfully. "By massaging you've spread the pus in his knee all over his leg," he said. "It has become a general infection now, and the lad may have to have his leg amputated, but maybe even that won't save him. He has to be taken to a hospital immediately." Orderlies from the railway workers' hospital came and took Leva away at once.

We were very upset over Leva's sickness. We were also tormented—I more than anyone—by the knowledge that on the instructions of the ignorant medical assistant, Leva's health had been seriously damaged. However, we were soon relieved to hear, via Artak, that although Leva had a blood infection and gangrene had set in, his condition had improved somewhat. Nonetheless, all the doctors in Kizyl-Arvat thought that he needed a major operation which could only be performed in Ashkhabad. Accordingly, a short time later Leva was taken to the prison ward of the regional hospital.

Chapter II

In Ashkhabad

On the road again—The situation in the Transcaspian region in 1918–1919—Suren Shaumian—In the prison hospital—The excesses committed by the British occupation forces—My attempted escape—Return to Baku.

We had been in the prison at Kizyl-Arvat for about three weeks when suddenly early one morning we were told that we were to be taken to the prison in Ashkhabad. We had not been able to understand why they had transferred us from the Krasnovodsk house of detention to Kizyl-Arvat in the first place, and now suddenly we were going to Ashkhabad. We were immediately beset by the same old doubts and fears, suspecting that this was only a pretext, the same kind of smokescreen that had been used with the Baku commissars, as once again the thought seized us that they were either taking us to be shot or were going to ship us to India.

I recall one incident in connection with our journey from Kizyl-Arvat to Ashkhabad. Armenak Amirov, who as I have said was a passionate billiards player and a hard drinker, said to me, "You have forty rubles. Let's buy some vodka and get drunk. It will liven things up and we'll be feeling good when they shoot us." I refused, since I wasn't a vodka drinker, and said I'd be perfectly calm without it. He tried to persuade me but I insisted that I was saving the money for an emergency. Then Amirov started to reproach me, calling me a skinflint, a stingy fellow, and so forth. At first I felt insulted, then I got mad, and pulling the forty rubles out of my pocket, threw them at him with a curse: "Here take the money and buy yourself your damned vodka!" He gave the money to the guard, who returned with two bottles of home brew. Amirov gave one bottle to the guard in return for the favor, poured himself a glass from the other bottle, and drained it in one gulp. Then he offered me a glass, but I flatly refused it.

I should note that up until then I had never drunk either vodka, home brew, or cognac, and I had had wine only twice before—a glass or two on each occasion: the first time on my twentieth birthday, and the second a year later. Surprising as it may sound—a man from the Caucasus who had never tried wine before his twentieth year—it is absolutely true.

I continued to refuse Amirov's offer, but he kept insisting. "What's it going to cost you, have a drink!" he said. "Just drink it down in a single gulp!" Finally I gave in to his urgings and downed a glass of home brew. As it hit my empty stomach I felt on fire inside, but my head was still clear.

I remember how they led us out of the prison onto the street and took us through Kizyl-Arvat to the train station, where they put us in a freight car. The men were put on the wooden benches on one side of the car and the women on the other, with the armed escort-guards in the middle. After a whle the liquor gradually went to my head and I became very excited, until I suddenly lost control of myself and began to curse wildly at the escort guards, calling them scoundrels and executioners. Moreover, my comrades told me later, I even tried to attack the guards. I don't remember anything after that. My comrades pulled me back down on the bench and I fell asleep until we reached Ashkhabad. That was the one and only time in my life when I got so drunk that I lost control over myself. By the time they led us under convoy guard through the streets of Ashkhabad, I was sober again.

They took us, as they had promised, to the Ashkhabad regional prison. It was a large, single-storied prison, laid out in the form of the Greek letter *pi*, that could hold a hundred people or more. They led us through the single entrance in the center of the building into the right wing. The cells in which they put us were located near the latrine from which an unbelievable stench wafted through the corridors and penetrated even into our cell. I am highly sensitive to smells, and I immediately felt sick and then nauseated.

In fact, our new cells was worse in every respect than the ones in the Kizyl-Arvat and Krasnovodsk prisons. It was very small, damp, and dark; somewhere beneath the ceiling there was a narrow slit of a window which let in the tiniest sliver of light. It was wintertime and we had no heat: we spoke feelingly and often of our "savioress"—the stove in the Kizyl-Arvat house of detention. Since only three iron beds could be squeezed into this cell for the four of us, Suren and I slept on one bed. The bed was so narrow that when Suren was lying on his back I had to lie on my side, and vice versa. We got so used to this that even in our sleep we turned automatically when one of us assumed the other position. We covered ourselves with Suren's military greatcoat. The beds could be folded up against the wall and even locked away during the day, but fortunately when we were there the guards didn't do this. There were no pillows but there were mattresses, and we were able to sit and lie on them during the day. Between the beds stood a small wooden table.

The Ashkhabad regional prison had one advantage over our previous prisons in that here, once a day on a regular basis, we were led outside for an hour's walk with the "criminals." I must say the criminals treated us "politicals" with great respect; they understood that we were not in prison for selfish deeds, but for political ideas—"for the people," as they put it.

The guards in the Ashkhabad prison were Russians, but the high, almost fortress-like walls surrounding the prison were manned by soldiers in the Anglo-Indian army—sepoys. When we were led out for our walk, they looked down at us from the prison wall, holding their rifles and machine guns at the ready. The soldiers were extremely tense, as if they expected us at any moment to mount an attack which would necessitate their fiiring on us. They had obviously been told many bad things about the Bolsheviks. It was the first time in my life that I had seen sepoys. Almost none of them appeared to be young; they had well-fed, even fat faces, or so they seemed, possibly because of their thick, black beards and the white turbans on the heads.

During our exercise hour we walked back and forth in groups around the courtyard, paying no attention to the rifle and machine gun muzzles trained on us. The prison was run on a precise, strict schedule. At eight o'clock in the evening the guard locked up the cells and gave the keys to the prison warden; nobody could open the cells until seven o'clock in the morning. A traditional closed-stool stood in the corner of the cell, but no one in our cell ever used it; as young, healthy people we had become accustomed to controlling ourselves. At seven o'clock in the morning the cell door was opened and we were let out to go to the latrine. Prisoners from other cells were let out at the same time. The latrine, in front of which there was always a line, was very rarely cleaned, so that it emitted a horrible stench.

During the winter of 1918–1919, while we were in Transcaspian prisons, we had lively discussions about the bits and pieces of information that filtered through to us on the state of affairs in the outside world. When we were still in Krasnovodsk we had learned about Turkey's capitulation to the Entente, followed by that of Hungary and Germany, and then we had heard about the end of the World War in November 1918. Lenin's announcement on behalf of the Soviet government of the annulment of the Brest-Litovsk Peace Treaty brought us great joy; it was a brilliant testimony to Lenin's genius in foreseeing the course of events. The treaty which had been forced upon us and which Lenin had insisted on signing while at the same time considering it obscene, had been in force for less than a year. It had been signed in March of 1918, when we were still very weak, but it had allowed us a breathing space in which to gather our forces and to organize the Red Army. Now, however, the defeat of Germany brought about a change in the alignment of powers in the world. Using this treaty as an example, Lenin had given us a superb and graphic lesson in the ability to take advantage of contradictions within the enemy's camp in the interests of socialism. In discussing the annulment of the Brest-Litovsk Treaty we recalled the heated arguments among Bolsheviks on the treaty a year earlier and saw that it had been a political blunder to have been against signing it at that time.

One thing we particularly hungered for was news of what had happened in Baku after the Turks had left and the British occupation forces had arrived for the second time. But it wasn't until January of 1919, after we had been in Ashkhabad prison for a while, that we received detailed and trustworthy information about that from comrade Topuridze, who had arrived from Baku.

The situation on the fronts of Soviet Russia also provided heated discussions among us. We were happy about the reports that General Krasnov's troops, which had been moving toward Tsaritsyn, had been stopped at the walls of the city, and then pushed back and routed, but we were depressed by the news of events on the Volga and in the Urals, which were connected with the rebellion of the Czechoslovakian Corps and the formation of a counterrevolutionary government in Samara. We were especially saddened to hear the news of the counterrevolutionary uprisings in Siberia and in the Urals, of the seizure of the cities by Kolchak's men, and of the revolt of the Orenburg Cossacks.

* * *

When the SR government came to power in the Transcaspian region, the life of the people not only didn't improve as its leaders had promised, but noticeably deteriorated. The British occupation forces could hardly be called excessively generous in the financial assistance they gave their "ally"; on the contrary, they literally cleaned out the storehouses in Krasnovodsk, taking the cotton, leather, wool, and other raw materials designated for shipment to Soviet Russia and sent them instead to the Persian port of Enzeli.

General Malleson wrote: ". . . in August of 1918 I concluded an official agreement, according to which I promised, on behalf of my government, to assist the Transcaspian government as far as I was able with troops and weapons, so long as it defended itself to the last." First Malleson sent an artillery detachment. "A detachment has been sent to Merv," he wrote, "and has dealt the enemy there heavy losses. The enemy, however, using its manpower superiority, has continued its movement towards Ashkhabad."

The British sent their sepoy troops, including the "crack" nineteenth Punjab infantry regiment, against the Turkestani Red Army, but could not change the situation at the front. The Indians fought unwillingly; anti-British demonstrations were already taking place in India itself.

The situation within the Transcaspian region changed rapidly, and not to the advantage of the counterrevolutionary government. The SRs, who had managed in the beginning to attract a significant number of blue- and white-collar workers under the banner of "constituent assemblies" and the false slogans of "democracy," quickly revealed their true colors. When Funtikov and Sedykh's band of right-wing SRs came to power, it immediately arrested and shot several Soviet commissars and active members of the left-wing SRs without bothering about investigations or trials. Hundreds of other Bolsheviks and left-wing SRs were arrested, the majority of them were exiled, with the aid of the British, to Enzeli and Petrovsk where the counterrevolutionary authorities were in control.

Since the British feared a Soviet attack from the sea, the leaders of the SR government promised, according to Malleson, to divert all ships and rolling railroad stock from Krasnovodsk, to make "possible the mining of Krasnovodsk harbor, to destroy water towers and refueling stations along the railroad line, to make the bridges impassable, and, if necessary, to remove the rails from sections along the line. A British engineer undertook an examination of the entire railroad line in order to draw up a plan for impeding the enemy's use of the railroad in case of an attack."

The fiscal economy of the Transcaspian region was in a state of chaos. Pay was regularly withheld from soldiers and white-collar workers, and on their part the population, particularly in the Turkmen rural areas, tried to avoid paying taxes in every way possible. The people's dissatisfaction continued to intensify until finally the government's order early in January 1919 of a new mobilization for the front was the straw that broke the camel's back. Feeling that they still had some influence, the SR leaders broke with the others and called a political rally at the Ashkhabad railway depot, counting on the people's support in the mobilization issue. Instead, at that rally in which Communist railroad workers in the underground took an active part, the majority came out against a new mobilization and refused to support the government. Seeing the handwriting on the wall, the British occupation forces sent their troops to the depot and broke up the rally before it ended. In the beginning, the British occupation forces had tried to operate behind the front of the Transcaspian government, but now, flushed with their victory in the World War, they became impudent and started to play openly with the fate of the Transcaspian people.

Soon after we had been transferred to the Ashkhabad prison we learned from the other inmates, and also from the local papers, which in one way or another found their way to us, that a change of government had taken place in Ashkhabad.

The regional government made up of twelve directors—right-wing SRs and Kadet elements—had been overthrown. Funtikov, Sedykh, and several other members of this government had landed in prison. The whole thing was affected by the British Expeditionary Command. Even Funtikov's government was now no longer to the liking of the British: not all members of that government had agreed with the policy of cutting the Transcaspian region and Central Asia off from Russia and forming a kind of separatist state under the direct protection of England. This scheme was part of the general Anglo-American-Franco-Japanese plan to dismember Russia.

The so-called Directory was instituted in place of Funtikov's government. At first, General Kruten was named head of the Directory, which was also called "The Committee of Public Safety"; but a few days later he was replaced by Semyon Druzhkin, the actual executioner of the Baku commissars. Druzhkin was a British agent from way back. The Directory, which consisted of six men, included two Turkmenians, chosen from among the reactionary beys. It is characteristic that the British ignored the nomination of one bey who had been proposed by the Turkmen leaders, naming in his place Oraz-Serdar, who had commanded Turkmenian and other units which had fought against the Turkistani Bolsheviks. Serdar had obviously been bought off by the British.

Most of the arrested members of the old SR government, with Funtikov and Sedykh at their head, were not put in our prison, but sent to the Ashkhabad house of detention.

Alaniya, who had arrested us in Krasnovodsk, received a promotion at about that time: as a person trusted by the British, he was made chief of the Ashkhabad municipal police. He was the person ordered to arrest the former leaders of the Transcaspian government (Funtikov and the others), an assignment he dishcarged on January 15, 1919. Later it was revealed that the arrested members of the Transcaspian government who were assigned to the Ashkhabad house of detention were not the responsibility of the local Directory, but of the British Command.

A short time later, an arrested colonel, the chief of staff of the Transcaspian army, was brought to the prison. They put him in the cell next to ours. It was the smallest, darkest, and dampest cell.

One day the head of the Directory, Druzhkin, appeared in our cell accompanied by an escort guard and some other people we did not know. We asked him why we were still being held as prisoners and what were the charges against us. "When will you release us?" Suren Shaumian asked him straight out. Druzhkin replied, "We're holding you as hostages. If the Bolsheviks in Tashkent shoot the right-wing SRs they are keeping prisoner, then we will shoot you."

* * *

We comrades lived on very friendly terms in our cell. Our rations were meager, since we no longer received any food packages from friends or transient prisoners. In the morning and evening they gave us kasha made of some sort of groats, and for dinner we got soup made out of the legs and heads of cattle. We called it "hoof and head" soup. You can get an idea of the kind of soup it was from the fact that although we sometimes found a vein or two on the bones, we never saw a single piece of meat. Later we found out that the Persian inmates who delivered the food would fish the bones out of the soup with their dirty hands, pick all the bits of meat off them, and only then take the soup around to the cells. We heard this from the other inmates. They brought us both the kasha and the soup in a single iron basin that contained four portions, instead of giving us individual bowls. For breakfast, dinner, and supper, they also brought a kettle full of boiling water and a few raisins. They gave us bread three times a day, but in limited quantities: we never had enough.

Once a scandal occurred in our cell completely unexpectedly. Suren and I were sitting silently on our bed, and Armenak Amirov and Samson Kandelaki were arguing about something. At first the quarrel was peaceable enough, but then the voices, Armenak's in particular, began to get louder, and a fight broke out. I wedged myself between the combatants, with the result that they rained more blows on me than on each other until Suren and I managed to separate them. The guard heard the noise and called the warden, who came into our cell and asked: "What's going on here, gentlemen?"

"It was just a noisy quarrel and nothing more," I said. "There's no reason for you to interfere."

The guard, however, insisted that there had been a fight in addition to the noise. The warden then demanded to know in no uncertain terms who was responsible for the fight, stating that the culprit would have to be put in another cell. I didn't know who had started the fight but I did know that Samson was a very restrained man, while Armenak was very much a bully and would certainly have been the first to pick a fight. Therefore, since I saw no way out of the situation, and realizing that it made no difference which of the two we would be forced to part with, I said that it was Armenak. Visibly insulted, he left the cell silently, without saying good-bye to us. When everyone had calmed down, Samson said that the reason for the fight had been a trivial quarrel and that it had got out of hand because of Armenak's hot temper.

I had already heard from other experienced prisoners that when people are cooped up with each other in a single cell for long periods of time quarrels often arise between them, with the result that even good friendships are ruined. During the time we spent in various prisons, the above incident was the only one of its kind. No quarrels ever arose between Samson Kandelaki, Leva and Suren Shaumian, and me, none of us ever raised his voice against another.

Samson Kandelaki was a very nice man. He was older than I and more experienced, with a longer record of service in the party. Heavy-set, although not fat, he was even-tempered and lacking in the usual Caucasian impulsiveness, although if provoked he could show his "Caucasian character" as he had in the incident with Armenak. He told us many interesting stories from his life, and we listened with the curiosity of the young to a more experienced, older comrade. He wasn't talkative, but he understood the value of humor and often had recourse to it himself, so that he wasn't offended when one of us poked friendly fun at him, nor were we offended by him.

One evening we were sitting on our respective cots, each thinking his own thoughts. I was thinking of my fiancée, wondering where she was, what she was doing just then, and how she had taken the news of our deaths. Involuntarily I fell into a reverie, then glanced at Samson and for no reason I can think of, the mischievous idea of teasing suddenly popped into my head. I said: "Of course, I realize, Samson, what a fix you're in. Your wife remained in Baku, but you're here in prison, and it's almost five months since she's had any news from you. She probably thinks you're dead, like our other comrades. Under the circumstances she's hardly going to keep on waiting for you." Samson realized I was joking, and smiling good-naturedly, said, "You've missed your mark, Anastas. That doesn't apply to me and my wife. We've been together for fourteen years, we have two children, and she loves me and will wait for me to come back." I continued to tease him, however. "Well, go ahead and calm your fears, if it makes you feel better," I said. "I'm better off than you are—I'm a bachelor, completely free."

How could I have known then that I would soon come to regret my joke? This is what happened.

A couple of days after we had returned to Baku after our release from prison and had become immersed again in party work, I realized that I hadn't seen Samson since our arrival. Since this absence from work was unlike him I was afraid that something must have happened to him and asked my comrades to try to find him. Two days later they unearthed him in a Georgian restaurant, drunk, and they brought him to my secret apartment. He looked awful—unshaven, exhausted, depressed, with red, swollen eyes.

"Samson, I don't understand you," I said curtly. "I've always had the highest opinion of you, but you've turned out to be an alcoholic rather than a revolutionary."

"Don't say such stupid things about me!" Samson snapped.

"Then what's wrong? Tell me what happened?" I said.

"Here's how it is," he replied. "I went home to my apartment, and saw my wife through the window sitting next to some man I didn't know. I went into the room; my wife, at a loss, jumped up. Seeing how upset and embarrassed she was, I asked her what was going on, who was this man? My wife answered in tears that she thought I was dead. 'We have two children,' she said. 'How could I bring them up alone? I met this nice fellow here. He gives me money to feed them. If I had thought for one minute that you were alive, I would never have become involved with anyone. Everyone said you had been shot.' I didn't say a word," Samson continued, "didn't even say hello to the children, just slammed out of the house and began drinking. I've been like this for three days now, spending my days and nights in a restaurant owned by a Georgian friend. I can't go home now, even though I love my children very much. My family life is shot to hell."

I immediately recalled my conversation with him in prison, and only then did I realize how thoughtlessly and cruelly I had acted toward my comrade. I still can't forgive myself my boyish prank!

I tried to calm Samson down, but seeing that I was getting nowhere, I changed tactics. "If that's the way things have turned out, Samson, there's only one cure," I said. "You have to put your whole mind into your work and forget everything else—otherwise, you'll become a drunkard and then you'll be useless to the party." To this Samson replied that he couldn't work in Baku: here, next door to his children and the wife who had betrayed him. I said that I understood, and advised him to go to Lenkoran.

At that time the Communists there, with the support of Azerbaijanian and Russian peasants, were attempting to take power into their own hands, and they had great need of strong cadres with leadership ability. I explained this to Samson. "If you go to Lenkoran," I told him, "you'll be in work up to your ears. You're a brave man, and courage is essential there."

Samson's face seemed to undergo a change as he listened to me, and after staring at me in silence for a moment, he agreed to go. He left for Lenkoran the next morning, after spending the night at my apartment.

A Soviet republic was soon formed in Lenkoran. We considered it our stronghold in the coastal area, and with the reopening of navigation we felt confident that the Soviet fleet would be able to bring troops there from Astrakhan. But in the end no military units were disembarked in Lenkoran, for the opposing forces turned out to be too strong. By the time Soviet power fell in Lenkoran, after having held out for several months surrounded by White Guard and Muscavatist counterrevolutionary bands, Samson Kandelaki had become one of the most well known leaders in the Lenkoran Republic. Along with several other Communists in the republic he perished heroically at the hands of the counterrevolutionaries.

* * *

During the time we spent in prison many different kinds of people passed through our cell. Some left a good memory behind them, while others were the sort you want to forget. But I shall always retain the fondest memories of the children of my friend, Shaumian.

While sharing a single, narrow iron bed with Suren Shaumian, I often thought what a remarkable fellow he was. Although he was young in years he conducted himself at seventeen like a mature person. How courageously he had taken the news of his beloved father's death, showing, in spite of his youth, exceptional courage and self-possession; he hadn't shed a single tear. I recalled that night when they took us to the same place they had taken the Baku commissars and we had lain next to each other on the damp floor of the railroad car, convinced that they were taking us to be shot, too. Suren and Leva had been prepared to share the fate of their father: their calm never left them. They endured the deprivations of prison life with great dignity, without whining or complaining.

Toward his comrades, Suren, the elder of the two brothers, conducted himself like an equal among equals and won their love and respect. His resemblance to his father was not only external, his manly calm and self-possession never left him, even in his most difficult moments. They were more than traits of his personality; they constituted his very essence.

While still a fifteen-year-old schoolboy, soon after Dzhaparidze's return from the Sixth Party Congress, Suren had actively participated in organizing the first Union of Soviet Youth with a Communist orientation in Baku. Several months before we wound up in prison Suren had already carried out important party assignments which his father had given him. In June of 1918, on his father's orders, Suren had delivered important documents to Lenin in Moscow, traveling via Astrakhan and Tsaritsyn. In view of the decrepit state of transportation and the dangers that threatened the mission every step of the way, the journey was a manifestation of real courage.

Suren's unforgettable meeting with Lenin was one of his most cherished memories. Lenin had received him, in his working office in the Kremlin, and after reading Shaumian's letter and questioning him in detail about the situation in Baku, had forthwith ordered the War Department to set aside the arms and materiel for Baku that was so vital for the defense of the city against the imperialist invaders. After the official part of their conversation was concluded, Vladimir Ilyich asked him how old he was. And when Suren replied that he was sixteen, Vladimir Ilyich smiled and remarked to one of his associates (V. D. Bonch-Bruyevich, I think), "You see what kind of children have already been born of our revolution! Soon we'll have grandchildren like this!"

On his return trip from Moscow Suren visited Stalin in Tsaritsyn with commissions from his father. It was through Stalin that the Baku Sovnarkom maintained contact with the government of the RSFSR. Suren and the comrades who had accompanied him, including Artak, Maro Tumanian, Leva, and others, arrived in Baku with the arms and materiel they had received from Lenin a few days before the temporary fall of Soviet power in the city.

My younger brother, Artem Ivanovich Mikoyan, also had very fond memories of Suren Shaumian. During his military service in the twenties, Artem Ivanovich was the Komsomol organizer in the Frunze Infantry Training School in Orel, where Suren was in charge. Suren was greatly loved and respected by students and commanders alike.

Sergo Ordzhonikidze also enjoyed warm relations with Suren. Sergo had made a trip to Baku in 1911 (at which time our party was illegal) to pick up some important party documents from Stepan Shaumian. Stepan called Suren and asked him to bring in the papers, and Suren went into the yard behind the house, dug up the papers which had been hidden in the ground, and brought them to his father. "That nine-year-old boy was already a revolutionary conspirator," Sergo said at the end of his story.

Suren Shaumian combined deep ideological conviction, spiritual lucidity, and unshakable principles with an exceptional responsiveness and goodwill toward all comrades who needed his support. His pleasant voice, the softly humorous speech so characteristic of him, are still alive in my memory. He possessed an outstanding mind, a businesslike character, and extraordinary organizational abilities. These qualities were developed further in the military, which he entered during the Civil War. His native inventiveness manifested itself in the way he managed to get to Moscow in the spring of 1919 after his release from prison.

In order to get out of the Menshevik port of Poti to Soviet-controlled Crimea and from there to Moscow, Suren was first forced to spend some time in Turkey. Disguising himself as the traveling companion of a certain "merchant," he and this "merchant" set out for Istanbul in a wretched little boat, ostensibly to sell gasoline which, as a matter of fact, the Bolsheviks in the Caucasus had asked them to bring to the Crimea for the Red Army. The boat's engine broke down while they were at sea and after hoisting their sail in the midst of a storm, they managed to make their way to the Turkish port of Sinope. There they had the engine fixed and they cut across the Black Sea toward Sevastopol, but Sevastopol was already occupied by the Whites, so they turned in the direction of Evpatorium. Still masquerading as merchants, they disembarked in Evpatorium and found the Whites were there, too, but Suren cleverly played the role of a businessman and anti-Bolshevik. Then he managed to make his way through Kharkov, which was also occupied by the Whites, and then on to Belgorod. There he bought himself some lieutenant's shoulder straps and disguised as a White officer, he headed for the front. To make his disguise more convincing he didn't shave and let his beard grow. He succeeded in getting across the front lines safely by hiding in a wagonload of hay, and a few days later he turned up in Moscow. It's true that this epic journey took a rather long time: Suren left Poti on May 9 and arrived in Moscow four months later, on September 16, 1919.

Suren described this journey very vividly in a small book published by Young Guard Press in 1930, entitled *Bolshevik Contraband*. I reread it not long ago, and I must say that it is very interesting even today, especially for young people.

In Moscow Suren entered the military academy and upon graduation was commissioned as chief of staff of the Azerbaijanian division. The commander of this division, Dzhamshid Nakhichevansky, was an old cadre officer in the Russian Army, of whom Suren became a great and trusted friend. Later Suren became the head of the Orlov Infantry Training School, which he soon reorganized into the first tank-training school in the USSR. Then he was transferred to the position of chief of armored tank divisions, first of the Byelorussian, then of the Leningrad military districts which had great strategic importance at that time. He became one of the outstanding commanders of the Red Army's armored-tank corps.

Brigade Commissar Suren Shaumian died in 1936 after a serious illness, at the age of thirty-four. Suren is an old Armenian name which means "military leader." It was a fitting name for our Suren.

* * *

During my stay in the Ashkhabad Regional Prison my gum disease became worse due to the absence of fruits and vegetables in the prison diet. Samson suggested that I ask to be transferred to the hospital. A day later my request was granted and I found myself in the regional hospital.

The prisoners' wing of the hospital did not resemble a prison at all. It was a typical hospital ward, relatively clean, bright, and warm, except for the bars on the windows. The change of environment in itself made me feel better, but the main thing was that for the first time in many months I washed with soap in a real bath, changed my dirty linen, and put on the clean hospital garments. They took my uniform away to be disinfected. After examining me carefully the hospital doctor had no trouble in diagnosing scurvy.

I found out that Leva Shaumian was in the same hospital as I, only in the general ward. The escort guards in the prison part of the hospital were very decent people (some of them were even refugees from Baku). They didn't keep too close a watch over me and even allowed me to visit Leva, as long as they went with me.

I found Leva in good spirits. He was the object of tender concern of the doctors and nurses; they had grown very fond of him there, and were doing everything possible to make him well again. The man most reponsible for his recovery was Dr. Petrov from Baku, whom I ran into some forty years later in Erevan where he was a professor and the director of an orthopedic institute. He performed several operations and got rid of the incipient gangrene which made it possible to avoid amputating the leg. In a short time the wounds started to heal and Leva began walking on crutches.

Leva and I were overjoyed to see each other again. He told me that they had taken him to the Ashkhabad regional hospital on the same train, but in a different car, on which they had transported us from Kizyl-Arvat to Ashkhabad. He also told me that Artak Stamboltsian had come to Ashkhabad from Kizyl-Arvat and had already visited him in the hospital, bringing him food.

Two days later Artak came to the hospital and was allowed to see me. We talked for a long time, during which he told me how he had made his way from Astrakhan to Krasnovodsk, from there to Kizyl-Arvat, and had now come to Ashkhabad after finding out that we had been transferred there. He survived by working at odd jobs, and had no contact with Baku. He gave me all the news that was current in Ashkhabad or that had been reported in the local yellow press. I was most interested in news about the situation in Baku and in the Caucasus, as well as in the Soviet Turkestan Republic and the Transcaspian region.

A few days later Artak came again, this time with Topuridze, a worker in the Baku Party Committee whom I knew well. He was a remarkable conspiratorial worker with the ability to organize the complicated, technical side of party business; arranging communications and the locations of secret rendezvous; obtaining illegal literature and carrying out other party functions which were very important at that time. He had been ordered to come to Ashkhabad by the already rebuilt Baku committee to ascertain our current situation and to give me their order to escape from prison and return to Baku. I learned from him that the Baku workers had moved sharply to the left and that the majority of them sympathized with the Bolsheviks. There were, however, few experienced leading party workers in Baku at that time and therefore the Menshevik leaders were still at the head of the workers' organizations.

Topuridze brought us thirty thousand rubles from the Baku Party Committee to be used on behalf of our imprisoned comrades and for organizing our escape, and he, Artak, and I discussed a plan for giving financial assistance to the men and women in prison. Topuridze also gave me a large number of Baku newspapers and supplements which I had sorely missed. After he left I avidly attacked those papers.

In view of what Topuridze had said, it was clear that whereas the British had come to Baku the first time "by invitation" and had operated behind the front of the "Tsentrokaspian Dictatorship," refraining from direct interference in internal affairs, the second time they entered Baku as genuine occupiers and victors. Typical in this regard is the first "Proclamation" issued by the Commander of the Army, General V. M. Thompson, on November 17, 1918—the day the British forces arrived by ship from the Persian port of Enzeli and entered Baku.

In this "Proclamation" General Thompson again tried in every possible way to mask the aggressive character of the British troops' presence in Baku. In an attempt to appear as the humane "savior" of the Russian Caucasus, he says that in occupying Baku he had received the consent of the "the new Russian government" (i.e., one assumes he means the puppet government of the White Guard Denikin) and of Turkey, although the whole world knew very well that Turkey had never had any legal rights to Baku.

I think it would be interesting to quote the text of Thompson's treacherous and false "Proclamation" which, it is easy to see, was calculated to appeal to the gullible inhabitants of the city:

> In entering Baku as the representative of the Allies, I make the following general announcements:
>
> Baku is being occupied by the forces of Great Britain on behalf of the Allies. I have been accompanied by representatives of France and the United States, and we are here with the knowledge and full consent of the new Russian government. In the armistice it concluded, Turkey also gave its consent to our occupation of Baku.
>
> In our hour of triumph we have not forgotten the great service given by the Russian people to the Allied cause during the first years of the World War. The Allies cannot return home until they have reinstated order in Russia and given her the chance to occupy her place among the other nations of the world.
>
> Peace has still not come to the Caucasus: my duty is to restore calm to the Baku area. The current devastation is entirely the work of our enemies. . . .
>
> There can be no question of the fact that the Allies have any intention of abandoning a single inch of Russian territory. Of this they have given their solemn promise to the Russian people.
>
> By the same token, the internal government of the country, or any part of it, is the exclusive affair of the Russian people, and the Allies will not interfere in it in any eventuality. We come to you with only one purpose: to reinstate order by removing the German and

Turkish hotbeds of discontent that are hindering the establishment of order and legality. Our treatment of all races and creeds will be absolutely equal. Provisional local and municipal administrative authorities shall be established by me, and all our efforts will be directed entirely to the reinstitution of law and order. In this matter I look confidently for the support of the sober majority of citizens, and suggest that they return to their responsibilities and fulfill their civic duty by upholding public order. For my part, the edicts I shall issue will place as few burdens as possible on the population.

In counting on the cooperation of all reasonable and sober-minded citizens of the city of Baku, I hope to accomplish as quickly and easily as possible the task levied upon me and my troops so that we can soon return to our homeland in the knowledge that we have helped Russia to take advantage of the fruits of victory over our common enemies.

The following necessary measures are in effect as of now:

1. Persons residing in the city, and therefore under the protection of the army, are obliged to hand in to the authorities all rifles and revolvers in their possession, at a time and place to be announced.

2. A reward will be given for turning in to the authorities any Turkish or German subjects discovered in any district of the city that has been purged by Turkish troops.

3. The city is under martial law, which will remain in effect until such time as the civilian authorities are sufficiently strong to be able to relieve the army of maintaining public order.

As is obvious from this "Proclamation," General Thompson did not consider it necessary to say on the basis of what laws governing international rights he, as an alien on foreign territory, could assume command there and replace the existing local administrative institutions (albeit "provisionally") with his own. He could hardly not have known that a national bourgeois government and municipal authorities existed in Baku at that time. The English general ignored this entirely, and Khankhoysky's Muscavatist government, which had grown used to submitting to the Turkish aggressors, merely grovelled before them in every possible way.

The following examples of "unburdensome" edicts issued by the British authorities for the benefit of the local population testify to the kind of "order" that General Thompson instituted in occupied Baku.

Those guilty of not handing in their weapons within the three-day period set forth in Thompson's orders were subject to: "a) corporal punishment; b) imprisonment; c) fines, or a combination of the above." Later, Thompson announced in his edicts that "meetings of more than ten people for whatever reason . . . are forbidden without express written permission. This also holds for street gatherings, which are unconditionally forbidden." People guilty of abusing the established rules were immediately subject to the most severe punishment. "Every strike committee member, rebel, or person guilty of inciting strikes or disturbances" was to be subject to such punishment.

In addition, the general declared: "Any person who does anything which I consider detrimental to the order, peace, or security of the citizens" shall be subject to the same punishment, and "more serious crimes shall be investigated by someone authorized by me to carry out swift and severe punishments, or by a special court, authorized to hand down the death sentence. . . ."

Here is where General Thompson revealed the true colors of a colonial gendarme. Unable to restrain himself for even two days after his arrival in Baku, and breaking all his promises not to interfere in internal affairs, Thompson came out in the local paper with a call to resist the Bolsheviks. "I summon you all . . ." he wrote, "to fight against bolshevism in all its manifestations . . . The Allies are the enemies of bolshevism!"

All these "orders," "threats," and "summonses" made it possible for the Baku workers to convince themselves at first-hand what "civilized" England meant in practice; to feel on their own backs the cruelty of a lawless and most capricious colonial regime.

The Baku proletariat, with its glorious revolutionary traditions, had gone through many economic and political strikes, had created its own workers' organizations in industry, and had frequently held mass meetings and demonstrations even during the worst periods of Tsarist autocracy, now found itself for the first time in a situation where even eleven workers could not meet and discuss their affairs without special permission. The entire Baku proletariat therefore immediately rose up in protest against these rules, and in spite of the Draconian negative sanctions, the workers of Baku continued to gather at their (separate) industrial plants and discuss the current state of affairs. The occupation authorities started arresting many of them for arranging or participating in meetings. This only added fuel to the fire and embittered the workers against the aggressors.

In December 1918 the British occupiers constructed gallows on the central square of Baku in order to frighten the population, and in the middle of the month they arrested a large group of well-known trade-union workers and political activists and threw them in jail. The Workers' Conference of factory and plant committee representatives met and decided to call a general political strike in protest. On December 24 workers from the oil fields, the oil-refining plants, and many other industries joined the strike.

Neither the Muscavatist government nor the British occupation authorities had expected decisive action by the Baku proletariat: they thought the Baku workers would be crippled without their experienced leaders. The strike, which had acquired a militant character, intensified and spread, lasting for more than four days. In the end the British command was forced to back down; they recalled the activists in the workers' movement who had been sent to Enzeli, and released the political prisoners from the Baku jails, and from then on they ceased to interfere with workers' meetings. It is ironic that the prisoners included SRs and Mensheviks, who were, for the most part, the leaders of the workers' organizations at that time, that is, the representatives of those parties who had invited the British to Baku as "saviors" only three or four months before! The Bolsheviks who had been under arrest included the young trade union activist Levon Mirzoyan, the somewhat older worker Anashkin, and a few others.

The victory achieved by the general political strike inspired the working masses and gave them faith in their own strength. The workers had in fact won freedom of assembly and the right to be active in their own organizations. As a result of the strike the Workers' Conference acquired great political authority and became the permanently functioning representative organ of the working class in Baku, which the British command and the Azerbaijanian bourgeois government had to take into account.

Immediately after their arrival in Baku, the British seized all naval vessels and announced that "all merchant vessels on the Caspian Sea and their crews are temporarily placed under the authority of the British military command." They also announced that Azerbaijanian's major natural resource—oil—was "under the jurisdiction of the British command," and a British administrative authority on oil was named.

Topuridze told us all of this, and it was corroborated in the newspapers he brought us. He also told us that the Baku Communists had brought a demand before the Workers' Conference for releasing the family members of the deceased Baku commissars, along with the other imprisoned Baku Bolsheviks from the Transcaspian regional prison, and that the Workers' Conference in its turn had brought this demand before the British command.

* * *

After the prison escort guard had looked me over and become used to me, one of the guards, taking a chance on trusting me, told me in private that he was a Communist. He was a Lett by the name of Otto Lidak. It turned out that he had been on the staff of the Red Army when taken prisoner by the Whites during a battle at the front. Not knowing that he was a Communist, they had brought him here to do garrison duty, and that's how he had become an escort guard. The clear gaze of Lidak's blue eyes, with their compelling honesty, immediately made me feel well disposed toward him, and his declaration of being a Communist made me very happy.* I asked him about Communist organizations in Ashkhabad, and whether he was in contact with them. He said that he didn't know of any such organizations. Then I asked him, as well as Artak and Topuridze, to try to get in touch at all costs with the underground Communist organizations in Ashkhabad, because I could not believe that such organizations did not exist.

Here is what Artak wrote about this in 1926 in his reminiscences:

> ... Mikoyan decided to gather the Ashkhabad organizations together. He told a certain comrade by the name of Otto Lidak to initiate the first organizational meeting. Lidak later was one of the organizers of and participants in the 1919 Lenkoran rebellion. I was told to maintain contact between the organization and Mikoyan. I had occasion to attend the first organizational meeting that took place in a sausage factory belonging to one of our comrades, who later was an active club worker in Baku. Unfortunately, our Ashkhabad comrades could not expand their work since the organization was in a state of disarray.

Meanwhile, I decided to carry out the instructions of the Baku Party Committee and make my escape. Such an escape was impossible from the prison, but among the guards in the hospital, besides Lidak, there was an Armenian soldier who had served in those units where I had been a commissar, and who knew and liked me, which made the hospital seem the more likely place to effect an escape. One day when Otto Lidak was on duty I asked him for permission to go to the municipal library without an escort. Lidak agreed. Of course, he was taking a great risk even though I promised him not to escape, but he trusted me.

At the library, to my joy and surprise, I happened on two books: one was called *The Trial of the 193* (revolutionaries and Narodniks [Populists] whose trial took place in 1878); and the other concerned the trial connected with the assassination of Alexander II (on March 1, 1881). Both books were official transcripts of court trials, and the one on the trial of the tsar's assassins made an overwhelming impression on me. I had already heard and read about Zheliabov, Kibalchich, and Perovskaya, but I hadn't known how courageously and heroically they

---

*Much later, after we were back in Baku, Lidak succeeded in getting out of the Transcaspian region and came to see me. Since he had had some military experience I suggested that he go to the Lenkoran Republic, and told him about the complicated and dangerous situation there. He agreed to go without hesitation. Later we heard that he had been elected secretary of the Lenkoran Municipal Party Committee. After the fall of Soviet power there, he and a group of comrades made their way in a small boat to the Transcaspian shore, south of Krasnovodsk. They crossed the Turkmen deserts and reached the vanguard of the Red Army, which had by that time occupied Nebit-Dag and were pushing through towards Krasnovodsk to complete the purge of the counterrevolution in the Transcaspian region.

had conducted themselves during their trial. And yet even a reading of the dry and confused transcripts of the Tsarist court recorders, who often clearly distorted reality in the interests of Tsarist "justice," allowed me to visualize the noble images of those revolutionaries, true giants of courage, revolutionary ardor, and faith in the cause of liberating the people, for whom they went to their deaths, heads held high.

I was moved and inspired to the depths of my soul by the nobility of Zheliabov, who not only conducted himself irreproachably at the trial, but had been tried of his own free will. The fact of the matter was that Zheliabov had been arrested several days before the assassination of the tsar and had no connection with the case. But when he found out that the person who had committed the straightforward terrorist act was facing trial and would most certainly be condemned to death, he demanded of the prosecutor that "he, Zheliabov, be included in the case of March first." In this way he expressed his solidarity with the assassins, declaring that he had previously made an attempt on the tsar's life and had not been with his comrades on March first due to a mere "quirk of fate." Therefore, Zheliabov wrote, he feared that for lack of formal evidence against him the court would fail to find him, a veteran revolutionary, guilty in this affair.

From the standpoint of revolutionary expediency, Zheliabov's behavior could be considered foolhardy. But in his view he could not have acted differently: for, if he had, a young man would have been condemned for carrying out his, Zheliabov's orders, which had been given to him on behalf of the People's Will party, while he, Zheliabov, would remain outside the legal process and court trial, at which he was, moreover, in a better position than the others to defend his party's line.

Nowadays every schoolboy knows that the Populists chose the wrong path to achieve their revolutionary goal of liberating the people, the path of individual terror. But the greatness of their souls, their all-embracing feeling of militant comradeliness, their ideological principles, their readiness to devote themselves to the cause of revolution, and to sacrifice themselves for the goal they served, could not help but evoke in me a feeling of deference toward them.

That book stayed with me for a long time. It seemed to me that it was just such revolutionaries—stripped, of course, of the Utopian illusions of populism and armed with Marxist principles—that Lenin had in mind when he wrote his famous words in *What Is to Be Done?*: "Give us an organization of revolutionaries—and we will transform Russia!" Lenin's dream in fact came true: the party of all-Russian revolutionaries that he nurtured accomplished the Great October Revolution.

When I remembered Shaumian and our other fallen Baku comrades, it seemed to me that they were, in fact, latter-day Zheliabovs who had mastered the laws of social development revealed by Marx and Engels and armed themselves with Lenin's revolutionary theory of the struggle for communism. I would like to mention in passing that both the film *Sofia Perovskaya* and the stage show, *Members of the People's Will* at the Sovremennik Theater, in my opinion, recreate the images of the populist-revolutionaries truthfully and in an emotionally satisfying way. Those images have been alive in my memory since youth. At that time I was young; one should not forget, moreover, that I read about them when I myself was a prisoner.

One day Topuridze came to visit me in the hospital and reported that Artak had come down with typhus and was in bed at the apartment of friends. Luckily, Otto Lidak was on duty in the ward that day and he agreed to let me go to see Artak. Although Artak had a high temperature, he was glad to see Topuridze and me and tried to raise our spirits by saying there was nothing terribly wrong with him. I arranged with Topuridze to see to it that a doctor visit Artak regularly and treat him at home. This was the second time I had left the prison wing of the hospital to go into the city, and I hurried to get back to the hospital, fearing a bed-check: I didn't want to get Lidak into trouble.

* * *

One day, after he had recovered, Artak came to the hospital and told me that a possible escape route had finally been devised. Through some local merchant he had located a Turkmen-bey who agreed, for a certain sum, to transport me illegally over the border into Persia, from whence I was to make my way to Baku along the Caspian shoreline. The possibility of escape lifted my spirits and my mood improved, although I knew that there would still be many obstacles on the path to freedom, and that danger lurked everywhere.

We decided on the route through Persia rather than on one through Krasnovodsk for the following reasons: It took about twenty-four hours to get to Krasnovodsk by train, and besides, it was hard to find a place to hide in Krasnovodsk if the necessity to do so arose, and it was even more difficult to get out of the port, which was crawling with guards. We counted on the fact that as soon as the authorities learned about my escape, they would immediately report it throughout the railway network and at the port. The Persian border, on the other hand, could be crossed on horseback in a few hours, during which time the local authorities would still not have managed to collect themselves.

The day of the escape was fixed to coincide with one of Otto Lidak's duty periods. It was arranged that he would flee with me. I was so overcome by the joyful prospect of imminent release that I couldn't stay calm for a minute. Then, the day before my planned escape, an escort guard came to the hospital for me in the evening, and I was taken back to prison. We could not figure out who had given our escape plan away. It's possible that no one did, and that the doctors, seeing that my health had improved, simply decided to release me from the hospital and return me to my prison cell.

With Topuridze's appearance in Ashkhabad our lives in prison improved significantly, thanks to regular food parcels, books, and local newspapers. In addition, we received notes from him and contrived to send him answers, managing also through him to establish contact with our female comrades, who were being held in the women's wing of the prison. Best of all, he told us that our comrades in Baku had intensified their struggle for our release. One memorable day word came that they had succeeded: on February 27, 1919, we were informed that we were being released from prison and would be sent to Baku. Our joy at this news was boundless.

* * *

The first day of our release from prison was something of an anticlimax, as we had to wait about twenty-four hours for a train, and we didn't know where to go. But the indefatigable Artak came to our rescue by arranging with an Armenian family, the Melkumovs, to put us up for the night. The Melkumovs had a small private cottage with a garden, typical of those owned by civil servants or merchants in those days. The lady of the house was very cordial and did everything possible to make us comfortable. After making sure that we all washed up, she served us delicious Armenian dishes—pilaf, *dolma, shashlyk*—that seemed to us the best food we had ever eaten. After dinner we slept peacefully and for a long time. The next day we boarded the train without incident.

We made the journey from Ashkhabad in a jubilant frame of mind, happy that we had left our trials and tribulations behind us. We men were particularly fond of the steadfastness our women had shown through the difficult trials they had undergone. The wives of Dzhaparidze and Fioletov, and Maro Tumanian won our special admiration. We knew that when Dzhaparidze's wife had been informed of her husband's death, she had been so spiritually shaken that for a time her mind broke down. She found the strength to recover, however, and on the train Varvara Mikhailovna Dzhaparidze was immersed in thoughts of seeing her daughters Elena and Liutsia as soon as possible. Satenik Martikian, too, dreamed of her reunion with her children Tatul, Azat, and Emma. Her husband, a close friend of Shaumian, was with the children somewhere in the Caucasus; he had managed by chance to slip through the net in Krasnovodsk where the rest of us had been captured.

We naturally talked a lot on the train journey with our women comrades, listening to their stories of how they survived in prison and telling them our own stories. But we also had time for more profound thoughts and reflections too. Only five months had passed, yet how much we had been through! And yet the difficulties we had overcome during that time somehow seemed to have lost their pungency and were already fading into a kind of pleasant memory of the past. It made us happy to know that we had had the courage to endure without losing one iota of party conscience and integrity. We felt ready to withstand any new trial and, in fact, we literally thirsted for revolutionary work.

I especially appreciated the enormous scope of the political work that awaited us in Baku, foreseeing its complexities, and realizing that it was up to us now to come to grips with these problems, and to work out the tactics for solving them on our own. At times I experienced doubts about our ability to handle things. After all, Baku was cut off from Moscow for the time being with no possibility of getting advice and instructions from the Party Central Committee. But my doubts passed; the solid Bolshevik traditions which we had inherited from our elder comrades, and our own limited, but nevertheless thorough experience in the revolutionary struggle reassured me. The main thing was that the Baku proletariat had itself passed through the difficult school of experience and had learned from its mistakes. By relying on the collective wisdom of the Bolshevik organization and the help of the working masses, one could count on success.

The journey home seemed to take forever. We wanted to plunge headfirst into the thick of the revolutionary struggle as quickly as possible. We had no escort guard en route; at least none was visible. But when we arrived in Krasnovodsk an escort guard appeared to take us down to the pier in the harbor, and after we had made our way up the gangplank to the ship, we found ourselves surrounded by British guards. Several soldiers with their carbines in a horizontal position—in complete fighting readiness, as they say—stood around on deck, but we tried to ignore them and walked back and forth around the deck while the ship was still standing at the pier. Suddenly, a British sergeant, obviously a senior (noncom), shouted at us to stop moving about. We gave him a look of surprise and waved our arms in an attempt to explain to him that we weren't doing anything wrong. Then he shouted something at us in English and made a threatening gesture with his carbine. Not paying any attention to him we continued to walk around the deck, laughing and waving our arms in incomprehension.

After some argument a British officer approached us, accompanied by a translator who conveyed to the officer our protests against being arrested, and then said the following: "The officer asks me to tell you not to get excited. You haven't been arrested at all, merely placed under the protection of His Majesty George V, the king of England." After saying this, they walked away. We didn't immediately grasp the hypocrisy and astonishing absurdity of the British officer's statement, but later we laughed when we recalled this incident.

I must say the way the British treated us during this entire time was more than incomprehensible. In Ashkhabad we had been free as birds for almost twenty-four hours. There had also been no restrictions on us during the train trip, and we had just begun to get used to the idea that we really had been released and were traveling to Baku as free people, when it seemed that we had been arrested again on shipboard and would be held under British military guard. "Which means that they aren't releasing us," we decided, "but transferring us to Baku under convoy guard in order to put us in the local prison." But this didn't frighten us since we were, in any case, relying on the Baku workers to force the British to release us. We knew that the Baku Committee had resolved to obtain the release of the families of the Baku commissars, as well as the rest of us. This resolution had found warm sympathy among the broadest masses of Baku workers, and under their pressure the Mensheviks, the right-wing SRs, and the Dashnaks had joined in with this demand. By doing so they were trying to demonstrate their humanity and even, in some degree, to escape responsibility for the deaths of the commissars.

While the ship was still at anchor we managed to get hold of a Russian-English dictionary which was being sold right there on the pier. From my schooldays I had a pretty good knowledge of German, but I didn't know English at all, so I read and pronounced English words with a German accent. What came out was just barely comprehensible, but it was clear that the British sergeant nevertheless understood something of what I said, probably because of the mimicry and gestures I used rather than because of my command of the language. Amirov's wife, who was of German origin, was with us. She was fluent in German, but didn't know English either, and like me, she was forced to communicate with our guards through gestures and reiterated phrases.

For one reason or another the sergeant not only calmed down, but even began to reveal an obviously friendly attitude toward us. He came closer, said something in English, laughed, and smiled; everything suggested that he was saying something nice to us.

When our ship hit the open sea a major storm began. We had food which we had bought while we were in Ashkhabad and therefore had no problem getting enough to eat, but the ship rolled and heaved so much that many of us lost our appetites.

Under normal conditions the ship journey from Krasnovodsk to Baku takes about twenty-four hours, but because of the storm it took us almost twice as long. Some comrades could not stand the rocking of the ship; long months of imprisonment had weakened them. One after another my friends left their cabins and went out on deck, tormented by seasickness. I, however, held out for a long time, managing to rise above the nausea.

I sat in the cabin with the British sergeant; he treated me to canned goods from his soldier's rations, which I was surprised to find rather well stocked. After a while I had the urge to try and make a convert out of my sergeant. Looking up the word for "king" in my little dictionary I said to him: "Konig George not gut!" and made an appropriate gesture with my hand. The sergeant flared up on the spot and shouted: "No, no, no!" and waved his arms. I was astonished that this working man, a Labourite and Socialist to boot (which he had told me he was), would defend the king so vociferously. I simply could not understand this strange "royal" Socialist. Recalling that Ramsey MacDonald had been praised somewhere in the party literature, I said to the sergeant: "MacDonald gut?" The sergeant nodded his head approvingly: "Good, good!" Continuing our "conversation," I asked: "Socialism gut?" He replied immediately, "Good, good!" Then I risked asking: "Capitalism—not gut?" The sergeant nodded his head affirmatively.

My "agitational activities" did not continue for long, however; it was clear that they weren't entirely successful, and besides I had exhausted my entire supply of English words. Also, by the end of the conversation my head had begun to ache; the rolling of the ship had got worse, and I went out on deck in order to breathe some fresh air. The first thing I saw was my comrades who had, as they say, been turned inside out by seasickness. This had a devastating effect on me, and I too finally became seasick. Anyone who has ever been seasick knows only too well how awful it is, especially if it lasts a long time. In any case, the memory of it stayed with me for years. Later on I traveled on boats, ships, and planes many times, but I was never so overcome with seasickness as I was then. It was obviously due to the general weakness caused by my stay in prison.

Gradually the sea began to grow calm; the rolling of the boat lessened. When our ship approached Baku, we were already back to normal, although the faces of many of us were still slightly tinged with green. The ship finally docked. We expected to be met by our comrades, but the British guards would not let us off the boat, and none of our friends could be seen on the pier. Not knowing what else to do we let out an earsplitting yell in the hope that the dock workers would hear it. When they looked in our direction we shouted at them to tell the Baku Communists and the Baku Workers' Conference that they had to rescue us because the British guards were holding us under arrest.

More than an hour passed. Then we saw that a most unpleasant fellow whom I knew, Rokhlin, one of the leaders of the Mensheviks, had appeared on the docks. At the same time three or four other people we knew appeared: Okinshevich, who was a public figure in

Baku and an old friend of Shaumian's family; the Bolshevik Voskanov, who was secretary of the presidium of the Workers' Conference; and one other person.

Rokhlin came up on deck and showed the ship's commander a document concerning our release (which, we learned later, the presidium of the permanent Workers' Conference had managed to extract from the British high command). We were released.

It was explained that because of the storm no one in Baku had known precisely when our ship would arrive. When the Mensheviks learned of its arrival from the dock workers, however, they did not inform the Bolshevik faction.

We dispersed to the old Baku apartments where our comrades were living. Okinshevich's wife took Leva with her from the boat in order to find a place for him in her husband's hospital—he was a well-known Baku surgeon. I don't think I have to say how happy we all were to be reunited with our close friends!

Artak and Topuridze arrived several days after our return to Baku. They had purposely not joined us in Ashkhabad so as not to attract suspicion to themselves and be arrested.

As I mentioned above, I had established rather good relations with a British sergeant on board the ship. In any case, he always had a pleasant, genial smile on his face when he saw me. It wasn't hard to explain psychologically: after all the intense anti-Bolshevik propagandizing the British command had done among their soldiers in which the Bolsheviks were painted as bloodthirsty villains and bandits, he saw that we were all completely normal people. As so frequently happens in such cases, his thinking underwent a sharp turnabout (all, take note, within the space of forty-eight hours) from suspicion and hostility to trust and goodwill.

One afternoon I happened to be walking down the street to the workers' club where our legal headquarters—the Presidium of the Workers' Conference—was located. Suddenly I saw a British military man decisively walking in my direction from the other side of the street, and my first thought was: perhaps he intends to arrest me? When he got closer I recognized him as the sergeant with whom I had become acquainted on the ship. He smiled broadly, pressed my hand in his two paws, and shook it in friendly fashion for a long time, saying something apparently very nice, as far as I could judge, in English.

Shortly after our return to Baku, some of the circumstances surrounding our release from the Ashkhabad prison became known to us. The right-wing SR Chaikin, a member of the Constituent Assembly, maintained in his speeches that we owed our release from prison not to the pressure exerted upon the British command by the Workers' Conference, but to his personal intervention on behalf of the Baku committee of right-wing SRs, on whose orders he had appealed to Minister Zimin in Ashkhabad with a demand for our release.

That Chaikin personally had taken steps in this regard was indeed borne out by the facts, but his references to the right-wing SRs and to a special role played by Minister Zimin were more likely dictated by Chaikin's need to "whitewash" the right-wing SR party in the eyes of public opinion. But Zimin could not have freed us without the direct order of the British. It was clear that he did so with the consent of General Thompson, because both the Directory and Zimin himself were totally subservient to the British high command at that time. Another document also makes this clear.

The day we were released a meeting took place in Baku between Churayev, the representative of the Baku Workers' Conference, and General Thompson. This meeting was reported in the Baku Menshevik newspaper for February 28. This is what General Thompson replied at that time to Churayev's demand that the British cease all further interference in the internal affairs of Azerbaijan and Baku: "But the release of women and children from the prison in Ashkhabad is, after all, also an internal matter, albeit one executed on the orders of the British command. Do you wish a halt to be called to the whole business?"

In his speeches Chaikin referred to one other most interesting circumstance. It turned out that after our release in Ashkhabad, Kun, the head of the local counterrevolutionary authorities in Krasnovodsk, had received instructions from the British command to detain us and hand us over to the jurisdiction of the British command as hostages—until the Soviet authorities in the northern Caucasus released several arrested British officers and other representatives of Great Britain.[10]

I want to say that the British evidently really did plan to let us go in exchange for their own people, but when they learned of the firm position of the Workers' Conference which was demanding our release, and fearing additional complications with the Baku proletariat, the British command, after putting us under guard in Krasnovodsk, was forced to release us in Baku. The leaders of the Baku Workers' Conference insistently demanded a warrant for our release from the British command. And such a warrant was issued to them.

Thus ended our wanderings through the prisons of the Transcaspian region during the period of the British occupation.

[10]See Appendix.

"I clearly recall my native village, Sanain, in one of the picturesque corners of Armenia." This is the way it looks now.

"There were five children in our family. Our relations with our father were very warm and affectionate; we loved our mother unreservedly. In the first row, from the left: eldest brother Arwand, mother Talita, father Ovanes, eldest sister with her son Serge Simonian (now a colonel in reserve in the Tank Corps), and myself. In the second row: youngest brother Artem, and youngest sister Asthik."

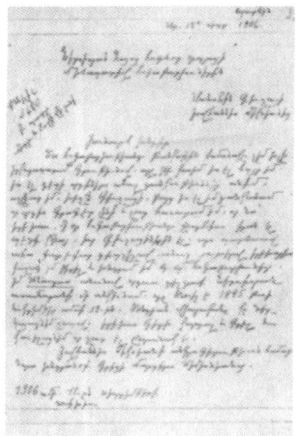

"Many years later my comrades from Armenia found in the seminary archives the original copy of my father's petition, written by Martiros Simonian."

"In Tiflis my father took me to the house of our relative, Aunt Verginia Tumanian. She was active in helping many revolutionaries."

A group of my classmates at the seminary: from left to right: M. Sarkissian, A. Shakhgaldian, O. Pogossian, B. Yeremian, A. Mikoyan, S. Sarkian.

"In the spring of 1914 we held the first conference of Marxist circle members in the apartment of Liusia Lisinova. She was an intelligent, well-read young woman."

My second cousin Ashkhen, after she became my wife.

"The supervisor of our seminary, Simak, future President of the Supreme Soviet of the Armenian SSR, was already a member of the Bolshevik party."

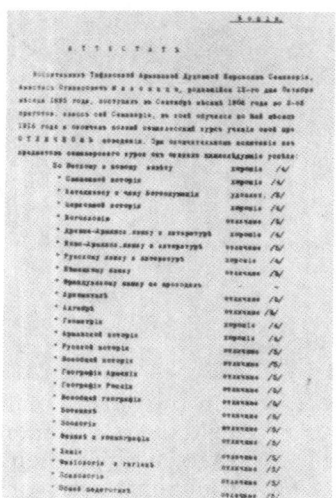 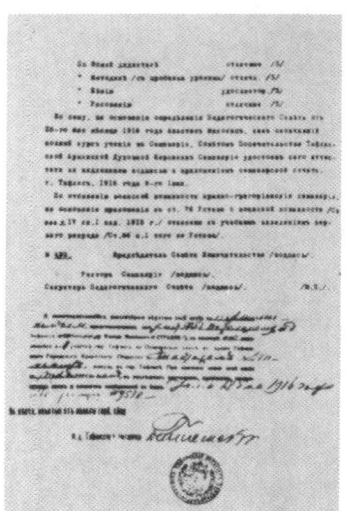

My graduation certificate from the seminary.

"Our professor, Ashot Ionessian, was an old-time Marxist."

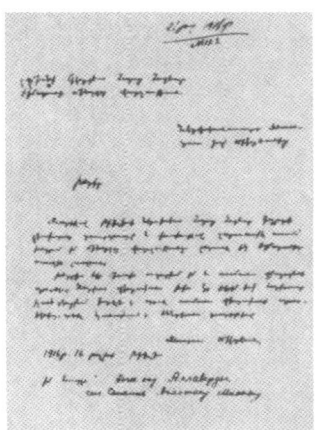

"My petition for admission to the Armenian Religious Academy, dated July 16, 1916."

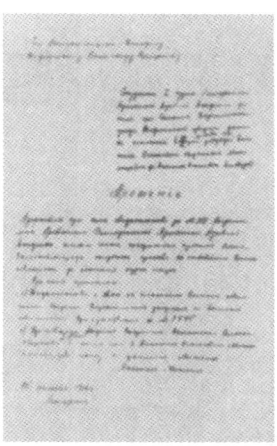

"I presented a petition to the Military Commander for deferment of military service until I completed my education." The text of the petition.

Seven delegates with deciding votes in the Fifteenth Party Congress were former students of our seminary. Six of them participated in our Marxist circle. First row, from left: A. Stamboltsian, S. Amirkhanian, S. Akopian, A. Mikoyan, G. Alikhanian. In the center: G. Gardashian. End right: V. Balian. Standing, from left: responsible party workers C. Amatviri, N. Andreassian. Seated in front of him: O. Bannayan. Standing: A. Shakhgaldian, A. Shaumian (he was not in the seminary), S. Markarian, A. Kostanian.

Among other active participants in the youth movement in our seminary, besides those in the group were:

K. Alabian

G. Voskanian

A. Iginazarian

S. Sarkian

E. Kochar

S. Parsadanian

Stepan Shaumian with his youngest son Serge in the Bailov Prison in Baku, 1916.

"I arrived in Baku at the end of March 1917." General view of Baku during this period.

"In the beginning of May 1917 there arrived in Baku from Petrograd Mikha Tskhakaya, who had great authority among the revolutionaries of the Caucasus. He was a truly heroic figure."

Philip Makharadze was an eminent Marxist writer. My acquaintance with him was a great pleasure to me.

"On October 2, 1917, the First All-Caucasion Regional Congress of the Bolshevik Party was convened. I participated in the work of the congress as the delegate of party organizations of Alaverdi, Manes, and Akhpat." An announcement of the congress.

G. K. Ordzhonikidze, 1920. "Never retreated, never surrendered. So he was in the years of revolution, in the years of civil war, and in the years of social reconstruction."

"In August 1918 Georgi Sturua, a well-known party activist, appeared in Baku, an unyielding and experienced conspirator."

"All our work was done in the Central Workers' Club in Baku."

Varvara M. Dzhaparidze

Olga Shatunovskaya, one of the active Communists in Baku, was the secretary of S. Shaumian. She stayed on in Baku after our arrest.

Old-time Bolshevik Maro Tumanian helped us find a place for our underground printing press.

Marusia Kramarenko

Ekaterina S. Shaumian

Suren Shaumian, a graduate of the Military Academy. He became one of the outstanding commanders of the Red Army's armored tank corps.

Lev Shaumian

A. I. Mikoyan, 1918

"The ship *Turkmen*, which was to evacuate us from Baku, was full of refugees and armed soldiers."

Victor Naneishvili, head of the detachment of the Red Army, which cleared out the reactionary bands, holding parts of Azerbaijan and the towns of Derbent and Petrovsk (now Machatchakala).

Group of party workers in an underground fighters unit, 1918. From left, standing: Sarkis, Artak, Babadjanov. Sitting: Mikoyan, Tovmosian, Ter-Serkissian.

Building occupied in 1918 by the Baku Committee of RKPLB.

"The prison in Krasnovodsk was near the harbor and we were taken there by foot under escort."

"The cell where they put us was narrow with wooden benches alongside one wall, cement floors, without mattresses or pillows . . ."

"We were delighted with your firmness and decisive political work." V. I. Lenin to Shaumian in 1918.
Mandate of Shaumian with signature of V. I. Lenin.

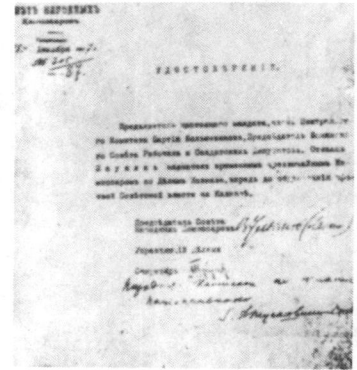

Laak Ter-Gabrielian was the President of the Extraordinary Commission on the Struggle with Counterrevolutionaries. Afterwards he became President of the Council of People's Commissars of Armenia.

"Not long before the shooting of the Baku commissars, S. Shaumian sent the People's Commissar of Education, N. N. Kolesnikov, to Astrakhan."

The People's Commissar of Justice, Artaches Korinian, was sent to Baku by S. Shaumian. Now he is an active member of the Armenian Academy SSR.

On the tragic day of September 20, 1918, by a cowardly plan formulated by SR henchmen, together with the English military authorities, they heroically died:

Stepan Shaumian

Prokofii Dzhaparidze ("Alyosha")

Meshadi Azizbekov

Ivan Fioletov

Mir-Gasan Vezirov

Grigorii Korganov

Yakov Zevin

Grigorii Petrov

Vladimir Polukhin

Arsen Amirian

 Suren Ovsepian
 Ivan Malygin
 Bagdasar Avakian
 Meer Basin

 Mark Koganov
 Fyodor Solntsev
 Aram Kostandian
 Solomon Bogdanov

 Anatolii Bogdanov
 Armeniah Borian
 Eyzhen Berg
 Ivan Gabyshev

 Tatevos Amirov
 Ivan Nikolaishvili
 Iraklii Metaksa
 Isay Mishne

Nariman Narimanov

Mir-Bashir Kasumov was, in 1926, among other distinguished Bolsheviks, a member of the court at the trial of those involved in the deaths of the twenty-six Baku commissars.

September 8, 1920, on Freedom Square, the funeral was held of the remains of the tragically killed twenty-six Baku commissars. The square now carries their name.

Among those Bolsheviks arrested by the English occupation forces in December 1918 was the worker I. I. Anashkin.

Bachram Agayev, Bolshevik, one of the leaders of "Adalet," the Communist organization of the Azerbaijanians, is of Persian origin.

A. G. Karayev was the president of the Workers' Conference in Baku.

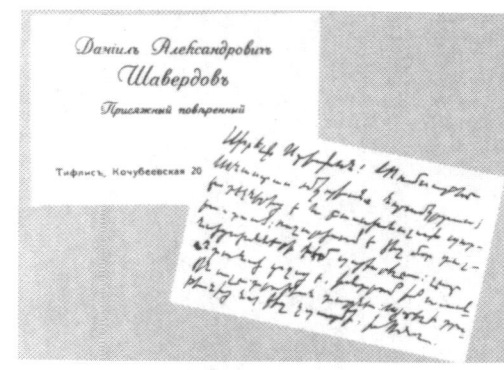

Danush Shaverdian, a person close to me who had an important influence on my political views and on my entering the party, was a lawyer in Tiflis in 1919. Above is his visiting card.

The first time I met Mamya Orachelashvrili was in October 1917. A physician by profession, he carried on an important work among the soldiers. S. M. Kirov, G. K. Ordzhonikidze, M. Orachelashvrili.

The boat *Turkmen*, one of twelve vessels on which the community of Baku transported oil to the Soviet Astrakhan. The journeys were perilous and many comrades perished on them. With great sorrow I remember them.

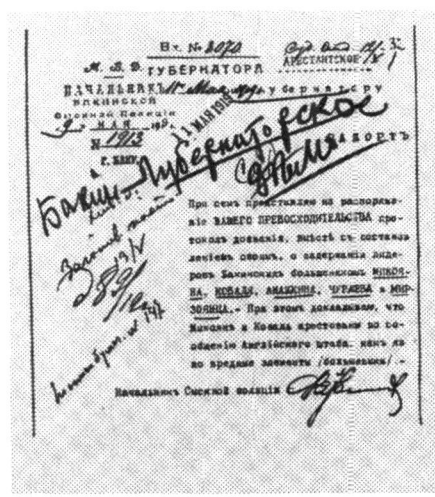

We were informed that we had been arrested by order of the British Headquarters. Report of the head of the Criminal Investigation Dept., where Mikoyan and other comrades had been brought after their arrest for organizing a strike in Baku in May 1919.

Commander Gubanov and Rogov organized the extraordinary maritime expeditions from Baku to Astrakhan in 1919. On one of these trips sailor Rogov tragically perished.

In October 1919 Fedya Gubanov, President of the Central Committee of the Seaman's Union, heroically died. He was tortured to death and then thrown into the sea by Denikin's men.

Comrade Mirzoyan, one of the leaders of the Baku strike in May 1919 was arrested along with us.

I knew Rukhulla Arkhundov from the time of the Baku Commune. Relations between us were close.

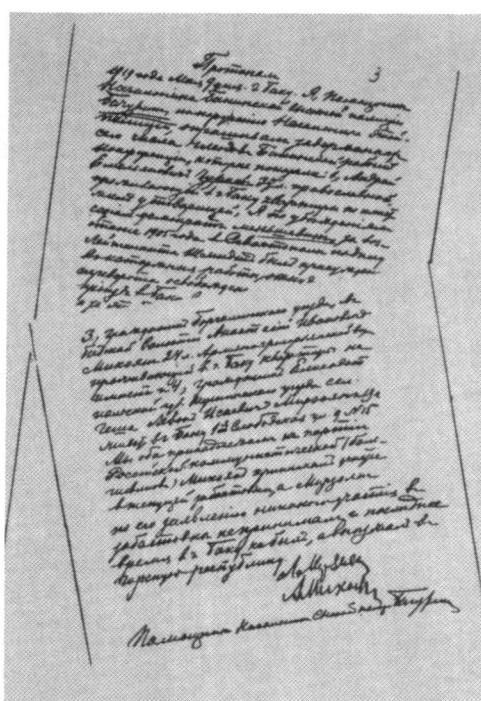

In the end the chief of the police called us in for interrogation. The affidavit of interrogation of the Baku Criminal Investigations Dept. According to deposition #913 we were sent to prison.

"Splendid!" Gorky exclaimed about Kamo. "This person was able to combine in himself an almost mythical courage with such invariable good luck, such unusual resourcefulness with such childlike simplicity of soul." S. A. Ter-Petrosian (Kamo), 1922.

In the archives of the Gendarmes Administration was found a photograph of the time we were in Tagiyev prison in the summer of 1919. From left to right, sitting: A. Mikoyan, Y. Figatner. Standing at right: B. Sheboldaev, A. Nuridzhanian, L. Gogoberidze.

From the beginning of 1919 the secret headquarters of the Bolsheviks in Baku was the apartment of the Kasparovs. The family was astonishing. Maria then worked as technical secretary to the Caucasian party organization.

In the autumn of 1919 Rosa Kasparova was arrested by White Guards. Her letters from prison are haunting in their courage and steadfastness. On the twentieth of February 1920 she was hanged. In a letter before her death she wrote to her mother: Happiness will return to us again like the bluebird.

Among those sent to help us from the Party Central Committee was Gamid Sultanov.

T. I. Otradner. He was a balanced and versatile person, with many years' experience in revolutionary fighting.

"S. M. Kirov at the time of siege of Astrakhan inspired the defenders, who were cut off on all sides," wrote Sergo Ordzhonikidze. S. M. Kirov and G. K. Ordzhonikidze.

V. I. Lenin in Red Square, November 7, 1919. In the autumn of 1919 I was sent by the Caucasian Party Regional Committee to Moscow to Lenin. That was my first meeting with Lenin.

From Samara to Tashkent we traveled more than a month with M. V. Frunze, who was then the commander of Red Army forces in Middle Asia.

Elena D. Stasova worked then in the Secretariat Central Committee of the party. She knew of my arrival from Kirov's telegrams. Businesslike, practical, and reserved was the impression I had of her. I was impressed by her unusual industriousness.

V. V. Kuibyshev. I remember how struck I was by his indefatigable efforts to gain a thorough knowledge of a situation, and at getting acquainted with people, and by his sober-minded approach to solving any problem.

N. F. Gikalo and A. Shazipov were leaders of mountaineer partisans in Chechnia in 1919.

A. Sharapov

M. K. Levandovsky, commander of the Eleventh Army. I remember him as a fierce commander. He looked stern, but had a good heart.

G. K. Ordzhonikidze, A. I. Mikoyan, A. Y. Belenky, and Serge Shaumian, 1927.

S. M. Kirov, G. K. Ordzhonikidze, A. I. Mikoyan, M. G. Yefremov, next to an armored train at the railroad station in Baku, 1920.

A. I. Mikoyan with family, 1929.

Anastas Ivanovich Mikoyan and Artem Ivanovich Mikoyan, 1941.

PART IV

THE BAKU UNDERGROUND DURING THE BRITISH OCCUPATION

Chapter I

The Struggle for the Masses

Debates about the strike—Fights at meetings of the Workers' Conference—At the Central Strike Committee—Memories of the fighters—We gather our forces—The Baku Conference.

And so, we were back in Baku. I have always loved this city of winds, "black gold," and eternal lights. Baku is the city of my revolutionary youth, holding the brightest and most vivid memories of those days. It was here that I met my famous friends—the future Baku commissars—here that I passed through the strict but priceless schooling of the Bolshevik underground and experienced that unforgettable time of struggle for Soviet power, for the triumph of our ideals.

After setting foot on the shores of the city we climbed into carriages and set out along the streets we hadn't seen for over five months.

I was happy to be free, to be in Baku again, and overjoyed at the thought that I would soon be among my close friends, the Baku Communists and workers. But my joy was overshadowed by the thought that Shaumian, Dzhaparidze, Azizbekov, Fioletov and our other deceased comrades were no longer with us and never would be again . . . I was worried about what awaited us here. I was anxious to find out what had gone on in Baku while we were away and what the situation was now.

Familiar streets flashed by before our eyes. Everything was the same as it had been when we left the city, only now one saw many more soldiers and officers in British military dress or in the plaid kilts of the Scottish artillery strolling through the city as if they owned it. It was depressing to realize that the forged boots of the British occupiers were still trampling our native land.

I stayed that night with a family of old friends. From them and from comrades who came to see me I learned the details of what had happened during the several months I had been away. It turned out that after all we had endured in the Transcaspian region, we were now forced to plunge into stormy revolutionary work "on the run," so to speak.

It was clear that the general situation in Baku remained both complicated and tense. The British military command of General Thompson had emerged quite openly as the de facto master of the city. The local bourgeois-gentry, nationalistic government of Khoysky and company danced to their tune. It's true that since the December, 1918, political strike, the permanently functioning legal organization of workers—the Baku Workers' Conference—had been coexisting with the British. Unfortunately, the majority in its ruling organ, the presidium, was composed of Mensheviks and SRs. It was headed by the capable worker Churayev, a good speaker who, although he was a Menshevik, had the trust of the Baku workers—mainly because of his glorious revolutionary past: he had taken part in the mutiny on the battleship *Potemkin*.

My friends said that the workers' economic situation was deteriorating day by day. The Azerbaijani capitalists, taking advantage of the support of the local government and the British occupation authorities, had once again seized the previously nationalized plants and oil fields, and reinstituted the old order of things there, destroying the terms of the agreement that had been won with such difficulty by the Baku workers back in the days of Kerensky's government.

In addition, the oil industry was undergoing a major crisis at that time. It was difficult to market oil, and since Baku had been cut off from Soviet Russia, the major importer of Baku oil, the wages of oilfield workers had decreased while the cost of living in the city kept on rising. The crisis also hit the maritime transport workers hard, with the majority of oil tankers in the Caspian fleet standing idle at their moorings. As unemployment rose, disturbances among the Baku workers became widespread.

In these conditions the oil-field workers, dock workers, and seamen insisted on going on strike. They remembered what they had won some three months' earlier during the political strike in December, and they hoped to be victorious this time, too. Even the appeasement parties favored a strike; and finally, the Mensheviks, although lacking faith in the effectiveness of a strike, did not dare to come out against it for fear of losing their influence over the workers.

About a month before our arrival in Baku, the Workers' Conference had passed a motion to initiate a general strike. A Central Strike Committee was formed. In those days the most popular words among the workers were *zabastovka* (strike) and *stachka* (strike). Thus, when we arrived in Baku on March 6, 1919, the first task on the agenda was to organize a general workers' strike.

On our first evening in the city there was a meeting of the Baku Party Committee, and the next day, March 7, a meeting of the Strike

Committee was called in order to resolve the question of the length of the strike.

We argued a great deal about this matter, with each person speaking several times. I thought it would be unwise and premature to initiate the strike under present conditions, but Levan Gogoberidze, representing our party on the Strike Committee, was especially set on the idea of an immediate strike. He argued heatedly that it was impossible for him as a Communist to come out in favor of a delay at the Strike Committee meeting. According to him, all the Strike Committee members favored starting the strike immediately; the workers were expecting it from one day to the next, and their patience had reached the breaking point. The speeches of other comrades reiterated this view. Attuned to the mood and demands of the Baku workers, they wanted to start the strike as soon as possible, and in addition, they considered themselves bound by the resolution which had been passed in this regard by the Baku Workers' Conference. Therefore when I, a neophyte in this matter, began to ask questions about the strike, several members of the committee became somewhat hysterical.

I asked, for example if, taking into account the present social, economic, and other conditions in Baku, my friends had confidence in the success of the strike? Wasn't it premature to go out on strike when the Bolsheviks were still an insignificant minority in the presidium of the Workers' Conference and on the Strike Committee? Would all the workers support us? Would the capitalists go for the idea of even partially satisfying the strikers' demands? I said that we could not ignore the dangerous possibility of the British occupation forces and the reactionary Azerbaijani government suppressing the strike, and at the same time disbanding the Workers' Conference too, whereas if we postponed the strike for two or three months, we would not only have time to become organizationally and politically stronger, but the whole situation would change: contact would be established with Astrakhan and the Red Fleet would be able to leave the Volga and occupy a dominant position on the Caspian Sea, thereby exerting an influence on the course of events here in the city.

The discussion stretched on late into the night during which time I managed to gain a deeper understanding of the situation, while on their part my comrades were able to see my reservations in proper perspective and to examine their own positions more critically.

As a result it became clear to us that the necessary conditions were still not ripe for initiating the strike immediately: not all of the Baku party organization had been reestablished at that time, and party cells had still not been set up at all major industrial plants; the Communists did not have a majority either on the strike committee or in the presidium of the Workers' Conference; and although the Baku trade unions, which were in the hands of the Mensheviks, had been organized, they were ideologically still very weak. It was impossible to begin the strike without firm party leadership: the Mensheviks, SRs, Muscavatists, and Dashnaks could capitulate in the course of the strike and cause the whole thing to fail.

I recall that after weighing all these circumstances, I introduced a motion which, while leaving in force the resolution of the Baku Workers' Conference to call a strike, proposed that we not be hasty in specifying when it should begin and end. The motion also proposed that we use what time remained to direct our broadest and most strenuous efforts toward preparing for the strike, strengthening the party organization by trying to win a majority in the Strike Committee, in the presidium of the Workers' Conference, and in the leadership of the trade unions. Finally, the motion called for delaying the strike until, at the earliest, the beginning of river navigation on the Volga, when we could realistically count on the aid of the Soviet fleet. In my opinion the strike would then become no longer merely economic, but political as well.

After lengthy discussion the Baku Party Committee approved my proposals. But the practical question arose: how were they to be carried out? The difficulty lay in the fact that the Baku workers were in favor of striking right away. Nobody wanted to even hint at workers' meetings of the possibility of delaying the strike. How were we, Bolsheviks, representatives of the most revolutionary party of the working class, to act in such circumstances?

We found the correct solution.

Levan Gogoberidze—the only Bolshevik representative on the strike committee—was told to expose the two-faced policies of the Mensheviks and SRs. On the one hand they were joining the strike committee and calling for a strike, while on the other they were holding behind-the-scenes talks with the British command and writing articles in the newspapers in which they undermined the workers' faith in the success of the strike. Gogoberidze was to demand an end to negotiations with the British command, and in the event this demand was rejected—to leave the meeting. Gogoberidze successfully fulfilled this assignment. The Mensheviks and SRs were stunned by such an unexpected manner of posing this question, and refused to accept the ultimata of Gogoberidze. At that point he demonstratively left the meeting of the Strike Committee.

Our most important tactical objective was threefold: to isolate the Mensheviks and SRs in the Strike Committee and in the presidium of the Workers' Conference, to attain for the Bolshevik party a dominant influence in these supporting workers' organs, and to unify the broad working masses around them.

We put forward the idea of organizing local workers' conferences and strike committees which would function as branches of the Central Strike Committee and would be in direct contact with the workers at industrial plants. This suggestion was supported by all the workers.

Looking ahead somewhat, I would like to say that during March and April we Bolsheviks who joined the presidium of the Baku Workers' Conference managed to accomplish a great deal of fruitful work by creating workers' conferences and strike committees in all the Baku districts, and in strengthening their organizational ties with industrial plants. Moreover, we managed almost everywhere to get Bolsheviks and vanguard, unaffiliated workers into leadership positions on the presidiums of the workers' conferences and on the strike committees.

All this helped to strengthen our party's influence on the working masses of Baku.

* * *

With March 11, the date set for the meeting of the Baku Workers' Conference, only days away, we decided to put the time to good use by preparing for the upcoming "battle" with the Mensheviks and the SRs. The Workers' Conference consisted of about 500 delegates who were elected by factory-plant, industrial, and marine committees, or directly by general workers' meetings. In addition, each of the seven political parties active in Baku at that time (the Bolsheviks, Gummet, Adalet, the Mensheviks, SRs, Dashnaks, and Muscavatists) could send one representative to the Workers' Conference, with the right under conference rules, of speaking for ten minutes.

In discussing our course of action at the upcoming Workers' Conference, we decided that on the first day of its session we would put several basic questions before the delegates. The first concerned the representatives of the Dashnak and Muscavatist parties at the conference. In view of the fact that neither one could be considered workers' parties in terms of their composition or program, we decided to bar their representatives from participating in the Workers' Conference. The second question concerned the presidium of the conference; we decided to introduce a motion calling for a reelection of the present composition of the presidium, since for the most part it was composed of people who were alien to the real interests of the working class.

We conducted negotiations with Karayev, the Gummet representative, who had joined the staff of the socialist faction of the Azerbaijani parliament. It was at that time that I became acquainted with him. In 1917–1918 he had been living in Tiflis and had joined the Tiflis Menshevik Gummet (there was no Bolshevik Gummet there). Karayev turned out to be politically well-disposed toward us; he agreed with our position and expressed his readiness to support the Bolsheviks. In addition, he told me that another Gummet member, Guseynov, shared his views.

We had the same sort of talks with Bakhram Agayev, a Bolshevik, one of the leaders of the Adalet. As early as 1918 Agayev and his organization had been on our side. In the present situation he could be relied upon as our comrade-in-arms.

After we had agreed upon these matters with Karayev and Agayev, we decided to nominate them as candidates for the new staff of the presidium of the Workers' Conference. As I've said, the worker Churayev was chairman of the presidium of the Workers' Conference, and we considered it possible to keep him on as chairman on the new staff of the presidium. I had a confidential meeting with Churayev in this regard, explaining that we wanted to keep him as chairman on the condition, however, that he not carry out the policy of the Menshevik party and its committee's instructions, but rather adhere to the decisions adopted by the majority of the presidium. He gave his consent to this, and we felt that one could have faith in him.

Keeping Churayev as head of the Workers' Conference was also very important in order to guard the conference against the persecution from the British occupation command and the Azerbaijan bourgeois government which would inevitably ensue if we elected a well-known Bolshevik as chairman of the presidium. Leaving Churayev in his post would have a certain "calming effect," and it would be much easier to conduct negotiations with the British command and the local government through him, since they had already had time to get used to him. Besides, Churayev could later turn out to be very useful to us in attracting workers, particularly those who leaned toward the Mensheviks: we never forgot the necessity of struggling for that segment of the Baku proletariat as well.

It was planned to have me become his assistant (his "comrade" in the nomenclature of that time) and I would defend the Bolshevik line in the presidium.

From among the SRs we decided to support Ilin as a candidate to the presidium. He was an SR "by accident" since he more properly belonged to the vanguard populist-intelligentsia. In any case, he was an utterly decent man. We put the same conditions to him as we had to Churayev. He willingly accepted them.

I remember that in discussing all these candidacies we thought about how we could later pull these people over to our side completely. It happened very soon in Churayev's case; he joined the Bolshevik party. Karayev also joined the party, but Ilin did not.

The following people were also mentioned as candidates for membership in the presidium of the conference; the young Communist Levon Mirzoyan, who at that time was a member of the Baku Trade Union Council and had joined the new staff of the Baku Party Committee; and also the unaffiliated worker Koval who had given a good account of himself in the position of chairman of the council and wielded great authority among the workers (a year later Koval joined the ranks of our party). In addition we considered three other comrades for membership in the presidium, including two Bolsheviks: Levan Gogoberidze and Ivan Anashkin.

Thus, according to our plan, six of the ten members and candidates for membership in the presidium represented the Bolshevik line, one had no party affiliation, two were SRs, and one a Menshevik. This composition secured us the necessary majority in the leadership of the Baku Workers' Conference.

Slightly more that 300 delegates gathered for the opening of the Workers' Conference. When the presiding officer Churayev declared the conference in session, a representative of the Muscavatist party arose and declared that the meeting could not be recognized as duly empowered since not all the delegates were present, but he was overruled by the overwhelming majority, and the conference began its work.

Levan Gogoberidze of the Bolsheviks spoke first, concerning the regulations governing the conduct of the session. First of all, he demanded that the representatives of the Dashnak and Muscavatist parties, as representatives of bourgeois-nationalist parties, be deprived of the right to attend and to speak at the conference. He added in this regard that membership in these parties must not hinder those who had been chosen as delegates to the conference from carrying out their responsibilities as delegates. Thus, Gogoberidze did not exclude from participation those delegates themselves who were affiliated with the Dashnak and Muscavatist parties. Gogoberidze's motion was passed by the conference. The Dashnak and Muscavatist representatives left the meeting of the Workers' Conference under protest.

Thus the first victory had been won. But very difficult battles still remained to be fought.

I had been assigned to make a major keynote speech at the conference on behalf of the Bolsheviks. I was to speak in detail of the circumstances surrounding the deaths of the Baku commissars and in doing so, to expose not only the role of the British command, but also the true colors of the SRs, Mensheviks, and Dashnaks in order to prepare the way for depriving these parties of their majority in the Presidium of the Workers' Conference.

I should mention that the Baku workers still did not know in full how the bloody tragedy of their commissars had come about and who had been responsible for it; the leaders of the appeasement parties had tried to deceive them by casting all blame on the British alone.

According to the established rules, I, like the representatives of the other parties, could speak for only ten minutes. Knowing the mood of the delegates, Gogoberidze made a motion on behalf of the Baku Bolshevik Party Committee that I be allowed a half hour for my speech since, as one who has just returned from prison where I had been with the Baku commissars, I was informed on what was going on. The reason behind Gogoberidze's motion itself evoked the intense interest of the conference delegates.

Friendly voices were heard saying: "Give him the floor! Let him speak! It's not necessary to limit the time of his speech!" The suggestion not to limit the length of the speech was accepted by the overwhelming majority of the conference delegates, a decision which in itself dealt the first serious blow to the Mensheviks and SRs.

Before broaching the current political situation and the tasks confronting the Baku proletariat, I dealt in detail with the circumstances surrounding the temporary fall of Soviet power in Baku last August and the subsequent deaths of the twenty-six Baku commissars. I exposed the treachery of the SR, Menshevik, and Dashnak parties which had dealt Soviet power a stunning blow and had gone over to the side of the British; I accused them—as well as the British command—of being the culprits responsible for the deaths of our comrades. When I declared that "the hands of the Menshevik and SR leaders sitting in this hall are crimsoned with the blood of the Baku commissars," a great hue and cry arose among the delegates, who stood up shouting, "Executioners! For shame! Down with the SRs, Mensheviks, and Dashnaks!" When this happened, the faces of many Mensheviks were covered with a deathly pallor. They were crushed by the widespread anger and indignation of the delegates.

Then I went on to analyze the state of affairs in Baku in light of the current situation of Soviet Russia, and I spoke of the tasks confronting the Baku proletariat. At that time we Bolsheviks did not even have our own newspaper in Baku. The SR paper, *Banner of Labor*, while refraining from comment on the politically pointed part of my speech (which lasted about an hour) published only a short summary of it.[11]

After I spoke, the floor was given to the Menshevik representative Rokhlin who called my speech an act of provocation that threatened to divide and demoralize the workers' ranks, etc. I must say that at first the audience listened to Rokhlin calmly enough, but gradually they became restless and dissatisfied and the noise they made became more and more frequent. One delegate shouted to the chairman: "Stick to the rules, his time is up!" The chair was forced to put to a vote Rokhlin's request for additional time. As the vote was being taken, Rokhlin, who had become overexcited, took off his jacket since he was sure they would extend his speaking time, but the meeting rejected a second proposal from the chair that Rokhlin be allowed to finish his speech. Perspiring, defeated, and pathetic, Rokhlin was forced to put on his jacket and step down.

The results of the discussion of the strike question came as a surprise to the Mensheviks and SRs. Gogoberidze spoke for the Bolsheviks and said approximately the following:

[11]See Appendix

> In view of the fact that the majority of the Central Strike Committee consists of Mensheviks and SRs who are, on one hand, conducting behind-the-scenes negotiations with the British command, and on the other trying to prove the impossibility of a general strike in their newspapers, thus undermining its success, the Bolshevik faction has been forced to leave the strike committee meeting.
>
> We are informing the Workers' Conference that as long as the Mensheviks and SRs are in control of the strike committee, the Bolshevik party will speak out against any strike led by such a strike committee. As long as agents of the bourgeoisie, who are playing along with the British command, are sitting on our fighting staff, any strikes are an act of provocation. Therefore, the Bolshevik faction demands immediate reelections of the strike committee and the removal of the Mensheviks and SRs from it.

A battle flared up as the result of Gogoberidze's declaration. In the end, by majority vote, the conference passed our motion to elect a new staff for the Workers' Conference presidium and instructed the presidium to form a Central Strike Committee composed of all members of the conference presidium, plus two representatives from each district. Each candidate on our list of nominees for membership in the conference presidium was voted on separately, and all were accepted. I also was elected to the new staff of the presidium.

At the concluding session, upon the suggestion of several delegates, the Baku Workers' Conference unanimously resolved to send greetings to the Second Congress of the Communist Internationale, which was in session in Moscow at that time, and to wish it fruitful work and success in achieving the victory of labor over capital.

This fact in itself was very significant. My Baku comrades told me that back in mid-January at a meeting of the Workers' Conference, the Mensheviks' motion (protested by the Communists) to send greetings to the Congress of the Second Internationale in Berne had been passed by a two-thirds majority. And now, a mere two months later the overwhelming majority of that same conference was sending its greetings to the Third Leninist, Communist Internationale!

I remember the enormous impression this all made on me. I thought about how much the political consciousness of the Baku workers had grown, how greatly their militant revolutionary spirit had intensified, how strong their faith in their own forces and in our Communist party had become! I could not help but recall Stepan Shaumian's confidence in the Baku workers and his conviction that they would soon realize, through experience, that the right-wing socialist parties were dragging them down into an abyss. He turned out to be right.

Another fact also illustrates the political clear-sightedness of the Baku workers. Back in December of 1918, in speaking at the Baku

Workers' Conference, a worker from the Elektrosil (Electric Power) plant had said: ". . . They invited the British in to maintain order, but the British are smothering us. We thought that the arrival of the British would bring about the flowering of genuine democracy in Russia . . . I suggest, comrades, that tomorrow we bring everything to a standstill and send a ship to Astrakhan for the Bolsheviks."

This speech was greeted by the conference delegates with applause that lasted for a long time. The worker from Elektrosil had precisely reflected the worker-delegates' general attitude to the British occupiers: a simple worker who had previously spoken out against the Bolsheviks' position and in favor of inviting in the British, had changed his mind after experiencing the occupation. And, what's more, the December 1918 Workers' Conference had been conducted under the aegis of the Menshevik leaders; only one Bolshevik representative had spoken at it—the worker Anashkin, a member of the Baku Committee of our party.

The speech of the Elektrosil representative hit the SR-Menshevik leaders like a clap of thunder on a clear day. They were forced to say the following about it in *The Strike Committee News:*

> At the last Workers' Conference one of the speakers made the mistake of saying that we ought to send a marine transport to fetch the Bolsheviks, a statement that won him applause. Why? Why did people applaud someone whom they would have greeted a few months ago with hissing and catcalls? Well, it's very simple. Our foreign-governors are doing everything they can to justify what the Bolsheviks have said about them. And regardless of the fact that other Bolshevik positions have in essence been rejected, the correctness of their judgment on the role and tasks "of the allies" arouses sympathy for them among a certain segment of the working masses. These distinguished foreigners came to us in order to stop the spread of anarchistic bolshevism, but are in fact busying themselves with other things . . . wouldn't it be better for them, the best proselitizers for the cause of bolshevism, to say straight out whose work they are doing—the work of democracy, the work of anarchy, or the work of international reaction? . . . We know the answer, but let it not be kept secret from others.

These voluntary or involuntary admissions reflected the process of reappraisal among the majority of Baku workers, including those who still supported the Mensheviks and SRs out of inertia more than anything else. Real life itself was facilitating this process daily.

Let's take just one example of the kind of treachery which was practiced by Khoysky's reactionary Azerbaijani government during the British occupation. On the day of the second arrival of the occupiers in Baku, the Khoysky government published the following declaration:

To the population of the city of Baku.

As a result of negotiations between the Azerbaijani government and General Thompson, the representative of the Allied forces at Enzeli, an agreement has been reached which provides for the entry into the Azerbaijanian capital on this morning, November 17, of a detachment of Allied forces under the command of General Thompson.

The entry of the Allied detachment is not an act of aggression which could destroy the independence and territorial integrity of Azerbaijan.

All state and public institutions will continue to function normally, as before.

The government has taken all measures required to maintain order in the city. The government calls upon all citizens to preserve calm and order and also to greet the Allied detachment as friends.

Chairman of the Council
of Ministers

F. Kh. Khoysky
Minister of Internal Affairs
B. Dzhevanshir.

Of course, the leaders of the government knew very well the conditions on which these British "saviors" had come. It was no accident that they had sent their representative to the British at Enzeli, where everything had been arranged beforehand. The British took control of all rail and water transportation, the oil industry, and the state bank. All this was done without any participation whatsoever on the part of the Azerbaijani government. With his edicts General Thompson dictated the daily routine for the entire population of Baku.

There is not even a single mention of Azerbaijan or its government in any British document. And the Azerbaijani government did not dare to debate or protest this. It is difficult to imagine how anyone could sink so low morally! The lies and deceit of this government, and the antipopulist regime of the occupiers gradually opened the eyes of the working class of Baku, evoked in it a storm of anger and indignation, and led to a general political strike in December 1918.

These events provided the Baku workers with excellent training in the class struggle.

The day before the presidium of the Workers' Conference convened, there was a meeting of the Baku Party Committee. At this meeting a preliminary examination was made of the question of the delegating responsibilities within the presidium of the Workers' Conference, a matter which was subsequently confirmed by the presidium.

We also discussed the question of what tactics we would adopt in the face of information that kept coming in from various industrial plants indicating the workers' intention to begin a strike immediately. We decided to use every means at our disposal to impede this grass roots movement. It was clear that the rift between our insufficiently organized preparations and the masses' growing desire for confrontation was getting wider. It was necessary to take emergency measures, and above all to create and strengthen the party organizations in the districts immediately. We decided that Sarkis Mamedyarov, Chikaryov, Tyukhtenev, Kasumov, Pleshakov, and Yakubov would concentrate on illegal party-organizational work. Responsible party-organizers were designated and confirmed for each district, and ways of combining legal and illegal party work were defined. We started from the premise that the organization of party work, and all our contacts, should be strictly secret, so that in the case of failure of the legal party workers, the illegal organization would remain intact. One of the immediate tasks of party organization work at that time was to gather together into a single unit all Communists who, for one reason or another, had still not joined in active party work. According to our estimates there were about 200 to 300 such Communists in Baku.

The second task was to attract into the ranks of the party nonparty workers, as well as former SRs, who had shown themselves to be staunch fighters against the British occupation and had actively participated in the strikes.

The question of work among Azerbaijani workers was the focus of particular discussion. It was decided to call a special conference on this question, in which Karayev, Guseynov, and Agayev would participate. The committee also examined concrete proposals concerning more active participation of Communists in nominating candidates for the leadership organs of the district workers' conferences and strike committees.

The day after the Workers' Conference we gathered for the first meeting of the new staff of the presidium. Churayev was elected chairman of the presidium, the SR Ilin and I were elected comrades (assistants) to the chairman. We immediately agreed that the chairman would not pass any resolutions on his own without first discussing them with the presidium membership. We also agreed that in the event any member of the presidium was arrested, he would automatically be replaced by one of the three standing candidate-members of the presidium, and in the event the chairman or assistant chairman was arrested, their places would be occupied by Gogoberidze and Anashkin respectively.

* * *

I must note that at that time the conduct of the oil manufacturers, the Azerbaijanian bourgeois government, and the British occupation command toward the workers was becoming more and more provocative. The industrialists not only rejected the workers' justifiable demands for improved working conditions, but stopped paying them their wages. The situation was red hot.

On March 15 a joint meeting of the Central Strike Committee and the district strike committees took place at which the district representatives told of the intolerable situation at the industrial plants and the indignation of the working masses. The representative from the Nobel Oil-Refining Works in Chornyi Gorod announced, for example, that they had already begun the strike and would not call it off until the workers were paid their wages. Several speakers condemned the Nobel workers for their separatist action, but at the same time they suggested that if the industrialists stuck to their guns, a general economic strike should be called within the next two to three days.

Ilin turned out to be more "Left" than anyone. Although he deplored the Nobel workers' separatist action, at the same time he himself proposed that political as well as economic demands be put forward in the upcoming strike.

Gogoberidze and Mirzoyan, speaking on behalf of the Bolsheviks, came out against combining economic and political demands, but supported the calling of a general economic strike if the workers were not paid their wages within the next two or three days.

The socialist members of the Azerbaijani parliament, Abilov and Pepinov, who had been invited to the strike committee meeting, also came out against a political strike while supporting an economic one. They were convincing in arguing that the withholding of wages was closely connected with the actions of the British command. In particular, Pepinov reported that the British had withdrawn one hundred million rubles from the bank for their own needs, and as a result there was no money left in the bank.

Pepinov and Abilov's positive response to the idea of an economic strike had great significance at that moment: it helped to quell the doubts of those Azerbaijani workers who were still under the influence of the Muscavatists, and had the effect of uniting them with the rest of the workers of Baku.

When my turn came I spoke out against a combined economic and political strike and proposed that a general strike be called only in the event of a continued refusal to pay wages.[12] An appropriate motion to this effect, introduced by the Bolsheviks, was passed by the strike committee.

All this immediately produced certain results. The Azerbaijani

---

[12]Brief notes of one of my speeches have been preserved in the archives, and present an idea of the debates at that meeting of the Central Strike Committee. See Appendix.

government immediately informed the presidium of the Workers' Conference that they had taken emergency measures. And, in fact, the wages were soon paid.

Three days later an emergency meeting of the presidium of the Workers' Conference took place. After hearing a letter from the Azerbaijani government to the effect that the workers had been paid the greater part of their wages and would be given the remainder in full in the near future, the presidium passed a resolution not to declare a general strike, but not to call a halt to the local strikes already in progress until the workers had received their wages in full. The conflict was soon resolved.

The issue of a collective bargaining agreement between the trade union and the oil manufacturers was also discussed at this meeting of the presidium. The person who reported on this issue, Koval, spoke of the substantial disagreements that had arisen during talks between the trade unions and the oil manufacturers. The latter accepted only seven of the forty-nine demands put forward in the new agreement. The presidium decided that before the final signing of the collective-bargaining agreement it would submit it to the workers for broad-ranging discussion. During the course of the meeting I made one additional proposal concerning the publication of *The Workers' Conference News* in both Russian and Turkish (Azerbaijani). My proposal was passed.

That same day all these questions were passed on for confirmation to the emergency session of the Workers' Conference, which approved the actions of its presidium and the Central Strike Committee, and confirmed the resolutions we had passed.

* * *

In two days, on March 20, six months would have passed since the tragic deaths of the twenty-six Baku commissars. Because of this the Baku Party Committee introduced a motion for discussion at the Workers' Conference to proclaim the twentieth of March a day of mourning, to call a daylong strike, and also to hold rallies and meetings in memory of the deceased comrades. The conference passed this motion unanimously. On that same day the Presidium of the Baku Workers' Conference issued an appeal to all the workers of Baku, summoning them to continue the struggle in which our comrades and leaders had lost their lives,[13] and the Central Strike Committee announced the plan for the one-day strike on March 20. It was envisioned that on March 20 all enterprises in Baku would cease work with the exception of the waterworks, hospitals, apothecaries, telegraph, and

[13]See Appendix.

some bakeries, and that the municipal electric power station would operate only at night.

The Baku Party Committee prepared and published in a special issue of a one-day newspaper that was dedicated to the memory of the Baku commissars.

In a lengthy article which I wrote for that paper at that time, the events surrounding the deaths of the Baku commissars were illumined in detail, and the criminal, treacherous role played by the leaders of the Baku right-wing parties, SRs, Mensheviks, and Dashnaks in this evil deed was exposed.[14] This issue of the paper constituted an indictment that laid bare the bestial countenance of the British occupiers and their assistants—the right-wing socialist parties.

It was from this newspaper that the broad masses of the Baku proletariat learned the details of the deaths of the Baku commissars: my speech, after all—made weeks earlier at a meeting of the Workers' Conference—had been published with this part missing, and the only people who knew about it were the conference delegates and the few workers whom they had managed to tell about it.

The workers gave their friendly support to our party's call for a one-day strike on March 20. On that day work stopped at almost all industrial plants.

The city was seething. Crowded rallies were in progress all over the place. Speeches were made by Bolsheviks, delegates to the Workers' Conference, and vanguard workers. On the streets, in theaters, and in cafeterias—any place where it was possible for the workers to gather—they gathered to demonstrate the unity within their ranks and, ultimately, within our Bolshevik party. It is characteristic that neither the British occupation command, nor the local Azerbaijani government dared to obstruct the strike or the political rallies in any way.

I had occasion to speak that day in Chornyi Gorod in the Nobel plant cafeteria, which held about a thousand people. After we were finished, the workers spoke. During the stories about how the Baku commissars had died, many wept. One had the feeling that the workers perceived the deaths of the commissars as a great personal loss. One old worker said: "We should blame ourselves for the deaths of our leaders; we did not defend them." He recalled the political meeting on July 25, 1918, on the eve of the fall of Soviet power, when in that same cafeteria, Alyosha Dzhaparidze had spoken for the last time. The old man remembered how they hadn't given him a chance to speak. "We were in despair and had lost faith in the Soviet's ability to defend Baku; we fell for the deceit of the SRs and Mensheviks, and voted in favor of the British against our leaders. Yes, we are guilty of their deaths by virtue of our blindness. But by the time we woke up and saw things clearly, our leaders were no longer among the living. Their

[14]See Appendix.

deaths opened our eyes and revealed the path of struggle that lies ahead. So we shall expiate our sins through unshakeable faith in the banner under which they fell." This speech made an enormous impression on everyone.

After the meeting was over I went to the Mailovsky Theater, where I spoke at another mass meeting. I arrived there with difficulty. There was a large overflow crowd of workers on the street who had been unable to push their way into the theater; in spite of the preannounced plan people had spontaneously streamed here from other districts of the city, and the adjacent streets were choked with people.

The rally was in full swing, with the workers already so excited that my first thought was that the leadership of the meeting was about to be snatched away from us. The excitement reached a fever pitch when one of the speakers, a maritime worker, reported that Druzhkin and Alaniya, the direct murderers of Shaumian, Dzhaparidze, and the other commissars, had been brought by the British command to Baku on their way to England via Batumi, and were at that very moment at the steamship pier. The thousand-voiced meeting began to buzz with calls for a march on the British command headquarters to demand that the murderers be handed over, and if they weren't, to tear the building down.

Realizing that such a spontaneous burst of enthusiasm would inevitably end in bloodshed and the rout of our forces (since it would not be difficult for the 15,000-man British garrison to shoot down an unarmed workers' demonstration), we announced firmly that such an action would only work to the advantage of our enemies. Workers' blood would be shed and would not achieve our goal. We reminded them that our fallen comrades, whose memory we were honoring that day, had always been opposed to such disorganized struggle on the part of the working class. We suggested sending a delegation from the rally to the British command with the demand that the murderers be brought to trial under the law. We also proposed leading an organized mass demonstration through the main streets of Baku in order to show both our power and our hatred of the enemies of the working class.

Although many still favored the idea of mounting an "attack" against British headquarters, the meeting accepted our proposal. One of the maritime workers, encouraged by all those present at the meeting, proposed that if the British wanted to send Druzhkin and Alaniya to Persia, the maritime workers should refuse to take them. The railroad workers made the same declaration.

A delegation to the British command was made up and dispatched, and the participants in the meeting, joined by the thousands of workers who had accumulated outside, marched in an impressive demonstration through the streets of Baku, singing revolutionary songs. The demonstration headed for the Workers' Club, the residence of the Workers' Conference and its presidium. I recall one typical incident.

When we marched through the streets with our heads bared and singing, "You fell as sacrifices . . . ," the people watching us from the sidewalks were also forced to remove their hats. One White officer standing on the sidewalk and looking at the columns of demonstrators passing by, did not bare his head, but threatening shouts ringing out of "Off with your hat!" forced him to remove his cap. The demonstration made its way peacefully through the main streets of the city and arrived at the Workers' Club where a rally took place.

The railroad workers kept their word: they made certain that none of the train crews would allow passage to Druzhkin—the man chiefly responsible for carrying out the execution of the commissars—or Alaniya, who had arrested them in Krasnovodsk.

The British command refused to hand them over. Afraid that Druzhkin and Alaniya would be torn to pieces on the train, the British outwitted us nonetheless, as we found out later, by sending them in an armored car via Shemakha to the station at Adzhi-Kabul where they succeeded in putting them on a train unnoticed and sending them to Batumi.

As a result of the mass demonstration, the day celebrating the memory of the Baku commissars became a day of triumph for the ideals of communism in Baku. We became convinced at first hand that the deeds and images of our unforgettable friends were alive in the hearts of the Baku workers. Their lives and great revolutionary deeds had inspired the workers to further struggle for the victory of communism. As before, they were with us, they lived on in our midst and were leading the Baku proletariat forward. Every one of us Communists could not help but be overjoyed.

Our enemies were also well aware of the fundamental changes that had come about in the arrangement of political forces in Baku. But we did not allow ourselves to become intoxicated with our initial success, realizing that the level attained by the workers' revolutionary movement also presented us with exceptionally lofty demands. We had to assume leadership of that movement.

* * *

In high spirits, we gathered that same evening in the Baku Party Committee where we shared our impressions, gave our assessments of the situation, and made predictions. It was clear to us that we had to expand our activities still more by creating strong party organizations in all the districts. It was necessary to secure a leadership position in the trade unions as well; and to put the workers' clubs in all districts under the leadership of the party organizations, thereby transforming them into bases from which to expand our mass political operations among the workers, and at the same time to use them as communications centers and cells for the district and industrial-plant party organizations. It was necessary at whatever cost to establish regular contact with Astrakhan, and through Astrakhan, with Moscow and the Party Central Committee.

Up until then we had received no instructions from the leadership either of the region or of Soviet Russia. Forced by circumstances, as the saying goes, "to be left to our own devices," we had had to define our tactics ourselves in response to developing events and to make independent decisions. Events were raising us up on a new significantly higher crest of the revolutionary wave.

In April, at one of the meetings of the Baku Party Committee, I learned to my surprise that my comrades had long ago created a fighting squadron under the aegis of the Baku Party Committee. On one hand I was happy to hear this. But on the other, I was astounded to hear what the members of this squadron, Comrades Gigoyan, Kovalyov, and Alikhanian, had to say at the committee meeting. They said that their squadron was well organized, that they had a solid supply of weapons, and that all that remained was to decide when to begin an armed rebellion in order to seize power in Baku. I was even more astonished at how many members of the party committee supported the idea of a rebellion. It even seemed to me that some members could vote in favor of this idea right there on the spot.

After lengthy debate the members of the committee appointed me to look into the situation more thoroughly and report my opinion at the next meeting of the committee. Alikhanian was the member of the military squadron I knew best. Two or three years younger than I, he was a close friend and former schoolmate, and as a dedicated revolutionary, he commanded our complete confidence. It is true that I knew Gevorg Alikhanian had been attracted to anarchism in his youth, and though that had been a "youthful passion," his character contained more than a little of the rebel. Gigoyan—a member of the VRK (Military Revolutionary Committee) of the Caucasian Red Army—had been arrested with the Baku commissars in Krasnovodsk. He had been released from prison sometime before our return to Baku, and when we arrived, he had been working in the underground party organization there for about two months. He was a few years older than I, had served in the old army, and was considered a rather experienced member of the revolutionary underground. The third member of the military squadron, P. Kovalyov, was older than all of us. He had come from Petrovsk where he had worked in the party committee. A former member of the party of Maximalists, he even now espoused a few anarchistic, ultra-Left views.

When I got together with this threesome and began to gain an understanding of what they were doing, I was amazed at how they could speak seriously of a rebellion in view of the limited quantity of arms and the tiny number of cadres they had at their disposal. I started to reproach them for having a frivolous, adventuristic attitude towards armed rebellion, and they began to justify themselves, saying they were placing their biggest stake on the Bolshevik-oriented sailors. But this argument did not stand up either, since at the time a large number of sailors were under arrest, or had been disarmed, and others were off in hiding somewhere. And there weren't that many "underground" sailors around who had not as yet been exposed as being in the navy. Although my threesome did not give in, everything nonetheless indicated that in their hearts they admitted the groundlessness of their positions.

At the Baku Party Committee meeting, in the presence of the threesome, I said, without revealing details, that the fighting squadron had no real power capability for armed rebellion, and also expressed my opinion on the political aspect of the question.

Even if, I said, we had enough weapons and fighting squadrons to organize an armed rebellion in order to seize power, it would still be adventurism that would end only in the complete rout of our organization. I pointed out that there were quite a few Mensheviks and SRs on the staff of the Workers' Conference, and that at the present time they also had control of the majority of trade unions, since party organizations in the districts had still not been created. It's true that the workers were in a militant frame of mind, but they weren't as organized as they should be. We had no contacts either with the Party Central Committee or with the Regional Committee in Tiflis; the northern Caucasus and the Transcaspian region were in the hands of the counterrevolution; only Astrakhan was under Soviet control, but we had no contacts with it. It was not in our power at the moment to overcome by ourselves the 15,000-man garrison of British occupation troops stationed in Baku, and several thousand Azerbaijani soldiers, and not one of us knew when Soviet Russia would be able to come to our assistance. These were the basic points covered in my speech.

Therefore, in conclusion I suggested that the fighting squadron continue its efforts to gather troops and weapons, but that the question of an armed rebellion be postponed to a later, more appropriate time.

It seemed to me that the members of the party committee not only agreed with me, but were even pleased by the way I had formulated the situation. Everything indicated that the militant threesome, made up of extremely energetic people who "rushed in where angels fear to tread" had, until then exerted a certain amount of pressure on their fellow members, and our conversation had helped the committee members to relieve themselves of that pressure.

As I think over what happened at that time, I am more amazed than ever at how prophetic many of Stepan Shaumian's pronouncements were. I recall as just one example the warnings he gave to the sailors in the Caspian Naval Fleet.

It must be said that the history of the Caspian Naval Fleet is in general rather interesting. In October 1917 the fleet, which sided with the Bolsheviks, played no small role in the revolutionary struggle for the transfer of power to the Baku Soviet of Workers' and Soldiers' Deputies. Rather important changes soon occurred in the fleet's personnel, however, when many of the most active, vanguard sailors left. This led to what happened during the critical period of the battles for Baku against the Turko-Muscavatist forces, when the sailors in the fleet, who had been deceived by the SRs, and through them partially bought off by the British, voted in favor of inviting the British into Baku.

Sharply criticizing the sailors of the fleet for that action Stepan Shaumian addressed them at that time, saying that they would very soon come to realize their mistake. The events which unfolded during the next few months in Baku opened the eyes of the deceived sailors. They realized the utter baseness of their action, regretted it, and the overwhelming majority of them reestablished their allegiance to Soviet Russia. In any case, after the British occupiers arrived in Baku for the second time, the sailors in the Caspian Naval Fleet joined ranks with the Baku workers and gave their friendly support to the common struggle against the occupying forces. My comrades told me that on the day of the victorious conclusion of the Baku workers' general political strike in December 1918 the artillery aboard all the ships of the Caspian Fleet fired a friendly salute in honor of the victory.

The Caspian Naval Fleet turned out to be a real thorn in the side of the British occupation command. It was mistress of the seas, and a mistress, moreover, who manifested openly Bolshevik sympathies at that time. The British had not yet managed to set up their own fleet to guard the city. A bitter struggle went on over the subject of the fleet. Some Mensheviks, in their speeches at the Workers' Conference, demanded that the Caspian Fleet be disarmed and turned into an ordinary merchant marine. Other Mensheviks, like Rokhlin, for example, proposed putting the entire fleet at the disposal of the Azerbaijani government. The British, for their part, dreamt of handing the Caspian Fleet over to the Denikinites. The Communists, however, succeeded in getting the fleet to depart to Astrakhan to be put at the disposal of Soviet power.

In February 1919 the British military command issued a disarmament ultimatum to the Caspian Fleet. This served as a signal to the fleet, in accordance with the decision of the Baku Party Committee, to weigh anchor and head out of Baku harbor for Astrakhan. Comrade Gogoberidze, as a representative of the Baku Party Committee, left with the fleet. But the ice on the Volga delta had still not melted, and there was no approach to Astrakhan. The fleet returned to Baku where there immediately ensued the disarmament of the battleships, the mass dismissal of sailors, and the arrests of "unreliable" Bolshevik-sympathizing sailors and commanders. The Baku Communists tried in every possible way through the Workers' Conference to help the sailors of the fleet and to prohibit the dismissals and persecutions of the sailors. But in the majority of cases their efforts did not yield the desired results.

With great difficulty some of our sailor-comrades succeeded in escaping the hands of the British. We soon passed a resolution through the presidium of the Workers' Conference to issue the British with an ultimatum demanding the release of the sailors. The British held out for a long time but were ultimately forced to give in. The sailors—there were about seventy of them—left prison. This success encouraged us and strengthened our position.

Meanwhile the Denikin counterrevolution was getting stronger and stronger. The Whites began to bolster their positions in Dagestan and on the Caspian Sea. Everything indicated that this plan had been put forward by the British, who wanted at all costs to have reliable support on the Caspian.

The British armed several merchant ships and tankers, such as the *Emmanuel Nobel*, the *Schmidt*, and others, with long-range guns and put them under their command. At the same time the occupiers set about organizing a Denikinite naval fleet under its joint command. With this goal in mind they transferred to Denikin a certain number of armed merchant vessels to be used for patrolling the northern part of the Caspian Sea. They were preparing to hand over to Denikin the entire Caspian Naval Fleet and all the military equipment remaining in Baku.

Seeing this, the presidium of the Workers' Conference made the following diplomatic announcement: in view of the fact that the military vessels and equipment are the property of the all-Russian government, and that this government has not yet been established in all the territory belonging to it, we cannot allow the transfer of this property to any person whatsoever. We demand the internment of all military vessels and equipment from now until such time as the all-Russian government shall have been fully restored within its boundaries.

Although the success of such a stall seemed doubtful under the circumstances, we nevertheless managed, with the support of most of the Caspian Fleet's personnel, to get the military vessels to remain at their moorings. It is true, however, that some of the military equipment fell into the hands of the Denikinites.

Not confident that we would succeed for long in keeping the military vessels at their moorings, we decided: with the help of our comrades—former sailors who had served on these ships—to remove the valuable machine parts from the ships and hide them in safe places. This operation was successfully carried out. It was a very important one: people who knew about such things maintained that the machine parts could be found or manufactured only in Petrograd. Thus, these vessels were prevented from being used by the counterrevolution.

In recollecting that time, I cannot refrain from telling how we helped to free a large group of worker-revolutionaries from the occupiers' prisons. When the Bolsheviks, who were operating illegally in the Transcaspian region, found out that after our return to Baku the Communists had assumed leadership of the Workers' Conference, they sent their representatives to Baku. They greeted the conference, expressing their enthusiasm over the Baku proletarians' heroic struggle, and declared the Transcaspian workers' solidarity with this struggle. At the same time they made the request that the Workers' Conference use its influence and appeal to the British occupation authorities with a demand for the release of worker-revolutionaries who were being kept in Transcaspian prisons.

The Workers' Conference greeted this request with wild applause. I was asked to speak in reply, on behalf of the conference. With the complete approval of the audience, I announced that the Workers' Conference would take all possible steps towards obtaining the release of its Transcaspian comrades.

Shortly thereafter, we sent the British command a written demand asking for the release of all political prisoners currently in the prisons of Krasnovodsk, Kizyl-Arvat, Ashkhabad, and Enzeli, and for their transfer to the jurisdiction of the Workers Conference. (It would have been absurd to leave our comrades where they were after their release, since they could again become victims of terror.)

In order to put some weight behind our demand we posed the threat of a general political strike. Unfruitful negotiations went on for about two weeks. In the end, however, the British command was forced to give in. All political prisoners were freed and transferred to the jurisdiction of the Workers' Conference.

Our numbers kept on growing!

* * *

In March of 1919 we managed to hold a Baku party conference—the first since the deaths of the Baku commissars. I remember that about thirty delegates from all the districts of Baku gathered for this conference. We listened to short reports about the political situation and condition of the party organization in the districts.

It was clear from the reports that a strong nucleus of Communists had been organized in almost every industrial district of Baku. They were under the leadership of Anashkin, Gogoberidze, Davlatov, Kasumov, Mayumedyarov, Mirzoyan, Plyashakov, Rodionov, Sarkis, Chukaryov, Yakubov, and other comrades. In Bailov, Bibi-Èibat, Balakhany, Surakhany, Zabrat; in Chornyi and Belyi Gorod, in the railway, municipal, and port districts of Baku; and at several businesses and plants, party cells were in operation, which, after this conference, were united into district organizations headed by regional party committees.

However, the condition of party operations was still not ripe enough for a general upsurge of the workers' movement. Many plants counted only one or two Communists among their numbers, and there were several plants, particularly small ones, where there was not a single member of our party. And although those Communists who had remained in Baku throughout everything were active party workers, they had not been sufficiently trained for carrying out the new, more responsible tasks now demanded of them, nor were they so well known to the workers as, for example, the Menshevik and SR representatives, who for the most part, had not been touched by arrests, executions, and exile.

Those who spoke at the conference cited many facts which demonstrated that the nonparty workers and former left-wing SRs who had been elected to the factory-plant committees and the workers' conferences were conducting themselves very well. Operations within the trade unions had begun to improve. At the conference our tactical approach was hammered out: to give active support to nonparty workers who distinguished themselves; to exert more influence on them in the realm of ideology; to show them confidence and all manner of support in preparing them for joining our party.

The conference passed the following resolution: in the course of the next few weeks to hold district party conferences, elect district committees, and immediately set about reestablishing party cells at the industrial plants. It was decided to accept into the party those nonparty people who had given a good account of themselves during the difficult events of the last several months—this was a better way of checking on potential party members than the usual candidate's service record.

The conference gave a great deal of attention to the issue of operations among the Moslem workers which were greatly in need of improvement.

The Gummet organization was in charge of operations among the Azerbaijani workers. Around that time it had undergone a significant change of personnel.

Perhaps it is worth recalling in this regard that the representative of the former Gummet (he was attached to the Baku committee of our party), A. B. Yusifzade, had taken part in the Sixth Communist Party Congress. In concluding his speech at that party congress, he said: ". . . I hope that the party will give us every possible support in our difficult and important work."

After this the chairman of the congress, V. P. Nogin, replied: "I shall permit myself, comrades, on behalf of the congress, to greet our Moslem comrades who are joining our party, and the delegate who has appeared for the first time at our congress. I worked in Baku in 1905–1906 and know what work our Gummet comrades have had to do and what great service they have given the party . . . I propose that the congress express its wish that the Central Committee give both financial and any other kind of support to Gummet."

Nogin's proposal was unanimously accepted by the congress.

It is well known that after the temporary fall of Soviet power in Baku the leaders and most outstanding activists in the Communist organization Gummet went to Soviet Russia. The Gummet-Bolsheviks who remained in Baku worked in the underground. When the Baku organization of the RKP (Bolsheviks) began to expand its activities, they took the most active part in it. At that time the leadership of the Gummet-Mensheviks moved from Tiflis to Baku. They were primarily members of the socialist faction of the former Transcaucasian assembly and then of the Muscavat parliament. They began their activities in Baku, as they had in Tiflis, under the name RSDRP—Gummet.

After events in Baku during the spring of 1919 Bolshevik influence in the leadership of Gummet began to get noticeably stronger, and the bitter struggle between them and the right-wing Gummet members intensified.

This was the question that confronted us: how could we use both the Adalet organization (which combined Azerbaijanis and emigrants from Persia) and particularly Gummet to intensify political operations with the Azerbaijanis? It was decided in this regard to hold a joint meeting with the Gummet-Bolshevik leaders and Adalet.

I remember three of the leaders of this organization very well—the Agayev brothers, gifted workers all. Although they had no education, these brothers were nonetheless naturally gifted leaders, possessed political experience, and were passionately, I would even say selflessly, devoted to the cause of communism. They took an active part in all meetings and functioned alongside the Communists.

There was not special report at the conference on the politics and tactics of our Bolshevik organization, although much was said on this score in the delegates' speeches. We had no differences of opinion on this matter since our line of attack had in essence been defined at the last meeting of the Workers' Conference.

At that time a particularly acute question was that of the necessity of establishing regular contact with the Caucasian Regional Party Committee located in Tiflis. Back in the fall of 1918, because of the Menshevik government's persecution of Bolsheviks in Georgia, the majority of the Regional Committee's members and the Regional Committee itself moved to Vladikavkaz (now Ordzhonikidze), where the Soviets were in power at that time. A group of Regional Committee members remained in Tiflis and acted as the center for Transcaucasia. After the Denikinites seized the northern Caucasus, leading workers from the Regional Committee made their way illegally from Vladikavkaz and

Baku to Tiflis and gradually began setting up operations in areas of Georgia and Armenia, although they did not get as far as Baku.

Because of this the conference ordered the Baku Party Committee to send its representatives to Tiflis in order to discuss with the Regional Committee all current issues, including the question of calling a meeting in the near future, of the All-Caucasian Party Congress not in Tiflis, but in Baku. The choice of Baku for the meeting was motivated by the fact that it was the scene of seething political activity at that time; whereas in Georgia and Armenia all was calm. Another important reason for holding the congress in Baku was also that in the prevailing circumstances not all our leading party workers felt they should leave Baku at the same time. The Baku conference elected Gogoberidze, Avis, Anashkin, Olga Korneyeva, Lidak, Sarkis, and myself as delegates to the All-Caucasian Congress. (However, as a result of insufficient preparation the Regional Committee cancelled its taking place at that time.)

The conference approved the policies which were being carried out by the Baku Party Committee and elected a new committee for the RKP (Bolsheviks). The Baku Party Committee which was in office before March was not, as usual, elected at the party conference, but was formed by means of what is known as co-optation. This had happened more than once in our party when existing dire circumstances made it impossible to hold either a party congress, or a conference. The newly elected Conference Committee was given the job of expanding mass operations among the Azerbaijani workers in all districts, by drawing active workers who had influence upon the Azerbaijanis into the staffs of the factory-plant and industrial committees and into the presidiums of the district workers' conferences; and by turning its main focus of attention to the work of the Gummet organization.

The next day we arranged an informal meeting with two representatives of Gummet—Ali Geydar Karayev, a member of the presidium of the Workers' Conference, and Mirza Davud Guseynov. One of the leaders of Adalet—Bakhram Agayev was also invited to this meeting. Those of us who took part in this meeting were Gogoberidze, Sarkis, Anashkin, and myself. We summarized our views on the general situation in Baku and said that we considered it of prime importance to focus attention on operations among the Azerbaijani workers and within Gummet. They agreed that at present a relatively small number of Azerbaijani workers were involved in the revolutionary struggle and some were still under the direct influence of the Muscavatists. They also agreed that Gummet was doing very little work among these workers. We called our comrades' attention to the fact that we could not continue to tolerate a situation where Gummet members were speaking out against our policies in several districts. Agayev completely agreed with us on this point. He detailed the plans of Adalet and assured us that the Persian workers would support all our measures and would join forces with the Russian and Armenian workers. He gave a correct formulation of the issue about whether members of Adalet should be allowed to be present at meetings of Communist cells, particularly in view of the fact that this was actually already the case in many localities. We promised to give our district organizations instructions on this matter.

Everything indicated that both Karayev and Guseynov had understood us correctly. But they spoke of the difficulties that confronted them. As for themselves they were personally with us all the way and said that we could rely on them as much as on actual Bolsheviks. They noted Mirfatag Musevi and Ashum Aliyev as people who could also be completely trusted. (Many of the members of the Gummet leadership adhered to the views of the right-wing socialists.)

Then we raised the issue of whether the time had not come for the Bolsheviks in Gummet to break away from the Mensheviks, since in the face of the current upsurge of enthusiasm in the workers' movement, and the working masses and significant shift to the Left, the existence of a united Gummet only worked to the advantage of the Mensheviks in Gummet and was hurting the Communists.

Because of his inherent softness, Karayev could not immediately agree to a split—for him it represented too sharp an about-face, but he promised to ascertain the feelings of his comrades in the Gummet leadership on this matter, and to learn precisely how preparations for the Gummet conference were going in order to insure a preponderance of Bolshevik supporters at it. Guseynov shared Karayev's views.

We agreed with them. All the more so since a certain amount of time was necessary for the masses themselves to become convinced, on the basis of their own experience, of the appeasement-oriented and treacherous nature of the Gummet Mensheviks, who were skillfully playing upon the nationalist sentiments of the politically backward laboring Moslem masses.

We were unanimous in supporting the idea that the Bolshevik Gummet members should remain in the parliament and use its tribunal for exposing the government's antipopulist policies.

On April 25, 1919 a Gummet conference took place in the Workers' Club at which a fierce struggle developed between the Mensheviks and the Bolsheviks. The leadership of the organization, comprised of eleven people, included seven Bolsheviks: Karayev, Guseynov, Musevi, Aliyev, Agazade, and others.

A short time later the question of splitting Gummet and creating a Communist Gummet organization actually came up before the Baku organization of the RKP (Bolsheviks). These things were soon realized.

About two days after our meeting with the representatives of Gummet and Adalet, Gogoberidze, Anashkin, and I left for Tiflis in order to take part in a meeting of the Regional Party Committee and to carry out the resolutions passed by the Baku Party Committee.

In proposing the convocation of a regional All-Caucasian Party Congress and in going so far as to elect our delegates to it, we were sure that our proposal to hold this congress in Baku would not meet with any objections on the part of the Regional Committee, since back in December 1918 the previous Regional Committee staff, then in Vladikavkaz, had issued an appeal to the party organizations in the northern Caucasus to send their delegates in January 1919 to the Regional Party Congress that was being convened in Vladikavkaz. At that time in Baku, and in all of Azerbaijan for that matter, party organizations had not yet been reestablished. The party organizations in Armenia and Georgia had been completely routed—as I mentioned earlier, some Communists were in prison, some were in hiding, and a great many had gone to the northern Caucasus where they took part in the Civil War, in the struggle for Soviet rule. Therefore, because of the current situation, many organizations were not represented at the Regional Party Congress, which nevertheless did take place in Vladikavkaz in January 1919 and which, in consideration of the enormous complexity of the situation, declared itself duly empowered. The Regional Committee of the party was elected at that congress.

Now, when the situation had changed, the convocation of a fully-empowered party congress had become both completely practicable and urgent. Many unsolved basic questions connected with the future development of the revolutionary workers' movement had reached the final stage of ripening; extremely critical problems had arisen concerning the unification and concentration of the forces of all party organizations, and the working out of a single set of tactics that took into account the general state of affairs in Soviet Russia. We needed close, daily contact with the Regional Party Committee, but it had been impossible for the simple reason that there had not been a single representative from Baku on its staff.

Chapter II

For the Unification of the Workers of Transcaucasia

Meetings with friends—Among near and dear ones—The Congress of Baku Unions—The issue of the Soviets—On the boats to Soviet Astrakhan.

After getting in touch—via the secret network—with comrades from the Regional Committee, we arranged a meeting with them. Luckily for us, almost all the members of the Regional Committee were then in Tiflis, and the few who were in Kutaisi also came to meet with us.

The session of the Regional Committee was summoned according to all the rules for clandestine operations. None of us had seen each other for a long time, hence the meeting was a warm and joyous one. We embraced each other in brotherly fashion and reminisced about mutual friends before settling down to a detailed discussion of the matters before us. The comrades from Tiflis waited with great interest to hear about the situation in Baku: they had heard many rumors—most of them positive—about what we were doing, but they had had almost no reliable information.

Our story aroused a lot of interest. The comrades from the Regional Committee were happy about the rapid regeneration of the Baku party organization and the expansion of its leading role in the workers' movement. We, in turn, asked the members of the Regional Committee to tell us what had been happening in Georgia and Armenia; how the committee's work was going; what forces the party had at its disposal in Tiflis; why the committee had not displayed sufficient zeal in establishing ties with Baku; and why it had not assisted us with cadres of leaders, of which there was such a shortage in Baku.

The following picture began to take shape after we heard the reports of the Tiflis comrades, and also that of Orakhelashvili on Kutaisi, and of Mravian on Armenia. Many of the Communists were in prison; some of the party members were, in fact, not working: they were afraid of reprisals. But the politically active segment of the party continued to work illegally, and underground organizations existed.

However, the Regional Committee had succeeded only recently in establishing regular contacts with the local organizations.

Then, on our suggestion, there was a discussion of the following issues: the summoning of an All-Caucasian Party Conference in Baku, and of an All-Caucasian Congress of Trade Unions. Formally speaking, there existed in Tiflis a Regional Caucasian Council of Trade Unions, but almost nothing was known in Azerbaijan of its work. True, that didn't distress us very much; the council consisted of Mensheviks; it had been elected no one knows when, and, naturally, it did not serve as any kind of guiding center for the regional trade union movement. We told our comrades that at the forthcoming Regional Trade Union Congress we would provide a sufficiently high percentage of Communists and their supporters from Baku. We could imagine, of course, how much more complicated things were in the trade unions of Georgia and Armenia, but if Communists and their supporters made up only a third or a quarter of the trade union delegates from these two Republics, then there was a real possibility of securing a leading role for the Communists both at the congress itself, and in the Regional Council of Trade Unions elected by it.

After the discussion the Regional Committee passed a resolution concerning the necessity of summoning a Trade Union Congress. Since the working class and professional movement had developed most extensively and successfully in Baku, it was decided to hold the Congress of Trade Unions in Baku. Therefore, the Regional Committee assigned us, the Baku comrades, the work of preparing for the congress. Many issues which were of great concern to us were resolved in the Regional Committee in just such a businesslike fashion.

In recalling how the discussion of these issues was carried out, I wish to make note of the high-level, party-conscious character and the comradely atmosphere that prevailed. We were particularly gratified that the Regional Committee members accepted our criticism of them in a partylike spirit, and tried to give us a frank explanation of both the objective and subjective reasons behind the deficiencies in their work. As a result of this meeting we formed good personal and comradely relations with each other, and these discussions not only enriched our knowledge of what was going on in the Caucasus but helped clarify the future prospects of our entire movement.

Later, as I strolled along the streets of old Tiflis looking at the buildings and store windows and gazing at the faces of the passersby, I realized an amazing thing: it looked the same as it did the last time I was there! Not one new building had been built or was under construction, the stores were the same as they had been before, only a little shabbier, and the inhabitants were as preoccupied as before with their own private concerns.

But there was one thing different about Tiflis: British officers and soldiers strolled leisurely down the streets, as in Baku, either in groups or alone. Frequently the soldiers would turn off into the yards of the houses and sell the local population what they hand in hand—odds and ends of some kind, or part of their quite abundant rations: canned goods, chocolate, condensed milk, etc. The appearance on the streets of British soldiers wearing exotic Scottish kilts was usually a cause for jokes and taunts among the onlookers, especially the boys, but the British behaved quite peacefully, even good-naturedly, and the population responded in kind. To some degree both the backyard buying and selling and the Scottish costumes became part of everyday life in Tiflis.

We could walk freely about the city at that time, and since we had come to the city legally, using authentic passports, we felt relatively at ease ourselves, although our visits to secret meeting places and underground apartments were, of course, another matter; here we firmly observed the precautions dictated by our illegal status as Communists. During my visit there I had a lot of meetings with different people. Especially pleasant were my meetings with the Tiflis comrades whom I had worked with earlier, in 1917.

The head person, if one may so express it, in the Regional Committee at that time was Philip Makharadze, for whom I had long cherished a feeling of deep respect. An old-guard, educated Marxist and a man of letters who had joined Marxist circles long before the formation of our party, Philip had a special kind of charm. He was of average height, had handsome features, blue eyes, and a long beard, and resembled the biblical prophets depicted in old paintings. He didn't have much to say in discussions, but the reports he gave were lengthy. In his relations with people he was polite.

I had frequent meetings with Danush Shaverdian, who was working in the Regional Committee at the time. As an able and experienced conspirator he enjoyed the special confidence of the party organization, even though he also had permanent legal status as a lawyer in Tiflis. For many years Danush was the treasurer and custodian of the most important secret archives of the underground regional center of the party. The party organization assiduously protected him from the police sleuths. No legal party work, which would be noticeable to the police, was given to him, and he was himself careful to the utmost degree. Danush enjoyed general love and respect.

Especially pleasant for me was my meeting with Comrade Mravian, on whose shoulders lay the main burden of leadership for all the Communist organizations in Armenia. I had known him before I met Makharadze. In the fall of 1915, after turning twenty, when I had joined the Bolshevik party, Mravian was already an old-guard party member and a professional revolutionary. He was a great help to me in the early stages of my active party life. He was working at that time on the legal newspaper that had a Bolshevik tendency, *The Struggle* (*Paikar*), where I met him in the editorial office on many occasions. After the victory of Soviet power in Armenia he became the People's Commissar of Education of the Republic, and was one of the prominent figures in the Communist party of Armenia.

My meeting with Comrade Lado Dumbadze also comes to mind. A worker in the main railway repair shops of Tiflis, he was also one of the old-guard members of the Communist underground. Well-trained politically, energetic, fearless, he inspired great respect because of his ability to defend his opinions boldly. Although the chief bosses in the railway repair shops were the Menshevik leaders, who turned these shops with their numerous body of workers into the mainstay of their support, Dumbadze nonetheless, along with a small group of Communists, had great influence and authority among the workers and took the wind out of the Mensheviks' sails on more than one occasion.

I had an extremely warm meeting with Amayak Nazaretian, a tireless member of the party vanguard and a Marxist man of letters. The first time I met him was in March of 1917 at the first legal meeting of Bolsheviks in the Zubalov People's Assembly Hall. He was then one of the three participants elected to the presidium of the meeting. Amayak was subsequently one of the outstanding figures in the Transcaucasian party organizations, and later he worked in Moscow in responsible party posts.

I was also pleased with my meeting with Mamiya Orakhelashvili, whom I had met for the first time in October 1917 at a session of the All-Caucasian Regional Congress of the party, where he represented the Vladikavkaz party organization. Now he was living in Kutaisi and held a legal position as an army doctor in the troops of the Menshevik government. A doctor by profession, he had served in this capacity in the Tsarist army while at the same time carrying out political work among the soldiers.

The Bolshevik Malakiya Toroshelidze was also a member of the Regional Committee of the party. I knew him well in 1917. He was considered a well-trained Marxist, and at the All-Caucasian Regional Congress of the party he had given a report on the nationality issue. Toroshelidze had a calm disposition and never got excited, and although he was firm in his convictions, in debates he never exacerbated the situation. In general he conveyed the impression of being a serious party worker. At that time he was serving as a member of the Central Board of Consumers' Cooperatives of Georgia (Tsekavshiri), on which he was the only Bolshevik member, all the rest being Mensheviks. They knew, of course, that Toroshelidze was a member of the Regional Committee of our party, but they didn't arrest him or relieve him of his post on the Consumers' Cooperatives. He carried out individual assignments for the Regional Committee in connection with the local party organizations.

I recall a curious circumstance connected with Toroshelidze: his wife was a Menshevik, and, moreover, a politically active worker who was a member of the Central Committee of the Menshevik party of Georgia. They lived in the same apartment. When I arrived in Tiflis and found out about this, I was utterly perplexed: can they really not ever discuss politics, I wondered, and if they do talk politics, how do they manage to conceal party business from each other? I was told that as a result of this situation certain members of the Regional Committee had expressed doubts concerning Toroshelidze. And, in fact, it seemed rather odd, not to say unnatural, for a member of the underground Regional Committee of the Bolshevik party and an active member of the Central Committee of the Mensheviks to live together as man and wife under the same roof. However, Toroshelidze made a good impression on me at this meeting, as he had before, and I couldn't share their doubts: I trusted him.

The comrades who had doubts about Toroshelidze's political integrity usually said: why are other Communists arrested and not he? They would then cite, by way of example, the fact that when mass arrests of Communists had been carried out in Tiflis, Toroshelidze had escaped arrest by leaving town two days earlier on a business trip for the cooperative, and they concluded that his wife probably had warned him beforehand. However, these were only conjectures which were provoked by Toroshelidze's unusual family situation.

After the final victory of Soviet power in Georgia, Toroshelidze continued to work effectively. In 1925 Stalin and I spent several days in Georgia, where Sergo Ordzhonikidze was then secretary of the party Regional Committee. Stalin invited Toroshelidze, as his old, prerevolutionary friend, to accompany us. We spent several days together, traveled to Kutaisi and from there to the Rachinsky district, and during this trip I came to know him even better as a completely worthy man, deserving of confidence.

During the time that I'm speaking of in my main narrative, two Bolsheviks—Mikha Tskhakaya, a prominent member of the Party Regional Committee and an outstanding party veteran, and the old-guard Bolshevik, Mdivani—were in prison in Kutaisi, arrested by the Mensheviks. After the establishment of Soviet power in Georgia, in February 1921, Mdivani became Chairman of the Revolutionary Committee, and then of the Council of People's Commissars of Georgia. I met him for the first time in 1920, when the Eleventh Army, at the request of the Azerbaijani Revolutionary Committee, came to Azerbaijan. He came along with the Army as a prominent political worker. I remember that he was in uniform, carrying a Mauser in a wooden holster and wearing a *budyonovka* (style of helmet worn by Red Army men during 1918–1921), and that he never parted with this "armor." It should be said that they were very becoming to him and he was proud of them. We became well acquainted at that time, and I liked him as a man who was candid, energetic, and witty.

I met Beso Lominadze for the first time at the session of the Regional Committee where he then represented the Tiflis Party Committee. He was younger than I; back in 1917 he still hadn't appeared in the political arena. Our report on the seething political life of Baku and the rapid growth of Communist influence made a tremendous impression on him. After the session he came over to see me, and we strolled about the streets of Tiflis for a long time while he questioned me about the situation in Baku. At times it seemed that he didn't quite believe what I was telling him. He was continually exclaiming: "Is that really true? Is it really like that? How did you manage that? That's astonishing! But see what's happening with us: Tiflis is one big swamp, with no movement, no strikes, no demonstrations, only illegal groups. Sometimes leaflets are distributed on different occasions, and we support our comrades who are in prison. That's the sum and substance of our work!"

Naturally, Lominadze somewhat underestimated what was going on in Tiflis at that time, but the contrast with Baku was, in fact, a striking one. He, on his part, was attracted by the general atmosphere in Baku. He asked me: "Will you take me to work in Baku if I made this request of the Regional Committee?" to which I replied that I would, with pleasure. "You yourself heard how we asked the committee to send us some experienced party workers who are well-trained Marxists," I said. "If the Regional Committee lets you go, we'll be happy to have you."

My meeting with Georgi Sturua was very cordial. So much bound us together! There had been our joint work in the commune and in the Baku underground; then prison; we had been together through many trials which reveal a man's character. We trusted each other completely, and loved each other like brothers. I remember how I even said how aggrieved I was that after all the Baku travails he had not returned to us in Baku, but had stayed to work in Georgia. He had an excellent knowledge of Baku, and the workers there knew and loved him. Georgi gave me his word then that he would return to Baku.

* * *

Aside from my meetings with the leading party figures in Tiflis there were also my reunions with former classmates, an overwhelming majority of whom had already become Communists. Apart from the friendliness and warmth of my relations with them, our conversations helped me to get my bearings regarding the current situation.

Do I have to add how happy my Aunt Verginia Tumanian was, and also her husband and children, that I had returned alive and healthy! They doted on me, and did everything to ensure that the three days which I spent in Tiflis were, in some sense, days of rest, despite the heavy amount of work that I had to do. I had very much wanted to go to Sanain and see my mother and my brothers and sisters, but such a trip would have required two days, and matters in Baku made it impossible for us to delay our return.

I didn't manage to see Ashkhen. She wasn't in Tiflis: she was near Sukhum, in a small Armenian village, where she worked as a teacher. I sent her a letter in which I asked her to return to Tiflis after the school year and to let me know when she was coming so that I could go there too. "That won't be difficult for me," I reported in my letter, "since my work requires me to go to Tiflis at least once a month to attend to various matters."

While I am recalling these far-off times, I want to speak in more detail of the Tumanian family.

My Aunt Verginia Tumanian had never been to school, but she could read and write, and was quite mature in a political sense. Naturally, she was not well versed in the fine points of politics, but she supported the revolution and the Bolsheviks with all her heart. Her husband, Lazar (on his passport—Gabriel) Tumanian, was a more educated person. He worked as a shop assistant, but dreamed of owning his own shop, and counted every penny until finally he was able to open a small shop where he worked without salesmen and assistants. However, this "business" didn't last very long. There was a big fire in his shop and once again Lazar Tumanian became a shop assistant. They lived in their own home in one of the most desolate sections of Tiflis, in the Surpkarapetsky ravine.

Lazar Tumanian was industrious, virtuous, and honest: it seems that these virtues were the major reason for his lack of success in "business." In contrast to his wife, he was not interested in the Revolution and socialism. On the other hand, he read the conservative Armenian paper *Mshak* every day from cover to cover, and, moreover, he collected every issue of this paper and bound them neatly together in one book—he had several such large bound volumes already. Politically he was limited, and believed only what was written in this paper. But his wife was a strong-willed woman and was, in fact, the master of their house. Her husband loved her and could not bring himself to contradict her.

I told Aunt Verginia about the events in my life in the past year and a half, and also, insofar as it was proper for her to know about them, about the development of revolutionary events in Baku. When I told her that we were holding secret sessions in Tiflis, she responded with approval.

One day Philip Makharadze happened to tell me that the Mensheviks had begun to follow him persistently, and even though he was constantly changing the location of his underground apartments, he was afraid, nonetheless, that he would end up in a Menshevik jail. I decided to have a chat with Verginia. The location of their house, plus neighbors who had no interest in politics, made her dwelling an ideal place for an underground apartment. I asked my aunt whether she would agree to give refuge in an apartment in her home to a prominent Georgian Communist, a very good comrade, whom the Mensheviks were after. She agreed without hesitation. Even after I cautioned her that this was a dangerous business—"If something happened to him," I said, "then you could all be in a jam"—she said that she wasn't afraid and that she was ready for anything. Then I asked her, "And what will your husband say about this?" "Don't worry," she said, "I'll have a talk with him, he won't object."

The following night I brought Makharadze home with me and introduced him to the family. My aunt showed him the room intended for him, and then gave him a delicious supper, leaving Philip pleased with the reception accorded him by the Tumanian family.

He lived there a fairly long time. He almost never went into the city, and when he had to meet with his comrades, they would go to see him, either alone, or in pairs. He would usually come into the sessions of the Regional Committee late in the evening, observing all the rules of conspiratorial security, and would return in exactly the same manner. From then on, every time I came to Tiflis I stayed at this apartment and had long chats with Makharadze. We got to know each other even better and became closer than ever.

I recall one amusing incident concerning both Makharadze and Mikha Tskhakaya that took place somewhat later. After arriving in Moscow for the Eleventh Congress of the party, we met with Makharadze the day before the congress was to begin in the Third House of Soviets, where the credentials committee of the congress was working. All of us filled out our questionnaires and received certification as delegates, as was the custom.

The old veteran Mikha Tskhakaya was with us. So that he wouldn't have any trouble filling out the questionnaire, the registrar decided to do it for him. "What year did you join the party?" he asked Mikha. "Write: a long time ago," answered Mikha. The registrar began to object: "That's impossible. I have to mark down the exact year." Mikha looked around him. "See that Komsomol member over there filling out his questionnaire?" he said, pointing to Makharadze. "See what year he puts down, and then make mine ten years earlier." All those nearby gaped in astonishment. We knew that Makharadze had taken part in the revolutionary movement since 1891; how long ago had our Mikha set out on the path of revolution!

Philip Makharadze continued to live happily in the Tumanian household while engaged in feverish revolutionary work until December 1919. I found out later that Philip had disregarded the rules of conspiratorial security and left the house during the daytime. He was immediately recognized on the street because of his beard (he couldn't part with it, even in the interests of security), and was arrested. Since

Philip had Lazar Tumanian's passport, the police discovered the Regional Committee's underground apartment at my Aunt Verginia's house, searched it, and removed a number of important party documents. Lazar Tumanian and Gaik, his seventeen-year-old schoolboy son, who did various errands for me and Philip, were arrested. But even after this, Verginia Tumanian continued to worry about Philip. With the help of my thirteen-year-old schoolboy brother, who also lived with them at the time, she sent messages to the prison to her husband and son and to Philip.

Perhaps this document, which an Armenian comrade recently found in the archives and brought to me, will be of interest. It seems that after the arrest of her husband and son, Verginia appealed to the Armenian Consulate in Tiflis with a request that it intervene in the matter and make every effort to secure their release.

As is evident from the document, Verginia exercised some subtle diplomacy in this matter when she didn't mention a word about Makharadze. She attributed all the Bolshevik papers found in her home to me, since at this time I was beyond the reach of the Menshevik government. I shall now quote this document.

> To the Attaché *sovetnik* of the diplomatic mission, Prince M. Tumanian. It is my honor to inform you that two days ago a resident of the village of Dseg (in Armenia-A. M.), Verginia Tumanian, came to see me and said that a search, headed by the commander of a special detachment, was carried out in her home on December 1 which yielded no positive results. The room of her neighbor was searched at the same time and various pamphlets and books of a Bolshevik nature were found. In this room there lived the thirteen-year-old student Anushavan Mikoyan, and his brother Anastas Mikoyan. The latter had come from Baku to stay for several days. The special detachment, not finding Anastas Mikoyan, to whom the forbidden literature belonged, arrested the husband of Mme. Tumanian—Gabriel Tumanian, sixty years old, and his son—Gaik, seventeen years old, a student in the seventh class at the gymnasium. In imparting this information, Mme. Tumanian humbly requests Your assistance in securing the release of her husband and son. The parties under arrest are currently in the Metekh fortress
>
> The Head of the Consular Department (signature),
> December 29, 1919

My brother remembered the details of the search in Aunt Vergush's house:

> That day I was sick, and was lying on the ottoman near the window. Dinner was being prepared. Suddenly the sound of boots

stomping over the cobblestones in the yard was heard—the house was being surrounded.

Auntie quickly stuffed some papers under my mattress and threw others onto the hot primus stove, and just then the gendarmes burst in. They rushed over to the stove, grabbed everything that hadn't had time to burn, and began the search.

They threw me off the ottoman. Then Auntie cried: "Don't touch the sick boy!" But that didn't help; they had already removed the papers from under the mattress and smacked me around. They hit me again because I wouldn't say where my brother Anastas was.

While this was going on, they were hitting and interrogating Lazar and his son Gaik in the other room. Then they were taken away to jail.

In trying to find a secret hiding place the constables began to break the room partition. But they still didn't find the hiding place since it was cleverly concealed by the china cupboard. (The cupboard covered an emergency subterranean exit to the garden in the event that the conspirators had to flee during the secret sessions.)

The messages which I carried to the prison Aunt Vergush cleverly hid under the pilaf: a letter was put at the bottom of a deep pot, and pilaf was piled on top. The constable poked a spoon around in the pot to see if there were any illicit enclosures, but whether the spoon was too short or the pot too deep, the letters managed to get to their addresses.

The Bolshevik Avetik Abovian was in the same cell with Lazar.

Verginia Tumanian was not only an intelligent person with politically advanced sympathies, she was also an excellent mother. She had seven children. Three of them died from infectious diseases, the remaining three daughters and one son joined the Communist party. Gaik Tumanian began his service in the Red Army after finishing Sverdlov Communist University. He graduated from the Military Academy. He was in Spain in the 1930's when the Civil War was going on, and during the Great Patriotic War he was at the various fronts as a member of the council of war of the armored forces, and in the Far East in battles for the liberation of Manchuria. Now a lieutenant-general in the Reserves, and a Candidate of Sciences, he teaches in one of the Moscow institutions of higher learning.

After I went to work in Moscow, Verginia Tumanian and her husband, along with my mother, lived with me for a very long time. Her husband died at the age of eighty. She and my mother lived together harmoniously, like sisters, and died the same week: Verginia was eighty-five, and my mother ninety-three. All were buried side-by-side in Novodevichy Cemetery.

* * *

On our return to Baku we reported the results of our trip to the Party Committee where the work we had accomplished met with approval. Now the resolutions which had been passed by us jointly with the Regional Committee had to be carried out; beginning, first and foremost, with the resolution concerning the Congress of Trade Unions. Moreover, during our trip to Tiflis quite a few assorted matters had piled up in Baku.

The Communists in the Baku Council of Trade Unions managed without any particular difficulty to put through a resolution concerning the convening of a congress of the trade unions of Transcaucasia, Dagestan, and the Transcaspian region. A suitable appeal to the trade unions of Georgia, Armenia, Dagestan, and the Transcaspian region was accepted on behalf of the Baku Council of Trade Unions, and the date of April 15 was set for the convening of the Regional Congress.

As I have said, the Caucasian Council of Trade Unions (elected at the end of 1917) continued to function in Tiflis, although in actual fact, it had neither the wish nor the possibility of extending its influence outside Georgia. The Baku Council of Trade Unions was the guiding center of all the trade unions in Azerbaijan. The Mensheviks and SRs who served on the council (occupying a rather large number of leading posts) agreed to the convocation of an All-Caucasian Congress. They assumed that with the support of the unions of Georgia and Armenia, where their influence was strong, they would be able to occupy the leading position in the congress itself and in the organs elected by the congress.

But we took the measures necessary to ensure victory for our party by seeing to it that in the election of delegates to the Congress of Trade Unions (first and foremost, in Baku) as many Communists as possible were nominated, also Gummetists and Adaletists, who were close to the Bolsheviks, as well as workers with a Bolshevik outlook, and Azerbaijan left-wing SRs, who then pursued a concerted policy with us. All this work was entrusted—according to the Baku Party Committee's policy—to Comrades Anashkin, Mirzoyan, and Poltoratsky.

The Georgian Mensheviks, who ran the Caucasian Council of Trade Unions, were taken unawares by the initiative of the Baku Council of Trade Unions. They didn't object to convening an All-Caucasian Congress of Trade Unions, but they disputed other issues related to the convening of this congress. In the first place, they spoke out against the participation of Dagestan and the Transcaspian region in the congress, insisting that the congress should have a purely "Transcaucasian" character, but the real reason for their objection was the fact that Dagestan and the Transcaspian region were pro-Bolshevik in their sympathies.

Secondly, the Mensheviks proposed that the congress be held not in Baku, but in Tiflis—which they said was the center of Transcaucasia—and the place where the First Congress of Trade Unions had been convened.

They were especially stubborn about the standards of representation to be applied, insisting that unions with up to 150 members send one delegate to the congress—on an equal footing with the unions having a membership in the thousands. The motive behind this was clear: small unions predominated in Georgia, but in Baku the opposite was true and the large unions with thousands of members were in the majority. This was not only unfair but had as its goal the undermining of the role of the Baku proletariat at the Trade Union Congress.

Finally, the Mensheviks demanded that no political issues be discussed at the congress. According to their logic, the congress should have a purely professional character.

In order not to give the Mensheviks grounds for wrecking the congress, we decided to meet them halfway and consent to the congress being held in Tiflis, but only if they withdrew all their other objections. This didn't help. After receiving the notification of the Baku unions of the All-Caucasian Congress in Baku, the Mensheviks called a Congress of Georgian Trade Unions in Tiflis at exactly the same time!

The only people who came from Tiflis to the Baku Congress were representatives from individual unions where the Bolsheviks and the Menshevik-internationalists constituted the majority of the leadership (the unions of tanners, pharmacists, and metal-workers). Included among the delegates from Tiflis were such Communists as Lominadze and Mzhavanadze, and such Menshevik-internationalists as Pirumov and the former member of the State Duma, Arshak Zurabov.

The Congress of Trade Unions opened in Baku on April 7, and on April 11, the Congress of Georgian Trade Unions opened in Tiflis. Meanwhile, negotiations still continued between Baku and Tiflis concerning the convening of an All-Caucasian Congress of Trade Unions. The Baku Trade Union Congress decided to send its own delegate to Tiflis to conclude the negotiations. My name was put forward. (I had already been invited on March 20 at a session of the Council of Trade Unions to participate in the work of the commission convening the Congress of Trade Unions.)

In the beginning of April I was sent to Tiflis by the Baku Party Committee to carry out negotiations with the Party Regional Committee regarding several issues which were of importance to us, and thus I was in Tiflis when the Congress of Georgian Trade Unions was taking place. On April 13 I addressed the delegates at the Congress of Georgian Trade Unions. Here is an abbreviated version of my remarks, which have been preserved in the minutes of the Congress:

The floor was given to a representative of the Baku faction, Comrade Mikoyan, who called upon the proletariat of Georgia to emerge from the narrow confines of their state and seek ways of uniting the entire working class.

The proletariat of Baku is striving for such unification in every way possible, and if it insisted on the congress being convened in Baku, it was only because it did not wish to leave the working masses of Baku without leadership at the present moment.

The proletariat and the bourgeoisie of Baku constitute two hostile camps, which spy vigilantly on each other and are ever ready for attack at a propitious moment. The absence of the leaders of the working class could prompt the aggression of the British command and be a decisive factor in the victory of the latter. Lack of funds also served as an impediment to a trip to Tiflis.

Despite all these considerations, the comrades from Baku would have come to Tiflis in the interests of unifying the working class, if there had not been major differences of opinion regarding the question of the agenda.

The speaker went on to speak of the necessity of a union between the proletariat of Tiflis and Baku despite the divergence of opinion between our parties, and he maintained that the working class of Baku, the Communists in particular, are not hostile to the independence of Georgia and are of the opinion that although going their separate ways, the democracy (the Georgian government) must defeat the common enemies of the working class together with them.

The Communists of Baku share a Marxist point of view and do not intend to carry Soviet power into Georgia on their bayonets, but they call upon the proletariat of Georgia to renounce their narrow nationalistic policy and unite with the entire working class.

The speaker concluded with a call to the democracy of Georgia to support the proletariat of Baku in its unequal struggle with the enemy, for the working class of Georgia must not remain a mere spectator while their class brothers are waging a life-and-death struggle.

Insofar as the Mensheviks' position regarding all this had already been decided beforehand, my appeal on behalf of the Baku proletariat naturally couldn't change anything.

The Congress of Trade Unions in Baku continued until April 15. It also witnessed the development of acute political struggle between the different tendencies within the workers' movement. My comrades had to struggle not only against the right-wing SRs and the Mensheviks, but also against the Menshevik-internationalists headed by Zurabov.

In contrast to the right-wing SRs and the Mensheviks, Zurabov adopted a kind of middle position, which was close to our party line in many ways. He set himself the task of reconciling the different factions within the working class and of passing a resolution at the congress which would unite them. However, this resolution turned out to be very vague politically. Under the circumstances, it could not have been otherwise.

The Bolsheviks presented a united front at the Congress, together with the Gummetists and the Adaletists. Another group to come out in concert with us with the Azerbaijani left-wing SRs, whose leader, Rukhulla Akhundov, said in a brilliant speech: "The slogans of communism are now being proclaimed in Red Moscow, Europe is in the throes of social revolution, but the wave of revolution has still not hit our shores. Socialists call upon the entire proletariat to unite; we must be untiring in our campaign for unity in the rural areas as well as in the cities. But whatever resolutions are passed, they will have no meaning until we achieve political power—the dictatorship of the proletariat. We have such power, but it was wrested from our hands. To prevent this happening again, we have to follow the true socialists and not those who summoned the British, and then the Turks, to come here and destroy Soviet power with their common forces. I appeal to you to be the equal of Red Moscow!"

Our comrades had to wage a prolonged, dogged struggle at the congress itself and also in its various committees in order to prevail. The fact is that some vacillating Bolsheviks were among those who succumbed to the "seductive" arguments of Zurabov, including, for example, the member of the printers' union, the Communist Khachiyev. In order to show some of the nonsense that cluttered the heads of such Communists, I shall cite a short excerpt from the paper *Nabat* (*The Alarm Bell*)—the organ of the printers' union:

> The Baku proletariat listens with great interest and secret anxiety to what is happening at the Congress of Trade Unions of Transcaucasia and the Transcaspian region. Despite the fact that the speakers of all three factions reached essentially one and the same conclusion, their talks provoked heated debates—and not over their basic points but rather because of their mutual recriminations for past sins, errors, and insults. And one has to be fair and say that the leaders of this unnecessary and harmful polemic were the Communist comrades, whose present-day psychology could be characterized as rapture with the taste of victory.
>
> One of the comrades expressed this sentiment: "We're familiar," they said "with your Menshevik, SR tricks: today you come over to us, but tomorrow, if the counterrevolution is victorious, you will again strike out against us, in unison with the enemies of the proletariat."

This statement was certainly not well thought-out, and one can only attribute it to the hot tempers of the Communist comrades. However, the congress was, by and large, free of such "temper" and expressed, through its votes, the will to form a united revolutionary front. Despite the demands of the Communist comrades to take their resolution as a foundation-stone (true, it was impressive, precise and well-defined), the congress found it necessary to concur with the suggestion of Comrade Zurabov: all three resolutions being essentially indistinguishable from each other, they should be handed over to the drafting committee to be drawn up into a single resolution. We say all this not to discredit the Communist comrades nor to extol the wisdom of Comrade Zurabov—that's a completely unproductive occupation. Rather, we have only one aim: that the Communist comrades curb their impulsiveness and abandon their "pose" as unyielding victors.

The efforts of the drafting committee to draw up a resolution which combined all the viewpoints represented at the congress ended in failure, as was to be expected. Then the three resolutions were put to a vote: that of the Communist-Bolsheviks, the SR-Mensheviks, and the Conciliators (the supporters of Zurabov). We were able to ensure that the congress adopted the resolution proposed by the Communists by an overwhelming majority.

The paper *Nabat*'s appeal to all the parties to join together in a united front and to put an end to party differences was rejected. In specific situations we were not opposed to unity of action with the other parties and with individuals belonging to them, when they displayed a readiness to fight for the cause of revolution, but we could *not* end the ideological struggle with the Mensheviks and SRs, nor submerge Communist ideology in vague socialist phraseology, nor could we conceal the gulf that separated the Communists, right-wing SRs, and Mensheviks or stop the criticism of their past treachery and their current policy of compromise with the bourgeoisie.

The Baku Party Committee could not remain indifferent to the incorrect line pursued by the paper *Nabat*, the organ of the printers' union (this newspaper came out weekly starting March 10, 1919, as "the organ of independent socialist thought").

The boss there was a member of the governing board of the printers' union, Arshak Khachiyev, who had been a Menshevik-internationalist (one of those associated with *New Life*) and then in October had come over to our party. Now Khachiyev was again having ideological doubts, and he had got the other Communists on the union board (they were in the majority there) to follow him in his pursuit of a policy of "independent socialist thought." This was proof of the fact that the Baku Party Committee still didn't exercise sufficient influence over all the Communists.

The position taken by *Nabat* became particularly intolerable in April during the time when the All-Caucasian Congress of Trade Unions was in session. The paper came out in support of Zurabov, calling for the unification of all socialist parties—from the right-wing

Mensheviks and SRs to the Communists—"mercilessly maligning party differences of all kinds, and explaining to the workers the destructiveness of their discord."

This exhausted our patience. We summoned all the Communist members on the board of the printers' union to the Baku Party Committee and explained to them that the policy their paper was pursuing was incompatible with the spirit of the party. We tried to persuade them, and then even threatened them, that if the paper's policy were not corrected, they would place themselves outside the ranks of the party.

The president of the union was then the worker-printer Poltoratsky. As a rule he was a good Communist, but he had displayed political shortsightedness in the leadership of the newspaper. Overloaded with union work, he had transferred power to Khachiyev, who began to take charge of all of the paper's editorial affairs. We did manage, however, to convince the comrades, and Khachiyev gave his word that he would submit to party discipline.

On May 1, *Nabat* began to come out as a workers' paper and as the organ of the printers' union. Lominadze became the editor. Thus our own party newspaper made its appearance, which was a great support to us in our political work.

* * *

Soon after my return from the Transcaspian region, I learned this bit of news: that the Mensheviks who were then head of the Baku Workers' Conference had raised the possibility (as early as January 1919) of the Baku proletariat forming a democratically elected organ of city self-government which, according to their plan, would henceforth replace, along with unions, the Workers' Conference.

The Mensheviks did everything they could to get rid of the Workers' Conference. The conference was a new form of organization which had been spontaneously created by the revolutionary masses during the occupation, it had been greatly strengthened during the general political strike, and had turned into a powerful organ of leadership of the Baku workers.

The Baku Communists had advanced the slogan at that time calling for the creation of a Baku Soviet of Workers' and Sailors' Deputies. They appealed to the workers with the call: "Demand the immediate convening of a Soviet of Workers' and Sailors' Deputies, rally closely around the sole defender of your interests—the All-Russian Communist party (Bolsheviks), which alone is capable of putting the Baku proletariat on the correct path of struggle for the power of the Soviets of Workers' and Sailors' Deputies." My comrades told me that the Bolsheviks' proposal to create a Soviet of Workers' and Sailors' Deputies was adopted on January 22, 1919, at a session of the Baku Workers' Conference to the applause of the overwhelming majority of the delegates.

When I learned this news, I was, of course, overjoyed, but at the same time somewhat puzzled. Some doubts and fears had sprung up, but at the time I didn't tell my friends about them. I had to do a little thinking myself, and check to see if my idea was accurate. When a definite opinion on this issue had completely taken shape in my mind, I spoke of it at one of the sessions of the Baku Party Committee. By that time, the leadership of the Baku Workers' Conference had passed into our hands, and I could see for myself what tremendous strength was hidden in this organization of the Baku working class which had already been formed and was recognized by everyone. Why should we now replace this organization with another? was the question I asked at the session of the Baku Party Committee. "Until we are certain that we will receive a firm majority in the Soviet, it's better to wait," I said. "The political situation in Baku continued to be a tense one. The British generals, the khans and the beks of the Azerbaijani government would inevitably see the Soviet as a real threat to the existing order and would not hesitate to destroy it.

"The Workers' Conference," I continued, "already exists in fact and unites around it the broad masses of the Baku proletariat as a strong, representative organization of the working class. It has already been recognized, and both the British occupation forces and the local government take it into account. They are forced to accept our letters and our notes and to reply to them. If a more suitable political situation should arise which would realistically allow us to raise the question of a seizure of power, then it would be a relatively simple step for us to reorganize the Workers' Conference in the shortest possible time into a Soviet of Workers' and Sailors' Deputies."

Without essentially disputing any of the issues which I had raised, the other members of the party committee expressed their reservations: how was it possible to renounce the resolution adopted by the Baku Workers' Conference, on the suggestion of the Communists themselves, concerning the formation of a Soviet; how could they cancel the party slogan relating to this?

I replied to this by saying that there was no need to cancel this slogan nor to renounce the resolution passed by the Baku conference. All that was necessary was not to force the issue or to insist on the early implementation of the resolution, and in the event that questions should arise, to explain that the resolution would be carried out under the proper conditions, after the necessary preparatory work. Meanwhile, I said, our major task is to strengthen the Workers' Conference, of which we are the leaders, and to strengthen the district workers' conferences. I was sure that in the course of carrying out this work the question of a speedy formation of a Soviet of Workers' and Sailors'

Deputies would not arise of itself. After discussing my thoughts on this subject, my comrades concurred with them.

At a subsequent session of the Workers' Conference the question again arose of creating a Soviet of Workers' and Sailors' Deputies. In a very calm tone we explained to the deputies that at the present time our main job was to strengthen the workers' conferences and that when the time was ripe, the Soviet would be created. This explanation turned out to be sufficient.

Some months later the issue was raised again, and this time, strange as it may seem, on the initiative of the Azerbaijani government: in July the government, via the Ministry of Labor, proposed to the presidium of the Workers' Conference that they organize a Soviet of Workers' Deputies.

This proposal astonished us. We began to wonder what dirty trick was concealed in all this, and we soon realized that the ruling party of Muscavatists (which was losing its influence over the Moslem workers with every passing day), deprived of its right to participate in the work of the Workers' Conference, evidently hoped to get represented on this Soviet. They would then try, as the elections were going on, by using any manner of manipulation, pressure, and repression, to turn the Baku Soviet into their own obedient tool, as had been done by the Mensheviks in Tiflis.

The presidium of the Baku Workers' Conference rejected this proposal of the Azerbaijani government.

* * *

Party work in Baku continued to develop in breadth and in depth. However, it suffered from lack of contact with Soviet Astrakhan, and hence with the Central Committee of the party. We didn't receive any of the central party newspapers and moreover the party coffers were empty, with no way of replenishing them. For these reasons, we began to look about for means of establishing contact with Astrakhan.

We decided to buy a fishing boat so that we could send our representative to Astrakhan with credentials to establish connections, but we didn't have the money to buy a boat. At this juncture a member of the board of the Caspian Cooperative Union, the Communist Shiga Ioanesian, came to see me. "I know that you're in need of money," he said. "The Workers' Conference also needs funds. I think I can help you in this regard." And he told me how, back in 1918, the Baku Soviet of Workers' Deputies had sent a large quantity of railroad ties, sleepers, and trolleys to Lenkoran to help carry export grain from the hinterland to the port in Lenkoran. But soon afterward, the Soviet forces collapsed in Baku and all this equipment remained unused and was at the disposal of a plenipotentiary of the Baku Soviet. Ioanesian told me that if the presidium of the Workers' Conference would pass an appropriate resolution, he was ready to organize, with the aid of the Lenkoran plenipotentiary of the Baku Soviet, the sale of this stockpile to private individuals, and to hand over the profits to the presidium of the Workers' Conference. This maneuver was, he announced, completely legal, since the only successor to the Baku Soviet was the Baku Workers' Conference.

After seeking advice we decided to give Ioanesian the necessary document from the presidium of the Workers' Conference. When I presented him with the authorization, I was, frankly speaking, doubtful of the successful execution of this scheme, which at the time seemed very dubious to me. But, strange as it may seem, about two weeks later the presidium of the Workers' Conference received the full value of the equipment sold in Lenkoran, thus providing us with considerable material support.

With the purchase of a boat now possible, that task was entrusted to Isai Davlatov—a very businesslike fellow, and an experienced conspirator—who managed to get hold of a fishing boat for us. Saraikin, a trustworthy seaman, and two other sailors were sent to Astrakhan to reestablish contact with the party organization there, and also to bring back some more party literature and "Nicholas II" banknotes, which had by that time been abolished in Soviet Russia, but which still had a considerably higher value in Transcaucasia than the Kerenki (Kerensky banknotes) and the emergency paper money then being issued in abundance by the bourgeois governments of Transcaucasia.

The boat made its way to Astrakhan, and four or five weeks later, in the latter half of May, it returned safely to Baku. They brought back some party literature (I don't remember exactly which) and some "Nicholas" banknotes. They also brought a letter from S. M. Kirov, who had gone to Astrakhan in the beginning of 1919 and was in charge of all the work there. Sergei Mironovich wrote that there was no gasoline in Astrakhan and that therefore their airplanes were idle while the Denikin forces were bombing Astrakhan with impunity. The same situation prevailed in the other districts of Soviet Russia.

We immediately began organizing the purchase of gasoline and boats: this would be the best assistance we could render the Red Army. Soon after, under the leadership of Davlatov, Comrades Gubanov, Rogov, and Saraikin organized an extraordinary maritime expedition from Baku to Astrakhan.

It was very difficult to stockpile gasoline in Baku; the British military command exercised vigilant control over its sale. In general, it was forbidden to export gasoline by sea (except to Persia). Boats setting out to sea were subject to inspection at the dock. However, we managed, nonetheless, to arrange both the purchase and the transport of the fuel. This was accomplished despite the many obstacles encountered. Each time we had to obtain permission for the export—now to

Lenkoran, now to Persia—of fairly substantial shipments of gasoline, it was necessary to get hold of boats, select special crews, and dispatch them in such a way that "there was nothing to find fault with." Thanks to the expertise of Davlatov and his comrades, we were able during the summer to send twelve sailboats to Astrakhan with gasoline.

If the acquisition of gasoline and boats, and the dispatching of them from Baku constituted a serious obstacle, then the very trip itself, from Baku to Astrakhan, was fraught with mortal dangers. The delta of the Volga was patroled by Denikin's warships. If the winds were favorable the sailboats could skirt quickly past Denikin's ships guarding the main channel of the Volga, and then make their way up the secondary branches of the river to Astrakhan. However, if the winds were unfavorable the boat was trapped in the open sea in plain view of Denikin's ships. One of our first boats fell into the hands of the Denikinites in just this way only a short distance from Astrakhan; its entire crew, except for Misha Sudeikin, was shot.

All those who sailed on these trips to Astrakhan were extremely loyal men, genuine patriots, and trustworthy Communists. There was not one incident of any crew member losing his nerve and making a mess of the undertaking, or turning out to be an agent provocateur. Misha Sudeikin's heroic behavior is well known. Brutally beaten, he lay in a faint among the bodies of his comrades who had been shot. When he came to, he managed to hide himself among a group of Denikin's wounded soldiers. Then, with the help of some sailors he knew, he made his way, poorly clad, barefoot, and hungry, to Baku. We greeted him as if he were someone from another planet. And even after this, not only did he not give up his work, he set out once again on a no less dangerous trip from Baku to Krasnovodsk, which was occupied by our troops, but which was surrounded on the sea by Denikin's ships. This time he fell under machine gun fire but again he barely managed to escape, together with Comrade Gogoberidze. On a third occasion (now with a large group) he set out in 1920 on the launch *Edichka*, on another trip to Krasnovodsk which was still blockaded by Denikinites. The renowned sailor Rogov, one of the leading crew members of these maritime expeditions, also performed many feats. He went to Astrakhan and back by himself in a boat. It was on just such a trip that he tragically perished.

Fedia Gubanov, the President of the Central Committee of the Union of Water-Transport Workers, perished in October 1919 when, because of unfavorable weather, his boat fell into the hands of the Denikinites near Petrovsk. He was tortured by them and thrown into the sea. Along with him, the Denikinites wiped out the boat's entire crew. Buniat Sardarov, a member of the party since 1906, perished in the same tragic way. The boat on which he was returning from Astrakhan to Baku with a group of comrades was seized by the British, and the entire crew was shot.

Of the twelve boats that we had sent out by the end of the summer, four were lost. And yet, every time, another boat would set out on the same dangerous journey, with a new crew, who were often going to face certain death. Many of our comrades died at that time. It is with great sorrow that I remember them.

By sending gasoline to Astrakhan, the Baku Communists rendered an invaluable service to the Red Army and to Soviet Russia during one of the most difficult periods of the Civil War. To say nothing, of course, of the fact that it was thanks to this that we managed to establish regular contacts with Astrakhan and Moscow, and with the Central Committee of the party, something which was of vital importance to us at the time.

Chapter III

The Birth of the Slogan: "For a Soviet Azerbaijan!"

Which decision is correct?—The May Day demonstration—Preparations for a general economic strike.

Meanwhile, the tempo of revolutionary events in Baku was accelerating by leaps and bounds. One after another, political problems of tactics and strategy arose to confront us—the young and still inexperienced leaders—difficulties which were compounded by the fact that we were cut off from contact with the Party Central Committee and its guiding directives. Our comrades in the Regional Committee were experienced, but since they were in Tiflis they had difficulty in grasping the situation in Baku and thought we were exaggerating the strength of our support. All this served to increase the profound sense of personal responsibility we felt for our decisions.

As we saw it, our main task was to strengthen the leadership of the workers' movement in Baku by preparing, arming, and training fighting detachments of workers for rebellion and the seizure of power. Our aim was to time the armed rebellion to coincide with the reopening of navigation when Soviet warships from Astrakhan would be able to come to our aid. However, we still had not reached an agreement on this question with Astrakhan because we did not know whether or not the fleet would set sail, and if it did, when this would happen. This held up the decision on when to start the rebellion: to begin it without having received a final answer from Astrakhan would be an act of political adventurism on our part.

We, the leading workers in the Baku Party Committee, were also concerned about what slogan to use as a rallying cry for the rebellion. The most obvious choice was the slogan of: "A Struggle for Soviet Rule!" But there were certain subtle distinctions to be made in stirring up rebellion within the complex national-governmental apparatus of Azerbaijan and Transcaucasia, where three national bourgeois states —headed by the Muscavatists in Azerbaijan, the Mensheviks in Georgia, and the Dashnaks in Armenia—were already in their second year of existence. What forms should a decision on the nationality question take under these circumstances, and how should this decision be combined with the slogan of Soviet rule?

I asked myself the following question: What concrete governmental form should Soviet rule take in the Caucasus; what should be its territorial boundaries? A resolution of the April All-Russian Party Conference had said the following on the nationality issue: "The party demands broad regional autonomy." A similar resolution had been passed by the Caucasian Regional Party Congress in October 1917. But, as yet, no one had answered the following questions: how should regional autonomy actually be effected in the Caucasus? Will there be a single Caucasian or Transcaucasian autonomous region, or should the old provinces with their current territorial divisions be preserved, or should some different principle of division be adopted?

We knew, for example, that in Tashkent the victorious Soviet government had proclaimed the existence of the Turkestan Autonomous Soviet Republic without making any changes either in the Central Asian provincial and district divisions then in existence, or in the internal structure of the region. Naturally enough, the question arose as to whether this example was applicable in our case. For me, personally, the answer was clearly "no." I realized that one had to take into account the very real existence of national states in the Caucasus, for although they were presently under the heel of the British occupation forces, they were formally independent, and it was commonly assumed that the occupation regime was provisional and transient.

It was clear, however, that it would be impossible to prevail, or to consolidate one's victory, without the active support of both the Azerbaijani workers, and the peasantry and the vanguard intelligentsia. During the brief period of the Azerbaijani bourgeois state's existence, the people of Azerbaijan had with their own eyes beheld their own national government for the first time in their history. On the other hand, however, the nationalist passions which had at first inspired the intelligentsia as well as many peasants and workers had gradually begun to cool down because the representatives of the Azerbaijani gentry and bourgeoisie in power had not only done nothing for the common people, but had actually made conditions worse. Goods were in short supply and the value of money had plummeted drastically. Industry, deprived of the Russian market for the sale of oil and the purchase of commodities, was in a state of crisis. The government had not opened any new schools or hospitals, and the land remained as before in the possession of the gentry, who upon coming to power, oppressed the peasants even more openly. All these things intensified the class-consciousness of the workers and peasants, and at the same time drew them and the vanguard intelligentsia closer together.

And so, in pondering these things, we came to the conclusion that in the upcoming revolutionary struggle for power, we, the Communists of Azerbaijan, had to perform the task of transforming the "independent" Azerbaijani gentry-bourgeois state into an independent Soviet socialist state that would maintain close ties with Soviet Russia and the Transcaucasian Soviet republics. We had to proclaim the slogan, "Long Live Soviet Azerbaijan!" and lead the masses to revolt under this slogan.

Once we had become convinced of the correctness of the slogan, "For a Soviet Azerbaijan!" we began sounding out various party comrades for their reactions before official discussion began in the Baku Committee. At first, the Russian Communists could not grasp the reason for this slogan, and they showered me with questions: "Does this mean that Azerbaijan will be directly joined to the body politic of Soviet Russia, and if not directly, then how will the two be connected?" There were many other questions as well. I replied that all nationalities residing in Azerbaijan would have equal rights and equal possibilities for developing their own cultures, but I tried to refocus such conversations on something else: "Is it possible for us to win the majority of peasants, workers, and members of the Azerbaijani intelligentsia over to our side if we tell them that with the victory of Soviet rule, a national Azerbaijani state will cease to exist?" This way of formulating the question made a strong impression. And indeed, if our victory resulted in the establishment of a Soviet Azerbaijani state, we would attract to our side those segments of the population which were at present still far from supporting us.

A few days later, after the ground had been prepared, the issue was put before a meeting of the Baku Party Committee. After lengthy discussion the committee approved the slogan: "For a Soviet Azerbaijan!" Karayev, Guseynov, and Agayev were very happy about our decision, for it made their struggle with the Muscavatists easier. It was decided to bring the question up for discussion at the Baku Party Conference, which was to be convened at the beginning of May 1919. The Baku Party Conference in turn approved the proposal of the Baku Party Committee, and the slogan, "For a Soviet Azerbaijan!" became the battle cry and program of the Baku party organization.

Shortly afterward, we went to Tiflis for the regular meeting of the Caucasian Regional Party Committee. When our Tiflis comrades learned about our decision they were extremely displeased. They continued to stand on the old positions and did not want to recognize any independent states in Transcaucasia, although these states were already in existence and it was impossible not to reckon with this fact.

I recall that my older comrade and friend, Philip Makharadze, and also a few other comrades, raised the most fervent objections to our decision. While they were unable to make us budge from our position, we promised, on seeing how persistent they were, to put this question up for discussion at the upcoming All-Caucasian Party Conference.

We understood very well that the question of a Soviet Azerbaijan would have repercussions that extended beyond the boundaries of the Azerbaijani republic. If we were not mistaken, then the question of a Soviet Georgia and a Soviet Armenia should also be raised.

With the approach of May Day, the workers' movement in Baku reached a new peak of enthusiasm. We were up to our necks in work, making decisions on a mass of questions that were coming at us from all sides, but in spite of this we began serious preparations for the upcoming May Day demonstration well in advance. We had to select as leaders for the columns of demonstrators reliable and experienced people who would be able to ensure order and forbid any excesses, as well as to safeguard the demonstration from provocations by the police and the occupation forces, and not give them an excuse to use their weapons against us. In addition, since our funds were meager, we had to get the industrial workers to prepare the necessary posters, slogans, and banners in the shortest possible time, under their own steam, and with their own resources. We were counting in this regard on the initiative of the workers' organizations. The Baku Party Committee, for example, made its own red flag; the district party committees, which had by this time been formed in all the districts of Baku, did likewise. The presidium of the Workers' Conference used its funds to make a score or so of fairly good posters on which the following slogans were written: "Long Live the Worldwide Republic of Soviets!"; "Long Live the Third Communist Internationale!"; "Down With International Imperialism!"; "Down With the Counterrevolutionaries!"; "Long Live the Merciless Struggle Against the Counterrevoluntion!"

Since the British military garrison in Baku was rather large, we made placards in English, which bore slogans proclaiming May Day as the international workers' holiday, as a day of solidarity between proletarians of all countries, and of proletarian friendship in the struggle against colonialism and for peace. We encountered great difficulties in making these placards, since it was extremely hard for us to find someone who knew English. The person we finally found was one of the coworkers at British headquarters who served there as a translator. But even he, I was told later, had made some mistakes in writing out the slogans.

In order to lend our column more "grandeur," we rented three cargo trucks. On one of them we mounted a five-cornered star about three-and-a-half meters tall; the second contained a tableau vivant illustrating the theme of the international spirit of the working class; and the third also had a tableau vivant of a blacksmith (worker) with a raised hammer. In addition, we organized a good-sized chorus made up of Baku musicians and singers who were sympathetic to us.

On the day of the demonstration the workers would gather at their respective factories, plants, and enterprises in their districts, hold political rallies there, and in organized fashion, by columns, to musical accompaniment and holding placards and banners, they would head toward the center of the city—to Freedom Square. Since there were no loudspeakers at that time, four platforms had been set up on the square in order to provide rostrums for the speakers. The columns of demonstrators, after marching through the square, were to set out along the main streets of the city in the direction of the embankment, and further, along Sadovaya Street, past the British headquarters and British barracks. Then they would walk down Nikolaev Street past the Azerbaijani parliament to the Workers' Club, where we envisaged holding a concluding rally, after which the demonstrators were to disperse to their homes peacefully and in organized fashion.

The Azerbaijani government and the British military command, foreseeing the impressive size and scope of the demonstration that was in the works, and fearing the influence of the "red plague" on the hearts and minds of their soldiers, forbade them to leave the barracks on that day. They simply locked them in there.

On the first of May, our newspaper *Nabat* devoted its entire issue to the May Day holiday. I remember that the following slogans were printed on its pages: "Long Live the International Proletarian Holiday of May 1!"; "Long Live the Worldwide Proletarian Revolution!" The paper also contained several articles devoted to May Day.

Toward the beginning of the demonstration, the presidium of the Workers' Conference, members of the governing boards of all the trade unions, and workers from the municipal district gathered at the Workers' Club, and we went as a group to Freedom Square.

At the square, which was already crowded with people, we assigned speakers to their respective platforms. New columns kept approaching from their districts, bearing the banners of their factories and industrial plants, and accompanied by music and singing. Many workers came with their families. The picture of the enormous mass of people which streamed into the square from all the streets leading into it was an impressive one. The columns did not pass on to the square itself, but moved in organized fashion alongside it, so that each column passed by all four rostrums, from which representatives of the Baku Bolshevik party organization, Gummet, Adalet, and the left-wing SRs greeted each district and industrial plant with salutations and speeches. All the speeches were short and inspiring; they contained revolutionary appeals and party slogans. The people responded to the speeches with a great upsurge of enthusiasm, spontaneous shouts of "Hurrah!" and thunderous applause. The square and the streets leading into it were filled to overflowing with a black, oil-like avalanche of workers who had come straight from the industrial plants.

Just as the columns of demonstrators were leaving the square and moving towards the embankment side, a small group of Muscavatist demonstrators—some three to four hundred people—unexpectedly appeared. They were carrying blue and green banners, waving them aloft with enthusiasm, but they failed to make much of an impression. The workers' parade stayed on course and passed by them calmly, led by the trucks with the *tableaux vivants* and the star. Marching in the first row of the lead column were the members of the presidium of the Baku Workers' Conference, the representative of the unions and the party. Nobody along the entire route taken by the demonstrators did anything to impede or hinder it. The police were nowhere to be seen, nor were there any British military patrols. We marched along, very pleased with the fact that our demonstration was progressing in such a calm and orderly fashion.

Suddenly, two British tanks appeared in the distance, aimed head on at our procession, moving slowly toward us along the center of the street. Each of us had the same questions: what does this mean? What should we do?

Talking the matter over quickly among ourselves we, the members of the presidium, decided not to retreat one step even if the tanks were to come directly at us. We joined hands, started to sing even louder, and continued marching straight ahead. We were naturally very agitated since we didn't know how it would end, but our perturbation was not outwardly visible. We marched calmly and confidently; whatever will be, will be The tanks were getting closer and closer; the moment of reckoning had come. And at the last minute, when only a few meters separated us, the tanks suddenly turned aside and stopped to let us go by.

We were proud of our self-possession and continued our march in still higher spirits. A short time later, a British military transport of sepoys appeared on the embankment, also moving in our direction. But in the end it, too, was forced to move aside and wait until our demonstration had passed by.

Thus, making our way up Sadovaya Street, we approached a large private house with a balcony, a residence that was very familiar to me since it was the house in which, a year before, on orders from the Baku Revolutionary Committee, I had arrested the well-known oil manufacturer, Tagianosov. Now it housed the British military headquarters, and from its balcony British officers gazed down at the demonstration.

After assessing the situation we decided on the spur of the moment to organize a political rally in front of British headquarters. I got up on one of the trucks and delivered a short speech directed against British imperialism and colonial oppression, demanding freedom for the peoples of the East from the military interventionists occupying Azerbaijan, and from the counterrevolution. While I was speaking I thought with vexation what a shame it was that the British officers did not know Russian and that we had not thought earlier of bringing a translator with us for just such a rally.

When I came to the end of my speech, a young man obviously

a student suddenly leapt up on the truck and asked permission to translate my speech into English. He proceeded to do so, speaking loudly, confidently, and with a conviction that had the demonstrators applauding him. The British, with their typical sangfroid, calmly heard out the speech, but seemingly they did not say anything in reply.

The demonstration moved on. Two or three blocks further on, when we passed the three-storied building which had been turned into a barracks for the British soldiers, the demonstrators expressed their sympathy and friendship toward the soldiers with shouts and friendly gestures, and our self-appointed translator carried on a conversation with them in English. It was here that our "visual agitational aids" came in handy—those placards which we had made up specifically for the British soldiers.

Then the demonstration approached the building that housed the Azerbaijani parliament where we also organized a political rally. I spoke first, then Karayev spoke in Azerbaijani. Addressing the members of parliament from the truck, I said, among other things: "Gentlemen landowners and capitalists! Know that by the next international May Day celebration, the Soviet of Workers' and Peasants' Deputies of Azerbaijan will be installed in this building." And indeed, just one year later, two days before May first, Soviet power was victorious in Baku, and the Revolutionary Committee of Azerbaijan was housed in the building where the bourgeois parliament had been. The May Day demonstration on that day was an unforgettably joyful celebration of the happy, liberated people of Azerbaijan.

The demonstration moved on to the building where the Workers' Club was located. There a rostrum had been set up on the balcony, and once again a big political rally was held which did not end until about six o'clock in the evening, when the happy and excited demonstrators dispersed to their own districts.

I must say that the demonstration, and in general the entire May Day celebration of that year made an enormous impression not only on the workers but on all the inhabitants of the city, to whom it seemed that "the Bolsheviks had already come." They were struck by the sight of the streets of Baku filled with workers marching under their own banners and slogans while the local bourgeoisie and the police hid in their homes and barracks, unable obviously to do anything about it. This realization represented a rather profound shift in the consciousness of all the citizens of Baku.

* * *

The next day, after we had gathered at the presidium of the Workers' Conference, a delegation of sailors from the merchant fleet unexpectedly appeared. They reported to us in agitated tones that they had received an order from the British command to make available eighteen boats for shipping troops and materiel to Petrovsk to aid the Denikin forces in suppressing a revolutionary uprising in Dagestan. Some of the ships were to make a stopover in Enzeli in order to take on a load of arms there and transport it to Petrovsk. The delegation asked us what they should do.

Naturally, we could not permit the British to help Denikin put down the uprising in Dagestan. We decided, however, to adopt flexible tactics in order to protect the sailors from possible sanctions, and instructed them not to openly defy the British order, but to say they could not carry out an order of the British military command without the permission of the Baku Workers' Conference, which they recognized as their leadership center. The sailors accordingly informed the British to this effect, and the next day a representative of the British military command came to us at the presidium of the Workers' Conference and said that they needed eighteen ships immediately, and were requesting the conference to deal with the matter of delivering them to the British command. We replied that we had to bring the matter up before the presidium of the Workers' Conference, and that we would not be able to give him an answer until the next day.

At the meeting of the presidium we resolved not to change our previous position and not to give the British the ships under any circumstances. The next morning when the representative of the British command came for our answer we told him that insofar as they needed the ships to transfer arms and troops in order to suppress a rebellion, that is, for the purpose of waging war, we, as opponents of war, could not satisfy the request of the British command. After receiving our answer, the British representative left, but he returned a short time later and said that they were prepared to limit their request to four ships, on which they intended to transport not arms and troops, but medical supplies and foodstuffs. To this we replied that the transfer of such cargo in the present circumstances also constituted an indirect support of the war, and thus we could not consent to it either. Our negotiations with the British command ended on this note, and it was clear to everyone that this was the actual starting point of the merchant seamen's strike.

I should mention that at just about this time a delegation of Baku railway workers called upon the presidium of the Workers' Conference to inform us that the British, after failing to get the arms transported by sea, were now trying to transport them to Petrovsk and Derbent by rail. The railway workers did not want to allow this, and were therefore asking our advice on what to do. We gave them the same advice we had given the sailors, maintaining our resolve not to allow the transfer of arms by rail under any circumstances.

Thus, we not only turned out to be on the verge of a partial strike initiated by the white- and blue-collar rail and maritime workers of Baku—which, I might add, bore a political character—but we had also entered into direct conflict with the British military command.

Reports from the various districts of the city gave evidence of a very tense situation: the workers no longer wanted to put up with the current state of affairs and demanded the calling of a general strike for a collective agreement which had long been in the offing. Moreover, workers at several plants had already begun economic strikes without waiting for further instructions from the strike committee. The Baku Party Committee was immediately convened to discuss the situation.

As before, we did not want to rush into a general strike, but now circumstances had taken such a turn that if a general strike were not declared the individual outbreaks in the local districts could be easily put down. In addition, since May 3 represented the expiration date set by the Workers' Conference for receiving an answer from the oil manufacturers, the government, and the British command in regard to a collective agreement and trade arrangements with Astrakhan, we came to the conclusion that it was necessary to call a general strike.

All this occurred at a time when we were awaiting the arrival in Baku of delegates to a previously designated illegal Caucasian Regional Conference of Communists. Because of the strike situation, suggestions were made to transfer the conference elsewhere and, consequently, to have the delegates return home to their republics, or to postpone the conference until after the general strike in Baku was over. However, the issue was complicated by the difficult conditions under which the Communists were then operating in Georgia and Armenia: the delegates could be arrested, and besides, like us, they were pressed for funds. Therefore, in spite of the unfelicitous circumstances, we decided to hold the party conference anyway.

On the evening of May 2, after a meeting of the Baku Party Committee, the All-Baku Party Conference took place. After hearing a report on the political situation in Baku, it approved the policy our party committee had worked out on this issue. But the most important question was the discussion of the slogan: "For a Soviet Azerbaijan!" The conference confirmed the appropriateness of the slogan. As a result we were able to go more confidently to the Caucasian Regional Conference, which was scheduled to open on May 6.

A meeting of the Baku Workers' Conference preceded it, taking place on May 4. We had conducted the elections of delegates to this conference according to a new system—one delegate was elected for every fifty blue- and white-collar workers. Previously the delegates had been elected to represent separate plants, with no account taken of the number of workers at each one.

As a result of the elections the Workers' Conference acquired a new staff. Therefore, we deemed it necessary to introduce a motion for reelecting the personnel of the conference presidium. The motion passed. Delegates were elected on an individual basis; each nominee was discussed. It is worth noting that all the candidates nominated by the Communists were voted in by an overwhelming majority. The following people were elected to serve as the new membership of the Workers' Conference presidium: Agayev, Anashkin, Azimzade, Gubanov, Ibragimov, Ilin, Mikoyan, and Churayev; the following were elected candidate-members: Gogoberidze, Kazbekov, Poltoratsky, and Sturua.

I was given the job of delivering a report at the conference on the conduct of the general strike. A brief summary of my report was published in the newspaper *Nabat*; therefore, I shall quote it:

> As you recall, comrades, in 1917, the Bolshevik authorities forced the capitalists to accept a collective agreement. Then the capitalists returned to power and refused to honor that agreement. It is up to the Workers' Conference, the guardian of the interests of the working class, to do the job of reinstating the collective agreement. The manufacturers wanted to limit the wage increase to three hundred sixty rubles and appease us with that, but the workers' commission drew up a collective agreement for the entire amount and spent three months making every effort to come to terms with the manufacturers; but even the most steadfastly conciliatory among us finally became convinced that it was impossible to reach an agreement with the capitalists by means of negotiations. It was decided to demand a final answer by May 3. And here is the answer: the proposed wage-rates have been turned down, and legal norms were not accepted. The workers cannot retreat, the gauntlet has been thrown down, and we have no choice but to continue the struggle.
>
> But the struggle must be viewed in a broader context. A collective agreement exists only as long as industry exists: if industry perishes, the agreement becomes a useless scrap of paper. On the Baku oil industry depends the existence of Baku itself, and of all Azerbaijan, for that matter.
>
> Everything here depends on oil. When oil is exported, money comes in; which means that there is something with which to satisfy the workers' demands. But this year two months have already passed since the reopening of navigation and nothing has been exported. And there are more than 200 million barrels of oil! In view of the current means of conveyance, as the oil manufacturers put it, it is possible to export each day only ten percent of the daily production. But there is nowhere to ship even that: Transcaucasia—Tiflis, Erevan, Batumi—do not require all that oil.
>
> A while ago British and other capitalists came here in order to organize trade relations for oil, but when they discovered how high prices here were, they said they would arrange to sell oil here themselves, and at lower prices.

One thing is clear: regardless of the political situation, Baku cannot exist without Russia, just as Russia cannot exist without Baku. This is not only the Communists' opinion, but that of the Azerbaijanian authorities as well. They need money, but with the value of their vouchers plummeting they do not inspire any confidence whatsoever, and the Azerbaijani government lives in daily expectation of economic ruin.

Nevertheless, the British, who are sitting pretty here, want to crush the Bolsheviks, although they know only too well that in the final analysis the Bolsheviks will win, and knowing this, they're trying to find any way at all of rescuing their money. But even they cannot find a way out of the difficulties of Baku and Azerbaijan.

Right now, for example, the question has come up: where can money be obtained to pay salaries? There's no money, no place to ship oil. As a result, life here will come to a standstill. This will have a disastrous effect on the entire population. Even if we set about exporting oil right now, we would not be able to export enough.

This means it is necessary to reduce output. But in the absence of exportation, even this won't save us: we are threatened with famine. Let anyone who doesn't believe (me) recall the time when black bread cost fifteen rubles here.

When the Azerbaijanian government agrees to open trade, the British stand in its way, and the impotent government is forced to give in to them. Their goal is to crush Soviet power. They are like the burglar who has broken into our apartment, and after robbing it, sets it on fire. The Baku proletariat itself has to chase these masters out!

We are obligated to step in and fight for trade relations!

A short time ago the representative of the Moscow Soviet on the National Economy said here that they have a vast amount of cotton textiles, footwear, etc., available to trade. This means that trade relations are an entirely real possibility.

The Baku proletariat must go to battle. Its interests are the interests of all enlightened (segments of) society, the interests of culture.

Don't be misled into thinking that our enemy—the British military command—is so frightening. After all, the British command is afraid of its own soldiers and doesn't trust them, and the soldiers don't trust their commanding officers. The best proof of this fear is the fact that they didn't allow the soldiers out of the barracks on May 1, and (yet) the soldiers willingly accepted our English-language leaflets: it is clear that a revolution is brewing even among the British, but they are too cut off here from their homeland and hesitate to rebel because they want to go home.

Not trusting their soldiers, the British are afraid to arrest us. The victory shall be ours. We have the sympathy of the whole laboring population on our side, because we are struggling in the name of everyone, and for everyone's benefit.

On behalf of the conference presidium, I propose today that we decide the question of whether to initiate an organized struggle—devoid of excesses—against those who will not allow us to establish communications with Astrakhan.

Naturally, this struggle will have unavoidable casualties. But it is better to sustain a few casualties than for everyone to perish from starvation.

Our sailor comrades have already begun the struggle. They refuse to allow the shipment of troops and supplies to assist the volunteer army which, along with the British, wants to crush both the peoples of Dagestan and the Bolsheviks. The conference presidium has ordered the sailors not to give up a single ship. The British authorities are threatening to dismiss all the sailors. (Mr.) Brown, the director of marine transport, has written to the Union of Water Transport Workers as follows:

"In reference to your order dated yesterday, I must inform you that if it is not rescinded, I shall consider the payment of wages on all ships other than postal vessels suspended as of May 2. In addition, if the ships *Van*, *Ekvator*, *Elbrus*, and *Zapad* do not put out to sea, as per yesterday's order, I request you to remove all commanding personnel from the above-mentioned vessels."

The sailors will be with us! The railway workers will support us!

The British want to crush us with starvation. So let it be not they who crush us, but we who shall crush them! (Loud applause.)

A lively debate arose around my report. Many people delivered speeches, and all of them, including SRs, Mensheviks, and also right-wing members of Gummet, were in favor of declaring a strike. One of the speakers proposed declaring a political strike; others—and they were in the majority—favored an economic strike. The railway workers declared that they were prepared to begin the strike immediately, and the sailors said the same thing.

The speech of one Azerbaijani worker by the name of Sumbat was worth noting. He said:

The Moslem workers have still not become completely aware of their class interests. It is possible that they will split from the rest, and in order to avoid such an eventuality, it is crucial that special work be done among them to clarify the aims of the strike. Moslem workers carry the double burden of foreign and local imperialists. Their liberation is possible only with the help of the all-Russian proletariat.

In speaking about the upcoming strike, Moriak Minayev concluded his address with the exclamation: "Long live Azerbaijani Soviet power!" These words, which had been addressed, moreover, to the Russian sailors, called forth wild applause from all participants in the conference.

In my concluding address I dwelt in particular detail on the question of a political strike. The Baku proletariat, I said, has had a rich tradition of general political strikes.

> We will keep this weapon in hand in the future and when a suitable moment arises, we shall take advantage of it. But the demand to declare the upcoming strike a political one is incorrect, and even harmful. In this strike we are advancing purely economic demands—for a collective agreement and the opening of trade. Our enemies—the Muscavatists and occupiers—will attach the label 'political' to our strike as an excuse for instituting repressive measures against the striking workers. Therefore, for the time being we will not advance a single political demand; with this strike we are attacking neither the Azerbaijani bourgeois government nor the British occupiers. Of course, an economic strike often does have political significance, but this does not mean that every strike must become political. In this instance we must adhere strictly to economic demands.

The conference unanimously passed the following resolution:

> After hearing the presidium's report on the collective agreement and the opening of trade with Astrakhan which are necessary for saving the Baku workers, as well as the entire population of Azerbaijan, from imminent starvation and complete financial and economic ruin; and in an effort to avoid stopping oil production, a measure necessitated by the impossibility of exporting stock-piled oil reserves, the Baku Workers' Conference has come to the following decision: to strive to obtain at all costs—by means of an economic strike—the acceptance of a collective agreement and for a decision concerning the export of oil to Astrakhan. The responsibility for determining the term of the strike, its initiation and its cessation, as well as the rules and regulations pertaining to its conduct, lies in the hands of the presidium and the Central Strike Committee. The duly elected personnel of the Central Strike Committee shall comprise the following: the presidium, two representatives from each district, two representatives from each trade union, one from the trade-union regional center, one from the sailors, one from the railway workers. Representatives of (political) parties may be co-opted into the staff of the strike committee, but they shall have consultative voting rights only.

I should note that by this time the Workers' Conference had begun to conduct itself completely differently with regard to the British occupiers.

I recall the following incident. During the May 4 meeting of the conference I left the presidium and walked out into the corridor where I happened to run into a tall, fat British officer who was on his way into the auditorium. Thinking that he simply did not realize that a meeting of the Workers' Conference was in progress in the hall, I told him that that was the case. I received, however, a totally unruffled reply spoken in good Russian:

"I know. But I've often attended the conference in the past."

The officer turned out to be Captain Walton, whom I had occasion to meet later on in different circumstances.

I said to him: "You're not allowed to be present there. But who knows, perhaps the conference will give you permission. Wait a minute, I'll find out."

I made an unscheduled announcement at the conference to the effect that a representative of the British command wanted to be present at the conference and that, according to him, the Mensheviks had allowed him to attend such meetings before.

The conference unanimously refused Walton's request, declaring the presence at its meeting of a representative of the British military command impossible. At the same time it addressed a motion to the British soldiers suggesting that they send their representatives to our meetings: the Workers' Conference was ready and willing to accept them as brothers, but it "could not receive the agents of imperialism at its house." This decision was adopted amidst wild applause and shouts of approval.

I reported the decision to Walton. He was very surprised. And I asked him on behalf of the conference to convey our request to the British soldiers.

On May 5 a meeting of the strike committee that had been elected by the Workers' Conference took place. It approved the decision to call a strike and sent the Baku proletariat's list of demands to the president of the Council of Ministers of the Azerbaijani government, to Central British Staff Headquarters (through Captain Walton), to the council of oil-manufacturers; copies of the demands went to all shipowners and industrialists, and also to the Minister of Labor. [The demands were as follows]:

> The Central Strike Committee, guided by the decision of the Baku Workers' Conference adopted on May 4 of this year, declares a general economic strike [to begin] on May 6 of this year, and presents the Congress of Oil-Manufacturers with the following demands:
> 1. Acceptance of the collective-bargaining agreement as worded by the Workers' Conference.
> 2. Establishment of trade relations with Astrakhan, and the settlement of several economic demands which have not been satisfied.

Once evidence has been given on your part of a willingness to satsify the above-stated demands, the Central Strike Committee is prepared to enter into negotiations.

The strike on sea and rail transport began on May 5. Its aim was to impede the loading of troop trains and military cargo being sent to Dagestan for Denikin's army which had already begun military operations against the revolutionary mountaineers.

Chapter IV

The First Transcaucasian Party Conference

Again the nationality issue—The strike—Our arrest—The recurrent treachery of the Mensheviks—Escape from the Baku prison—The lessons of the strike.

Members of the Regional Committee and delegates of the Transcaucasian Party Conference met in Baku on May 5. A session of the Regional Committee took place during which we familiarized the newly arrived comrades with our situation. They approved of our decision concerning the general strike and agreed with us that, despite the complex state of affairs in Baku, the Regional Party Conference should not be postponed. Those of the Baku comrades, myself included, who were involved in the leadership of the strike received permission not to attend all the sessions of the conference.

Some twenty-five to thirty people, along with the recently arrived delegates, took part in the work of the Regional Conference. Of the delegates I can recall Makharadze, Misha Okudzhava, Orakhelashvili, Nazaretian, Shaverdian, Toroshelidze, Mravian, Lado Dumbadze, Fyodor Kalantadze, Kakhoyan, Georgi Sturua, and Beso Lominadze. The names of the others are lost to me. There were also delegates from Dagestan and the Northern Caucasus. The conference took place at the Kasparovs' apartment. I remember that we began the first session on May 7 at nine o'clock in the morning, having resolved to work through the day and late into the night, with short breaks. Misha Okudzhava was elected chairman of the conference, and Gogoberidze and Mirzoyan were elected secretaries.

The conference began with reports from the rural locales. In addition, there were two financial reports: one from the Regional Committee in Vladikavkaz on the operation in the Northern Caucasus, given by Philip Makharadze. The report from Tiflis was given by Sarkis Kasyan. I didn't hear these reports—arriving only in time for the discussions afterwards, since I was attending sessions of the Central Strike Committee.

The comrades came out with sharp criticism of the weak points in the Regional Committee's operations both in the northern Caucasus and in Transcaucasia. I was told later that Makharadze was continually trying to explain and "justify" the deficiencies in the Regional Committee's work, and it was felt that Kasyan's report, on the whole, wasn't very interesting either. Without denying the weak leadership of the Regional Committee he tried to attribute it to the difficult conditions created by the Mensheviks in Georgia and by the Dashnaks in Armenia, and also to the shortage of leading cadres resulting from the departure of many Communists for the northern Caucasus.

My report threw light on the state of affairs in Baku and Azerbaijan.

It was clear from the speeches given at the conference that, despite continuing persecution in Georgia and Armenia, the party's work there had picked up noticeably in the last few months, with the Communists' influence among the workers increasing and in some districts among the peasants as well. In a number of districts in Georgia and Armenia preparations were already being made for armed rebellion. The authority of the petit bourgeois highland government in Dagestan had spread only to Temir-Khan-Shuru and Derbent; in the rest of Dagestan rebel bands held sway which were led by the Dagestan provincial committee of the party.

Bitter fighting had broken out between those revolutionary rebel bands and the Denikinites, who occupied Petrovsk, and then took Derbent, too. We knew that there were partisan detachments in the mountains of Chechen led by Nikolai Gikalo and Aslanbek Sheripov. We had been kept well informed on the situation in Dagestan since March by regular visits of comrades from there. The conference passed a resolution to give active support to struggling Dagestan.

At the conference Lominadze criticized the Party Regional Committee fairly justifiedly for its recall from the Tiflis Soviet of Worker's Deputies of a group of Communist deputies. By doing this we had deprived ourselves of an admittedly unsatisfactory but nonetheless legal platform from which a Communist's voice could be heard.

The nationality issue was a prime subject of debate at the conference. It began when Philip Makharadze criticized the Baku resolution concerning an "independent Soviet Azerbaijan." He did not take into consideration the conditions which had been created by life itself, and he continued to be a dogmatic defender of the old policies. Naturally we, the Baku comrades, mounted our own criticism of Makharadze.

We said, in particular, that "the Muscavat party's strength derived from the fact that it marched under the slogan of an independent state. If we adopt this slogan for our own purposes, but invest it with our own Soviet meaning, we said, then we will disarm the Muscavats in the eyes of the workers, who were following their line, contrary to their class interests. Once disenchanted with the Muscavatists the workers would go where their class interest dictates, immune to the false lure of the nationality issue. Now, I said, in order to counteract Makharadze's argument, a completely different situation exists than the one in 1917, although even then we decided the issue incorrectly and did not support Stepan Shaumian's idea of dividing Transcaucasia into three regional divisions, thus creating three autonomous national units. Now everything has taken a different turn. For more than a year there has existed, in actuality, three independent, national, bourgeois states, and while their independence is illusory, with the British Military Command in the driver's seat, it is impossible not to reckon with the very real existence of these national states. If we followed Makharadze's proposal we would inevitably alienate the petit bourgeois, nationalistically inclined strata of the Azerbaijani populace, and without a close bond with the bourgeoisie and the peasantry Soviet power cannot hold out and be victorious in Transcaucasia.

The prolonged debates on this issue revealed that we could not find a common ground with Makharadze and those comrades from Georgia and Armenia who supported him. Then we announced categorically that we would not renounce the slogan of an independent Soviet Azerbaijan, and the only compromise we would make would be to agree not to use this slogan in reference to Georgia and Armenia, on the condition that our Georgian and Armenian comrades did not come out against us. On this point we concurred, having agreed that the question would be resolved once and for all by the Central Committee of the Workers' Communist party (Bolsheviks).

The conference continued its work until late into the night. I remember very well that as I was leaving this session, it was already light out. In the morning there was a final session at which a new Party Regional Committee was elected. The conference delegates then dispersed, with Sturua, Makharadze, Mravian, and Lominadze staying in Baku for the strike.

At our suggestion the newly elected Regional Committee passed a new resolution creating a Baku Bureau which would consist of those Regional Committee members who worked in Baku. We were given broad powers to make decisions regarding all issues pertaining to Azerbaijan, which would then be followed up by a report concerning them to the Party Regional Committee. Such a reorganization of the party structure was necessary if we were to invest greater dynamism into the leadership of a changing situation. The Baku Bureau became, to all intents and purposes, the organ which directed all the Bolsheviks' activities in Azerbaijan.

* * *

The Central Strike Committee, as I have said, had designated May

6 as the date for the general strike, having arranged that the signal heralding the strike would be the cutting off of all electric power by three o'clock in the morning. As I recall, all the businesses and institutions of the city were divided into three groups. The first included the electric-power stations, the railroad, and water transport; the second included all remaining industries and organizations; and finally, there was the third, reserve group—the waterworks, slaughterhouses, hospitals, and cooperative societies.

Knowing that the members of the Central Strike Committee could be arrested, despite all precautions dictated by secrecy, we decided to appoint a backup committee ahead of time. There were five members on the main strike committee: Gubanov, Karayev, Mikoyan, Sturua, and Churayev. We appointed Gogoberidze head of the backup committee. In addition, and just to be on the safe side, we organized yet a third strike committee, to be headed by Anashkin. It was resolved that while the main strike committee was still functioning, those comrades on the second and third strike committees would not take part in the leadership of the strike, but would work at their posts in the presidium of the Workers' Conference. In the event of the arrest of the main committee, the second committee would automatically take its place, and at the same time the candidate-members of the presidium of the Workers' Conference would then automatically become members of the presidium. The backup strike committee was empowered to select, in coordination with the districts, emergency district representatives to the Central Strike Committee. Leadership of the conduct of the strike itself at the different industrial enterprises was entrusted to the factory and production committees. A resolution was also passed to enlist the services of twenty responsible workers for the Central Strike Committee who would carry out various assignments for the committee during the strike. The youth committees were given responsibility for organizing security for the strike committees: they had to have people ready to stand guard outside the buildings where the strike committees were holding their sessions, warn them when an appearance by the police was imminent, and also carry out the various technical assignments required by the Strike Committee.

The conditions under which the strike was to be implemented required that our work be carried out in strictest secrecy. It was therefore decided that the leaders of the strike would not show up at the Workers' Club, the legal center of our movement; and that the Strike Committee would hold daily sessions of about two hours in length which would be held in a different place each time (the locations of the sessions were, as a rule, secret). The representatives invited from districts usually went to the Workers' Club, and from there our young conspirators took them in small groups to the place where the session was being held. No one was allowed to leave the premises until the session was over in order to prevent any agent provocateur who might have penetrated our midst from being able to leave before the session ended and give away our meeting place. In order to maintain contact between the members of the central and the district strike committees, secret meeting places were set up at the Workers' Club, and in the event of its being closed—at the Printers' Union and in the Union of the Oil Industry Workers.

As had been arranged, at exactly 3:00 on the morning of May 6, 1919, the lights went out. The strike had begun. The railway and water transport workers had begun the strike the day before. Now they were joined by the industries and organizations in the second grouping. But would the workers of all the industries go out on strike? This question had, naturally, worried us, and we waited anxiously for morning when reports from the districts would come in.

Afraid that the government, expecting a strike, might have arrested the workers at the electric power plants that night and sent strikebreakers in to put the stations in operation again, we sent a special person around to the power plants who was instructed to remove basic parts and components from the generators and to conceal them in a safe place. We were especially anxious about the Bailov electric power station, where the SRs influence was still quite strong.

The Baku electric power stations—and there were three of them: in Bailov, Belyi Gorod, and Balakhany—were to play a key role in the strike. In the first place, there was the profound psychological effect which the onset of total darkness would have on the population, and secondly, the shutdown of electric power would prevent any strikebreaker from doing his job.

In the morning it became clear that every industry in the city which had figured in our plan had gone out on strike. Arrests of the leaders among the workers were carried out immediately, especially among the Azerbaijani workers, where our influence was still not strong enough. The authorities used deception, lies, and intimidation and, as a result in some places where the majority of workers were Moslems, the strike didn't begin decisively enough, and in some instances the workers even agreed to go on working. The plenum of the Strike Committee, which met on May 6 to consider this issue, listened to reports on the state of affairs in each district.

On the whole, however, the strike went successfully and was well-coordinated. The government meanwhile continued to take repressive measures by arresting the most skilled workers at the electric power stations and making them work by force. Several people in Bailov were detained in this way. The Strike Committee immediately ordered that the skilled workers of the power plants be sent to other districts for the duration of the strike so that the police couldn't find them. This was done. Then it turned out that our comrades at the Bailov power plant hadn't managed to conceal the machine parts on time. In short, quite a few mix-ups occurred. I even had to go to Belyi

Gorod myself to see to it, with the help of our comrades, that the decision concerning the stoppage of the power station was implemented.

When they didn't find any workers at home, the government officials rounded up their wives and children and evicted them from their apartments. The repressions got worse. If, however, they exerted an influence in individual, isolated cases on certain Moslem workers, they certainly failed to achieve their goal with other, politically conscious workers, who were only embittered even more by all this.

The government, the British Military Command, and all the national groups of the bourgeoisie which were usually at odds with each other now joined together in a united front against us. They made wide use of provocations and slander. The gendarmes, the police, and the government agents spread all kinds of rumors among the Moslem workers; namely, that the strike had been organized not to improve the living conditions of the Azerbaijani workers, but that it had been organized specially by the Russians and Armenians against Azerbaijani in general and the Moslems in particular, and that the Russians were armed and ready to carry out a slaughter of the Moslems.

During this time only one newspaper, *Azerbaijan*, was not on strike; reactionary Muscavatists had been specially chosen as typesetters in the paper's printing shop. From beginning to end, the paper was filled with abuse and slander directed at us.

On May 8, in particular, the paper published a so-called governmental appeal, concocted by order of the British, in which the Azerbaijani government misrepresented and falsified the meaning of the strike and its goals, and tried to rally around itself public sentiments favorable to it and the invaders. The appeal called upon the workers of Baku to continue working, and identified the general strike as a political one, which "pursues goals that are directed against the foundations of the Azerbaijani Republic and the state system, and being led by Bolshevik-leaning elements and agents of the Council of Deputies." In the appeal it was announced that the prime minister was instructed "to discuss jointly with the entrepreneurs and the workers those parts of the workers' economic demands which are just." However, not being sure that the workers would respond to their summons, the authorities threatened to settle accounts with them if they didn't get down to work. They said straight out that "those who interfered with the normal course of operations in governmental, public, and private concerns would be punished according to the martial laws that were in force."

In addition to his governmental appeal, the paper also published a front-page appeal from the Muscavat party's Baku committee. Without beating around the bush it said right out: "The Bolsheviks are deceiving the workers, pushing them on a dangerous course by inciting them to an adventuristic strike. And with it all, they act as if the strike is purely economic. That's a lie! This strike has a political character and pursues Bolshevik aims . . . Comrades, don't believe the disguised, masked adventurists and liars."

Our opponents tried to make it appear as if the Bolsheviks—those agents of the Soviet government—wanted to cause a civil war, slaughter the Moslems, and so forth. On May 8 the same paper, *Azerbaijan*, in an article entitled "The Truth about the Strike," prophesied that "the Bolsheviks would immediately introduce into the well-regulated life of Azerbaijan the bloody nightmare of interethnic and irreconcilable political dissension, the horrors of which are already well known to the Baku proletariat . . . In order to achieve this goal, the political maniacs from the Workers' Club have resorted to the demagogy of the most transparent lies, and to the senseless intimidation of the working masses with threats of future unemployment and starvation."

There was no way to respond to similar provocations of the bourgeois press. Realizing that agitation in the press could play a most powerful role in the continuance of the strike, we decided as early as the first session of the Strike Committee to put out *News of the Strike Committee*. This task was entrusted to Arshak Khachiyev, a member of the governing board of the Printers' Union. Despite the persecution of the police and other obstacles, we managed to put out two issues of the *News*. However, soon after, the police occupied all the printing houses and made it impossible to continue publishing the paper. Then we resorted to putting out short proclamations.

The presidium of the Workers' Conference was instructed to prepare a special appeal to the British soldiers, explaining the economic and peaceful character of our strike and calling upon them not to interfere, since it was the same kind of strike their own British workers organized at home when fighting for their own economic interests. We told them in advance that their officers would attempt to foist upon them the role of bloody executioners.

I remember that when the text of this appeal was ready and had been translated into English and printed on leaflets, we gave the job of distributing them among the British soldiers to the young women of Baku—members of the International Union of Working Youth who had helped us several times before in various party assignments. We felt sure that not one British sailor or soldier would raise a hand against the girls for distributing leaflets, whereas men could have been arrested for this. You should have seen the enthusiasm and the joy with which the girls later reported to us the completion of their assignment!

The strike's progress was painful and difficult. Although it was a general strike, we had to keep up the struggle in literally every place of business and every district against the numerous government agents, police officers, and gendarmes who tried to hinder us by ringing false alarms at the factories and by intimidating, arresting, and terrorizing the workers. This, of course, had an influence on a certain segment of the workers, who had begun to go back to work in some places a few days after the strike had started. However, the majority of the workers held out heroically.

The police carried out an intensive search for the members of the Strike Committee. We worked day and night. Sometimes we were forced to change the location of the executive committee sessions of the strike force and the plenary meetings of the Strike Committee several times in the course of the day. Members of the local strike committees were continually being arrested in the districts. But, despite this, the leading organs continued to function, with reserve troops of activists entering the lines in place of those who had been arrested.

On May 8 the Central Strike Committee, via the Worker's Club, received an offer to enter into negotiations from the Ministry of Labor of the bourgeois Azerbaijani government. We set a preliminary condition that all those who had been arrested be released, that the repressions and provocations come to an end, and that our delegation receive immunity. The Ministry agreed to our demands, and the three representatives chosen by the Strike Committee—Karayev, Churayev, and I—appeared at the Ministry of Labor building that evening. There, at a large conference hall on the second floor the Minister of Labor, Safikiurdsky, occupied the chairman's seat at the table. We sat to the right, at his invitation; across from us sat representatives of the capitalists: three stout fellows of an advanced age. One of them was a Russian, the other an Azerbaijani, and the third was an Armenian. A little distance away sat the factory inspectors, along with Abilov, a member of parliament. The representatives of the British Command, Captain Walton, kept aloof from the proceedings.

We had agreed beforehand that Churayev and Karayev would play the principal roles in the negotiations. Churayev particularly was more informed than any of us on the conditions of a collective agreement and was in the know regarding all the details of our struggle for it, while it was Karayev's job to hammer away at the importance of exporting oil to Astrakhan, and to insist on the implementation of trade relations with Soviet Russia, emphasizing their importance not only for the Baku workers, but for the entire economy of Azerbaijan.

I must say that they substituted our demands very thoroughly, illustrating the hopelessness of the Baku worker's plight and demonstrating the inevitable ruin of the oil industry and the resultant starvation that would befall the population of Baku should the export of oil to Astrakhan not be set in motion immediately. The factory inspectors, Abilov, and the British representative were silent the entire time and took no part in the negotiations.

The capitalists took turns speaking and argued heatedly against a number of the agreement's important points they considered unacceptable. They announced that if the export of oil to Astrakhan were not authorized, they would refuse to sign any collective agreement whatsoever. Thus we found ourselves face to face with the British Command, which had prohibited the export of oil to Astrakhan. There existed no technical means for transporting it elsewhere and such means could not be set up in a short period of time. The oil industry in Baku was at an impasse.

While the negotiations were still going on I went over to Captain Walton and asked him if he had given the British soldiers the invitation of the Worker's Conference to send their own delegates. He said that the invitation had been passed on to the British Command which would most likely inform the soldiers about it. Then I asked him how he had liked our appeal to the British soldiers regarding the strike. Walton said that unfortunately, since our May Day handbills and our appeal had been written in very bad English, it had been difficult for him to read them. I could not let this remark go by, and replied, smiling, that after Soviet power had been established in Baku, we would invite him to be our translator, and then our appeals and proclamations would be more grammatically correct. He responded with a rather sour smile.

It must be noted that the role of the Labor Minister Safikiurdsky at this conference was rather a wretched one. He alluded to the fact that the British Command had prohibited the export of oil to Astrakhan, since it regarded this as aid to the Bolsheviks, and said that the Azerbaijani government would not consent to this either. In regard to the collective agreement he said that he would make every effort to get the industrialists to sign it, with the exception of a number of legal points, concerning dismissal from work, right to work, etc. In addition, he proposed that the wage increase we demanded be reduced.

The discussions lasted a long time, about three hours. When it became clear to us that the negotiations would yield no practical results, I made the following declaration to the minister:

"It is our opinion that the export of oil provides support not only for the workers of Baku, but for Soviet Russia as well, and we are in favor of such support because Soviet Russia, not the British imperialists, is the best friend of the Azerbaijani people. At the same time, we have nothing against the independence of Azerbaijan, nor will Soviet Russia attempt to overthrow the present government of Azerbaijan, if it consents to sell Russia oil. The economic interests of both Azerbaijan and Soviet Russia are mutually served by this. This is evident from the fact that in your government's declaration there was a statement concerning the desirability of initiating trade relations with Astrakhan. Now, however, you are opposed to this because the British Command is opposed. Where, then, is the independence of Azerbaijan? Whose interests are you protecting—those of the British invaders or the Azerbaijani state government? Give us a straight answer, Mr. Socialist Minister."

Safikiurdsky muttered something incomprehensible about the fact that "one had to take the actual situation into account," and proposed that the parties concerned express their final opinion.

Churayev said that our delegation didn't have the authority to retract any of the demands put forward by the Strike Committee without reporting back to it, since only the committee could make the final decision. "Tomorrow we'll convene the Central Strike Committee, discuss your proposals, and give you our answer."

The capitalists, on their part, promised to report to the Union of Oil Manufacturers concerning our negotiations. At the end of the meeting the minister announced that for the duration of the negotiations the promise of the government not to arrest the strikers had remained in force. On this note the negotiations came to an end.

We had no sooner come out of the doorway and taken a few steps along the street when five policemen surrounded us and announced that we were under arrest. No matter how we tried to explain to them that we had been at the Ministry of Labor at the government's request to take part in negotiations, and that we had been guaranteed immunity, nothing helped. Evidently, Safikiurdsky, scoundrel to the bitter end, had simply duped us, and hadn't kept his word. Karayev, as a member of parliament, began to insist that the police go with us to the Minister of Labor, who was still in his office, and have him verify our word. Fortunately, the fact that Karayev was a deputy of parliament made an impression on the policemen and they did as we asked.

At the minister's office we confronted Safikiurdsky. "Didn't you have the courage to keep your word of honor as a Socialist Minister?" we asked. Safikiurdsky, pretending to be astonished and embarrassed, angrily ordered the policemen to let us go. After we had left the building we quickly dispersed in different directions down the dark alleys so that we wouldn't fall into some other trap.

On May 9 the British general Milne announced with characteristic bluntness that the British troops had occupied Transcaucasia in order "to maintain law and order until a final solution of the territorial disputes had been reached at the current peace conference . . . I, the undersigned, as the commander-in-chief of the British forces in Transcaucasia, make the following announcement to them: it is the duty of residents to conduct themselves in a peaceful manner and to go about their ordinary business insofar as it is possible . . . However, should they refuse to discharge their duties in this regard, the necessities of wartime demand that punishment be meted out . . . In general, the violation of any of the instructions listed in this proclamation will be brought before a military court and will be punishable by death, or by some lesser punishment, depending on the individual circumstances."

This is the way the British colonizers had acted over the centuries, everywhere they went: in Africa and in Asia. This time, however, the gallant general forgot that these were different times and different people.

At first we kept up the conspiracy in exemplary fashion. The young people set up communications brilliantly and in so doing, displayed resourcefulness, loyalty, and genuine heroism. But with every day it became harder and harder to observe the precautions of conspiratorial secrecy. It was very difficult to find new locations for our meetings, especially for such large meetings as the plenums of the Strike Committee. Workers' meetings were called in all districts on the morning of May 9, for not only did information have to be gathered on the course of the struggle, but the workers had to be given support and their fighting spirit raised. It was decided that all the members of the executive committee would go out into the districts and speak at the worker's meetings.

Churayev and I went to Chornyi Gorod, to a meeting at the Nobel company's cafeteria where the place was filled to the rafters. The workers who hadn't managed to get into the cafeteria were standing in crowds on the streets, where rows of policemen were also lined up. The speeches were fairly short, but an extraordinary surge of spirits was felt among the workers of Chornyi Gorod. The local youth had done a magnificent job of organizing security for us. When we came down from the podium at the end of the meeting and began to walk through the crowd, the young workers who accompanied us changed our caps several times to confuse those trying to follow us. They led us down back alleys which were known only to them and through various gates out onto the street, where they put us in a carriage. From there we went to the regular session of the plenum of the Strike Committee which was taking place at the oil industry workers' union hall.

On this occasion the rules of secrecy had been violated, for we were meeting in the same location for the second time. However, a reliable guard of young men was placed all around. After we had heard the reports from the districts, we handed out instructions to the district leaders. Word had got out that the British Command had stationed a British battalion of engineers at the Bailov electric power plant and was attempting to put it back in operation. (Not only did they not succeed in doing this, but due to their unfamiliarity with the way the system operated, two men were killed.) The most difficult situation of all was on the railroad in Baladzhary, where the government had managed to send a train.

Even at the beginning of the strike a number of SRs and Mensheviks, who had been discharged from their electoral duties, offered us their services for various tasks. Some SRs were already carrying out such assignments for us, and since we had no work of our own in Baladzhary, we sent the SR Rebrukh there as our representative. It should be said that the SRs who offered us their services carried out the Strike Committee's assignments quite effectively in the beginning, but the more intense the struggle became the quicker they were to propose anarchist acts. For example, in Baladzhary they suggested tearing up the rails and blowing up the track in order to impede the movement of the trains, which the government had already managed to put in operation with the help of strikebreakers, and they also proposed blowing up the electric power lines into the city, and so forth. It's understandable why we rejected all these suggestions of violence.

During a session of the Central Strike Committee I stepped out into the large corner room adjacent to the union of oil industry workers' office, and suddenly I saw policemen aiming their Mausers at me from all sides. The shout "Stop!" was heard. I then rushed back to the room where the session was being held and told my comrades that we were surrounded by the police. "Destroy all papers!" I shouted, just as a police officer appeared in the room with two gendarmes.

I should add that while this was happening everyone present displayed composure. We were ordered to step into the big room, but our first thought was that we had to manage to destroy all papers that could compromise us in any way. I personally had quite a few of them. We began to tear them up into little pieces. Luckily for us, we managed to destroy all papers during the confusion of the first few minutes.

All in all there were about forty of us: in addition to the members of the Strike Committee there were also representatives from the districts, young communications workers, and those who worked in the office which we had by then acquired for the publication of leaflets. They led us out onto the street, surrounded by a large police detail, and we walked down the streets of the city to the police station with our heads held high, singing revolutionary songs. The passersby stopped, watching us as we went by both with curiosity and with obvious sympathy.

At the police station the business of straightening out our names began. On the way we had agreed among ourselves that the leaders of the strike—Churayev, Anashkin, and myself—would take all the responsibility on ourselves. Koval and Mirzoyan, whom the police knew well by sight, would give their real names, but would refuse to admit that they had taken part in leading the strike. The majority of the comrades whom the police hardly knew by sight, or did not know at all, would conceal their real names and deny their participation in the leadership of the strike. All the rest, who were not threatened with any particular danger, would give their real names.

When the checking of names began, Gubanov, who as a member of the strike force's executive committee was by agreement to conceal his name, began to waver for some reason and gave his real name. Georgi Sturua, who had recently arrived from Tiflis, was with us. The police knew that he was a member of the strike force's executive committee but they didn't know him by sight, and when the head of the criminal investigation department asked, "Who here is Sturua?" we replied that Sturua had already returned to Tiflis. The police therefore never found out that Sturua was one of our number, for he used the name Akhabadze during the interrogation. Karayev, a member of the Strike Committee was also arrested, but he took advantage of his position as a deputy, demanded his immediate release, and was in fact immediately let go.

After the identification procedure had been completed, the head of the criminal investigation department announced that five of us: Churayev, Mirzoyan, Koval, Anashkin, and myself—were to be isolated from the rest of the prisoners. Furthermore, Koval and I were informed that we had been arrested by order of British headquarters for trying to instigate an armed insurrection and that we were therefore under the jurisdiction of the British Command. An interesting document in this regard has been preserved—a communiqué from the head of the Baku Criminal Investigation Department to the governor of Baku.

Then the five of us were taken to the Criminal Investigation Department, while the other comrades remained at the police station, not knowing what would happen to them next. While we five were being led down the street, a large group of young people—members of the Internationalist Union of Working Youth—followed to see where they were taking us.

We were put in a dark, dirty hole of a room in the department with broken windows. We were all so tired that we lay down right there on the floor to get a little sleep, having been able to snatch only a few hours' sleep during the previous few days. But we didn't manage to get any sleep here either, for we had no sooner lay down when horrible cries and moans were heard from the next room where policemen were giving a merciless beating to some prisoners. Naturally, these cries had an effect on our state of mind. We thought that it would soon be our turn.

Finally we were called to the office of the head of the Baku Criminal Investigation Department, Fatalibekov. He began by asking us who we were. After we gave him our names he began to interrogate Churayev and Koval, attempting to wound their self-esteem so that they would admit they were Bolsheviks. "You cowards, you never confess!" he shouted. "We're sick of this long drawn-out affair!" He was especially fierce in his attack on Koval, who stubbornly maintained that he was not a Bolshevik. Evidently, Fatalibekov had instructions from the British to arrest Koval on no matter what charge, irrespective of his involvement in the strike. We stuck up for Koval, saying that only three of us were Bolsheviks: Mirzoyan, Anashkin and me. Churayev was a Social Democrat Menshevik, and Koval had no party affiliation.

Fatalibekov still didn't believe that they weren't Bolsheviks, and he tormented them with his questions for a long time until we demanded that this senseless interrogation be terminated. Then he took a signed statement from us saying that we three were Bolsheviks, but that Koval and Churayev had nothing to do with the Bolshevik party. I have to admit that at this point we made a mistake: instead of writing simply that we were Communists or Bolsheviks (in this case we could have somehow taken evasive action if we had to), we wrote that we were members of the All-Russian Communist party (of Bolsheviks), completely losing sight of the fact that this party was outlawed in Azerbaijan, and membership in it alone could land you in the clutches of the British invaders.

Finally, around two o'clock in the morning the long ordeal came to an end. Bachurin got in touch with British Command headquarters and asked what should be done with me and Koval, who were under the direct jurisdiction of the British Military Command. Where should we be sent—to the warships or to prison? The order was received: for the present we should be sent to prison. Then he drew up a draft order signed by Fatalibekov, arranging for the five of us to be sent to Baku Central Prison. This document has also been preserved in the archives.

Between three and four o'clock we were taken to the central prison and put into a room on the fifth floor. This was in the cell block for the lifers and those condemned to death. The cells here had thick steel doors, with small openings for peepholes. Our request that we be given cots was turned down. We made places for ourselves on the floor, and slept the sleep of the dead.

The next day the vice-governor came to our cell. We again requested cots, and also asked that we be moved from this cell block to another one; after all, we were neither on death row nor were we lifers. By the way, I recall that as a joke I then said to the vice-governor that it was in his best interests to have the prison equipped as well as possible because, who knows, one day he might find himself there instead of us. I don't know whether it was this argument that had an effect on him or whether he was simply a reasonable person, but in any case we were given cots.

Soon we began to receive messages, via the prison grapevine, from comrades who had been arrested for active participation in the strike; there were about 200 of them in this prison. They informed us that they had presented the prison administration with an ultimatum; move us in with them, and if their demand wasn't met by a certain time, they'd begin to break down the doors. That same day we were moved to the second floor where our comrades were. Evidently the prison authorities were afraid to use arms against political prisoners. The conditions in our new quarters were incomparably better. Our cell on the fifth floor had been kept locked all day, here only the door leading to the entire cell block was locked. The doors of the individual cells were open and the prisoners could circulate freely among each other's cells and walk about in the corridor.

To celebrate our reunion an impromptu concert was organized with singing, dancing, and declamations. Everybody had a good time, although naturally, we were all anxious about what was happening on the outside. None of us had any information about how the strike was going, and whether it had been broken. A short time later we received, along with provisions from the outside, a letter from our comrades and a leaflet which had been printed by the new Strike Committee (it seems that within an hour after our arrest they had managed to put out their first leaflet in which they informed the workers that the members of the Strike Committee had been arrested, but that a new Strike Committee had taken its place and the struggle was continuing. "Long live the victory of the Baku workers!" was the way the leaflet ended.

We could not imagine a greater joy. But we were continually thinking about what policies our comrades would carry out next, and when and how the strike would end. It was hard for us, isolated in prison, to give any advice to those on the outside. Soon, along with provisions, we were given the newspaper *Azerbaijan* for May 10. In it we read the article entitled, "The Detention of a Barge Loaded with Weapons," which contained a false account of the detention in the naval port dockyard of a barge called *Belaya* on which there had been discovered "four long-range guns, plus a lot of shells, ammunition, machine guns, a large supply of rifles, and around two million rifle cartridges. According to reliable information, this barge was left over from the former 'Tsentrokaspii' and in connection with the latest events was to be used for arming the striking Bolshevik workers."

We requested permission from the prison authorities to write a statement of protest to the socialist-faction members of parliament to be read aloud in parliament and to appear in print. Permission was received, and we wrote a text which exposed all the provocation and slander that had been directed against us by the paper *Azerbaijan*. Preserved in the archives is a letter from the public prosecutor of the court to the warder of the Baku Central Prison, dated May 13, in which he wrote: "I suggest that the enclosed declaration by the prisoners Churayev, Mikoyan, Anashkin, and Koval be returned to the writers, and that it be explained to them that I cannot pass on such a communiqué because it contains expressions insulting to public officials."

A short time later our declaration nonetheless managed to find its way into the Menshevik paper *Iskra* (*The Spark*), where it was published on May 20. This issue of the paper has been preserved.[15]

As we learned from the paper *Azerbaijan*, the head of the government made this announcement at a session of parliament: "The government is thoroughly convinced that this strike does not have an economic character, but rather is pursuing exclusively political goals.

[15]See Appendix

We have facts which enable us to assert that the local leaders of the Workers' Conference are acting on orders of the Council of Deputies . . . The agents of the All-Russian Council of Deputies want to set the local workers against the Azerbaijani Republic. In order to insure the success of their criminal plans, they have involved the workers in a strike, concealing their political aims with economic demands. . . . If it is forced to, the government won't hesitate to deport the obvious agents of the Council of Deputies beyond the borders of Azerbaijan . . ."

After the speeches of the representative of the parties, a resolution was passed by the parliament in which the strike was declared to be "a political act of aggression against the Azerbaijani system of government," and the "strong measures of the government" met with its complete approval.

The socialist members of parliament did not speak out against the government's declaration concerning the character of the strike or against threats to use repressive measures against the strikers. True, they didn't vote for this resolution, but proposed their own formulation: "After listening to the government's explanations and approving of its measures to preserve order and peace, the parliament turned to the next items of business."

Such a formulation skirted the issue of the character of the strike in cowardly fashion; on the other hand, the measures taken by the government "to preserve peace and order" met with open approval. Thus the socialists approved the government's repressive measures against the participants in the strike. This was an act of self-exposure before the workers on the part of the right-wing socialists.

Word reached us in prison that among certain comrades who had remained at liberty a bitter debate was raging concerning the question of whether to stop the strike or prolong it. The sailors, the workers form Chornyi Gorod, the majority of the railway workers, and the Russian workers, for the most part, would not even consider ending the strike.

However, in Balakhany, Surakhany and in the other oil districts, the Muscavatists had managed, by means of deceit, false promises, and repressions, to force part of the workers to go back to work at the plants and to get some of the trains moving again. Such a campaign was aided, first of all, by the fact that the workers had been deprived of their genuine leaders: during this time many of the worker-activists (including Azerbaijanis, too) were in jail. In Baku more than 200 of the leaders of the workers' and party organizations had been arrested.

The number of Moslem workers who took part in the strike also declined at this time. Under these conditions it was difficult to count on a prolonged and successful struggle.

In the face of this first set of difficulties the Mensheviks and SRs, who had come out earlier in favor of the strike, were now beginning to agitate for the cessation of the strike in their locales. However, they encountered opposition from the progressive vanguard of the workers. For example, the Menshevik Rokhlin, thinking that he would earn himself a pretty penny for his role in halting the strike, asked permission to speak at a meeting of the workers of Chornyi Gorod on the necessity of calling off the strike. Knowing the mood of the workers in that district our comrades allowed him to make the speech. And what happened? The workers responded to Rokhlin with indignation, and he, having disgraced himself, left without getting anywhere.

On May 11 the Mensheviks, the SRs and the right-wing Moslem socialists conferred among themselves and convened a so-called interparty conference. This little assemblage was aimed at assuming leadership of the Baku proletariat behind the backs of the legally elected Central Strike Committee, and capitulating to the enemy. A representative of the Baku Party Committee was also invited to this conference, but as soon as he realized that this was the usual treachery, our representative refused to take part in any discussion of the strike issue. After declaring that the conference had no authority and that the legally elected Central Strike Committee alone could concern itself with the fate of the strike, he left the conference in a showy flourish.

This self-styled conference nevertheless discussed the progress of the strike and passed a resolution to call off the strike immediately, without any stipulations or conditions. The organizers of the conference were able to popularize this resolution quite broadly among the workers of Baku, and some of them took the bait.

After taking into account this extremely complicated situation, the Baku Party Committee, with the participation of Philip Makharadze, decided after long debates to withdraw the basic demands of the strike. It did, however, attach the necessary condition that all those who had been arrested be released immediately, and future repressions be curtailed. Having received the Azerbaijani government's consent to this demand, the Central Strike Committee passed the resolution on May 12, "to call off the strike, and to go back to work at twelve noon on May 13."

Ending the strike, however, turned out to be a far from simple task, with the sailors and workers of Chornyi Gorod protesting most energetically. As Sarkis told us later, a large mass meeting took place on Freedom Square, comprised of over 6,000 workers and a large number of sailors. Since they were all in a fighting mood and wanted to continue the strike, our party speakers had a hard time convincing them of the necessity of ending the strike. However, all the participants of the meeting resolved: "to continue the strike until those arrested have been released."

The government had no sooner succeeded in getting the Central Strike Committee to call off the strike than it began to delay carrying out its side of the bargain; namely, the simultaneous release of the prisoners. In any event, the twelfth and thirteenth of May passed, and no one had been released. Only then did it become clear that the resolution passed by the participants in the Freedom Square meeting concerning the termination of the strike *after* the release of the prisoners had been more correct than the resolution of the Central Strike Committee, which took the government at its word.

* * *

In the main block of Central Prison No. 1 where all five of us were now transferred, there were thirty-seven men who had been arrested with us on May 9 at the session of the Central Strike Committee. The only woman—Vera Gorodinskaya—was held in another place.

Thus there were forty-two men in all interned in the main prison block. Among us there were representatives of many nationalities and various parties. The majority were Communists and worker-activists without a party affiliation, but there were also members of the SRs and Mensheviks.[16]

In order that our stay in prison would not be wasted we decided to use the time for our own political education. Taking advantage of the fact that the prison allowed prisoners to move around within the block, we assembled them in the corridor and organized lectures for them. I remember the topics of our first two lectures: "The Nationality Issue and the Communist Party," and "Historical Materialism."

I gave the lecture on the first topic, and Lominadze spoke on the second. At the same time we asked Sherin and Rebrukh, who were educated, knowledgeable SRs, and the Menshevik law student Popoviants, who were all present at the lectures, to give their own supplementary presentations. After they spoke a general discussion began. Both the listeners and we, the lecturers, made speeches.

It happened that Lominadze and I each spoke several times. I must say that we managed to deliver sharp and convincing criticism of the policies of the SRs and Mensheviks, demolishing their arguments completely. All those present at the lectures—not only the Communists, but those without party affiliation as well—could sense the force of the Communist ideas and the weakness of the policies of our adversaries.

Our colecturers were so depressed by all this that they refused to participate in the discussions the second day. Our three opponents—the two SRs and the one Menshevik—went back to their cells. But we continued propaganda work for several days more.

On May 14 the administration of the prison informed us that soon all the prisoners, except ten of them, would be released. Included in the ten were members of the presidium of the Worker's Conference, two of whom—Koval and myself—were accused of planning an armed

---

[16]See Appendix.

revolt, and, as especially important prisoners, were under the jurisdiction of the British Command. By the way, I should add that the accusation against Koval was a complete misunderstanding. A renegade from the anarchists who became our fine and close comrade, Koval had never been a Communist and never contemplated an armed revolt, being occupied strictly with union concerns. Evidently they had confused him with Kovalev, who really was a Communist and who, as I recall, had even joined the military commission of the Baku Party Committee, along with me and several others.

All the time I thought: on the basis of what factual evidence can they accuse me of planning an armed rebellion? What evidence do the invaders have against me? Our party was, in fact, preparing for an armed rebellion. We had had a top secret military commission functioning all along which had acquired arms, stockpiled them in special storehouses, and was preparing cadres, but the police hadn't known anything about this, otherwise the other comrades more directly involved in this enterprise than I would have been arrested too.

Then I remembered that several days before the strike three workers had come into the Worker's Club and told me in the presence of Kovalev that there was a barge in the port with 3,000 rifles and other weapons on board. They claimed that the government didn't know that the barge existed, and proposed that since it was not guarded, we could help ourselves to the arms. At the time we were amazed at this information and asked them several times whether it was really true that there was no guard on the barge. They claimed that they had been watching the barge for several days and hadn't seen any guard.

All this had seemed improbable to us. We couldn't believe that the Muscavatist government and the British invaders who had been in charge of the port for half a year knew nothing of the existence of the barge. On the other hand, the information was cause for rejoicing. Could it really happen, just like that?

However, at the time we couldn't do anything about the barge. The information about it had been given us two days before the start of the general strike when it would be extremely dangerous to undertake anything of this nature. We decided, therefore, that if the report could be corroborated we would attempt, after the strike, to send some arms to Dagestan and also to Lenkoran for the revolutionary units which were operating there under the leadership of our party.

When, soon afterward, the government published the report in the paper that the barge carrying munitions belonged to the Workers' Conference, we realized that there was obviously one agent provocateur among the three workers who had come to us with the information, sent to us by the military commander of the city, Grigoryev, who wanted to bury us.

Now, in prison, and trying to think of any concrete evidence that could be used to substantiate the charge that I had been planning an armed rebellion, I suddenly remembered the whole story of the barge and decided they wanted to connect Koval and me with the barge in some way, so that they could concoct a "case" against us and hand us over to the British military court, or deport us via Persia to India, which was equivalent to death in those conditions.

All these issues were debated very heatedly among the narrow circle of Baku Party Committee members who were in the prison. Different plans were drawn up for our escape. The mastermind and spirit behind all these plans was the indefatigable, persistent, and very resourceful Sturua. First he proposed organizing an attack on the prison by a detachment of our comrades from the outside. Then he got another idea: during the inspection when two or three guards usually came to our cell, we would disarm them and flee the prison. Finally, a plan was proposed to file the bars in the cell window and escape to freedom. Technically speaking, all these schemes were very difficult to execute. All of them would, most likely, fail, but we nonetheless made preparations to carry them out.

With the help of the worker who adjusted the lamps in the cells, we could make contact with comrades on the outside and get hold of files, and also a rope for the ladder which we would need to get down to the ground from the window. This was especially difficult to accomplish since sentry guards were continually walking beneath the windows. We would have to pick a time when the guard was looking in another direction and fall upon him, or we would have to get out some other way and have a carriage in readiness somewhere nearby.

All this, of course, was extremely chancy and entailed a high risk of failure. When the prison authorities announced the imminent release of the majority of the prisoners in our group, the plans for escape were put aside.

But mortal danger continued to threaten both me and Koval. We all expected that sooner or later the misunderstanding regarding the charge against Koval would be cleared up and nothing terrible would happen to him, but my situation was a lot more complicated. Sturua therefore put forward a new plan for my liberation; to leave the prison disguised as another comrade. To exit from the prison in this way didn't appeal to me personally since it would mean that someone else would have to suffer for me, but my comrades didn't allow me to debate the subject. Sturua claimed that the one who stayed in my place wouldn't experience any particular punishment and would only spend a few extra weeks in jail. "But you, Anastas," he said, "are threatened with the death penalty."

The decision was made. Quite a large number of volunteers were found for the job, but it was necessary to select someone who closely resembled me. Thus the choice fell on Grisha Stepaniants, a technical worker on the Baku Party Committee, who asked impassionedly to be left in prison in my place. His candidacy was approved and Sturua then set about altering my appearance. With an old pair of blunt scissors that he found somewhere he trimmed my beard and moustache, literally pulling out the hairs one by one. It was very painful but I had no choice but to endure it. Then they dressed me up in Stepaniants' clothes and gave my clothes to him.

We had never expected that they would release us so soon. One evening, around ten o'clock, after the cells were usually locked until morning, several members of the prison administration suddenly entered our block. Naturally, we all immediately poured out of our cells and gathered in the corridor. One of the officials announced that all the prisoners, with the exception of ten whose names would be read out, would be released forthwith from prison.

We were astonished. Usually people were released from prisons in the mornings, but for some reason they had decided to release us at night. Later we found out that, starting in the evening, groups of women from the Tagiyev Textile Factory had gathered around the parliament building with their children and begun to demand our release; among the prisoners there were several Azerbaijanis who worked in this factory. Once before the strike I had been at this factory and had spoken at a meeting which, as I recall, stretched out late into the night. Since the factory was far from town the workers had urged me to spend the night there. I had made friends with many of the workers during the course of the evening, and especially with the president of their factory committee, Kafarov, a Communist at whose home, by the way, I spent the night.

The commandant of the prison began to read off the names of those who were to be released. Those who were called had to respond and then step over to a designated place near the door. When Stepaniants' name was called, I answered, stepped forward, and took my place among those being released. I was carrying a large pillow in my arms which covered almost half of my face (among the workers employed at the prison there were some I had associated with when I wrote our collective letter to the parliament in the prison office; these people might know me by sight, and then the entire plan for my escape could be burst like a soap bubble).

When the reading-off of names had concluded, all of us who were being released were removed from the prison block and taken to the office. The comrades who walked beside me tried to arrange it so that I didn't stand out and would not catch the eye of the prison officials. The office was poorly lit, with a single lamp hanging over a table whose light fell only on the table, leaving the rest of the room in semi-darkness. Seated behind the table was the warden's assistant with a thick book containing the prisoner's names lying open before him. He began to call on us one by one for interrogation, and then he checked the answers with the information that appeared in the prison register (surname, first name, patronymic, place of birth, province, district, and township).

Knowing approximately what sort of questions were usually asked in such circumstances I had gotten the information I needed from Stepaniants that afternoon and had memorized it. Therefore, I awaited my turn with confidence—until I suddenly heard the warden's assistant ask the first comrade called to give his Baku address. I realized that I could not answer this question, since I had forgotten to ask Stepaniants his address—this was due to the fact that when I had been arrested I had refused to give my address and this psychological moment had had an effect on my conversation with Stepaniants.

I immediately whispered the problem to Sturua. He didn't lose his head, but walked over to the table where the warden's assistant was sitting and tried to peek into the prison register to find out what address was written there for Stepaniants. But this turned out to be impossible. What could we do? Was there a way out? One of our comrades who realized what had happened suggested that I say I had come to the prison to write a statement and not to conceal, therefore, that I was Mikoyan. But this was quite improbable and naive, if only because I had not requested any such permission, and, in addition, at night, when prisoners were being released, would not be the time when statements would be written.

Then the thought flashed through my head that since my apartment was far away and was locked at night, I could ask to spend the night in my cell, and be released in the morning. This is what Sturua and I decided to try. When my turn came to be questioned I put the question to the warden's assistant: "Couldn't I stay here in my cell overnight?" He turned the electric lamp light on me so that he could see me better. "What's your name?" he asked.

"I'm Stepaniants," I answered, and the thought occurred to me: now he will recognize me, and . . . But it was dark in the office and he didn't recognize me.

"Wait a minute," he said.

I was elated. It was a pity, of course, that I wouldn't be able to leave the prison, but at least the escape attempt would not be discovered. Then something else occurred to me: if I were the only person to stay overnight, then I could easily attract the guards' attention. In order to "shield" me, then, three or four of our sailors asked if they too could stay overnight, saying that their ships were at sea and they had nowhere to spend the night. They were given permission, much to my relief.

But the indefatigable Georgi was still not content, and went on trying to think up another way out of the situation that would get me out of prison, no matter what. A few minutes later he came over to me and said quietly: "Popoviants looks a lot like you. They've only just checked him out. I'll persuade him to stay in your place, and you can go free tonight." This suggestion came as such a surprise and seemed so unlikely to me that I couldn't even find anything to say.

Without even waiting for my answer Sturua went over to Popoviants and began to try and persuade him "to become" Stepaniants and stay overnight in prison. At first Popoviants didn't want to do it, and was even indignant, but Sturua's inspiring arguments convinced him. "Mikoyan is threatened with death, but nothing is threatening you, even if you spend the night in prison under Stepaniants' name," he said. "Maybe you'll spend an extra week in jail, but that's all." Besides, he said, since Popoviants was a Menshevik, it wouldn't be hard for him to get released.

But then Popoviants said: "How can I say that I'm Stepaniants when I just said that I'm Popoviants?"

"Very well," answered Sturua, "then someone else will answer for you."

Sturua demanded that I play out this little scene, so when the name Popoviants was called out I stepped forward and took my place among those about to be released. When "my" name, "Stepaniants," was called out, Popoviants didn't move and another comrade answered for him, and he stayed with those five men who had been permitted to spend the night in prison.

A few minutes later those of us who were to be set free were led out onto the terrace where from minute to minute we expected to be released, but still nothing happened. One of our number then asked that he be given his passport, since it was forbidden to walk about the city without a passport. That only caused yet another delay, and in the meanwhile the prison officials might discover the trick any minute, and I would be detained. But the irrepressible Sturua took charge of the situation by walking over to the one who was raising a fuss and whispering sternly in his ear: "Stop fussing about your passport—you'll get it tomorrow." He obeyed. Finally, before the prison gates were opened for us, we were counted one more time, "for good measure." Everything was in order, the doors were opened, and we went out to freedom. It seemed too good to be true.

Everyone went off in different directions so as not to fall into the clutches of the police again. And it was the right thing to do since, as it turned out, one of our group of comrades was again detained by the police on suspicion of a jailbreak. He was brought back to the prison and they looked into the whole matter before they released him again.

Sturua and I stayed together, choosing the most out-of-the-way back alleys in our flight to the Kasparovs' underground apartment. Here's what Maria Kasparova, the oldest daughter, remembered about that night:

> I remember how at midnight Anastas and Georgi appeared after

their flight from prison. It was a magnificent escape for its boldness and success! Not long before this the Azerbaijani government had arrested Andzhiyevsky—one of the leading party workers in the northern Caucasus—and had handed him over to the British who had in turn, handed him over to the Denikin authorities, who hanged him on top of Mashuk Mountain. In view of this we did everything possible to insure that nothing similar happened to our imprisoned comrades.

We waited agitatedly in our apartment till late at night, expecting Georgi to appear any minute. When the bell finally rang, all of us—Rosa and the others—rushed to the door. When Georgi walked in we surrounded him, too excited at first to notice the young man standing behind him. It was Anastas Mikoyan. The unexpected appearance of someone whose fate had made us all so anxious filled us with joy. We made him sit down while we gathered around him and listened spellbound to the story of their escape, which had ended happily.

We found out later that our five comrades who had asked to spend the night in prison had returned to their cell, and after waiting about an hour (as we had earlier agreed among ourselves) until they felt we were out of danger, they had summoned the guard and told him that Mikoyan was not among them. The guards became all confused, ran to the office and began to check the books. It turned out, according to the books, that I hadn't left the prison, that Popoviants had been released, and that Stepaniants had been in the office but had returned to spend the night in his cell.

The guard returned and began to question how all this could have happened. On being questioned Stepaniants said that he hadn't even left his cell because he had received permission beforehand to spend the night there and to go free in the morning. Popoviants said that he had been in the office, had requested permission to spend the night in his cell, and had been given such permission. There was no possible way to disentangle this complete rigamarole. Then the enraged guard left, saying that none of the five would be released.

The following morning our friends went to the prosecutor and demanded the release of those comrades who had stayed in prison overnight, because "they could not and should not answer for the escape of the other prisoners." In the end they were released, and the warden of the prison was arrested.

In the archives of Soviet Azerbaijan there is a report drawn up on May 15, 1919, by the warden of Baku Central Prison No. 1, describing the events surrounding my escape from prison. Fairly recently another discovery was made of two other interesting documents in the archives also relating to this event which concern the warden's assistant Bairamov (his full name was Abas Kuli bek Bairambekow), who suffered on account of me. I shall take this opportunity, long overdue, to express my apologies to Bairamov who, nevertheless, had to spend a whole month in jail on my account. (It is curious that I, the prisoner, on whose account he suffered, spent a total of only five days in jail!).

* * *

The next morning we got in contact with our comrades and met at a session of the Baku Party Committee. We had to evaluate the situation, plot our future course of action and, first and foremost, find a way to release those of our comrades who still remained in jail.

We were in good spirits, although we were all naturally distressed that the strike had to be ended without all our demands being met. One comrade even wondered if it hadn't been a mistake, in general to have declared a strike? This question had to be answered, especially since the Mensheviks were shouting from the housetops that the strike was simply a Communist adventure even though they had voted for it when the decision to strike was being made.

I spoke at the session and reminded everyone of what had preceded the strike. We, the party's leading workers, had been fully aware of how difficult it would be to force the British who were blockading Russia to give Astrakhan access to oil. Breaking the blockade was an unlikely feat. But despite that we did the right thing when we declared the strike, I said. First of all, we had to take into consideration the mood of the workers, who were anxious to go into action. Secondly, we had to be at the head of this struggle, and in the course of the bitter class fight, test the organizational strength of the Baku working class and the influence of the Communist party. The majority of the workers had showed their worth admirably during the strike, I pointed out, and our party organization also turned out to be up to the mark. During the strike there was no question but that it had increased its authority.

I conceded that a certain segment of the Azerbaijani workers still remained under the influence of the Muscavatists and the right-wing socialists, but the strike had also revealed that many of the Azerbaijani workers fought heroically hand in hand with the Russian workers against the British Command and the Azerbaijani bourgeois government. A fairly large number of leading Azerbaijani cadres were tempered in the fight, an important achievement in itself. Only individual isolated industries were under the influence of the Muscavatists, and because of the repressions carried out by the government, strove to end the Strike Committee. There was no doubt, in any case, that the strike was both begun and ended in orderly fashion, under the Central Strike Committee's leadership. Of course, it was a pity that it ended before all those in prison had been released, and it was our task, therefore, to obtain their immediate release.

"The very fact," I said in conclusion, "that we have disengaged from the battle in an orderly fashion, even though we did not win all our demands, does not testify to our defeat; on the contrary, we were tempered in the struggle, maintained and strengthened our organization as well as our class consciousness and unity. Above all, the enemy was not able to break us. And the fact that certain plants called off the strike without authorization is a result not only of the government's repressions and Muscavatists' false propaganda, but also of the treacherous behavior of the SR and Menshevik leaders. Our primary task, therefore, is to expand the fight against the Muscavatists, especially those who were active among the Azerbaijani workers, and also to unmask in any way that we can the perfidious role played by the SRs and Mensheviks."

A resolution was passed at this session of the Baku Party Committee to convene the Workers' Conference the following day to discuss the results of the strike. After a broad exchange of opinions we were unanimous in our assessment of what should be done despite shades of difference in our evaluation of events. Immediately after the meeting Gogoberidze and I began drafting a resolution for the Workers' Conference (this resolution was also to contain the theses of Gogoberidze's report at the Conference).

The Workers' Conference met on May 16 under the chairmanship of Karayev. I was not present, since I had to lie low after my escape from prison. A proposal of Karayev's was adopted to make Churayev, Mikoyan, and Anashkin who, as was announced, were under arrest, honorary chairmen of the conference. Gogoberidze then gave a detailed report on behalf of the Central Strike Committee. The discussion of his report was a stormy one.

The representative of the Adalet party, Agazade, who aligned himself completely with Gogoberidze, declined to take the floor, and demanded that the right-wing leader of the Gummet party and member of parliament, Pepinov, explain the unworthy actions of his ideological accomplices in the party during the strike. Pepinov began to cover up the role of his parliamentary faction and the right-wing segment of Gummet, in every way that he could.

The Mensheviks made a shrewd move at the Conference when they chose as their speakers Shtern, Popoviants, and Rebrukh, who had been arrested along with the Central Strike Committee members and were now free, feeling that these speakers would be listened to attentively and would produce the desired impression. Their speeches were directed at that part of Gogoberidze's report which had exposed the Mensheviks' treacherous actions during the strike. These speakers claimed that the strike had failed because of poor organization, and that since it was the Bolsheviks who had led it, responsibility for the failure must rest on their shoulders. The workers' delegates repeatedly interrupted the speeches of the representative of the right-wing parties with indignant shouts from their seats.

After Gogoberidze's brilliant summations, one of the delegates, the Muscavatist Zulfugar, shouted a malicious rejoinder from his seat: "The agents of the all-Russian Council of Deputies have convened here," and with that he left the meeting in a demonstrative fashion. By a majority of 194 votes, with 3 opposed and 24 abstaining, the resolution proposed by the committee was adopted:

> In its first confrontation the Baku proletariat found itself facing a united enemy—international capital and its local villains: the khans, beks, capitalists, and their parties of Cadets, Muscavatists, and Dashnaks. The Muscavatist party in particular stands out for its provocational zeal, interminable lies and shameless slander, aimed at provoking national persecution and bloody slaughter.
>
> With the help of its numerous agents it (the Muscavatist party) declared the peaceful strike to be an armed rebellion, directed against the Moslems and the "independence" of Azerbaijan. Bayonets and whips did the trick where lies and provocations failed.
>
> The so-called right-wing socialist parties, or rather, the decaying corpses of these former parties, who were for the strike in the beginning, showed their treacherous colors during the bitter struggle and instead of coming to the aid of the fighters, they swooped down on the field of battle like a flock of black kites to end the struggle and they set up a funeral parlor—the interparty conferences.
>
> The parliamentary block of Moslem pseudosocialists betrayed the proletariat the very first day by expressing its confidence in the government and by approving the bayonets, the whips and the Muscavatist provocations. The Baku proletariat did not gain a victory, but it did manage to leave the first battle in an orderly manner, having preserved its proletarian army and enriched it with the experience of a first engagement for battles to come.
>
> The workers of Baku consider the past strike to be not the end, but the beginning of the struggle, the first battle where the Baku proletariat received its baptism of fire, and they are aware that the causes of the strike have not been eliminated but, rather, have become exacerbated. Therefore, they have not laid down their arms, not given up the struggle, but are preparing successfully for a new battle, by organizing and closing their ranks, paying particular attention to their politically uninformed comrades—the Moslem workers—and trying to tear them away from the bourgeois and petit bourgeois traitorous parties and with the aid of the Communist groups Adalet and the left-wing Gummetists to organize and unite them around the supreme proletarian organ of Baku—the Workers' Conference.

This resolution of the Workers' Conference which was passed by the motion of the Communists, threw the SR and Menshevik pen-pushers into a frenzy. Day after day they began to revile the Communists, trying to discredit them with every means possible, going into tirades about the defeat of the working class, the failure of the Workers' Conference leaders, etc., etc. Heaping all kinds of abuse on the Communists they tried with all of their strength and with every means possible to restore the influence of their parties and their leaders, representing them—in counterbalance to the Communists—as the only intelligent, faithful, experienced, and trained leaders of the working class.

They were unable to achieve any real success among the working class, but among that part of the intelligentsia whose understanding of events they had distorted, they had considerable success.

Unfortunately, our paper *Nabat (The Alarm Bell)*, the editor of which was then Lominadze, did not carry out an active struggle against this newspaper campaign. I spoke about this to Lominadze once and suggested that he write some sharp articles which would set forth our principles and unmask in print all this SR-Menshevik hullabaloo and give a correct appraisal of the reasons behind the strike and its outcome. I told him that we had to unmask the Mensheviks' claims concerning the so-called defeat of the Baku proletariat. We had to raise the workers' spirits, increase their faith in their own strength, and summon them to prepare for a new battle which would take place when the propitious moment arrived.

Lominadze was a very candid person. He told me frankly that, in his opinion, the strike had ended in defeat. I argued with him, trying to prove the opposite, but that time I couldn't convince him completely. Naturally, I was displeased by this, and told him sharply that if he didn't want to write articles, I would do it myself.

I had in mind two articles: one devoted to a criticism of our opponents' policies and a substantiation of our policy up until the declaration of the strike, and the other devoted to the strike itself and its results. I wrote the first article right after Lominadze had left me, still under the influence of my heated argument with him. I never managed to write a second article, however, being swamped with pressing business.

The text of the article "The Recent Strike," which was published on May 23, 1919, in the paper *Nabat* has been preserved. This article, I feel, portrays the situation in detail, and also the policies of the representatives of the different classes and parties. Their real surnames are used.[17]

The major conclusion which we drew for ourselves regarding the outcome of the strike—and this was particularly marked in the resolution of the Workers' Conference—was the necessity of doing our

[17]See Appendix

utmost to intensify our work among the Moslem workers, trying to tear them away from the perfidious bourgeois and petit bourgeois parties, and with the help of the Communist groups of Adalet and the left-wing Gummetists to rally them round the Communists and the Workers' Conference. With this goal in mind the Baku Party Committee convened a special expanded conference, in which the leaders of the left-wing Gummetists and Adalet would participate.

At the conference a lot was said about the mood of the Azerbaijani workers after the strike. For as a result of the Muscavatists' treacherous policies and the government's cruel repressions, revolutionary sentiment began to appear even among the most backward segment of the Azerbaijani workers. This movement toward the left was assisted a good deal by the unseemly behavior of the right-wing Gummetists during the strike and particularly after its conclusion. However, this leftward movement among the Azerbaijani workers had to be consolidated, which was why many of those who spoke at the conference said that the time had come for the left-wing and right-wing Gummetists to sever relations with each other and purge the Gummet party of the Mensheviks, thereby turning it into a genuinely Bolshevik organization.

Karayev and Guseynov, who had spoken out two months earlier against such a split as untimely, agreed now that the time for a split had indeed arrived. An agreement was reached as to how the Communists for Gummet could take advantage of their majority in the leadership of this organization and bring about the practical realization of such a split in the near future. The split took place in July 1919; when the Gummetist-Communists broke decisively with the right-wingers, and the latter left the organization. This was an important victory for our party.

At the same conference the issue arose in the Baku Party Committee of the expediency of participating in the Azerbaijani parliament. This issue was raised by Karayev, who even before the conference had asked permission of us on more than one occasion to leave the bourgeois parliament of which he was deputy. He was indignant at the actions of the parliament during the strike and declared that he could not remain in that "swamp."

However, we all spoke out in concert against such a move. It was impossible for us to refrain from utilizing the legal parliamentary platform for our political struggle and the propagation of our views. We recalled that the successful combination of legal and illegal forms of party work was one of the fundamentals of Bolshevik Leninist tactics.

We managed to convince Karayev, and he reconciled himself to the necessary continuation of this work, which was extremely unpleasant for him. And I have to say that he carried it out very successfully, skillfully using his position as a deputy of parliament in the interests of the party.

Chapter V

A Letter to Lenin

Messengers to Moscow—Tigran and Shura's visit with Lenin—Assistance from Moscow—Events in Lenkoran—Akhundov.

It was the year 1919. The situation of the young Soviet Republic was difficult. At precisely this time, on July 29, 1919, Lenin observed worriedly: "Murman in the north, the Czechoslovakian front in the east, Turkestan, Baku and Astrakhan in the southeast—we see that almost all the links in the chain forged by Anglo-French imperialism have become hooked up with another."

With each passing day we felt more acutely the necessity of establishing direct comunications with Moscow and the Party Central Committee, and above all, with Lenin. We couldn't resolve the most important policy issues on the basis of our own thoughts and surmises. It was impossible to continue working, particularly in the complicated circumstances then prevailing in Baku and Transcaucasia, without the advice and directives of the center and without an exchange of experience. We realized, of course, that it was no easy matter to establish contact with Moscow. Distance was not the only thing that separated us—the counterrevolutionary front stood between us—but in spite of all obstacles it was imperative that communications be established.

Back on the eve of the strike a young man had turned up at my apartment, introducing himself as Tigran Askendarian. He said that in 1918, when he was a seventeen-year-old youth, he had become a Communist and had joined a party military unit, and had later been evacuated along with the Soviet troops to Astrakhan. From there he had gone to Kizlyar with a group of comrades on a party assignment, ending up in Astrakhan along with the retreating Red Army, and now, on assignment from Paderin (head of the political department of the Eleventh Army) and Kolesnikova (chairman of the Astrakhan provincial party committee), he had come to Baku to reestablish communications with us, informing us of the state of affairs in Astrakhan, and returning with information about our situation. As I recall, he had not brought any written documents or other items with him; all he had was a certain sum of money.

His appearance did not provoke suspicion. Quite the opposite—he won me over. He was a decent, sympathetic fellow, and gave ever indication of being an enthusiast. He gave me his Baku address. When we checked up on him we found out that his parents had perished at the time of the Turkish seizure of Baku, and that his neighbors spoke well of him. I told him to take a rest, warning him that he would soon have to return Astrakhan.

The letter I wrote to the Party Central Committee and Lenin containing information about the situation in Transcaucasia and our tasks there, was printed on cloth. In it I said that due to Transcaucasia's year-long isolation from the center, we had been deprived of directives and assistance in our struggle. "And you," I wrote, "are either completely unfamiliar with the actual state of affairs in Transcaucasia, or have been misled by incorrect information given to you by irresponsible people, and therefore you are not guaranteed from making serious mistakes in external policy directed at our region. Taking advantage of this small opportunity, I wish to inform you . . ."

My letter-report to Lenin was rather broad-ranging. It made mention of the situation in Baku and Transcaucasia, the occupational regime, and the role and politics of the Azerbaijani nationalist-bourgeois government. It spoke of the fact that in Azerbaijan there was more "combustible material, more socially acute contradictions, more of a class basis for proletarian revolution, more dissatisfaction about and hatred of the existing government" than in either Georgia or Armenia, and that there existed in Azerbaijan a well-organized working class which was under Communist influence and leadership. The letter also reported that "the Baku organization has recognized an independent Soviet Azerbaijan with close political and economic ties to Soviet Russia," and that "this slogan is very popular and can unify around itself the broad masses of laboring Azerbaijanis and lead them to revolt." The letter contained descriptions of our Workers' Conference; of the beginning and outcome of the Baku workers' general strike; of the growth of our party's influence among the Baku workers; of the debates with our Tiflis comrades on the subject of an independent Soviet Azerbaijan; and of the situation in Georgia and Armenia. It reported on the joint reign of the British and White Guards in Petrovsk, on their rout of the Dagestan Party Committee, on the revolutionary mood of the Dagestan peasants, and the organization of rebel detachments in Dagestan.

We told Dagestan that we were preparing an armed rebellion, creating rebel strong-points in the area, and that we would be able to act as soon as we knew that units of the Red Army and Fleet were ready to assist us. We requested Lenin and the Party Central Committee to send experienced party workers (especially Moslems) to

Baku, and also to send party literature written in Russian, Azerbaijani, and Armenian.

And so, when it came to deciding whom to send as our first messenger to Moscow, we unanimously hit upon Tigran as our man. Not that we didn't have enough dedicated party men of our own, but we understood very well that dedication alone was not enough in this matter; we needed someone who possessed special adroitness and experience. Tigran know the road from Astrakhan to Baku very well, and it would be easier for him to get his bearings on the return trip and slip through all the barriers.

It was decided to have him accompanied by Shura Bertsinskaya, a twenty-year-old Communist and technical secretary in the Bureau of the Regional Party Committee who was a native of Baku, and had worked in the northern Caucasus in 1918. After the fall of Soviet power there, Shura had made her way to Georgia along the Georgian-Military highway, and then had returned to Baku. We had complete confidence in her.

She was short, delicate, pretty, and looked younger than her twenty years, giving the appearance of being no more than a high school student. But she had already been involved in various very complicated situations through taking part in the struggle against the counterrevolution in the northern Caucasus.

So the two of them, Tigran and Shura, were assigned to take our letter to the Party Central Committee and deliver it personally to Comrade Lenin. (I should note that, just to be on the safe side, we created an "understudy" for Tigran and Shura and sent a third messenger to Moscow—the experienced Communist, Khoren Borian—with the same letter.)

We were afraid that the letter might not reach Lenin directly, since I didn't know who would receive it there, whose hands it would fall into, or where they would send it. We very much wanted Lenin to see it, and it seemed to us that once the Central Committee understood our situation, we would receive directives on all the matters that were worrying us. Therefore, we underlined the names of the addressees: To the Central Committee of the Workers' Communist party (Bolsheviks) and Chairman of the Sovnarkom, Lenin.

The story of how our delegation made its way through the territory that had been seized by Denikin is told by Shura Bertsinskaya in her memoirs:

> When it was proposed that I accompany Tigran to Moscow, I joyfully agreed to go. To travel and see Soviet Russia with my own eyes, to cross borders and front lines, to be in danger and overcome obstacles—all these things were so satisfying to my desire for heroics and great deeds that I could hardly wait for the journey to begin.
>
> At that time Tigran was eighteen and I was twenty.

Finally at the end of May 1919 the day we had been waiting for arrived. Anastas Mikoyan's report on the situation in Azerbaijan and Transcaucasia had been printed on a piece of cloth; our orders for delivering it to the Party Central Committee now in Moscow were printed on the same sort of fabric. We began our preparations for departure by finding hiding places for these missives. Tigran had a service jacket which served as an outer garment. We ripped out the breast pocket and sewed up Mikoyan's report to the Central Committee in its place; we managed to "install" our orders in a corner of my jacket. Both of us had already received identification papers under assumed names. I had taken the name of my deceased friend, Zina Yeremenko, and Tigran that of a pupil at the modern school, Grisha Zorabov.

And so, finally, carrying only a small traveling bag each, we bought tickets to Petrovsk at the station and left Baku on the evening train. We purposely waited till the last minute to board the train and sought out a car which was dark except for the intermittent flickerings of lantern lights.

After Khachmas Station we heard a noise in the car and a few minutes later several people came into our compartment and demanded our documents. After examining Tigran's papers and insistently shining the lamp in his face, they abruptly ordered him to wait outside in the corridor, whereupon they themselves exited.

Tigran quickly took off his service jacket and handed it to me. In his service jacket Tigran looked like a rather grown-up young man, but in the blue-striped shirt which he had on underneath and in his student's cap with its modern school cockade he could easily have passed for a boy considerably his junior. On leaving the compartment, he caught sight of a guard on the platform who was obviously waiting for him to appear. Instead he made a quick turn and exited from the opposite side of the car into the lavatory. Then, unrecognized by the guard he passed through into the adjacent third-class carriage, which was very crowded, and lay down in an upper bunk.

I, of course, didn't know anything of this. Suddenly I heard something—a commotion of some sort had begun. People with lanterns came into our compartment, examining the faces of the passengers and whispering among themselves about something. They left and came back again. Obviously they were looking for someone.

What had happened to Tigran? No, I couldn't let it appear that I knew him. The responsibility for getting Comrade Mikoyan's report to Moscow now lay with me alone. And that was the main thing. I heard what the passengers were saying to each other: "They probably took that young man off the train. It happens a lot here—they take them off and shoot them!"

Morning, Yalama Station. I didn't feel like sitting still, so I stepped off the train, hoping that I might manage to find something out.

Marvelous countryside, the fragrant southern forest, transparent dew, cool air and freshness . . . And suddenly—there was Tigran! He greeted me joyfully, pretending he'd just bumped into an acquaintance by chance. In a whisper he said that during the night they had shone the lantern straight into his face several times, but still did not recognize him as "the" young man in the service jacket.

We arrived safely in Petrovsk and left the train. All over the city cavalry men could be seen riding about with white ribbons attached to their Caucasian fur caps; anti-Soviet posters and caricatures of OSVAG* hung on the walls. The city had just been occupied by the White Guards the night before. No exit from the city was permitted without special passes.

We went at once to the pier where we tried to persuade one of the boatmen to take us to Chechen Island, promising to pay well (from the supply of money we had been given for our journey). No one would agree to do it. Only the day before it would have been possible, but that day an order had been posted: "No exit without special passes." Just to see what would happen we went to the command post to get a pass. We made up a story about wanting to go from Chechen Island to Staro-Terechnaya in search of our parents, who had fled there from Baku during the Turkish invasion. Looking at Tigran, the officer said: "A mobilization has been announced, and you, young man, must present yourself tomorrow to be drafted; you shall join the army. You're just the right age."

So we had really gotten ourselves into a bind, with the way back blocked and any move forward extremely dubious. We reached a decision: Tigran would return to Baku, even if he had to go on foot, and I would get to the island by myself, since I had not been refused a pass.

We were wandering around the city when suddenly whom did we see coming down a hill but Khoren Borian. We knew that Khoren was on the same assignment we were, but he was to have gone to Soviet Russia via the northern Caucasus, whereas we were enroute to Astrakhan by way of Chechen Island. Khoren reported that he had managed to get hold of a remarkable document belonging to the Armenian committee, fully authorizing it to inquire into the situation of Armenian refugees in the northern Caucasus. He too was traveling to Chechen Island, going by a cutter which was due to depart that evening, and he advised us to try to get on board too.

We stopped off at a hotel for a while, and then went down to the pier late in the afternoon. The cutter was at the dock and the embarkation of passengers had already begun. Most of the travelers were sack-bearers carrying flour. We got on board and went down to the hold for "security," but about fifteen or twenty minutes later

*OSVAG—Denikin counterintelligence.

everyone was ordered above for a document check. We went out on deck and saw that the commander was standing a short distance away on the pier, next to him there was someone in a military uniform. One by one the passengers got off the cutter, showed their documents to the commander, and returned on board. Suddenly Tigran nervously squeezed my arm and whispered. "Look. That's Andrea Miakov who's standing next to the commander in the military uniform. Miakov was in the same class as me at school!"

What were we going to do? If Tigran showed his student card made out in the name of Zorabov, who was also a student in the same class, then everything would be revealed in no time. Besides, Miakov knew that in August 1918 Tigran had volunteered for a party unit and had joined the Red Army. Should we leave the cutter, forget about the trip? But where could we go? The pier was closed in by military personnel, one of whom was Miakov.

There were fewer and fewer uninspected passengers remaining. Meanwhile, Tigran had his eye riveted on Miakov's movements, and suddenly he saw his chance. Someone had come onto the pier to talk to Miakov, and at the very moment that Miakov walked over to meet him, Tigran hurried up to the commander, displayed his student card with its two-headed eagle insignia, and in a minute was back on the cutter again. Miakov turned back from the man he'd been conversing with, and without noticing anything, strolled back to the commander. It was split-second timing, but Tigran pulled it off.

Still shaking from our close call, we went below and entered a small cabin where a group of officers and their ladies were seated. The officers were kissing the ladies' hands, paying them compliments, and amusing them with stories about the "horrors" of the Council of Deputies, frightening them with the fearsome word, "Bolshevik."

Finally, the whistle blew. The cutter cast off from the pier and a few minutes later we were at sea. For the first time that day we began to breathe easier. We returned to the deck and gazed at the light of Petrovsk fading away in the distance. Our sleepless night and the excitement of the long day caught up with us then, and on finding a tiny open space on deck near the steam whistle, we lay down and fell sound asleep. We slept so soundly that Tigran, whose arm had fallen across the steam-operated safety valve of the whistle, did not feel the steam burning his arm all night. He had a scar on his arm for the rest of his life.

In the morning we arrived at the desert island of Chechen and set off for the Bolotins' house which was at the very end of the settlement. Ivan wasn't home; his wife said that he had gone off somewhere, but should be back soon. His brother Fyodor, however, was at home, and like everyone on the island he was a fisherman. That meant he knew how to handle a boat, which meant he would be able to get us to Astrakhan.

Nevertheless, we waited for Ivan. Two days passed, and still he did not come back. On the third day his young wife ran into the house, sobbing bitterly, and told us that the Denikinites were ransacking the huts. They had captured her husband, she said, and were torturing him, and had already shot someone.

Tension filled the air as the entire settlement seethed with anxiety. We realized we couldn't wait much longer. As we sat in the hut like trapped animals waiting for evening, we talked with Fyodor, telling him that we had to leave that same night at any cost. Fyodor agreed to help us and left on "a reconnaissance trip" in search of a suitable boat. He had to be careful since about 200 meters from us there was a lighthouse around which a sentry marched slowly, with measured gait. He was an Indian/Hindu soldier from the British and Denikinite base on the island.

Fyodor returned eventually with the news that he had found a boat, a very small boat, but, "It's May now," he said with a smile, "and it will get the three of us to Astrakhan."

The sun finally set and night came. We crept out of the house and found that the only things visible in the distance were the lights of the British escort vessel and the passing silhouettes of the marching Hindu sentries on the strip of shoreline. Suddenly Fyodor said in a soft voice: "Stop!" We took cover. We heard the sound of a boat scraping on the sand. Someone had come up on shore and passed by not far from us, and then once again, there was silence. Fyodor and Tigran carefully climbed into the boat. Then Fyodor returned, picked me up and carried me over to the boat, pushed it to a place where the water was a little deeper, put up the sail, and we set off.

It was such a lovely night, we had been so lucky to get hold of a boat, the sea was so calm and caressing and the fair wind so reliable, that we all slept soundly—even Fyodor—who seemed to our eighteen- and twenty-year-old eyes to be a rather mature person (he was only about thirty).

At dawn Fyodor anxiously awakened us. We looked around and saw nothing, but the distant barking of dogs could be heard, and we knew we were quite close to a shore settlement. "The Village of Staro-Terechnaya, a nest of White Guards," Fyodor said. A boat was moving away from the shore in our direction. The wind had long since calmed down and our sails hung like lifeless rags. "Take down the sails!" Fyodor commanded. "Shura, take the rudder, Tigran and I will get the oars. Row as hard as you can!" The chase began.

A little less than an hour passed. The distance between us and our pursuers increased steadily; when it began to get wider and wider we realized that they had stopped rowing. The settlement was already hidden from view; but Fyodor and Tigran kept on rowing for the air was motionless and there was no hope of using the sails. Three hours passed. "I can't row any more." Tigran was the first to give up. "Okay, you can stop," Fyodor said.

Thus, for three whole days and nights, catching the tiniest bit of wind, we slowly progressed toward our destination. Then, on the third day we suddenly heard the roar of artillery fire. "There's a sea battle going on somewhere," Fyodor said. A cutter passed by close to us; on it waved a proud flag that caused wild excitement among us, for it was imprinted with the initials: RSFSR.

Hurray! . . . It meant we were already home!

* * *

When Tigran left for Moscow, I had given him strict orders to bring back at any costs the shorthand report of the Eighth Party Congress which had taken place in the last half of March 1919. We knew very little about the congress: newspapers from Soviet Russia did not reach us; our only news of what happened at the congress was culled from rumors and from reports in the bourgeois press.

Tigran and Shura had been ordered to hand the letter to Lenin personally. They carried out this assignment successfully, and by July 1919 returned safely to Baku via Astrakhan and the Caspian Sea with a group of responsible party workers, so necessary to us at that time.* They included several well-known Baku workers. Naneishvili, Buniatzade, and Gamid Sultanov.

In his meetings with simple people Lenin would leave them enraptured by his warmth and cordiality. Below is a page from Shura Bertsinskaya's memoirs describing her and Tigran Askendarian's visit with Lenin when they delivered our letter:

> Cordially extending his hand and smiling Vladimir Ilyich rose and came forward to greet us.
>
> Tigran, who was filled with excitement, exclaimed, "Greetings to the leader of the worldwide proletariat from the workers of Baku!" and out of the fullness of his heart he embraced Vladimir Ilyich.
>
> "Sit down, comrades, and tell me the news," Vladimir Ilyich said gently, trying to calm him down. He was interested in every detail of what we had to say.
>
> "And where is the Lenkoran of which Comrade Mikoyan writes?" (He pronounced the "r" in Lenkoran softly, so that it sounded like "Lenko-an.") "Comrade Mikoyan writes about a peasant congress in Karabakh. Where is Ka-abakh? I looked, but I couldn't find it."
>
> A large map of Transcaucasia was hanging right there on the wall

---

*After the final victory of Soviet power in Azerbaijan, Tigran and Shura were sent away to school and became first-rate metallurgical engineers.

of the office, and Tigran pointed out to Vladimir Ilyich the places where important events were taking place for Transcaucasia at that time. On another wall there was a large map of European Russia, with little flags on it indicating the front lines of the Civil War.

We told him that Sergo Ordzhonikidze was in Baku in deep hiding, and the comrade Mikoyan had asked us to tell him that Ordzhonikidze wanted to get to Russia. What's the best way of managing that? we asked.

Vladimir Ilyich was overjoyed. "Sergo is alive, then, in Baku!" he exclaimed. "It is absolutely necessary to get him out of there."

"It might be a good idea, Vladimir Ilyich, to send a submarine for him," Tigran said.

"But how deep is the water at the shores of the Caspian Sea in the vicinity of Baku, would a submarine be able to get through?"

Tigran replied that not far from the Derbent shoreline the water was sufficiently deep for a submarine. One could get to the submarine in a (row)-boat.

"It's an idea, we'll think about it," Vladimir Ilyich concluded.

In describing the conduct of the British troops in Baku, we noted that the British soldiers were busy selling cigarettes, chocolate and (outer) clothing.

"Well, that means that soon they'll be selling rifles too," Vladimir Ilyich said with a laugh.

"Is Lenkoran a real state?" he asked. "It has an army; does it have its own money?" And then he said: "How did you manage to get through Moscow so quickly?"

We were about to reply in a word or two, but Vladimir Ilyich was interested in hearing the details. "And were there dangers?"

We related our misadventures. He laughed when we told how Tigran had outwitted the White Guard sentry that had been posted for him on the train.

"Just don't tell anyone how you got here, no one, not a soul," he warned us insistently. And suddenly, he surprised us by saying: "What are you thinking of doing now, perhaps stay here and go to school?"

"No, Vladimir Ilyich," we replied with all the decisiveness of youth, "we promised Comrade Mikoyan to return."

"And do you know English?"

With regret we answered "No" to his question.

"Well, if you have already decided to go back, we shall give you leaflets in English, printed on onionskin, and money for the party organization. They'll put it all in a suitcase for you. You can go get the suitcase from Comrade Klinger at the Comintern."

And once again he said: "Tell no one about your trip, not a single person, not even your close friends!"

Vladimir Ilyich conversed with us so cordially and naturally, and questioned us in such a fatherly way that we forgot Comrade Stasova's warning to keep our report down to ten or fifteen minutes. And the words we had prepared in advance in making our report completely vanished from our minds. We spent forty minutes with Vladimir Ilyich, after which, happy and proud, we left to carry out his instructions.

* * *

Toward the end of March 1919 several party members arrived in Baku from Lenkoran. Since we lacked contact with Lenkoran we knew very little about the real state of affairs there. They told us, for example, that although Lenkoran had been included in the composition of Azerbaijan, the power of the Azerbaijani government did not extend to it. Ever since the overthrow of Soviet power last summer the region had been governed by counterrevolutionary officers and representatives of the Lenkoran kulaks. The British command, in declaring its "friendly relations" with the Azerbaijani government, could, of course, have crushed "the Lenkoran incident" and made this rich region submit to its ally, but the Lenkoran rulers were direct agents of Denikin, and thus were even more to the liking of the British than the Azerbaijani nationalists.

The comrades from Lenkoran told us that they had succeeded in gathering the fragmented forces of the local party organization and in winning influence and authority among the laboring peasantry. They also reported that among the military personnel in Lenkoran there were a rather large number of former members of the Red Army who had been sent there during the time of Soviet rule and had remained after its fall. The sympathies of the local Azerbaijani peasants were also against the Muscavatists and White Guards. In additrion, there were rebel detachments of Azerbaijani peasants with which the local party organization was in close contact. In view of all these circumstances, the Lenkoran comrades affirmed that it was completely realistic to think in terms of seizing power in the Lenkoran region.

This communication was so important that once we had meticulously checked our information, we had to give serious thought to our comrades' suggestion.

It became clear that we did have the possibility of seizing power in Lenkoran, but to hold on to it without help from the outside was another matter entirely. And in general, there wasn't much sense in creating such a small "Soviet oasis" without the hope of expanding it, or even preserving it intact. However, within the larger framework of our hope that, with the opening of navigation, the Red Fleet would begin operations for seizing Petrovsk and Baku, the Lenkoran issue acquired concrete significance. Located on the shore of the Caspian

Sea and possessing a fairly good harbor, Lenkoran would make a good base for the Red Fleet. Besides that, it was rich in grain and people, and could become a real force in our subsequent struggle for Soviet power in the rest of Transcaucasia.

In view of this it was decided to overthrow the rule of the White officers and to establish Soviet power in Lenkoran.

As far as I can remember, between the fifteenth and twentieth of April, Gogoberidze was posted to Lenkoran as a fully authorized representative of the Baku Party Committee. He was accompanied by Agazade, Gubanov, Starozhuk, Kanevsky, and others. In the course of several days they succeeded, by relying on local cadres, in effecting a coup, chasing out the White Guard regional head-of-state, and creating a revolutionary committee in Lenkoran, headed by Bolsheviks.[18]

In the middle of May in revolutionary Mugan a special congress took place which lasted for four days. The congress proclaimed Soviet power in Mugan, established the Mugan Soviet Republic, and elected a regional Soviet of Peasant, Worker, and Red Army Deputies.

In a word, things had got off to a good start in Mugan. Gogoberidze told us about it in detail when he returned to Baku.

As before, the first item on our agenda at that time was the matter of establishing regular contact with Astrakhan. We tried to set up radio communications with Astrakhan, but for technical reasons nothing came of the attempt. One of our comrades, an experienced sailor and Bolshevik by the name of Kozhemiakin, took it upon himself to make his way to Astrakhan by sailboat. His assignment was to tell them there about our situation in Baku, and to return with relevant information.

In spite of the difficulties, this operation turned out to be a complete success. A month later Kozhemiakin returned on the same sailboat with two comrades sent by Kirov. One of them was the sailor T. I. Otradnev Uliantsev, the chairman of the Revolutionary Military Soviet of the Eleventh Army.

The *Revvoensovet* of the Eleventh Army and the command of the Red Fleet in Astrakhan showed particular interest at that time in strengthening Soviet power in Mugan. As we had surmised, they intended to use Lenkoran and all of Mugan as a beachhead for naval operations, first and foremost for landing an amphibious force. Therefore, Otradnev had orders to go to Lenkoran and get the necessary work underway there.

From our conversations with Otradnev we established that doubts had arisen in Astrakhan concerning the information they had received from our representative. Apparently Kozheniakin had exaggerated somewhat in his account of our forces and successes in Baku. They did not totally believe what he said and had ordered Otradnev to go

[18]See Appendix

to the scene, recheck everything and report on the state of affairs in Baku, but to stay there himself in order to work in Lenkoran under the leadership of the Baku Party Committee.

We were very happy about our meeting with Timofey Ivanovich Otradnev. He was thirty-one at that time, having joined the party in 1909 as a worker. In 1916, while serving in the Baltic Fleet, he was arrested for doing underground revolutionary work among the sailors, and sentenced to hard labor. After the February Revolution he became a member of the Kronshtadt Bolshevik Party Committee. He was a delegate to the historic April Conference of the RSDRP (Bolsheviks), elected to membership in *Tsentrobalt*, and was president of the socio-democratic Naval Court. In 1918 Otradnev was the military commissar for the organization of the Red Army in the Stavropol area. Then, at the suggestion of Kirov, he worked in the Astrakhan military field-court tribunal.

He was a well-built man, with an open Russian face, large, intelligence eyes, and a pleasant smile. His smoothly combed, parted hair and short clipped moustache lent him a certain imposing air. Energetic and full of initiative, he was at the same time composed and reasonable, and made a good impression on us.

We decided to send him to do leadership work in Lenkoran and granted him extraordinary authority, just one indication of which can be seen from the content of the special mandate he was issued in May 23, 1919.[19] These orders have by chance been preserved.

After arriving in Lenkoran, Otradnev very quickly won the trust and support of the local comrades, and under his leadership Soviet power there was strengthened rather rapidly.

For our part, all necessary measures were taken to give supplementary assistance to Lenkoran, most importantly in the form of experienced fighting men to strengthen the armed forces of the small Soviet republic. And we had such fighting men at that time. After the fall of Soviet power in Baku, several thousand Red Army men overcame many hardships to make their way to Georgia. Once there, however, they lived in poverty, for the Menshevik authorities treated them with great suspicion and prevented them even from getting work. Gradually they began to drift back to Baku where the Workers' Conference gave them what help it could. We selected the most reliable among them and began to send them to Mugan in small detachments in order to fill out the ranks of Red Army.

It should be noted that after the proclamation of Soviet power in Mugan the British Military Command and the Azerbaijani government declared a blockade of Lenkoran. No ship, not even a single sailboat, was allowed to enter or leave Lenkoran. The strictest controls on movements in or out of the area were established on land and sea.

[19]See Appendix.

However, we kept on sending small groups of fighting men—twenty to thirty in a group—to Lenkoran in fishing boats and sailboats. I remember that on four or five occasions (perhaps even more) our boats fell into the hands of the police and our comrades thrown in jail. There were cases where those who had been arrested were killed.

But this terror did not weaken in the slightest either our energy, or the ardent desire of the workers and the Red Army men to get through to Lenkoran. We succeeded in transferring about 500 experienced, reliable troops there, which constituted a significant support force for Lenkoran.

Otradnev brought a code from Astrakhan for secret correspondence, plus a large sum of money (Nicholas II notes) to meet the needs of Lenkoran and Baku. Thus for the first time we were able to create groups of party workers in the districts who were completely free of all other work responsibilities. Moreover, we repaid the money we had borrowed from the Workers' Conference to buy a boat, and we put a certain amount into the financial aid fund for needy Red Army men who had come from the northern Caucasus. We also had the opportunity now to publish party literature ourselves since, despite our requests, very little literature (even in Russian) had trickled down to us from Russia. In order to publish literature in Azerbaijani we enlisted the aid of the Bolsheviks in Gummet (when it was unified) and of the workers from Adalet, as well as of Azerbaijani left-wing SRs, one of whose leaders was Rukhulla Akhundov.

I had known Akhundov since 1918, from the time of the Baku Commune. At that time he was the editor of the newspaper, *News of the Soviets of Workers' Deputies*, which was published in the Azerbaijani language. I liked him very much. I got together with him frequently, and we established a good relationship of mutual trust. Our friendship grew even stronger during the period of the Baku underground under the British occupation. Rukhulla Akhundov's political point of view actually differed in no way from the policies we were pursuing. In spite of his youth—he was twenty-two at the time—he was a firm, principled man, deeply devoted to the cause of revolution. Of medium height, and thin, he was always smartly turned out, punctilious in keeping his promises, perhaps his only weakness his excessively hot temper.

I knew that Akhundov had an excellent command of the Azerbaijani literary language. He had a sharp pen, and since he also had a good knowledge of the Russian language and literature, we wanted him to translate the works of Marx and Lenin into Azerbaijani. Besides, he might be able to draw a large group of Azerbaijani intellectuals, including some from his own party, into this project, which was something we were interested in having happen at that time.

In talking things over with him, I told him that the Baku Party Committee would undertake to publish the translations. He agreed to the project, but wanted his translations and their publication to be under the sponsorship of his party while having the financing come from our Baku Committee, since his party did not have sufficient funds at its disposal. Understandably, I rejected this suggestion and, as I recall, even expressed bewilderment: how could the left-wing SR party publish Marxist literature on funds provided by the Communist party?! "Your way of doing it is absolutely unacceptable," I said, and asked him to think about my suggestion again and let me have his answer.

A few days later he sent me the following brief note.

> To Comrade Mikoyan:
> You have often suggested that I leave my other duties behind and begin work in regard to publishing Turkish books and brochures. Up until now, owing to circumstances beyond my control, I have been unable to leave the press. Now, however, in view of a conflict which has arisen between the manager and myself, I have left the press.
>
> If you are still interested, I can undertake the project. If your answer is affirmative, we can then talk about the details and terms of the job.
>
> With comradely greetings, Rukhulla.
>
> P.S. By the way, I should note the fact that the other day I was considering the possibility of taking a position at another place. Please send me your answer immediately, so that if the work you promised is no longer available, I might not lose the position.
> Baku, July 2, 1919.

It must be said that when I got the idea of suggesting to Akhundov that he do the translations, I had the thought in the back of my mind that in the process of doing them, and once he had penetrated to the essence of Marx's and Lenin's teachings, he would liberate himself from SR ideology and become a Marxist. That was precisely what happened.

After I received Akhundov's note, we got together immediately. He agreed to my suggestion, and shortly afterwards became a Communist. In the summer of 1919 the Azerbaijani left-wing SRs merged with the Gummet Bolsheviks.

To jump ahead somewhat, I would life to say a few more words about Akhundov and about how he later matured and became a prominent figure in the party.

After the victory of Soviet power, when Kirov was Secretary of the Central Committee of the Azerbaijani Communist party, Rukhulla Akhundov, who was greatly respected within the Communist party of Azerbaijan, became the Central Committee's Secretary for Cultural Affairs.

The publication of books in the Azerbaijani language was the most important condition for effecting a cultural revolution in the republic. It was Akhundov's ardent desire to expand the development of the publishing business in Azerbaijan, and he accomplished a great deal in this area. With exceptional zeal he busied himself first and foremost with the publication of Marxist-Leninist literature in the Azerbaijani language. He considered the publication of Lenin's opus his life's work.

At the end of 1924 the presidium of the Central Committee of the Communist party of Azerbaijan passed a resolution calling for the modernizing of the printing industry in the Republic, and in 1925 Rukhulla Akhundov was sent abroad to acquire polygraphic equipment, especially a linotype that had typeface in the new script. (At issue here was the introduction of a new, improved printed version of the Arabic alphabet; the old alphabet was to continue in existence alongside the new for a short while longer.)

Lev Lifshits accompanied Akhundov on his trip. In Germany they found some typeface that could have been purchased, but it turned out to be unsuitable. Discovering that new typeface at that time was manufactured only in America, Akhundov and Lifshits informed Kirov of this fact and the Central Committee of the party decided to send them to America.

They left by ship for New York, where something unforeseen happened to Akhundov upon his arrival. During the health inspection he was discovered to have trachoma and was sent to the "Island of Tears" (Ellis Island) where he was put uder quarantine as a person suffering from a disease which forbade entry into the country.

In spite of all their petitions, Amtorg—the Soviet trade delegation in New York—failed to receive permission for them to enter New York. There was only one thing to be done—return home. But Akhundov was unable to return empty-handed. "It has never happened," he said, "that I've failed to achieve what I set out to do."

In a meeting with the Amtorg representative Akhundov stubbornly insisted that his work be concluded despite the circumstances. The Amtorg people warned him that that would entail great hardships for him. "We didn't come here on an excursion," Rukhulla said, "but to do work. I am staying here until the work is accomplished."

Then Amtorg and Lifshits searched out the company they needed in New York, and the company's representative began travelling to the "Island of Tears" to negotiate with Akhundov. The negotiations went on for more than a month, under very difficult conditions, until the parties finally defined the specifications for the needed equipment. Akhundov got his way.

At that time Vladimir Mayakovsky arrived in America. According to Lifshits' account, Mayakovsky intended to visit the "Island of Tears" out of (simple) curiosity, but when he found out that the Secretary of the Central Committee of the Azerbaijani party was in quarantine on the island and refused to leave until he had achieved his goal, he became doubly interested in making a trip there and meeting Akhundov personally. But somehow it happened that Mayakovsky never got to the island and the two men did not meet. Nevertheless, the stories he had heard about Akhundov caused Mayakovsky to form a very high opinion of him.

In 1928 Mayakovsky's poem "The Big-Wig" appeared in print. The poem had come to be written on account of a newspaper story which reported that a member of the Central Executive Committee of the USSR by the name of Akhundov had slapped the face of a passenger in a dining car. Mayakovsky could hardly have realized that the newspaper story concerned the same Akhundov who had interested him so much when he was in America. And, besides, in the poem itself he writes:

> I have no idea,
> who I'll run into.

In the poem Mayakovsky painted a colorful portrait of a Soviet big-wig but it bore no resemblance to the culprit in the incident described in the paper. The incident, however, actually took place. Akhundov had taken some remark made by his travelling companion as an ethnic slur, and, unable to restrain himself, had slapped the man.

It was the newspaper report, of course, that had motivated Mayakovsky to write the poem. Akhundov himself later felt very badly about the incident, the only one of its kind that occurred in his life.

Both Sergo Ordzhonikidze and Kirov valued Akhundov highly and considered him one of the best and most well-trained Azerbaijani Marxists.

Chapter VI

The Knights of the Revolution

Sergo Ordzhonikidze—His arrival in Baku—My acquaintance with Kamo.

I had first heard about Sergo Ordzhonikidze in the spring of 1917 from Stephan Shaumian and Alyosha Dzhaparidze, who had worked with him in the party organizations of Transcaucasia even before the revolution. They spoke of him as a principled and courageous revolutionary, a tireless organizer of the masses, so that from the very start my mental image of Sergo exuded an aura of revolutionary romance.

He had joined the party as a seventeen-year-old youth in 1903, the year of the Second Congress of the Russian Social Democratic Workers' party (RSDRP), and devoted his entire life to its cause. Sergo carried out selfless underground work in Baku. His meetings with Lenin in 1911, and Lenin's lectures at the school for party workers in the small town of Lonjumeau near Paris, were very important in Ordzhonikidze's evolution as a professional revolutionary. In her memoirs Krupskaya says of Ordzhonikidze: "From that time on he became one of our closest comrades."

Sergo played a prominent role in the overall party arena in the period when preparations were being made for the Prague Party Conference of 1912. As one authorized to organize the conference, he visited Petersburg, Moscow, Kiev, Rostov, Ekaterinoslav, Tiflis, Baku and other industrial centers of the country. On these trips at a time when the most intense struggle was being waged with the Mensheviks, Trotskyites, and Conciliators, he displayed great skill in his efforts to rally the local party organizations around the Leninist idea of a general party conference. The conference chose Ordzhonikidze as a member of the Central Committee, and of the Russian Bureau of the Central Committee. When he returned to Russia he did a great deal of work to implement the resolutions of the conference.

Of his fifteen years of underground work, eight were spent in prisons, penal servitude, and exile. The prisons of Tiflis, Sukhum, and Baku, the Schlusselburg fortress, and exiles in Siberia and Yakutsk did not break Sergo's iron constitution, but were the universities that trained him in the struggle and strengthened his ideological convictions. His character was such that not once did he back down or bow under during the Revolution and the Civil War—in Petrograd, the Ukraine, Byelorussia, the northern Caucasus, and Transcaucasia—or during the years of socialist construction.

I'd like to depart a little from the chronological framework of my memoirs and speak of Sergo Ordzhonikidze not only as an important party and state figure, but also as a man with whom I worked for many years and remained friendly with until his death.

Sergo was a very deliberate person. His emotional makeup, his opinions—both political and philosophical—his actions and his way of life, all formed a single, compact whole. He was knowledgeable about the most complex political and economic issues, and expert in party politics and tactics, and in the party's and the working class's methods of struggle. The question arises: how did he get so knowledgeable? After all, by education he was only a doctor's assistant.

It is here, it seems to me, that one of the most striking characteristics of this outstanding man reveals itself. Gifted from childhood, Sergo was always studying—in the midst of revolutionary events, in the course of the bitterest struggle between the then existing parties, while overcoming inner-party dissensions, in fierce interparty skirmishes, in underground circles, in the dogged work of self-educations, in personal contacts with all sorts of people, at meetings, sessions, conferences, congresses. He swallowed up knowledge avidly, making it a permanent part of himself. Prison and exile became for him a magnificent university of life and learning; I was always amazed by the number of books which Sergo had read while he was in the Schlusselburg fortress.

Sergo avidly loved life, people, and the struggle. He was a man of active temperament, interested in everything. His political and theoretical views, therefore were closely tied to direct action, which he pursued with great enthusiasm, and it was for this very reason that Sergo showed his worth so brilliantly in the most difficult years of our party's struggle.

Let us recall the year of 1918. Soviet power has collapsed in Baku. German-Turkish troops have entered Transcauscasia: three bourgeois national states have been formed which are hostile to Soviet Russia. In the northern Caucasus the Germans and the Denikinites have opened up a front against the Bolsheviks. The Red Army, lacking supplies from the center, experiences great hardships; its ranks are falling away, people are dying of typhus, of malnutrition. The army is forced to fall back to Astrakhan where there is Soviet power, and supply operations are still functioning; from there it is possible to continue the fight with the enemy.

Ordzhonikidze was then in Vladikavkaz. He was Special Commissar of the South of Russia and the head of the Committee for Defense of Tersk Province. He was confronted with this question: should he retreat with the troops to Astrakhan, or stay behind with the local partisan detachments of workers and mountaineers and continue the struggle where he was? Sergo decided to stay and fight to the end, even though there was little hope of victory, with Menshevik Georgia to the rear, hostile to Soviet Russia, and brutal bands of White Guards on the offensive.

However, revolutionary honor inspired Sergo; we have to stand our ground, he said. He sent a telegram to Lenin: "Rest assured that we shall all perish in an unequal fight, but we shall not disgrace the honor of the party by fleeing." These words are expressive of the whole man.

A half year passed before Sergo could return to Russia and display his talent on the other fronts of the Civil War. When, roughly a year later, he returned to the northern Caucasus with units of the Red Army the mountaineers greeted him as a national hero. Their trust in him knew no bounds.

Sergo was not a born orator, but he had an exceptional talent for immediately entering into direct emotional contact with his audience and winning them over with his sincerity, directness, and simplicity. His emotional makeup was such that although he was usually calm and self-possessed, he became unrecognizable when he came up against lies, intrigue, or rank injustice. Then he would explode with indignation and rage and was capable of things he would later sincerely repent.

Once one of the local deviationists, a certain Kobakhidze, indulged in a coarse attack against Sergo, practically accusing him of political corruption. He cited as evidence the following "fact": when Ordzhonikidze returned to the Caucasus the mountaineers had presented him with a saddle horse as a token of their special affection and gratitude. Sergo accepted the gift in accordance with Causcasian custom, but not considering the horse to be his personal property he put it in the stables of *Revvoensovet* (the Revolutionary Military Council), using it mainly for riding in ceremonial parades (in those days not only commanders but also members of the revolutionary military councils formally reviewed at parades). As a member of *Revvoensovet*, Ordzhonikidze had a complete right to a horse provided at public expense, and when he left Tiflis this horse stayed behind in the stables of *Revvoensovet*.

Kobakhidze twisted these facts to suit himself.

To accuse Sergo—an impeccably honest man—of corruption was more than monstrous. Therefore, when he heard about it, he blew up, and not being able to control himself, he slapped his slanderer in the face. A legal "case" sprang up. Dzerzhinsky, who was then empowered by the Central Committee of the party to take care of such matters, reached the conclusion that Ordzhonikidze was not guilty.

Ordzhonikidze's behavior, however, made Lenin indignant. He wrote: ". . . Comrade Ordzhonikidze has to be punished as an example to others (I say this with regret because I am one of his personal friends and have worked with him in exile abroad) . . ."* Lenin condemned Dzerzhinsky's judgment of Ordzhonikidze's action as incorrect and explained with that adherence to principle so characteristic of him:

"Ordzhonikidze was an authority with regard to all other citizens in the Caucasus. He does not have the right to indulge in that hot temper, which even Dzerzhinsky makes reference to. Ordzhonikidze is, on the contrary, obliged to conduct himself with a degree of self-possession not required of an ordinary citizen . . ."** (Incidentally, here is a perfect example of the high standards Lenin always set for Communists occupying important posts in the party and state.)

Sergo was sincerely repentent over everything that had happened and could not get over it for a long time. Without minimizing his guilt in any way he wrote in the interests of explaining the truth that the incident had occurred not because of his political quarrel with Kobakhidze, but had been prompted exclusively by the personal insult which had been dealt him.

There was another trait characteristic of Ordzhonikidze. We all knew that he had kidney trouble. It was accepted at that time that one turned for help to foreign specialists in particularly complicated cases. Lenin insisted on this more than once, and so Sergo went to Berlin. The famous Professor Borkhart confirmed the opinion of the Soviet doctors: one kidney had to be removed immediately.

It was suggested that Sergo have the operation in Berlin. He flatly refused, declaring that he would be operated on at home by his own doctors, with whom he had become very close. He was especially fond of Fyodorov (he had frankly "fallen in love" with him, as he himself would tell us) a court physician of the former tsar in Leningrad. At that time Fyodorov was considered a truly odious figure, the physician of the Royal Court who had been present even at Tsar Nicholas Romanov's abdication from the throne! However, Sergo knew Fyodorov very well personally, believed in him, and agreed without the slightest hesitation to have him perform the operation.

The operation was successful, making everyone very happy, including the doctors, who had taken a liking to Sergo for his open, expansive nature. (It should be noted that it was because of Sergo that the country recognized Dr. Fyodorov as an outstanding physician when, in 1933, he was awarded the Order of Lenin.)

*V. I. Lenin. *Complete Works*, Vol. 45, p. 361.
**V. I. Lenin. *Complete Works*, Vol. 45, p. 358.

The doctors expressed their feelings for Sergo in a very original way; while he was being treated by them, a new medicine had been discovered which yielded excellent results in kidney treatment. They named the medicine "Sergozin"—in honor of Sergo and his wife, Zina. As far as I know, this medicine is still being used.

I'd like to call attention to another of Sergo's characteristics: while he was unable to conceal his anger, he didn't bear grudges or take revenge on anyone, but he could never bear an insult in silence. I recall an episode which took place before my eyes in 1921 during the Tenth Party Congress in Moscow at which I myself was a delegate. Sergo was not at the congress, being prevented from attending it by a very strained political situation which had arisen in Transcaucasia concerning the Sovietization of Georgia.

I should add that our party had been racked for more than two months prior to the Tenth Party Congress by a discussion, initiated by the Trotskyites, about the trade unions; factional groups were being formed with their own platforms, thus threatening the unity of the party. Ordzhonikidze, who was in Transcaucasia, came out clearly and loudly on Lenin's side on this issue. In April 1921 at the Third Congress of the Communist Party of Azerbaijan he had given a report on the trade union discussion during which he brilliantly clarified the basic issues and revealed the antiparty quality of Trotsky's and Shliapnikov's position. In the end the overwhelming majority of the Congress of Bolsheviks of Azerbaijan supported Lenin's platform.

As I said, Ordzhonikidze did not attend the Tenth Congress, but when his name came up in connection with nominations to the Central Committee some military delegates from the northern Caucasus unexpectedly began to shout out objections from their seats to Ordzhonikidze's candidacy. These delegates were sitting in the back rows and they caused a stir throughout the hall. One of them got up on the podium and said that Ordzhonikidze shouted at everybody, ordered everyone about, and didn't take the local party workers into consideration, and that for these reasons he shouldn't be on the Central Committee. This demogogic address affected the mood of the delegates, many of whom, moreover, did not know Ordzhonikidze.

The one who had spoken out was the political worker Vrachov. He sticks in my memory very well, because his face was bandaged in white gauze, evidently because of a toothache. Some years later when he spoke at the Thirteenth Party Conference Vrachev admitted that his challenge to Sergo at the Tenth Party Congress had not been done on his personal initiative, but under heavy pressure from the Trotsky faction, with whom Vrachev was then affiliated. He told how during the Tenth Party Congress he had taken part in a secret meeting of Trotskyite delegates at the congress where a resolution had been passed to reject the nomination to the Central Committee of a number of comrades who were supporters of Lenin, Ordzhonikidze included

(Vrachev subsequently moved away from Trotskyism, admitted his mistakes, and adopted the correct party position.)

Stalin spoke out in defense of Ordzhonikidze at the Tenth Party Congress. He spoke calmly, in a soft voice. He cited biographical data, told of Sergo's work in the underground and on the different fronts of the civil war, and recommended that he be elected to the Central Committee. It was clear, however, that he hadn't managed to convince the delegates, since they continued to make a fuss.

Then Lenin spoke out in defense of Sergo's candidacy, to the following effect:

"I have known Comrade Sergo for a long time, from the underground days, as a dedicated, energetic, fearless revolutionary. He proved his worth admirably during exile abroad, played a prominent role in organizing the Prague Party Conference in 1912, and was then elected a member of the Central Committee. He carried out active work in Petrograd before and during the October Revolution. In the civil war he showed himself to be a courageous, talented organizer. But there is one correct observation regarding Sergo in our comrades' criticism of him, and that is that he shouts at everyone. That's true. He talks loudly—even when he's talking with me, he shouts the same way—but you, most likely, don't know what the problem is. Obviously, it's because he's somewhat deaf in his left ear, so he shouts because he thinks that no one can hear him. But this deficiency shouldn't be taken into consideration."

This made the delegates smile and even laugh good-naturedly. It was clear that Lenin's support of Sergo's candidacy had repulsed the opposition's attacks against him (although it should be observed, parenthetically, that Lenin never excused Sergo for his individual outbursts and the disruptions he caused.) In any case, the seriousness of any misgivings among the delegates would be reflected in the number of "no" votes cast against Sergo's candidacy. After Lenin's speech a secret ballot was taken, and Ordzhonikidze received an overwhelming majority of the votes. Like Dzerzhinsky, he received 438 out of 479 votes (only Lenin had been chosed unanimously)—more votes than a number of other prominent members of the Central Committee received at that time.

I remember how astonished I was at Vladimir Ilyich's powers of observation, for while I had also noticed that Sergo's hearing was worse than others', I hadn't known that it was, in fact, his left ear that was affected.

Soon after the congress delegates had departed for their respective locales, Ordzhonikidze found out from V. A. Sutyrin (who was then Assistant Head of the Political Administration of the Causasian Army) about Vrachev's speech at the congress. However, as to be expected, he didn't harbor any feelings of malice against him. On the contrary, the opposite occurred.

In the summer of 1922 Sergo was sent by the Central Committee, together with Eliava, and someone else whose name I can't recall, to Central Asia to look into the state of affairs there, particularly the struggle with *basmachestvo* (an anti-Soviet movement). There he met Vrachev, who was then a member of the Military Council of Turkestan.

As Vrachev related later, Ordzhonikidze's commission familiarized itself objectively with their work and did not criticize him personally. Moreover, at the end of 1922 Sergo raised the question of transferring Vrachev to Tiflis as a member of the Revolutionary Military Council of the independent Caucasian army, of which Ordzhonikidze himself was a member.

Vrachev said that shortly afterwards he was summoned to Moscow to the party's Central Committee where he was received by Stalin. When he was asked whether he objected to being transferred, Vrachev said he had no objection. Then, without any irony, Stalin asked him if he would be making any more challenges against Ordzhonikidze. Vrachev replied that, naturally, he would not, since now he knew Ordzhonikidze very well and had a deep respect for him.

I should add that, as a rule, the only people who responded badly to Ordzhonikidze were those who did not know him personally or knew him only slightly, and who gave too trusting an ear to the hostile or run-of-the-mill gossip that frequently circulated in certain circles about prominent party workers.

That same Sutyrin mentioned above told about the incident in the mid-twenties when he rode to Moscow in the same train compartment with Sergo. At one of the stops Sutyrin stepped out onto the platform and bought a copy at the bookstand of Serafimovich's *The Iron Flood*, which had just come out and was still relatively unknown. He asked Sergo if he wanted to read it, saying that he had read the story himself and had liked it very much. When he heard the author's name, Sergo became somewhat irritated (as it then seemed to Sutyrin) and said that he didn't intend to read that book. When Sutyrin asked him why, Ordzhonikidze replied: "Because I consider the author of the book a bad person."

It turned out that in the early twenties Serafimovich had been sent to the Causasus as a correspondent for *Pravda*. At the time Sergo was working in the Caucasus as a representative of the center, which had wide-ranging authority and was fully empowered by the Soviet government.

The situation in the Caucasus at that time was a very complex one. Without being properly informed about the situation and the subtleties of the most bitter of class struggles which was then being waged in the Caucasus—a large multinational territory—Serafimovich had succumbed, unfortunately, to the influence of gossip, and without checking any of it personally, had slandered Sergo to Lenin in a letter sent to Moscow in which he attributed actions to Ordzhonikidze which the latter had not committed. As a man who was honest and upright to the extreme, who was very sensitive and hated unfairness and slander, Sergo was naturally genuinely hurt by the author's treatment of him.

Fortunately, this conflict was resolved a short time later, to the mutual satisfaction of both sides. When he saw how universal love and respect for Sergo was growing in the country, Serafimovich realized that he had been profoundly mistaken in his hasty conclusions regarding Ordzhonikidze. At the end of 1928 he addressed this letter to Sergo:

> Comrade Ordzhonikidze.
> Dear Comrade,
> Several years ago I did you a great injustice.
> Everybody whom I have spoken to about you—even those who have disagreed with you—is unanimous in their response: Comrade Ordzhonikidze is a most dedicated revolutionary, an exemplary party member. I have kept a constant eye on your political work, and more and more I am convinced that I was unfair to you and made a complete mistake.
> Not long ago I met you in the Kremlin, and again it hit me painfully; here is a comrade who has been treated with great unfairness on my part which was undeserved.
> I have to tell you this.
>
> With Communist regards,
> A. Serafimovich
> December 7, 1928

This letter is a fitting description not only of Ordzhonikidze, but of Serafimovich himself. The honest man could not rest until he had written to Ordzhonikidze and admitted his unwitting mistake. After this, the past was forgotten.

While I was working in Nizhny-Novgorod I often met Sergo at party conferences, at plenums and meetings of the Central Committee, and at the Congress of Soviets. In March 1922 we met at the Eleventh Congress of the Workers' Communist Party (Bolsheviks, RKP). Vladimir Ilyich still took part in the work of this congress, but his health had deteriorated so much by that time that the doctors demanded treatment and rest. Lenin himself was very aware of this fact.

Taking the state of Lenin's health very much to heart, Sergo tried to persuade him to go to the Caucasus to rest and undergo treatment. And it seems he succeeded, for plans for such a trip were arranged with Ilyich. Several of Lenin's notes to Ordzhonikidze apropos of this have been preserved. In one of them he wrote:

> . . . Write me a few lines, please, so that I shall know that every-

thing has been completely arranged and there won't be any "misunderstandings." My nerves are still in a state, and my head pains won't go away. In order to give medical treatment a fair chance, I have to make the trip one of complete rest. . . . It is very likely that in view of how busy you are, you will not be able to carry out the tasks we talked about yesterday by yourself, nor is it rational, of course, that you should undertake such a thing. Find somebody who is efficient and dependable and attentive to small details and delegate the job to him (then, by the way, it'll be more agreeable for me when *you're* not the one I have to criticize).*

Lenin also said that in a month's time he would expect to receive from Sergo "a detailed map and information concerning appropriate places—their height above sea-level, their seclusion, etc., etc.; also descriptions of these places and the districts and provinces in which they're located."** In other notes to Ordzhonikidze apropos of this same subject, Lenin said that Kamo (S. A. Ter-Petrosian) wanted to go with him and that he had agreed,*** and he expressed his concern over whether the place that had been chosen would suit not only him but Nadezhda Konstantinovna, and that from that point of view "Borzhomi water was ideal."**** However, Vladimir Ilyich never managed to make the trip to the South.

My meetings with Ordzhonikidze became more and more frequent after I moved to Rostov (in May 1922) where I was named Secretary of the South-East Bureau of the Central Committee of the RKP (h). Naturally, I was interested at that time in Sergo's work, and in the operations of the Transcaucasian Communist parties in general, insofar as we "lived" alongside each other; many of the problems encountered in the northern Causasus had points in common with analagous problems in Transcaucasia. In order to bring about practical solutions to many of these problems, it was important for me to study the experience of those working in the Transcaucasian republics. My meetings and conversations with Sergo were especially useful to me because he knew the northern Causasus and Dagestan better than I did. Usually, when we had to go to Moscow to attend plenary meetings of the Party Central Committee and congresses of the Soviets, Voroshilov (who was also working in Rostov at that time) and I would join up with Ordzhonikidze and Kirov, who were traveling via Rostov. We would ride in the same train car and would also make the return trip together. During those trips we engaged in the friendly conversations and lively exchange of opinion that takes place between close comrades who work together.

*V. I. Lenin. *Complete Works*, Vol. 54, p. 229.
**Ibid.
***V. I. Lenin. *Complete Works*, Vol. 54. p. 230.
****Ibid., p. 242.

At that time many complex problems were emerging in the northern Caucasus concerning nationality, class, and other things which gave rise to sharp conflicts and frictions. It was quite difficult for me to find my bearings in these issues, especially in the beginning, whereas Sergo, who had worked in the northern Caucasus during almost the whole period of the civil war and its victorious aftermath, had amassed a wealth of experience, and knew the local cadres very well, especially those comrades with whom he not infrequently came into conflict. Sergo's authority among the northern Caucasian and the Soviet party workers was great, and they heeded whatever he had to say. I would take every opportunity, therefore, to meet with Sergo in order to consult with him regarding these issues.

In July 1926 at a plenum of the Central Committee of the All-Union Communist party (VKP, Bolsheviks), Ordzhonikidze, Kirov and I were elected candidate-members of the Politburo. Up until this time we had all worked in the Caucasus. Immediately after the plenum Sergei Moronovich Kirov was sent to Leningrad as Secretary of the party *Obkom*; for Zinoviev could no longer be tolerated as head of the Leningrad organization since he had become part of the vanguard opposition movement within the party.

The appointment of Kirov was a very good one, for, as later years showed, he was fully able to cope with the tasks imposed upon him. He managed to put party life in Leningrad back on the correct path and to retain for the party many thousands of Communists who had been thrown into a muddle by Zinoviev. Once again the Leningrad organization became a true support of the Party Central Committee, and the standard-bearer of Leninist ideas.

Ordzhonikidze also played a positive role in the struggle against the Zinoviev opposition. During the Fourteenth Party Congress some of the members of the Central Committee, including Sergo, Kirov, and myself, went to Leningrad, where we spoke out in defense of the policies of the Central Committee and against the opposition.

Soon after the July plenum (1926) of the Central Committee a resolution was passed to appoint me People's Commissar of Internal and External Trade, thereby relieving me of my duties as Secretary of the Northern Caucasian Regional Committee of the party. (The story of how I became a People's Commissar is quite interesting, but more about that later.) A short time after this the Central Committee passed a resolution to transfer Ordzhonikidze from Transcaucasia to the northern Caucasus, to the post of Secretary of the Northern Caucasian Regional Committee of the party. A protest against his transfer was submitted to the Central Committee from members of the Transcaucasian Regional Committee, who insisted that Ordzhonikidze be kept in Transcaucasia. Stalin, however, insisted on having his own way, and the transfer went through. As I found out later, Ordzhonikidze himself also thought that he should continue his work in Transcaucasia, but did not lodge a protest with the Central Committee, and after the decision regarding his appointment was made he set out for Rostov.

It was obvious even then, that if Sergo had to be transferred from Transcaucasia, then naturally it should be to a leadership post in all-union work, for which he was then fully qualified. And that's what happened. Before months had gone by the Politburo passed a new resolution—to move Ordzhonikidze to the post of People's Commissar of the Workers' Communist Inspectorate, and Chairman of the Central Control Commission (CCC) of the party. This post had, in fact, been vacant for three months, since after Dzerzhinsky's death in July 1926, the People's Commissar of Rabkrin, V. V. Kuibyshev, was made Chairman of the Supreme Economic Council (VSNKh).

Ordzhonikidze's appointment to the post of Chairman of the CCC was confirmed at a joint plenum of the Central Committee and the CCC in November 1926. Despite the pressure of many members of the opposition, including their leaders, no one lodged a protest against Sergo, and he was appointed Chairman of the CCC with only one vote against, and six abstaining.

This new appointment was a very felicitous and useful one for the party. Ever a consistent advocate of Leninist politics and a resolute fighter of the opposition, Sergo was able to show the necessary tolerance for those who had erred. Following the spirit of Lenin's ideas and directives concerning the unity of party ranks, Sergo sought to examine each issue as objectively as possible, trying whenever he could not to sharpen antagonisms, nor permit diverse platforms to take shape, fighting against a split in the party with all his might.

The general membership of the party was pleased to have someone like Ordzhonikidze at the head of the CCC. At that time there was a tendency to expel from the party ranks in too-hasty fashion even old-guard Communists who had been led astray in supporting the opposition. Sergo Ordzhonikidze and the entire CCC (the presidium of the CCC consisted of old tried-and-true Bolsheviks, who were devoted to the cause of party unity) spent hours trying to convince such Communists of their error. In the majority of cases they succeeded—true, of course, only in regard to the middle- and lower-level party workers, since as far as the leaders of the opposition were concerned, it had already proved impossible to "make them over."

Ordzhonikidze fought passionately against the opposition. At the Fifteenth Party Congress he said: "The Revolution, comrades, is no laughing matter. As soon as you begin to shatter the party that led the revolution, then you can ruin any revolution. Vladimir Ilyich has spoken of this many times. We have no right to permit our party and our Revolution to be destroyed, but this is precisely what the actions of the opposition are leading to."*

*"The Fifteenth Congress of the All-Union Communist Party (Bolsheviks) December 1927." Stenographic Reports, Vol. 1, Gospolitizdat, 1961, p. 436

In 1929 at the Fifth All-Union Congress of Soviets the first Five-Year Plan for the development of the economy was ratified. The task now was to mobilize all forces to carry out the plan for the country's industrialization. A new job awaited Sergo; in November 1930 he was made Chairman of VSNKh.

As the Five-Year Plan was being implemented, the complexities and difficulties of managing a rapidly growing industry in totally new conditions became clearly evident. A decision was made to section off from the VSNKh two separate independent people's commissariats, one for light industry and the other for the timber industry, and also to transfer the food industry to the people's commissariat of produce. What remained of the VSNKh—i.e., all the branches of heavy industry—became its People's Commissar.

The country set about the construction of huge ferrous metallurgy plants, gigantic mines, and new machinery-construction factories. The most important task became the training of cadres of industrial-technical intelligentsia, who would be able to master the advanced technology and provide the economic leadership needed by these industries.

Sergo contributed a great deal to the solution of these problems. His authority as a party and government leader which he enjoyed throughout the party, as well as his personal charm, contributed in many respects to that great upsurge of creative energy shown by the workers of heavy industry, which created the preconditions for the fulfillment and over-fulfillment of the Five-Year Plan. It should be added that the workers of heavy industry were very proud of their People's Commissar. Soon after this the party awarded Sergo the honorary title of Commander of the Army, Commander of Heavy Industry. His right to this title was demonstrated by everything he did. Sergo was able to pick talented people, particularly from among the youth, whom he would support in every way, and then correct and take in hand when they made mistakes.

It was on his recommendation that Zabeniagin, a young engineer-Communist, was made head of the vast metallurgical complex at Magnitogorsk. Later Zabeniagin became Vice-Chairman of the Council of Ministers. Included among those whose careers he promoted was the then young engineer-metallurgist Tervosian, whom he appointed to the post of head of the Spetsstal (Steel) Trust, and who subsequently became Minister of Ferrous Metallurgy and also Vice-Chairman of the Council of Ministers of the USSR.

Sergo also took note of Likhachev, a talented organizer, who became head of the Moscow automobile factory, and then a People's Commissar.

Sergo attached a great significance to the rationalizers, and always brought striking initiative to the all-union arena. His role in the Stakhanovite movement is well known. This movement, which began in the coal industry, spread to all branches of the economy. Sergo knew how to raise the enthusiasm of the popular masses and channel it toward the solution of concrete, practical tasks.

Although he devoted his main attention to the development of heavy industry, Sergo did not lose sight of the people's need for consumer goods. At that time the food industry was undergoing a radical technical transformation and was in dire need of appropriate equipment and machinery. Ordzhonikidze rendered invaluable service in this regard and I, who was in charge of the food industry in those years, had occasion more than once to take advantage of his energetic support in the practical solution of the food industry's problems.

Sergo worked tirelessly to inspire everyone with the belief that everyday work at the factories was not a commonplace matter, but, rather, something majestic and inspirational. He fought stubbornly for a production standard, for cleanliness in the factory yards and inside the shops, for optional organization of the work site, for the introduction and development of cost accounting, and for raising the production efficiency. In the remarkable initiatives of our current day innovators you can see traces of undertakings begun by him, to which he devoted his heart and soul, as well as his organizational talents.

To conclude, I would like once again to note Sergo's exceptional qualities as a comrade. He didn't make friends easily, but when he did, it was a genuine friendship, and he always knew how to show a true concern for his comrades.

Not long ago I got to see a heretofore unpublished letter which Ordzhonikidze had sent on February 13, 1928, to K. E. Voroshilov, who was then People's Commissar of Defense. In this letter Sergo requested, in connection with the forthcoming awards to be given to veterans of the civil war, that eight veterans of the struggle in the Caucasus be recommended for awards, since in his opinion they deserved to be decorated but had never been honored in this way.

Included among them was Sergei Mironovich Kirov, to whom Sergo devoted these lines in the letter:

> . . . He is as well known to you as he is to me. Nevertheless, I shall say a few words about him. During the siege of Astrakhan, Kirov was the one who inspired the defenders of the city, who were cut off on all sides. In 1920, by order of the Revolutionary Military Council of the Caucasian Front, Kirov flew from Astrakhan to Sviatoi Krest (Holy Cross) in our old wreck of a plane and advanced with the troops to the northern Caucasus. In 1921 during our advance into Menshevik Georgia, Kirov himself accompanied us through the Mamissonsky Pass, where the troops cleared a path for themselves through nine feet of snow. It was only Kirov's presence that could have inspired the almost half-naked and barefoot Red Army men to such a fear.

From this document I also discovered that it was Ordzhonikidze who had included me among those comrades to be awarded the military decoration of the Red Banner for my part in the armed struggle for Soviet power in Baku. Neither Sergo nor my other comrades had ever told me this. Ordzhonikidze himself was awarded the military decorations of the Red Banner in the spring of 1921, when the civil war fighting had come to an end and the country was beginning peaceful socialist construction.

Many examples could be cited which would corroborate what an outstanding political and governmental figure on the Leninist model that Ordzhonikidze was. All this is fairly well know to everyone. In my memoirs, therefore, I would like to stress that Sergo was also an outstanding man. To quote Gorky, a man with a capital M.

He was a dedicated, truly noble Knight of the Revolution, a confirmed standard-bearer of Bolshevik truth. He loved to say: "party spirit—first of all, and first and foremost." When he addressed the young commanders of our industry at that time he taught them "never to rest on their laurels, never to be conceited, because conceit, comrades, is merely a sign of ignorance."

The name Sergo Ordzhonikidze always enjoyed tremendous authority and popularity in the party and with the people. It is an illustrious name in the history of our Communist party, and in the history of the Soviet state.

* * *

Now I shall return to my memories of 1919.

In May of that year we, the Communists of Baku who were working in the underground, received a communication via Tiflis from the Transcaucasian Regional Committee of the party that Ordzhonikidze intended to make his way to Moscow. There was only one route from Tiflis to Moscow at that time, and that was via Baku and then by sea to Astrakhan, where there was Soviet power and where Sergei Mironovich Kirov was head of both the political and military leadership.

Ordzhonikidze found himself in Tiflis after our Eleventh Army had been forced to retreat in January 1919 after having waged fierce battles in the northern Caucasus with the superior forces of Denikin's "volunteer army." The main units of the Red Army's Eleventh Army, led by Levandovsky, escaped into the Kalmyk steppes and to Astrakhan. The other units, headed by Ordzhonikidze, fought heroically in the Caucasian foothills until the last bullet had been fired and then headed into the mountains. In the mountain villages Sergo formed partisan detachments consisting of Ingushes and Ossetians, and in early May he made his way to Tiflis over the Caucasian range and the almost impassable Khevsursky mountains.

Sergo arrived in Baku with his wife, Zinaida Gavrilovna, his inseparable companion at the civil war fronts and in later life. They had married during exile in Yakutsk, and their close friends called her Zina. The wife of the slain Baku commissar Alyosha Dzhaparidze—the Bolshevik Varvara Mikhailovna—also came with them, on her way to Moscow. She had languished in Transcaspian prisons for more than half a year with our group of Baku comrades. She was going to Moscow, where her two daughters—Elena and Liutsiya—were. They had been evacuated from Baku together with the wife and younger children of Stepan Shaumian, not long before the fall of Baku in 1918.

The legendary Kamo (S.A. Ter-Petrosian), a professional revolutionary, who was going to Moscow to see Lenin, also arrived with them. He had many plans for military operations he wanted to show Lenin for his approval. He dreamed, for example, of making his way with a group of comrades into Denikin's headquarters and blowing it up. Ordzhonikidze loved Kamo and was deeply affected by his death three years later. When he spoke at Kamo's funeral on July 18, 1922, on behalf of the Central Committee of the RKP (Bolsheviks), Sergo became so upset that he could barely manage to speak, and could only record a few words: "Dear Kamo!" said Sergo, "I met you eighteen years ago. I was young. You considered it your duty to explain to me how to become a Bolshevik, how to fight for the interests of the proletariat . . ."*

Ordzhonikidze's stay in Baku was arranged according to all the rules of conspiratorial secrecy. His bodyguard was made up of the most tried-and-true members of the Youth Union, and every evening meetings were arranged in secret apartments between Sergo and a small circle of party workers where we gave him detailed information about the current state of affairs and about our plans. He showed a keen interest in everything, wishing to gather as fresh information as possible for his report to the Central Committee and Lenin on the situation in Baku and throughout Azerbaijan. It was also very important for us to get Ordzhonikidze's opinion about our ongoing political debate with the Tiflis Communists concerning the establishment of Azerbaijan, Georgia, and Armenia as national states. Ordzhonikidze listened to our discussion with great attention and approved our position, which especially pleased us since he was on his way to Moscow and could defend our point of view there.

I asked Sergo whether Georgian Mensheviks, who were in power then, knew of his visit to Tiflis and whether they might not have tried to arrest him. He smiled and said that naturally they knew about his

*G. K. Ordzhonikidze. Articles and speeches, Vol. 1, Moscow, Gospolitizdat, 1956, p. 241.

visit. What was curious was the fact that the leader of the Mensheviks, Noi Zhordaniya, whom Ordzhonikidze had known very well even in the days of the tsar, had deemed it necessary to inform Sergo in a roundabout way that he shouldn't appear on the streets since he could be arrested by the British. This meant that the Mensheviks didn't intend to arrest him themselves, but neither would they stand in the way of his arrest.

Sergo recalled with pleasure his first trip to Baku for party work, back in 1907, when he had turned twenty-one. He had thrown himself whole-heartedly into party work among the factory plant workers, combining that with a job as a doctor's assistant at the first aid station of the Asadulayev works. Having shown his colors as a true Bolshevik in the stormy revolutionary struggle of the Baku proletariat, he had always preserved a warm affection for this workers' center.

He told us how after his return to Russia from abroad Lenin had given him the task of calling an All-Russian Conference of Bolshevik Organizations in Baku, which was to select an Organizational Commission which would help convene the Prague Party Conference. He recalled Stepan Shaumian's active participation in this conference with special warmth. After the arrests in Baku, Sergo recalled, the All-Russian Conference was moved to Tiflis, where it reconvened in the apartment of E. D. Stasova. It was there that a resolution was passed regarding an organizational commission which would help convene the Prague Party Conference, and establishing a procedure for elective delegates to this conference. From Tiflis Sergo returned to Paris as a plenipotentiary of this organizational commission, and his report at the Prague Conference of the work it had done received the approval of all the delegates, and of Lenin in particular.

I remember how in answer to our questions Sergo told in detail how the defeat of the Eleventh Army in the northern Caucasus had occurred. He also related the story of the temporary fall of the Soviet power there at the beginning of 1918. At the time a feeling of modesty prevented Sergo from telling us how difficult it had been to fight in the mountains of Ingushetiya and to cross the Caucasian range. But we soon realized ourselves what this had cost him.

One evening we arranged a meeting between Sergo and a group of Azerbaijani socialists who had their own delegates in parliament. I recall that Ali Geidar Karayev, Mirza Davud Guseynov, and Samed Agamali-ogly attended this meeting. These comrades had done a great deal of work on behalf of our party, and we held them in high regard. The comrades familiarized Ordzhonikidze with the situation in parliament, as well as in the Muscavat government, and the districts of Azerbaijan, and they told him about the work they were doing in accordance with a plan worked out with the Baku Bureau of the Party Regional Committee. Sergo was very pleased with this meeting.

He spent four days in Baku, and then on the afternoon of June

13 he and a group of comrades left for Astrakhan on a fishing boat. We had decided to entrust Sergo's fate to Rogov, an experienced courageous sailor, and a well-trained Communist as well. I should add that by that time we were less afraid that these trips would come to grief, since these "sea expeditions," which Dovlatov was in charge of were now very well organized.

Nonetheless, we were actually on edge until we heard—more than four weeks later—that they had arrived safely. It had taken them nearly ten days to reach Astrakhan on account of gale-force winds. From a security point of view, such weather was most advantageous since Denikin's ships, which were based in Petrovsk, sought shelter in port in foul weather. During calm seas our boats had to hide from the Denikinites in the secluded coves along the Transcaspian coast.

Rogov told us later that Sergo, who was susceptible to seasickness, had had an extremely hard time during the storm, to the point where he had considered disembarking, but his comrades had persuaded him that that was out of the question: White Guardsmen were the masters of the coastline.

* * *

Kamo came to see me several times during those days. It was then that I became closely acquainted with him, although I had already known about some of his heroic feats, and of Lenin's confidence in him. In life Kamo was an extremely modest person who behaved simply and didn't like to talk about himself.

Not long ago, while I was rereading Gorky, I took note of his magnificent sketch of Kamo, whom Aleksei Maksimovich (Gorky) has known personally. Even before he met him Gorky had heard legendary stories about him. During the heroic days of the first Russian Revolution, when Gorky had first learned of Kamo's existence, he had found it hard to believe that this man "was able to combine in himself an almost mythical courage with such invariable good luck, such unusual resourcefulness with such childlike simplicity of soul." Gorky, who had seen a lot in life and was endowed, moreover, with a rich creative imagination, always felt that if he wrote everything that he knew about Kamo, then "no one would believe in the actual existence of such a man, and the reader would take the imago of Kamo to be a belle-lettrist's fabrication."

Of working-class origin, and without much theoretical training (and always painfully conscious of that fact, by the way) Kamo was a natural-born revolutionary, and moreover an unwavering one, devoted to his party duty until the end. Revolutionary work, the people who knew him would justly remark, was as necessary to him as "bread and water."

Kamo's courage knew no bounds. L. Krasin often said of him: "Sometimes it seems that he is spoiled by his successes and is being somewhat silly and playing the fool. But that is due . . . not to the frivolity of youth, or to braggadocio or romanticism, but to some other source. He is silly in a very serious way, but at the same time, as if in his sleep, having no thought of reality . . . "

Many examples could be cited of Kamo's daring courage. But perhaps the most astonishing of his feats was the feigned insanity to which he resorted after he had been in Berlin on a party mission and had been handed over by Wilhelm II's government to the gendarmes of the tsar, shackled by them, and put in the Tiflis Psychiatric Hospital. He simulated his so-called insanity for almost three years, deluded the most experienced doctors and psychiatrists, and finally, after deceiving all his guards, escaped from the hospital.

Gorky gives a very faithful portrait of Kamo, whom he saw in 1920:

> A strong, powerful man with a typical Caucasian face and a fine, very penetrating and austere expression with soft dark eyes, he was dressed in the uniform of a Red Army man . . . It was immediately evident that questions about revolutionary work got on his nerves and that he was wholly absorbed in something else. He was preparing to enter the military academy. . . . It was absolutely impossible to connect everything that I knew about the legendary daring of Kamo, his superhuman will, and his amazing self-possession, with the man who sat before me at a table laden with textbooks. . . . It was incredible that he should have endured such a prolonged exertion of strength, yet remained such a kind, unpretentious comrade, and retained the freshness, the strength and the spirit of youth.
>
> He still hadn't overcome the young man in him and was romantically in love in a youthful way with a fine woman. . . . He spoke of his romance with that lyricism of passion available only to young men who are healthy, strong, and chaste. . . . He spoke of the necessity of his going abroad to work with the same kind of passion that he evidenced in his love for the woman. "I asked Ilyich (Lenin) to let me go," he said. "I'll be a useful person abroad."
>
> "No," Lenin said, "study!"
>
> "Well, what could I say? He knows best. Such a man! He laughs like a child. Have you heard how Ilyich laughs?" he smiled brightly and then darkened again, complaining about how hard it was to comprehend military science. . . . He was handsome in his own way, with a beauty that was unique and not immediately evident. . . .

"For me," wrote Gorky further on,

> Kamo was one of those revolutionaries for whom the future is

more real than the present. This doesn't mean that they are dreamers, absolutely not, it means that the strength of their emotional, class-conscious, revolutionary spirit has been organized so harmoniously and so solidly that it nourishes their mind, provides soil for its growth, and marches as if in front of it, as if showing it the way.

Aside from their revolutionary work, all the reality in which their class lives seems to them like a bad dream, a nightmare, whereas the real reality in which they live is the socialist future.

I apologize for such a long quotation, but I feel that Aleksei Maksimovich Gorky has captured Kamo's personality here, and that it would be impossible to improve upon it.

During our meeting at that time Kamo was very interested in how we had managed to organize a Soviet Republic in Mugan. He was always questioning: "How can you trust all these comrades who are surrounded by White Guardsmen, they could betray you, turn out to be traitors, turn coward?" During one of our conversations he even said to me: "Do you want me to stay here and not go with Sergo? Send me to Lenkoran to check out the leading cadres." And he proposed this plan: he would go to Lenkoran as our plenipotentiary with a group of comrades, but he would be dressed up in a White Guard uniform. At night he would seize the leading Bolshevik comrades in Lenkoran and lead them out as if to be shot. If any of them began to beg for mercy and turn traitor, he would shoot them, but those who stood firm he would spare; that way, said Kamo, you could be absolutely sure of those who had passed the test, knowing they wouldn't let you down.

I didn't want to start an argument with him, although I rejected outright such a "method" of testing people. I said to him. "I know all these comrades well. We trust them completely, especially since there have already been quite a few instances when they have shown themselves to have a fighting spirit and to be staunch Bolsheviks."

Kamo was obviously disappointed that he wouldn't be able to carry out his plan. He then told me about many other plans he intended to propose to Lenin. One concerned having his own detachment to carry out diversionary activity at Deniken's rear, and he had a ready plan for this "operation," apropos of which he said to me: "I'm going to Moscow for a blessing and support."

Although on that occasion I hadn't agreed with his plan, we remained good friends. Kamo didn't argue with me very much at that time, since he was profoundly convinced of the correctness of his proposal. And he was genuinely hurt that I hadn't understood and accepted his proposal. Subsequently, when Kamo received consent in Moscow to organize a detachment to carry out diversionary activity against Denikin and his staff, he selected the men he needed and then, wanting to test them, applied the same "method" which I had rejected.

A special group of Kamo's trusted followers, who were dressed in the uniform of White Guardsmen and their officers, attacked his detachment somewhere in the forest during their training exercises. The soldiers of the regiment were disarmed and lined up—with the idea that those who turned out to be Communists would be shot. But those who "repented" or declared themselves to be enemies of the Communists were promised mercy.

It turned out, however, that there were no cowards in the detachment. But, on the other hand, one did announce that he was an agent of Pilsudsky, ripped open the lining of his service jacket, and removed an appropriate document. Kamo was pleased that he had been able to discover the traitor in this way. When it was all over and he had explained what was going on, he began to embrace the remaining members of the detachment as true friends and comrades. However, this entire "operation" had such an adverse psychological effect on one of the men that he became seriously ill—that was Fyodor Alliluyev, the son of the prominent Bolshevik, Sergei Alliluyev.

When Lenin found out about this "method" of testing people, he became very angry at Kamo.

At the end of September 1919 we received a letter from Kirov in code, in which he informed us of an impending visit from Comrade Kamo. As we found out later, he was coming to make plans to blow up Deniken's headquarters. Our job consisted in making the necessary preparations during the next two weeks to insure the safe reception of his group. I remember that this information delighted all of us, and we immediately set about making the necessary arrangements. It turned out that all the ways and means that had been employed previously were inadequate for the reception of such a large group, and especially for such a large quantity of explosives.

A new plan was proposed by a member of the Party Regional Committee, the worker Chikarev, who together with Dovlatov took care of similar matters for us. He had a lot of contacts among the fishermen and his plan consisted in the following: forty *versts* south of Baku there was a small island called Bullo, where no one lived and only occasional fishermen came to fish. Chikarev would borrow a boat and fish on this island, which was to be the place where Kamo's detachment disembarked. His men would dress themselves in fishermen's clothes and move off the island in groups, and the dynamite would be hidden on the island, to be delivered to Baku in separate loads. We duly informed Astrakhan of this plan.

Kamo's arrival was delayed however, and we began to worry. I found out the reason for the delay in October when I went to Astrakhan and wrote to my comrades in Baku: "Kamo's group was delayed because of the Mamontov raid. But there is a report they're leaving Moscow. We have to wait for them." Soon afterward all of Kamo's group arrived on the island and were conveyed safely to Baku, but the dynamite fell into the hands of the police. With the help of some agents, however, who were connected with the police, our comrades managed to save the dynamite and recover all of it for themselves.

Chapter VII

The Tactics of a United Front

The first Soviet ambassador in Persia—We buy arms—The fall of Lenkoran—The balance of power in Transcaucasia—At workers' meetings—Our underground apartment—Death in the heroic style—A trip to Sanain—Two attempts at arrest—Debates with the Mensheviks in Tiflis—Revolutionary Dagestan.

I cannot refrain from saying a few words in these memoirs about Ivan Kolomijtsev. Our first acquaintance with him was in April 1919 when he made his way to Baku from Persia, via Lenkoran, and showed up at the Baku Party Committee. He was in his early twenties, the same age we were—a gifted man, and a fearless, courageous fighter, utterly dedicated to the cause of revolution.

As a lieutenant in the tsar's army Kolomijtsev had been assigned to General Baratov's forces in Kermanshakh, Persia, serving as the head of counterintelligence. As the influence of bolshevism within the soldiers' ranks intensified, changes came about in Kolomijtsev's thinking, too, and he went over to the side of the Bolsheviks. He was elected a delegate to the Second Regional Congress of the Caucasian Army, held in Tiflis in December 1917 where he joined the Bolshevik faction. From the spring of 1918 Kolomijtsev was a member of the Military-Revolutionary Committee of the port of Enzeli, where on Shaumian's orders, he directed the evacuation of Russian troops and materiel.

In the summer of 1918 Shaumian sent Kolomijtsev at the head of a Soviet mission to Teheran, in Persia. After the temporary fall of Soviet power in Baku the White Guard officer element routed the Soviet mission in Teheran in November 1918. It was only by a miracle and with the support of progressively inclined local Persian residents that Kolomijtsev managed to get out alive.

When he appeared at the Baku Party Committee in April 1919 he provided us with detailed information about the situation in Teheran, northern Persia, and Lenkoran. He very much wanted to go on to

Moscow, and we succeeded in sending him by boat to Astrakhan, where he arrived without mishap.

In June we received two letters from Kolomijtsev reporting that he had carried out our request, and was also sending money with Kanevsky and Starozhuk, via Mugan. He wrote that the money was being sent mainly to permit the Baku Party Committee to purchase arms for the purpose of creating an army in Mugan. Further on in the letter he informed us that there was no Azerbaijani literature to be had in Astrakhan, and he promised to bring several comrades back with him to work among the Moslems. It was especially pleasant to find out from his letter that as a result of conversations he had had with Astrakhan party leaders, he had become confident that "the issue concerning Azerbaijan will be decided in the way you have resolved it—in spite of the opinion of Tiflis."

After he had familiarized himself with the political situation in Astrakhan, and with the course of the struggle against the Mensheviks and SRs, who at that time had closed ranks with the Whites and were actively struggling against Soviet power, Kolomijtsev wrote to suggest that we change our tactics in regard to the Mensheviks and SRs. "It's not worth catering to them in Baku. In my opinion you must put the question point-blank: either they're with us, all the way, without any reservations, or they're definitely against us . . . I don't think it's worth your while to make a fuss over Churayev either. Let him either come over to your side, or go to hell!

"I'm not writing all this simply for the sake of writing, to get it off my chest, but after making a long and careful examination of policy at the center, insofar as it is being forged here, in Astrakhan. And since your policies should not diverge from the center's, I recommend the above . . . I'm becoming convinced that you haven't been firm enough in discrediting menshevism and SRism, and that you've been too lax in granting them entrée here and there, and in occasionally retreating from your positions."

We had a detailed discussion of Kolomijtsev's letters and his criticism of our policies at a joint session of the Baku Regional Committee Bureau and the Baku Party Committee. We did not agree with his point of view. The fact of the matter was that Kolomijtsev was not taking into account the different stages the struggle had reached in Russia and in Baku; he wanted to transfer automatically to Baku the approaches that were being used in Soviet Astrakhan, where the situation was entirely different. We felt, moreover, that our policy was correct and had completely justified itself in practice, and we decided not to change it.

In regard to the state of affairs in Astrakhan, Kolomijtsev reported that there "the entire titanic workload rests on the shoulders of three-fifths of the people, who are very intelligent and totally dedicated . . . The situation in Astrakhan itself is rather unstable. Among the workers we find all kinds of riffraff who are undermining the ruling authorities by playing on the shortage of bread . . . "

Knowing how we hoped that with the coming of spring the Red Fleet would enter the Caspian Sea and aid us in the armed struggle, Kolomijtsev wrote that the fleet "would most likely be busy in the Tsaritsyn area for a long time." We were particularly distressed by his warning not to expect any military aid from Astrakhan in the near future. "Hold firmly to a course that will enable you to cope without any aid whatsoever," he wrote. "The basic task should be to reconnoiter the enemy forces and to send to Astrakhan the most detailed information concerning the numbers and armaments of the enemy, especially ships . . . " Further on he pointed out that we had to buy up arms: "All that it's possible to get in Baku (in the province) and in the Persian towns bordering Mugan."

We had already begun the purchase of arms in Baku, but now we redoubled our efforts, aided by our party "businessman," Dovlatov. He selected suitable people and opened a restaurant in the city especially for British soldiers. Taking advantage of the vulnerabilities of the British soldiers, who were exhausted from their campaigns, and exploiting their predilection for alcohol, he set up his restaurant operation in such a way that his patrons never had enough money to pay their bills. In place of money he willingly accepted weapons. It was clear that the restaurant was to the taste of the British soldiers: they came there in droves, and every day Dovlatov was left with a rather large number of weapons given him in payment for spirits.

The purchase of arms in Mugan was undertaken by a special group of local comrades. Upon receipt of additional substantial sums of money we instructed them to buy even more arms.

* * *

By June the military situation in Mugan had become extremely complicated. By that time gangs of Persian robbers, plus detachments of White officers who had been chased out of Lenkoran, as well as bands of Muscavatists, had stepped up their military operations against Soviet power. There was bitter fighting all over the Mugan area, with numerical superiority on the side of the enemy, and units of the Mugan Red Army in retreat.

At the end of June the situation had become critical. The White Guards broke through into Lenkoran, and bitter fighting broke out in the streets. In one of the street battles Otradnev was fatally wounded and died a few hours later in the hospital; the heroic life a true son of the revolution had come to an abrupt end. However, Soviet units managed at that time to press the enemy and defend the city. After being defeated in that battle, the White Guards gathered their forces again, and in the second half of July fierce fighting broke out.

At the beginning of July we received news from Astrakhan that a Red flotilla had suffered defeat at Fort Aleksandrov (now Shevchenko) in battles with the superior forces of British and Denikinite ships; some vessels were sunk, others damaged, and the rest had returned to the Volga estuary. Thus, there was no longer any chance of our getting a fleet and troops from the north that year.

The Baku Party Committee was keeping a close watch on the current situation, particularly in Mugan. It was decided to inform our comrades there that, in the event of a special emergency, they should use the ships and fishing boats they had to evacuate Mugan. The first to be evacuated should be those whose lives would be in danger if captured. We advised them to set sail either for Astrakhan or the Turkestan coast in order to meet up with the Red Army, which at that time occupied Ashkabad and was pressing forward in the direction of Krasnovodsk. Taking into account that strategical communications with Baku and Astrakhan had been made difficult, we suggested to our Lenkoran comrades that they make final decisions on their own in response to the situation.

We found out later that Kolomijtsev, who had been named to the post of fully empowered RSFSR representative in Persia, had arrived in Lenkoran on a motor cutter from Astrakhan with a group of co-workers, had made contact with the Persian consul there, and completed the necessary formalities for his trip to Teheran; the duties which had been assigned to him made it mandatory that he reach his new post as soon as possible. However, the situation in Lenkoran had become so acute that he was unable to leave. Whether because of difficulties in getting out of the city, or because he did not want to leave his comrades in the lurch, he stayed in Lenkoran and assumed leadership of the struggle against the White Guards, self-sacrificingly fighting to the last day.

The forces were clearly unequal and the counterrevolutionary bands soon seized Lenkoran. The Red Army men who survived tried to make their way to the Turkmen coast of the Caspian Sea in boats and launches, but British warships and the armed vessels of the Denikinites seized some of them at sea and turned them in the direction of Petrovsk. Kolomijtsev, Kandelaki, and other of our leading comrades managed to escape to shore, but were captured by the White Guards and killed.

Kolomijtsev—the first fully authorized representative of Soviet Russia in Persia—was shot by White Guards near the Persian city of Bendegryazi. He was the first Soviet diplomat to give his life for the power of the Soviets.

In Petrovsk as well, many Communists from Mugan were shot. Only two small groups of Lenkoran Communists managed to escape.

One of them, headed by Otto Lidak, made its way through the Turkmen deserts to the city of Kizyl-Arvat where Soviet rule had already been reestablished. The second group, headed by Vakhram Agayev, took the land route and reached Baku with the help of local Azerbaijani peasants.

That was the tragic end of the epic of the Mugan Soviet Republic, which we had hoped would become a base of disembarking units of the Red Army and expanding the subsequent struggle for a Soviet Transcaucasia. The main cause of its fall was a shift we had not foreseen in the alignment of forces, a shift that worked to the advantage of the counterrevolution.

* * *

The situation of Soviet Russia on the southern fronts became more and more precarious. With Anglo-French aid Denikin had managed to bolster his army significantly and had speeded up his movement north, threatening the Donbass area and ultimately, Moscow.

Quite another situation had arisen in the eastern part of the country, however, where our troops were battling successfully with Kolchak's army. In Turkestan the Red Army, taking the offensive, pressed hard on the counterrevolutionary forces of the "Transcaspian government" and was closing in on Krasnovodsk in earnest. The possibility of a complete liberation of the Transcaspian region was at hand. Naturally, this made us very happy, although we also realized it was impossible to expect any real assistance from the Red Army at that moment since the sea coast was blockaded by British and Denikinite warships.

After assessing the situation and taking into account first and foremost the lessons we had learned from the defeat in Lenkoran, we decided not to organize new seats of rebellion, but insofar as possible to support those already in existence, principally those in Dagestan and Chechnya. Denikin simply could not suppress these rebellions. In Zangezur and Karabakh, and also in the Kazakh area of Azerbaijan, the rebels held out for a long time. In these regions Azerbaijani and Armenian peasants joined forced amicably under the banner of Soviet power.

We were confronted at that time by extremely complicated tasks: we had, first of all, to forbid uncoordinated and premature political actions by the Azerbaijani working class, while we built up our forces and worked to bolster the party and our underground armed contingent in organizational terms. Meanwhile, it was necessary to refrain from open, armed outbreaks until a crisis was reached in the situation on the Denikin front and was resolved in favor of the Red Army.

Secondly, we were obligated to increase our aid to Soviet Russia by arranging for supplementary shipments of gasoline to Astrakhan, and by expanding our military intelligence network in the Denikin-occupied territories and organizing the transmission of valuable military information to the Red Army command through our agents, who crossed the front lines on a regular basis.

Thirdly, we established strong ties with the party organizations in several cities of the northern Caucasus occupied by Denikin, and mobilized them for organizing demolition work in the rear of the White Guards.

Finally, we saw to it that in Dagestan and Chechnya the raids of mountain partisans on the rear sections of Denikin's army were carried out with greater frequency and substantial force.

At the end of June, Boris Sheboldaev, who had been Assistant People's Commissar for Naval Affairs during the period of the Baku Commune, arrived in Baku completely unexpectedly. It turned out that during the evacuation of our armed forces from Baku to Astrakhan in 1918, when our ships had been stopped at Zhiloy Island, he and two comrades had managed to go ashore, seize a fishing boat, and in the face of great difficulties, make their way to Fort Aleksandrov. From there Sheboldaev had gone to Astrakhan by boat, and had then been sent to the Kizlyar district with the goal of subsequently getting through to Dagestan and making contact there with the local rebels. Displaying exceptional resourcefulness, Sheboldaev had managed to carry out his assignment, working in the Dagestan party *Obkom* until he left here to come to us in Baku.

We were very happy at this addition to our ranks of an older comrade with a good deal of military experience, and his arrival at this time when we were attempting to organize military intelligence was particularly fortuitous. By decision of the Party Regional Committee Sheboldaev was made head of our intelligence operations, with the right to select needed agents personally. We gave meticulous attention to the organizational side of our intelligence operations, dividing up the entire northern Caucasian rear of Denikin's army into military districts—Rostov, Krasnodar, Armavir, Grozny; in each of these it was decided to have a chief resident with a group of intelligence agents who were provided with the necessary codes and means of communication. Sheboldaev's headquarters were also put in charge of the intelligence network in Enzeli and certain other centers in Persia where the British were in control at the time. He was to receive from local party organizations in Transcaucasia all information of a military nature which could be of interest to the Red Army Command.

* * *

Denikin's victories confronted us with several tactical problems, posing a threat not only to the revolutionary proletariat of Russia, but to the existence of the Transcaucasian national republics as well. The Mensheviks, who hated the Bolsheviks, were overjoyed at Denikin's victories, but a complete victory for Denikin would not have been convenient for them, however, since the White Guards would hardly be likely to treat the Mensheviks with excessive consideration. The Azerbaijani bourgeois government was also worried, and its anxiety was reflected in the press of its ruling parties and in the speeches of individual government representatives. It should be noted that with every victory the Denikinites behaved more and more impudently toward this ludicrous government.

Above all, the victories of Denikin's army alarmed the laboring masses of Transcaucasia, disturbing both the vanguard workers and peasants who supported Soviet power, and the nationalistically inclined section of the population of Transcaucasia which saw in Denikin's victory a threat to their national gains. Since only the Bolsheviks and Soviet Russia were fighting against Denikin, distrust toward the Bolsheviks was gradually replaced by a certain sympathy.

Apparently, at the beginning of July, the head of the Menshevik government, Gegechkori, in his speech at a meeting of the Tiflis Soviet, threw out the remark that the Mensheviks also wanted to fight against "General Denikin's counterrevolution," and that "to unite all the forces of democracy would not be a hindrance in this affair."

We knew very well, of course, that Gegechkori had no serious intentions of uniting with the Bolsheviks and waging a real struggle against the Denikinite threat. He wanted to win authority among the revolutionary masses. He also wanted to make an impression on the British military command, and on Denikin himself, by way of warning him that if he mounted an attack on Georgia, the whole nation would rise up against him to the last man. After the Mensheviks, Muscavatist leaders spoke out in similar terms.

At a joint meeting of the Baku Bureau of the Regional Committee and the Baku Party Committee a resolution was passed: to propose to the Mensheviks and Muscavatists the formation of a united front against Denikin of all democratic and national forces. The members of the Regional Committee in Tiflis agreed to this. On the basis of this resolution the presidium of the Workers' Conference proposed to the Tiflis Soviet of Workers' Deputies the creation of a united front of laborers in the Caucasus against the counterrevolution.

We, the Bolsheviks, then openly declared that while we did not support the policies of the existing governments of the Transcaucasian republics, we were prepared to support any actual struggle directed against Denikin on the part of separate parties, groups, and even governments, and with these aims in view, we proposed that a Transcaucasian workers' congress be convened in order to discuss these questions and to create a Transcaucasian organization which will occupy itself with the practical mobilization of the workers against Denikin. In addition, we said that we were ready to mobilize the Baku workers for this cause, and to put both them and our entire party organization under arms—even if that meant including them in the armed forces of bourgeois governments.

Following our orders, Gogoberidze thereupon announced the proposal at a meeting of the Workers' Conference where it was approved unanimously, to the accompaniment of a great upsurge of enthusiasm. A telegram containing relevant details was sent to the Tiflis Soviet. Their reply informed us that they accepted our proposal, and requested permission for representatives of the Tiflis Soviet to come to the Workers' Conference in Baku in order to discuss the question of convening a Workers' Congress. For their part, they invited representatives of our Baku Workers' Conference to Tiflis for talks. Gerasim Makharadze, Urushadze, and a third representative whose name I have forgotten, came to us from Tiflis at that time.

Knowing that the Workers' Conference was completely on our side and that the Mensheviks could accomplish nothing with their speeches, our only fear was that the workers might disrupt their speeches and in doing so give the Mensheviks an excuse for discontinuing discussion of a united front. Therefore, we issued a warning to all comrades to listen patiently to everything the Mensheviks had to say and allow them to express themselves in full.

We realized that through the tactics of a united front we would gain access to the proletariat in other republics of Transcaucasia. Such tactics would be to the liking of the workers, would bolster the authority of our party and facilitate unification of the forces of all laboring people in Transcaucasia against Denikin, and in the final analysis, would constitute serious support for the proletariat of all Russia in its battle against the counterrevolution. Realizing this, we decided to focus our propaganda on this occasion not only on the differences of opinion which divided the Transcaucasian proletariat, but rather on their common, and at the time, main task—the struggle against the Denikinite counterrevolution.

Philip Makharadze, the official spokesman for the Communists speaking under a false name was unable, however, in addressing the Workers' Conference, to confine himself to the main point. Although we had cautioned him about the necessity for self-restraint, his speech was harsh and inflammatory, leading one to the foregone conclusion that there was no longer anything for the Mensheviks and us to talk about. Fortunately, the overall conduct of the conference did not give the Mensheviks any reason to rupture negotiations, and the workers, after calmly hearing out the speeches of the Georgian Menshevik delegates, unanimously passed the resolution proposed by the Communists.

Then the Menshevik delegation requested permission from the

Workers' Conference to speak at one of the mass workers' meetings, and they were accordingly permitted to address a large political meeting of workers in Balakhany. Rather than have the Baku Party Committee's official spokesman speak at this meeting, we decided to give the workers themselves a chance to debate with the Mensheviks, which they did so well that the Mensheviks left Baku extremely depressed. To their surprise the masses of workers demonstrated complete unanimity with the Communists; as a result the Mensheviks were convinced at first hand that they would not have any success among the Azerbaijan workers, and moreover, they began to fear that we might be a "pestilence" that could infect the working class of Georgia.

With regard to a united front policy, we foresaw several possible outcomes. If the Mensheviks caused the policy to break down, we would in fact have lost nothing while boosting our authority among the masses even higher. In the event of a successful outcome, we hoped for the convocation of a Transcaucasian workers' congress, which would in itself represent a major victory for us, since we would undoubtedly be in the majority at the congress by having the support of the working class of Azerbaijan, the railway workers of Armenia, representatives from Dagestan, and also some of the workers of Georgia. Thus it was clear to us that if the Workers' Congress succeeded in making the struggle with Denikin a reality, then we, even while joining the personnel of the national armies, would achieve a great deal.

After the delegation of the Tiflis Soviet had departed from Baku, the delegation of the Baku Workers' Conference, composed of Sturua, Gubanov, and myself, left for Tiflis.

Prior to our projected talks with the Tiflis Soviet about the united front, the Menshevik government of Georgia had instituted mass repressions against the Georgian Communists, imprisoning many of them and causing those who had still not been arrested to operate under cover, since there was no possibility of operating legally. When our talks about a united front began, however, the Mensheviks ceased the arrests and even permitted the Communists to publish their own legal newspaper.

In Tiflis we immediately made contact with the Party Regional Committee and discussed the prospects for organizing a united front.

Many Tiflis comrades were pessimistic, among them Philip Makharadze and several others said straight out that the Mensheviks were just "playing at diplomacy," and since they did not intend to wage any sort of struggle against Denikin, they naturally would not permit a Transcaucasian Workers' Congress to be convened. They knew "their own" Mensheviks better than we did, but nevertheless we decided to pursue our policy, citing the fact that in principle the Mensheviks had agreed to the creation of a united front.

Our delegation was received in the former palace of the Tsarist

Governor-General of the Caucasus, where the government of Georgia and the Executive Committee of the Tiflis Soviet were located. We were greeted by Gerasim Makharadze, who had just made a trip to Baku. Outwardly he greeted us very politely and conversed calmly with us, but none of our attempts to ascertain whether the leadership of the Tiflis Soviet agreed to the convocation of a Transcaucasian workers' congress were successful. Makharadze would not say either "yes" or "no," and when he could avoid an answer no longer he said that two opposing opinions on the matter existed within the Tiflis Executive Committee, and only the Soviet as a whole could decide the issue once and for all. Gerasim Makharadze himself was the chairman of the Soviet. The promise to call a meeting of the Tiflis Soviet suited us just fine: we would get a rostrum from which to speak before a broad workers' audience.

After this informal meeting, we telegraphed the presidium of the Workers' Conference in Baku that "the Bureau of the Tiflis Executive Committee has split into two camps. One is in favor of convening the congress, the other opposes it. They have not come to any decision. There will be a meeting of the Soviet in the near future. The Tiflis representatives who attended the Baku Conference—members of the organizational commission—oppose the congress. The Soviet has the floor."

The regular publication of a local party newspaper and our legally authorized arrival in Tiflis had noticeably enlivened the activities of the Georgian Communists; their influence among the Tiflis workers grew stronger. Since the delegates from Baku had diplomatic immunity, we decided to use the time before the convocation of the Tiflis Soviet for making public appearances. We attended a meeting of the Trade Union of Apothecary Workers and the Union of Curriers; the governing boards of which contained many Communists. Sturua and I delivered reports at the meeting about the success of the workers' movement in Baku and the aims of the united front against the Denikinites, calling upon the Tiflis workers to join the people of Baku at this time of grave danger for the Russian Revolution and for the peoples of Transcaucasia.

I must say that at all these meetings the workers gave us a very fine welcome, listened to us with great attention and were very receptive to the resolutions we proposed.[20] We also held a meeting at the House of the People at which a paper was presented on the subject of "The Gentry-Bourgeois Counterrevolution and the Transcaucasian Governments." The Mensheviks did not put forward any official opponents at this meeting and the paper was very enthusiastically received. One sensed an enormous sense of animation among the

[20]See Appendix

participants, who hadn't heard such open speeches from the Bolsheviks in a long time.

Meanwhile, the Mensheviks were delaying calling a meeting of the Tiflis Soviet. Obviously, they were discussing how they could wreck the meeting to their own advantage and in doing so, put the blame on us while they came out unscathed. It became known in the Regional Committee that the Menshevik Central Committee had passed a resolution to block the formation of a united front. Nonetheless, a week later the meeting of the Tiflis Soviet took place. The Mensheviks put forward three speakers to act as their spokesman: Gerasim Makharadze, Dzhugeli, and Arsenidze. These speakers committed slander by implying that the Bolsheviks were playing the hypocrite in their proposal for a united front, and that in Bolshevik Russia, workers and peasants were being shot, revolutionary democracy was being persecuted, etc.

Dzhugeli's speech was expecially offensive in this regard,[21] while in contradistinction to Dzhugeli, Gerasim Makharadze conducted himself more subtly, avoiding crude attacks. In general, Arsenidze spoke most politely of them all. Without provocative statements or undignified attacks, he openly declared that our paths had diverged and could never come together.

With the next two speakers, the Mensheviks descended to farce. The first was Voytinsky, a prominent right-wing Menshevik and member of the Menshevik Party Central Committee who had fled from the Russian revolution. He tried as hard as possible to put himself out on behalf of Georgian democracy and praised it to the skies. He was followed to the podium by one of the former prominent leaders of the Baku Mensheviks, the inveterate anti-Communist, Ayollo, whose hands were steeped in the blood of the twenty-six Baku commissars. I knew him personally and detested him. Besmirched with political dirt from his head to his heels he had in his time hot-footed it out of Baku, found shelter with the Tiflis Mensheviks, and was now speaking as a "connoisseur" of the Baku proletariat.

It should be noted that the Tiflis Soviet had not had a reelection for more than a year. When the Bolshevik-Deputies had walked out in protest, the Soviet had been restocked with Menshevik civil servants who were now in the majority. I recall that "workers' deputies" wearing cockades and shoulder-straps predominated in the hall, and we were struck by the sharp contrast between this audience and the Baku Workers' Conference. In spite of the Mensheviks' efforts at "filtering" the crowd, however, a rather large number of workers had found their way into the gallery. Therefore, while savage hatred for the Bolsheviks seethed down below on the floor, almost the entire gallery was on our side. Whenever the hall burst out into applause, the gallery drowned

[21]See Appendix

it out with whistles and cat-calls, and when applause rang out from the gallery noise and whistles resounded below.

I was the first of the Baku representatives to speak. I had planned to speak calmly, and without making any personal attacks to substantiate our position and refute everything that had been said by the Georgian Mensheviks. But Ayollo's speech had upset my equilibrium, and I decided to settle accounts with him by uttering one pointed sentence. I remember very well how I said: "Before answering the objections against a united front that have been expressed here, and before stating our platform, I must say that I consider it beneath my dignity to reply to the hooliganish speeches of the base provocateur, Ayollo—the monster who was long ago thrown overboard from the revolution by the Baku proletariat."

I had hardly had time to finish the sentence when an incredible din arose in the hall, accompanied by applause and yells of approval from the gallery. Several people in the first rows of the hall rose from their seats shouting "Provocateur! Liar! Give him a good beating! Kill him!" and they threw themselves upon me, some with raised fists, a couple even pulling out their pistols. Unbelievable chaos ensued before the members of the presidium decided to calm down their party colleagues who had become so enraged. On behalf of the presidium the same old Dzhugeli spoke and said that although "everything Mikoyan has said here is a lie and slander," that I should nevertheless be given the chance to say my piece. "Later we shall answer all their "slanderous accusations," Dzhugeli said. The audience calmed down somewhat.

In calmer tones I began to refute, one after another, the arguments of the Menshevik speakers. Sweeping aside their attacks on the Bolshevik party and Soviet Russia, I tried to focus attention on the fact that at that very moment the Denikinite threat hung not only over Soviet Russia but also over the peoples of Transcaucasia. I said that although Denikin was winning at present, his victories were short-lived, and he would inevitably fail in the south of Russia just as Kolchak had failed in the east, although he had succeeded before his defeat in occupying all of Siberia, the Urals, and a part of the Povolzhie. I accused the Menshevik leaders of double-dealing, and alleged that while holding talks with the Baku workers about forming a united front against the Denikinite threat, they were simultaneously conducting secret negotiations with Denikin's representatives in Tiflis.

The atmosphere in the hall was tense, to put it mildly. Hostile rejoinders and shouts continued to be heard from the first rows, while in the gallery people applauded as before. In spite of the noise, however, I said everything I had wanted to say.

Our next speaker was Georgi Sturua. His fact-filled, calm speech called forth new applause from the gallery and hostile shouts from the floor, since he exposed the treacherous policy of the Mensheviks. During his speech I wondered with regret how it happened that I was the one who had fallen victim to Ayollo's provocation, and due to my impulsiveness I had given the Mensheviks an opportunity to create a scandal right at the beginning of our speeches.

The speech of Arshak Zurabov, a former member of the Menshevik faction in the State Duma, left a painful impression. In 1917 Zurabov had made his way to the Constituent Assembly on the Menshevik slate. He was a good speaker; I heard his paper in Baku on May 2, 1917; at that time he was developing a sort of "intermediate" position, not fully justifying defensism, and at the same time criticizing the provisional government. Somewhat later, still not completely sharing the political views of the Mensheviks, Zurabov joined the Menshevik-International group.

In speaking here at the meeting of the Soviet, Zurabov expressed himself rather straightforwardly in favor of a policy of consensus with the Bolsheviks in the struggle against the White counterrevolution. It must be said that Zurabov spoke convincingly—he was an experienced and able speaker whose ideas could easily be understood by the Menshevik workers. Precisely for that reason his speech was immediately perceived by the presidium and part of those present as hostile; again there was noise and malicious shouts, not only from the first rows of the hall, but also on the part of the presidium. I remember that Dzhugeli hurled the rejoinder, "Political chameleon!" at Zurabov, while the Mensheviks shouted from their seats: "Get out, your place is not in Georgia, but in Dashnak Armenia!"

The expression on Zurabov's face testified to his pitiable and at the same time tragic position: insults rained down upon him from his own party comrades of the day before, and obviously hurt him very deeply. In all conscience I must admit that I felt sorry for this man, who had made mistakes, frequently shifted his position, and now was crushed and rejected by everyone.

Speaking briefly for the second time, Dzhugeli took issue with my statement that the Menshevik government of Georgia was conducting secret negotiations with Denikin's representatives. "We are not conducting any such negotiations," he said. "We are holding talks with the British Command." This "refutation" of his was more like a confirmation of the information that we had. It meant, we reasoned logically, they really were conducting negotiations with Denikin's representatives through British intermediaries and, obviously, they had already achieved something. Otherwise, why would it have been necessary for them to interrupt so abruptly our talks about a united front which had been going on for more than a month It was clear that the whole point of "negotiating" with us was only to frighten the British command and the Denikinite representatives and, as the saying goes, to sell themselves to the highest bidder.[22] At the end of the

[22] See Appendix.

meeting of the Tiflis Soviet the Mensheviks proposed their own resolution and got it passed by majority vote.

And so, the picture was clear. Even before the end of the meeting we had agreed to delay our departure from Tiflis in order to make some additional open appearances among the workers for as long as the Mensheviks would tolerate them. We had consulted on this matter with comrades from the Party Regional Committee, who supported us. It was decided to deliver speeches at several meetings as representatives of the Baku workers; such speeches, in the opinion of the local comrades, could have a major influence on Tiflis workers.

We decided to hold the first political meeting at the Central Tiflis Railroad Shops. As far as I can recall, something like three or four thousand workers were employed in the shops. At that time the Mensheviks headed the Union of Railway Workers, and we could not even think of organizing the meeting with the help of the Trade Union Committee; they simply would not have let us into the shops.

The Tiflis Party Committee arranged through one of its comrades —Abakidze—who was a member of the governing board of the Railway Workers Union, that the scheduling of the meeting would be announced in all sections of the factory five minutes before the end of the working day and that the workers would gather in the locomotive shop. We managed to gather something like a thousand people. The members of the Trade Union Committee spoke out against opening the meeting to us, however, declaring that the meeting had not been prepared in advance, and that many workers were not present, and proposed that the meeting should be postponed to the next day when it would be possible to gather all the workers.

After Sturua spoke, expressing himself in favor of holding the meeting immediately, the overwhelming majority of workers voted to begin the meeting.

Sturua and I both spoke at the meeting. We spoke about the struggle of the All-Russian proletariat and the Denikinite threat to the working class not only in Russia, but also in Transcaucasia; we exposed the policy of the Mensheviks who had wrecked the formation of a united front and had refused to convene a workers' congress. On behalf of the Baku proletariat we declared the readiness of the workers of Azerbaijan to struggle to the end against our common enemy—Denikin. "At a time," I said, "when our best comrades are sitting in prisons, and when yesterday, and the day before, still more hundreds of workers and revolutionaries were arrested; at a time when the veteran of the revolution, Mikha Tskhakaya, leader of the Caucasian workers, has been thrown into the Kutaisi Prison—at the same time Denikinite generals and officers are openly strolling through the streets of Tiflis and organizing counterrevolutionary bands against the workers of Russia." An explosion of indignation and shouts: "Shame on the Mensheviks, down with them!", "Free our Tskhakaya!", "Long live the united workers' front!"—rang out in answer to [my] words.

Our speeches were followed by many others made by Mensheviks from the trade union, members of the railway workers' union, and local Communists. The situation was clearly shaping up to our advantage. And indeed, when it came to a vote, the enormous majority of participants in the meeting accepted our proposed resolution which condemned the Mensheviks and accused them of breaking up the united front, persecuting revolutionaries, and giving aid and protection to the Denikinite counterrevolution.

As soon as the meeting was over, we tried as quickly as possible to get out of the factory shops unnoticed in order not to fall into the hands of the police who had arrived just in time. We managed to slip away in good time.

With this, our attempts to create a united front for the struggle against the Denikinite element in Transcaucasia came to an end. There were no more talks between us and the Georgian Mensheviks. In Georgia itself the Georgian Bolsheviks' bitter struggle against the Mensheviks continued. At that time the Dashnak government of Armenia held a "neutral" position on the question of a united front, and therefore, we generally did not conduct any negotiations with it.

As for Azerbaijan, when the immediate danger of a Denikinite invasion hung over Transcaucasia the Musavatist party raised a great hue and cry about it, and even began a campaign to prepare for struggle against the Denikinite threat. Representatives of the Baku committee of our party, as well as of Gummet and Adalet, met with the Muscavatists several times. At first the talks went satisfactorily. We were conducting a large-scale mobilization campaign in the working-class districts, particularly among the Moslems. The Muscavatists tried to take the initiative into their own hands; they even intended to organize a mass demonstration of workers. The issue was discussed at an inter-party conference. Our comrades proposed delaying the beginning of the demonstration somewhat in view of the fact that we were not sufficiently prepared for it, but the Muscavatists did not agree to this.

Disagreements broke out among us. Some proposed not participating in the demonstration at all and boycotting it. Others, although admitting that we were poorly prepared for it, nevertheless thought that it would be politically incorrect to boycott such a demonstration and that we had to take a most active part in it.

The latter point of view was recognized to be the correct one. There were few Communist supporters at the demonstration, but our speakers Karayev, Gogoberidze, and Agayev had a big success at the political rally, especially among the Moslem workers.

The subsequent course of events revealed that the Muscavatists, like the Georgian Mensheviks, did not intend actually to wage a struggle against Denikin, but rather to talk about a struggle, with the aim of encouraging the British command to protect them from Denikin. All their talk was simply an attempt to rehabilitate their party to some degree in the eyes of the workers. It was this secondary problem that they failed to resolve, however. They ceased even the verbal campaign against Denikin when the British command informed them that there would be no invasion if Azerbaijan cooperated with Denikin by furnishing him with oil and other supplies, and by not selling oil to Soviet Russia. This unmasked the Muscavatist party even more in the workers' eyes, while in their turn the Baku organization of the Workers' Communist party (RKP, Bolsheviks), and the Gummet and Adalet Communist organizations strengthened their own positions among the Moslem workers even more.

Thus, although the tactics of a united front against the Denikinite threat was not in fact victorious, our struggle for the creation of the united front had yielded undoubtedly positive results.

* * *

It is hardly necessary to mention what an important role in the underground conditions of that time was played by the secrecy, reliability, and dedication of the people at whose homes we met, through whom we made contact with each other, and whose apartments frequently became the battle headquarters of our underground Bolshevik organization.

From the beginning of 1919 the Kasparovs' apartment became the main secret apartment of the Baku underground Bolsheviks up to the very day of the reestablishment of Soviet power in Azerbaijan at the end of April 1920.

Maria Kasparova, the daughter of the owner of that apartment, recalls:

> The three-story building in which we lived was located in the very center of the city, in a bourgeois district (on the corner of Karantinnaya and Krasnovodskaya streets). Our apartment was on the third floor; the owner of the building, Adamov, lived below us. All the tenants were well off. These things could not have been better suited to our conspiratorial plans. And, in fact, it was difficult to believe that it was in this building that the headquarters of the underground Communist organization in Baku was located.
>
> My dear, clever mother immediately consented to having the apartment used for revolutionary purposes. Secret meetings and conferences of the leaders of the Communist party were held at our apartment for more than a year without a single mishap. The following people were often there: S. Ordzhonikidze, A. Mikoyan, Karayev, Philip Makharadze, Vano and Georgi Sturua, Sheboldaev, Mamiya

Orakhelazshvili, Kamo, Gamid Sultanov, Buniatzade, Mirza Davud Guseynov, Levan Gogoberidze, Sarkis, Georgi Alikhanian, Beso Lominadze, Isay Dovlatov, and others. Some of them hid there for a rather extended period of time. The Caucasian Regional Conference of the Bolshevik Party took place in that apartment in the spring of 1919.

But, nevertheless, one time the police paid a visit to our apartment. This was the situation: someone warned us from the courtyard that the police were on their way. The older members of my family were in the apartment at that time, and when the police opened the door into the dining room they saw seated peacefully around the table my grandmother, Mama, my aunt, and her children drinking tea and eating yogurt. Seeing such a peaceful family idyll the police withdrew without even stepping into the room, offering extensive apologies. Meanwhile, gold and paper money which had reached the Baku party organization from Astrakhan via the Caspian Sea was secreted in an adjacent room, as well as caches of weapons.

The Kasparov family was amazing. Still hale and hearty, though no longer young, Tatiana Kasparova was a good mother and a kind and hospitable hostess who took little part in political conversations, but was wholeheartedly dedicated to the cause of revolution. We were always the beneficiaries of her unchanging cordiality and concern, and while she never showed the slightest doubt or hesitation about continuing this extremely dangerous game, she knew very well what risk it entailed for her and her family. She was fearless.

She had five children. The three oldest had already joined the ranks of our party and were actively assisting us, especially in arranging secret contacts and meetings. Her oldest son, Vanya, was a young doctor. (Later, when Kirov was already in Leningrad, Vanya worked as the head of the Leningrad party *Obkom's* organizational department). Her daughter Maria was a technical secretary of the Caucasian Party Regional Committee. Her second daughter, Rosa, was a third-year student at the Petersburg Institute of Medicine. The third daughter, Grachiya, was still a schoolgirl at that time; later she joined the Bolshevik party. Their younger brother Leva, still a little boy then, also became a Communist later on.

I would like to speak in particular about Rosa Kasparova, who returned to Baku from Petersburg in the spring of 1917. She read Marxist literature and never missed any meetings held by the Bolsheviks, and even before the October Revolution, in August 1917, Rosa had joined the ranks of the Baku organization of Bolsheviks.

In March 1918, at the time of the Muscavat rebellion against Soviet rule, Rosa worked in a field hospital. People related how heedless she was of the enemy's bullets as, standing on a medical truck in a white coat, she rode along the entrenchments on the city streets and tended to the wounded Red Guards, evacuating the seriously wounded. I met her for the first time at the field hospital where she bandaged my wounded leg. Beautiful, cheerful, solicitous, always wearing a smile, she captivated the hearts of the wounded fighters, delighting them with her very presence.

At the end of the summer of 1918 when Turkish forces had reached the walls of Baku, she volunteered for the front-line positions, where, under enemy fire, she carried the wounded to safety. Everyone fell in love with her. She remained at the front lines alongside the fighters until the last day of the defense of Baku and, after the fall of Soviet power, Rosa worked in the Astrakhan underground.

I remember that when we were choosing the most trustworthy party people for operations at the rear of Denikin's troops, Rosa Kasparova insisted that she also be sent to do this dangerous work. Her mother raised no objections although one could imagine how she felt about it. Rosa was accompanied by another girl, eighteen-year-old Katya, who had just married my close friend, the Baku Communist, Kostya Rumiantsev. Kostya had been sent at the head of a group of comrades to the Rostov and Novocherkassk regions. He was arrested, did time in a Denikin prison, and escaped certain death only because of the Red Army offensive.

Rosa's regular reports to the Caucasian Regional Committee of our party, which contained a concrete political and economic analysis of the state of affairs in Denikin's camp, demonstrated how greatly she had matured in the complicated conditions of her work. "Her letters to our mother," writes Maria Kasparova,

> were full of concern for her and imbued with the desire to calm her fears somehow. "Don't worry too much about me, I am alive and well," she wrote in one of them. And in order to calm mother down once and for all, she added, "Send me my blouses, collars, and lilac-colored beads."

In the fall of 1919 we received the news that Rosa, Katya Rumiantseva, Suren Magauzov, and some of our other comrades had been arrested in Armavir. For me personally this was a terrible blow, filling me with such guilt that for a while I avoided going to the Kasparov's apartment. When Rosa's mother learned how I felt she communicated her surprise through my comrades at my long absence from her home, and I began coming to their house again, where I witnessed the amazing steadfastness and courage of this woman and mother. Without giving any outward sign of her grief she suddenly left for Armavir where her daughter was in prison.

After making her way with difficulty to Armavir at the rear of Denikin's troops Tatiana Kasparova assumed the care not only of her own daughter, but of her daughter's comrades as well. She went to the prison every day, delivered parcels, and established contact with the families of the imprisoned comrades.

Soon all the prisoners were transferred to Pyatigorsk. In spite of severe beatings and tortures, the White band still had failed to extract a single word of confession from the young Communists, and had not succeeded in breaking their spirit. Their friends were extremely anxious about the situation of Rosa and her comrades. A member of our underground committee, Boris Sheboldaev, who was himself in a Baku prison at that time, wrote to Maria Kasparova on the outside:

> Why don't you write me about how things stand with Rosa and how it is likely to end? I want so much to say something nice, warm, and comradely to her so that she might feel my support and not have gloomy thoughts.

One of the young Communists arrested with Rosa was Tamara Nadzharova, who managed to escape. Here is her account:

> In prison I first became acquainted with Rosa and Katya. Rosa made a great impression on me. She was an intelligent, cheerful girl with large, beautiful, sparkling eyes. We read a great deal, usually aloud. Rosa read beautifully, and we would crowd around her and listen to her, enraptured. I remember a small volume of Kuprin's stories in which Rosa particularly liked *The Garnet Bracelet*. She and I passed many a sleepless night together since almost nightly comrades of ours were shot in the prison courtyard.

Rosa's letters from prison to her family and friends are models of courage and fortitude. After the tortures and beatings she wrote: "We have hope of getting well again; just try to do something so that they won't hang us, all the rest is a mere trifle." In another letter to her mother Rosa wrote,

> I so passionately don't want to die without having lived! After all, I've hardly lived yet, and suddenly—death! Well, down with gloomy thoughts, or they'll think I'm afraid of death. It's all nonsense! Since the day of my arrest I haven't cried once, or even gotten a tear in my eye, and that's the way it will be to the end . . .

To her eight-year-old brother Levochka, whom she loved ardently, she wrote,

> I kiss you tenderly, tenderly, many times. In the evenings I spend hours thinking about all of you, and till next time I want you to be a good boy, don't be lazy, and don't make ink blots with your pen.

Lelenka! I want to write so many things to you, but right now you wouldn't understand. Just make sure that you always remember me.

To her brother Vanya she wrote, "Everyone has gone to bed, but I am writing, tomorrow is my birthday. Like a little child I'm writing and thinking what our people will bring me tomorrow. It's funny!"

Beneath the blows of the Red Army the routed Denikinite units retreated from Pyatigorsk. During the retreat the Denikinites transferred Rosa, Katya, Suren and the other comrades to Grozny. Rosa's mother received a good-bye note from her daughter in which she reported that they were being taken to be executed. Trying to comfort her mother and cheer her up, Rosa wrote: "Dear Mama, happiness will return to us again like the bluebird." She asked her to bring up her beloved Leva to be a stalwart, dedicated Bolshevik.*

When she received Rosa's note, her mother again set out to follow her and by chance turned up on the same train which was transporting the condemned. She did not manage to see Rosa again, however. Shortly before the arrival of the Red Army, on February 20, 1920, Rosa and Katya were hanged in Grozny. Suren Magauzov followed them to the gallows.

A few hours before he was executed, Suren managed to send two letters to friends on the outside. They were written on scraps of material he had torn off of his shirt. Fine letters, full of strength of spirit and courage!

I shall quote both of them.

> Dear Tamara,
> I am sending you my last comradely greetings. It's twelve o'clock. I am awaiting death, but I feel cheerful. Life—in the fullest sense of the word—is unknown territory for me. I didn't have time to realize my last wishes. I'm in a cell apart from Rosa. She is in a heroic frame of mind. All of us have been sentenced to be hanged—sixteen people. Regards to my comrades. I kiss you all. Suren.

In the other letter Suren wrote:

> Dear friends!
> It was my fate to be a witness to the death of my glorious comrades. It's painful for me without my dear, splendid Rosa. She died the death of the brave: she went to the gallows boldly, without grumbling, and without fear. The young worker Katya also perished.
> May their bright memory live forever! Give my regards to Rosa's mother and comfort her. The mother of such a glorious, fearless

---

*The Communist Lev Kasparov was at the front from the beginning of the Great Patriotic War and died a heroic death in battle with the Fascist aggressors.

comrade should not grieve or shed tears. Happy is the mother who has such daughters. I am in a secret cell and have been awaiting my fate for three days. Seven of the fifteen revolutionaries will probably be hanged today. I shall follow the example of Rosa and my other comrades and go boldly to the place of execution. Good-bye, be well, Suren.

Thus did these brave young heroes, selflessly devoted to Lenin's cause, lay down their heads for the victory of the revolution. And there were many like them.

I have permitted myself this short digression from the main subject of my memoirs because I feel a moral obligation to say a few words about these very young people whom I had occasion to meet on my life's path. Their lives can and should be a good example for the youth of today.

* * *

It was at the end of June 1919, I think, that having arrived in Tiflis for a meeting of the Regional Party Committee, I decided to make a trip to my native village and see my family, if only for two or three days. A civil servant whom I did not know was traveling with me in the same compartment. Everything indicated that he was on his way to his place of work, the copper smelting factory that was located next to our village. The whole district was part of the so-called neutral zone between Armenia and Georgia, and at that time it was governed by civil servants appointed by the British Military Command.

My traveling companion was a Georgian. He asked me where I was going and who I was. I replied that I was going to the village of Sanain to visit my mother, and that I worked in Baku as a clerk at the oil fields. I gave my real name, feeling there was no point in hiding it since it would be easy to find it out, and besides, I figured that the local administration had no ground for picking on me. In conversing on the most varied topics, however, I avoided political arguments or even mentioning politics, and he did not thrust them on me.

The train entered a narrow gorge with steep cliffs closing in on us from both sides as the railroad tracks twisted their way along the short of the swift-coursing Debet River. It was a small river, but unusually forceful and noisy. I admired the beauties of my native countryside, which touched me much more then than it had in my childhood.

I had not been home for more than a year, a year full of stormy events and tragic as well as joyous experiences, so that one could say that I was returning to these parts a different person. Everything around me seemed exotic and uniquely beautiful: nowhere had I seen, and nowhere would I ever see, beauty that was so close to my heart, such forest-covered mountains, such wild cliffs and turbulent rivers.

Around midday we arrived in Alaverdi. It was here at this station that the old, French-owned, copper smelting factory was located. While we were still en route, my neighbor had said that there were fewer workers at the factory than before, that the factory was not working at full capacity due to technical inadequacies and poor organization of operations, etc. I have already said that an underground party organization existed at the factory: in 1917 I had spoken there at a meeting of Bolsheviks who elected me as a delegate to the Caucasian Regional Party Congress. However, taking into account that my neighbor in the compartment belonged to the civil service world, I naturally could not and did not show any particular interest in the factory, in particular in its workers.

After getting off the train I ran into a fellow villager whom I knew, and the two of us set out together for our village which, like many others in that gorge, was located on a high plateau. As we walked along the steep footpath, so familiar to me from childhood, making our way along the cliffs to a height of 300 meters above the river flowing through the gorge, my companion told me about life in the village. He asked where I had been and how things were with me. He was, in fact, very surprised to see me, since rumor here had it that I had perished. Along the way we met another fellow villager whom I knew—I should mention that in our village it was the custom to greet every passerby whether one knew him or not, and so while this fellow villager had nodded to us in passing he clearly did not recognize me. When my companion chided him, "Don't you recognize him?" the other man looked me over intently, took a few steps back, and then threw up his arms and exclaimed, "It's yet to be seen that a corpse should come to life!" Then he embraced me—we had once been good friends.

Although it was less than three kilometers to the village it took us almost a whole hour to get there because of the steep ascent. When I entered our house my mother and I threw ourselves into each other's arms, tears ran down her cheeks. During the whole time she had thanked God that He had kept her son alive. My father was no longer alive, he had died the year before, but my innumerable relatives crowded into the house and plied me with questions, and I was forced to answer them all.

It happened to be dinner time, and I ate everything my mother put before me on the table with great pleasure. She was at the height of bliss: she sat next to me and kept looking at me as if she couldn't feast her eyes on me enough. Since she didn't understand politics, I didn't talk about political matters; I said that I was living well, working in Baku, and that there was no need for her to worry about me. She began to ask questions: "What do you mean? After all, there was talk that you had been arrested by the British and killed." After dinner I took a walk through the village to see what had changed. Since we all knew each other in the village, my walk turned into a series of conversations. I made my way to the cemetery, to my father's grave. I had a guilty conscience about him. In the spring of 1918, when rail communications between Tiflis and Baku were very unreliable and threatened to be cut off entirely at any moment, I had received a telegram from the village telling me that my father was gravely ill and wanted me to come to say good-bye to him. My duty as a son obligated me to go immediately, but that meant that I would be unable to return to Baku; I felt I could not leave my revolutionary work in Baku at the risk of not being able to return. Besides, there was no certainty that I would find him alive.

Now, standing before my father's grave, I mentally asked his forgiveness.

At that time my older brother was supporting our family. Two years before the First World War he had been called into the army where he served for more than six years, returning to his homeland at the end of 1917. Now he was working as a carpenter—in the same profession as my father. In terms of the times, the family was not badly off. When my father was alive our household had only two goats; now a cow had appeared as well. Around that time my younger brother Artem had finished the four-year rural school, and I had decided to arrange for him to continue his education in Tiflis, hoping for the hospitality of my aunt, Verginia Tumanian. From September 1919 my brother began living at her house in Tiflis and attending the Armenian school.

In the evening our whole family sat down to supper on the porch. During the meal a neighbor's son, with whom I had been friends as a child, came over and shook hands with me, and then he took me aside and told me that he had served in the militia of the "neutral zone" and had come to warn me that his commanding officers had decided to arrest me. He suggested I flee that very night. I thanked him, but didn't say a word to anyone else, and we continued with our supper.

When everyone had gone home I told my mother and my older brother that I would have to leave that same night. They were very hurt and surprised, and I was forced to explain that if I didn't leave that night I would be arrested. I had already consulted with my younger sister's husband, Akop, a copper smelting worker, on how best to flee. The only, and very ancient, stone bridge over the river was the most dangerous place to get past, but luckily for me, nobody turned out to be in the vicinity of the bridge; we crossed it safely, and set out in the opposite direction from Tiflis. It was clear that I could not wait for the train at Alaverdi station so we set out for the Sanain station, but decided that I shouldn't appear there immediately either.

Akop accordingly went there alone and spoke to our relative who had a small stall at that station which, under the circumstances, was very convenient for us. Akop gave him money to buy a ticket for me and got him to agree that when I reappeared at the station he was to hide me in his stall until the train came. Having several hours yet to kill before train time, Akop and I then went to the nearby village of Odzun where my aunt who lived there gave us a warm welcome.

It might be worthwhile to relate at this point one amusing incident that took place then. As I recall, in conversing with us, my aunt mentioned that my mother was trying very hard to get my older brother married to my aunt's neighbor's daughter. The girl's mother had agreed to the match, but her father was still hesitant. My aunt suggested we pay them a visit. "You'll have a chance to get acquainted with the girl and her parents," she said, "and at the same time perhaps you can help me arrange the match."

The neighbors in question greeted us cordially and extended us their hospitality. I liked the girl who was intended to be my brother's fiancee and I also liked her parents. When we had a chance to talk I asked the girl's father straight out whether he consented to give his daughter in marriage to my brother. He looked my in the eye attentively, thought for a moment, and then after a short pause he said, "I agree to give my daughter's hand, but not to your brother—to you!"

This unexpected turn of events left me momentarily speechless. I managed to stammer that for the present I was not intending to marry. "And besides," I said, "You know very well that according to the old customs, I can't marry before my older brother." With that our conversation came to an end, and my chance mission as a matchmaker ended unsuccessfully.

When we set off again in order to be at the station before the arrival of the train, it was already dark and Akop went on ahead to check the situation. He returned satisfied that everything was quiet. Then we went to the station together and I waited in the proprietor's stall during the fifteen or so minutes before the train pulled in, and then boarded it unnoticed, and took a corner seat. It seemed an eternity before the train started to move. The next stop—three minutes at the platform in Alaverdy—was my most anxious moment, as I kept expecting that the police would suddenly enter the car and begin checking documents, but that didn't happen. I got by the station safely, arrived in Tiflis and from there went on to Baku.

* * *

In July of 1919 I again journeyed to Tiflis for a meeting of the Regional Party Committee. The meeting place was on the city outskirts in the last row of houses on the slope of Mama-Davyd mountain, at the home of a tailor named Razhden, who was a reliable, but by no means a particularly well-known Communist. The police almost never appeared in this district, and it seemed to us that it was a good safe place to meet. Each of us would come to the meeting separately, while a comrade stationed at some distance from the house had the job of making sure that none of us had acquired a "tail."

On my way to Razhden's I decided to check and see if anyone were following me before I had reached my destination, so I went into a store for two or three minutes, pretending that I wanted to buy something, and then went out to the street. Not having seen anything suspicious, I continued on more confidently.

At Razhden's house I found two comrades already there—Georgi Sturua, and Slavinsky, who was from Vladikavkaz. Several minutes passed, and suddenly two policemen entered the house. "Are you Mikoyan?" they asked me. "Yes," I replied. "You're under arrest." "For what?" "I have orders from Kediya, chief of the Special Section." (The Special Section was a Menshevik group organized for the purpose of tracking down Communists; it had a very bad reputation.) My two other comrades were also put under arrest.

Georgi Sturua, ever resourceful in such situations, immediately struck up a conversation with the elder of the policemen, whose name was Lipartiya, trying to talk him out of having me arrested. He spoke first in Georgian, then switched to Russian: "You can arrest me and this other comrade here—we won't protest, we'll go to the prison with you, but you mustn't arrest Mikoyan under any circumstances." He explained that Mikoyan was being pursued by the Denikinites and the British command, and that if the police arrested him, the Menshevik government would hand him over to the British, the British would hand him over to Denikin, and he would be executed. "And surely you know," he said, "that in a short time, perhaps even within six months, the Bolsheviks will be victorious in the Caucasus. And then it will become known that you, Lipartiya, were responsible for the arrest and death of Mikoyan. That will be the end of you. Therefore, think of your own self first."

The policeman began to explain that he had to carry out the orders of his superiors unquestioningly. "All the more so," he said, "since, as Mikoyan was walking along the street Kediya recognized him and ordered me to go after him and arrest him. How can I return empty-handed?"

Sturua came up with additional counterarguments which I now can no longer remember, and finally Lipartiya said, "Well, all right, I'll let you two go, and I'll take Mikoyan. I can't appear before my boss without him." But Sturua refused to agree to that so in the end all of us left for the police station together. Even out on the street Georgi kept up his argument that I shouldn't be arrested, and that the policeman would regret it later.

Finally, some signs of hesitation began to creep into the policeman's words. "I have a family," he said. "How am I going to support them if I'm fired from my job?" Sturua immediately picked up on this and replied that he needn't worry about that since he would be given the necessary financial assistance. Lipartiya fell silent. "You and your friend will get five thousand rubles," Sturua said. "Tonight at the Kazyonny Theater a girl will be standing outside. Go up to her, tell her your name, and she'll hand the money over to you."

We were approaching the courthouse building, not far from Kediya's headquarters, when Lipartiya finally agreed. We three comrades immediately dispersed in different directions and made our way through various back streets to our respective secret apartments.

We kept our word. That same evening Lipartiya received the sum he had been promised, the money handed over to him by my fiancée Ashkhen, who was then spending her summer vacation in Tiflis. It's interesting that about fifteen years later I received a letter from Lipartiya. He described everything that had happened, and asked me to certify in writing that he had really released me from arrest—he needed it in order to receive his pension.

* * *

I have already noted that Dagestan was one of the main centers of revolutionary events of that time, with a rather strong party organization taking shape there, and the rebel movement developing successfully under the banner of Soviet power. The bold raids carried out by the Dagestan partisan detachments on the rear sections of the Denikinite army inflicted very substantial casualties, and in their own way the railway workers of Dagestan and the dockworkers of Petrovsk were also waging an active revolutionary struggle.

The Dagestan Bolsheviks were in constant contact with the Baku Party Committee and relied on its support, and the Bolshevik organization of Azerbaidzhan sent arms, money, literature, and manpower to their Dagestan comrades. The special role played in all this work by Famid Sultanov, a member of our regional committee, should be noted. Among the vanguard Baku workers and Communists there were many from remote mountain villages in Dagestan who maintained contact with their countrymen and had an active influence on them. I, for example, have a vivid memory of the colorful figures of two outstanding Dagestan workers—the prominent Communists Kazimamed and Aydinbekov—who were of enormous help to me in party work when I first arrived in Baku in 1917.

Dagestan was in contact not only with Baku, but with Astrakhan as well. Kirov kept a close eye on developments in the Dagestan political struggle and it was due to his involvement that in the beginning of 1919 the prominent Bolshevik Buinaksky was dispatched to Dagestan from Astrakhan with a group of military workers. Buinaksky played a major role in the overall strengthening of party work in Dagestan and in the establishment of close contact with us.

By the summer of 1919 the influence of the Communists in Dagestan had grown so much, especially in the mountain villages, that the question of an armed rebellion in May could be raised in practical terms. However, the start of the rebellion was delayed for some reason, and during that time the Mountain counterrevolutionary government decided to take the offensive and put an end to the "sedition" once and for all. Having traced the time and place where the Bolshevik leaders had gathered for a secret meeting, the police seized them and threw them in jail.

On July 19, 1919, I reported to the Party Central Committee concerning this matter: "In mid-May the government of the Mountain Republic, in an act of open betrayal, called a halt to the war against Denikin and concluded an agreement with him after rearresting all our comrades (up to thirty-five men)—leaders of the planned-for rebellion—and handing them over to the Petrovsk volunteers. They have not as yet been shot, since Denikinites arrested in Lenkoran were declared as hostages for them. A short time ago a military field-court trial took place. As a result, five of our best comrades—Ullubi Buinaksky, Abdul Barab Hadji Magoma-ogly, Abkurakhman Izmailov, Saib Abdul Khalimov, and Madzhit Ali-ogly—were sentenced to death."

After the deaths of Buinaksky and his comrades in the fall of 1919, when the Mountain government had been driven out by the Denikinites themselves, Comrade Kazbekov was sent from Baku to Dagestan where he became active in the work of the Dagestan party *Obkom*. He was well known in Dagestan: in 1918 he had been a member of the Dagestan Revolutionary Committee, and after the temporary fall of Soviet power there he had worked in the Baku party organization.

From June of 1919 Boris Sheboldaev, whom Kirov had sent from Astrakan, played a most active part in preparing the Dagestan workers for a July anti-Denikin rebellion. The Bureau of our Regional Party Committee gave all possible assistance to preparations for the rebellion and to the struggle against Denikin's army.

A few words should be said about the role of the local clergy in the struggle. The clergy was split into two groups: the reactionary clergy headed by Imam Gotsinsky, which reflected the interests of the exploitative bosses, and those clergymen who stood on the side of the revolutionary poor, and often even coordinated their operations with the Communists.

During the years of the Civil War the Dagestan Bolsheviks often made use of two prominent figures in the Pan-Islamic movement, Ali-Hadji Akushinsky and Uzan-Hadji (who wielded great influence among the masses at that time), in order to weaken the positions of the internal counterrevolution and to organize the struggle against the interventionists and White Guards. In particular, they assisted the Dagestan comrades a good deal in drawing broad segments of the poor into the rebel detachments. It was said, for example, that Ali-Hadji Akushinsky made an active contribution in the Dagestan mountain villages by mobilizing the poor for the struggle against Denikin.

The rebel movement drew in mountaineers from many districts of the northern Caucasus. Battles were fought with varying degrees of success. But in all cases the mountaineer partisan raids inflicted serious casualties on the White Guard units, and most importantly, deflected part of the Denkinite armed forces from the front-line struggle against the Red Army. This in itself constituted important support for the Red Army during that critical period.[23]

---

[23]See Appendix.

Chapter VIII

New Arrest

Interrogation—In the Baku prison—From one prison to another—We are released.

I was now forced to work in Baku under strictly secret conditions. I did not take part in any legal meetings whatsoever, attending only the secret plenary meetings of the Baku Party Committee, which were usually held in working-class districts. As a rule these meetings took place backstage in clubs while rehearsals of amateur productions by workers' groups were in progress. The meetings of the Bureau of the Baku Party Committee, which had a small number of members, took place in secret apartments in the city, either at the Kasparovs, or at the Chernomordik's, or at the home of one or another of our reliable comrades.

At that time I was living at Chernomordik's, whose apartment was vacant while the owners were away at their dacha during the summer. One day Levan Gogoberidze dropped by with two comrades. One was Avis Nuridzhanian, who was well known to me from work in Baku, and the other was Yuri Figatner, with whom I had recently become acquainted. Figatner had come to us in Baku illegally, via Georgia, from the northern Caucasus where he had been a People's Commissar of the Tersk Republic before the temporary fall of Soviet power, and he had just recently returned from Dagestan where we had sent him to enlighten us on the general situation there. He had come to tell me about his trip. Both Nuridzhanian and Figatner were good and loyal Communists.

We spent a long time discussing current politics, talking a great deal about the situation in the districts and the activities of the Workers' Conference, and sketching out several proposals which we intended to put up for discussion before the Baku Party Committee and the Bureau of the Regional Committee.

The day was very warm, and after a while the heat became unbearable so that it was difficult to breathe in the apartment, at which point Gogoberidze suddenly suggested: "Let's go for a swim in the sea. I don't think there's any danger, Anastas, since I doubt that the police will be waiting for you on the beaches." We had become so wilted by the heat by now that I agreed at once to this tempting suggestion.

It was late in the afternoon, drawing towards evening, and we made our way safely through the back alleys to the municipal beach. The sea was very calm and we swam around with great pleasure in the warm water, feeling so refreshed and cheered that we completely forgot about the heat which, in any case, had begun to diminish somewhat as the sun went down. By this time we had begun to feel hungry, however, and since neither I nor my friends had anything to eat in the house, the resourceful Gogoberidze suggested going to a restaurant and having a good meal for a change. I felt that it would be better to go to a cafeteria, but Gogoberidze rejected this idea; police plainclothesmen were in the habit of keeping watch over the workers' cafeterias, he pointed out, and they might recognize me there, but nobody would expect to run into Bolsheviks in a fancy "bourgeois" restaurant.

Persuaded by this reasoning we followed Gogoberidze to the Tilipuchur restaurant, where he had dined several times before. The restaurant was brightly lit and crowded with well-dressed men and women who were sitting around, laughing and drinking. Gogoberidze led us through the room to an empty little table in a somewhat secluded corner where he ordered some steaks "rare," and, as I recall, some Cahetian wine.

We ate with great relish and were in such good spirits that we entirely forgot about danger. Having finished the steak, we were thinking about having another when a police superintendent and two regular policemen suddenly appeared in the dining room. It became clear that we had fallen into a trap. As the superintendent was making his way across the room to our table, I remembered with a sinking heart that I was carrying several secret documents in my pocket. There was just enough time before he reached our table for me to whisper to Gogoberidze to start arguing with the superintendent about our being arrested.

While Gogoberidze was protesting our arrest in the sharpest terms, on the grounds that he was the chairman of the Baku Workers' Conference and therefore immune from arrest, I was thinking feverishly about how to get rid of the papers. The superintendent, of course, was looking at him, and the two policemen looked first at the superintendent, then at Gogoberidze. In fact, no one was paying the slightest attention to me; even the surrounding public was looking at Gogoberidze and at the superintendent of police. As soon as I realized this I took the documents out of my pocket and thrust them under the tablecloth. No one had noticed what I had done.

Only then did I breathe a sigh of relief, for now I could act the part of whomever I wanted. Among the documents hidden under the tablecloth was an official order (mandate) written on a small piece of white linen in the shape of a theater ticket of the present day. It contained instructions issued to Comrade Eshba, the Party Central Committee person in charge of initiating party work in the Caucasus. The mandate had been signed by the secretary of the Party Central Committee, Elena Stasova, and stamped with the seal of the Central Committee of the Workers' Communist party (Bolsheviks). Eshba had in fact just arrived in Baku the previous evening and had left the mandate with me at my apartment. I had put it in my pocket and hadn't had time to hide it anywhere. Like Nestor Lakoba, Eshba was the generally recognized leader of the Communists of Abkhazia.

Finally, interrupting his altercation with Gogoberidze, the police superintendent turned to me and asked who I was. I replied, "I'm a teacher. I have no party affiliation, and have come from Tiflis to look for work. My name is Ter-Israyelian (I had a passport made out in this name.) I was at a conference with the chairman of the Workers' Conference, to whom I had come with a request to find me a position. He promised to help me, and invited me to join him for dinner with two other comrades, who were also conferring with him on the question of getting jobs." The other comrades, conducting themselves quietly and calmly, gave the police superintendent similar replies. They led us out of the restaurant and escorted us to the nearest police station.

On the way there I thought frantically about what apartment I should give as my residence when I was interrogated. It couldn't be Chernomordik's apartment, since important party documents were kept there. It would be best to give an apartment where nothing secret was being kept and where the owner had been forewarned that, in case of emergency, I would say that I was a temporary boarder at his place: this was the secret apartment we kept in reserve for such emergencies.

It was dark outside on the street by now but there were still many people about. Completely by chance we ran into Olga Shatunovskaya, who was walking in our direction, and was very surprised when she saw us under police escort. Under cover of the noisy chatter on the busy street I managed to whisper to her to warn the comrades and to put the apartment in order which I planned to give the police as my residence.

At the station a different police superintendent began the interrogation. Gogoberidze repeated everything he had said at the restaurant and again demanded his release, without saying anything about us. The three of us, each one separately, told the superintendent our alibis, which did not seem to elicit any suspicions on his part, he took everything we said on faith. We asked to be released since we weren't guilty of anything, but they didn't let us go. We spent the night and half the next day at the police station, a stay brightened by our receiving a basket of food from one of Levan Gogoberidze's fellow countrymen who was the owner of a Baku restaurant where Levan ate rather often. Somehow he had found out about Gogoberidze's arrest—possibly he had been told about it by people who were in the restaurant at the time. In any case, in the morning the guard brought us a large basket from him containing some excellent *shashlyk*, fresh bread, and a large watermelon—an unexpected and pleasant surprise.

At midday they put us in two phaetons and drove us under guard to the Bailov Prison on the outskirts of Baku where they put the four of us into a small single cell. The cots contained only bare boards with neither mattresses nor pillows. But that didn't particularly upset us since we were used to sleeping on boards, and we had no need of blankets in the summer's heat. We were fed very badly for the first two days until our comrades arranged for food to be sent in Gogoberidze's name. In addition, Gogoberidze succeeded in getting one of the guards to agree to transmit a letter to a certain address on the outside, having promised to pay him well for the favor. Since the letter could fall into the hands of the prison administration, we wrote it in Aesopian language. The main point conveyed in the letter was to have them let us know, via the person who had passed on the letter, what steps were being taken to obtain our release.

Boris Sheboldaev, who had been arrested ten days earlier, was, we knew, in the same prison. Sheboldaev was by nature a calm and restrained person who never got angry or lost his temper in arguments; he was a man of few words who spoke in a well-thought-out manner, excellent as a man and as a Communist. We could not even guess how he had come to be arrested, but supposed at the time that he must have been betrayed by one of the people who were assisting him in obtaining documents from British headquarters in order to make a trip to Denikin's camp disguised as a British officer. I recall that I even ventured the thought that perhaps Sheboldaev had been arrested not immediately, but after his connections had been ascertained, and thereby after our spy network had been partially uncovered. That was extremely dangerous both for Sheboldaev himself and for many other comrades.

In any case now, after we had ended up in the same prison with Sheboldaev, we decided not to be in a hurry to make contact with him. There was no way we could help him (or, for that matter, that he could help us), and it would be only too easy to complicate the situation further: all we had to do was show that there was a link between Sheboldaev and ourselves, and things could turn out badly for all of us.

The guard of whom I've spoken turned out to be a decent man. He not only delivered our letter on that occasion, but even brought us an answer. And he did this more than once.

Our comrades on the outside came up with two plans for getting us released. The first hinged on their sending us saws to cut through the bars on the window facing the street at a predesignated time of night when the sentry would have already been taken care of, and a phaeton waiting to take us to a prearranged spot. The second plan consisted in getting pistols to us via the guard. On the appointed day, just before they locked the cells for the night, we were to disarm the guards, lock them in our cell, and make our way through the corridors out to the street where our comrades would be waiting for us.

Both plans were very risky, with failure possible in either case; it was impossible to foresee all possible complications in advance. We told our comrades to prepare for both plans until we reached a decision on which of them to use. At the same time we asked our friends to have Karayev, as a deputy in the parliament, to take energetic steps for the release of Gogoberidze who, as chairman of the Workers' Conference, had been arrested illegally. With Gogoberidze released, everything could be resolved more simply.

It was unbearably hot in the cell and became more and more difficult to breathe. Besides, being forced into complete idleness straight from a period of intensive work left us feeling particularly oppressed, so that an enervating lassitude began to make its appearance. I was so tormented by the thought that it was I who had broken the rules of secrecy by agreeing to Gogoberidze's suggestion and permitting carelessness, thus unwittingly driving both myself and my comrades into prison. It was all the more irritating to think of the heavy load of party work that was waiting for us on the other side of the prison walls. There are some mistakes which cannot be rectified, and this was one. I could not really reproach Gogoberidze, since a much larger share of the blame lay on me: I was older than Levan both in terms of age—I was twenty-four—and in terms of experience in political work, and therefore should have been more circumspect.

Gogoberidze himself, however, made all my oppressive thoughts go away. Cheerful and happy, he always enlivened things with his gaiety and his jokes, and in this way boosted our morale. On the other hand Avis Nuridzhanian felt very depressed. He was constantly singing some mournful song in Azerbaijani, the words to which, as I recall, did no more than confirm that

> In my cell it's hot and close,
> And my heart is sorrow-laden . . .

Avis was a sensitive and impressionable person. He had been a member of the party for less than a year and it must be said that at that time joining the party often brought with it the real possibility of being put behind bars and thus being deprived of normal life if not being deprived of life itself. However, to the end he held firmly to the positions of a Communist. To jump ahead somewhat, I shall note that about a year later Avis turned out to be one of the active leaders in Armenia of the rebellion and struggle for the victory of Soviet power there, and showed himself to be a devoted and self-sacrificing fighter. His sorrowful songs, however, annoyed Gogoberidze, who would constantly ask him to be quiet and not get on our nerves.

Our third cell-mate—Yuri Figatner—a serious, intense person, didn't joke around, and didn't like, and perhaps didn't understand, jokes. Lying on his cot and gloomily training his gaze at the ceiling, he would brood about something for long periods of time. It seemed to me that he feared he would share the tragic fate of a friend of his, the Communist Andzhievsky, chairman of the Pyatigorsk Soviet of Workers' Deputies.*

Newspapers sent by our comrades on the outside began to reach our cell. They brought some life into our monotonous existence, and we began to catch up on current events and to discuss them. Thus, about a week passed until one evening, around ten o'clock, the senior prison guard opened the door of our cell and informed us that we were to be transferred from Bailov to the Central Prison. This news upset us very much. We knew that usually around eight o'clock in the evening the keys to the prison cells were handed over to the superintendent, and only at seven o'clock the following morning were they returned again to the guards so they could open the cells. And now, suddenly, late at night, the keys were in the guard's possession, and they wanted to transfer us to another prison.

We all had the same thought: it can't be that they simply are going to send us over to the Central Prison—and to a man we refused to leave our cell at night. The guard said that a group of high-ranking authorities had arrived at the prison office to carry out our transfer, but we said to tell them that we categorically refused to leave.

The guard left, and we talked worriedly about what this new turn of events could possible mean, coming to the conclusion that this was not a transfer to Central Prison, but a ruse to get us out of here in order to put us on a ship later that night and hand us over to the British. We decided to refuse unequivocally to be transported at night, reasoning that during the day we would manage to make contact with the outside and perhaps take certain measures.

A short while later the chief of the gendarmerie appeared at our

---

*Andzhievsky was a talented speaker and popular activist whose name at one time resounded all over the northern Caucasus. In the spring of 1919 he had made his way with a group of comrades through the Caucasian mountain chain and Georgia to us in Baku, where he became involved in party work. I met him several times. He gave the impression of being an enthusiast, an energetic worker and dedicated revolutionary who could not sit idly by, but had to act, work, and struggle. Andzhievsky was soon arrested by the Muscavat authorities and, obviously at the Denikinites' demand, handed over to the British who in turn transferred him to the northern Caucasus—to the Denikinites. All this was done so quickly that we didn't even have time to intercede. I remember with what great sorrow we experienced this loss.

cell door, backed by guards and many policemen who filled the corridor. To their demand that we immediately vacate the cell Gogoberidze replied: "We cannot comply with your demand. We will not go anywhere at night. Tomorrow, during the day, you can take us anywhere you wish, at your pleasure." The chief of the gendarmerie again repeated his demand, saying that if we would not agree to go of our own free will we would be taken by force.

We realized that they very likely could use force, and give us a pretty good beating besides. We looked at Gogoberidze. After thinking for a moment, he said that we protested the action, but would have to submit to force. The policement then crowded around us and led us through the corridor to the prison office, where, a few minutes later, Boris Sheboldaev was also brought. This was another unnerving surprise, since we had been careful not to have anything to do with him, and the police had never even questioned us about him or even mentioned his name in conversations with us.

We were now more than ever convinced that far from simply transferring us to another prison, they wanted to hand us over to the British military, together with Sheboldaev. When Boris was led into the office, he had a rather sleepy look about him. Glancing around, he recognized us, and letting out some incomprehensible sound, passed his hand across his face and said: "Ah, now I understand everything!" It was clear that he too had suspected that something nasty was afoot. However, we made no response to his words and did not even greet him, pretending that we didn't know him.

At that moment a guard entered with a long, thick rope in his hand and asked the authorities: "Will this do?" I could not restrain myself and asked jokingly, "So, gentlemen, are you getting ready to hang us?" "The fact is," the prison superintendent said, "that at the moment we don't have any handcuffs. Therefore, we shall tie you up together with this rope." And that's what they did, putting our arms behind our backs and tying them with the rope and then tying us up to each other. Then we were led single file, under heavy guard, through the dimly lit corridor and out to the street. There a truck stood ready to receive us. Policemen climbed into the back of the truck with us to make sure we didn't escape.

It was a warm, windless, moonlit night, and it was pleasant to breathe the fresh air after the cell, but we weren't in very good spirits. We didn't try to talk amongst ourselves in the presence of the policemen, and to be frank, there wasn't anything to talk about. Everyone was busy trying to figure out what might be awaiting us at our destination.

The road we drove along wound along the shoreline. When our truck approached the pier we expected it to stop at any moment and have them load us on a boat. When we passed the pier without slowing down and continued in the direction of the Central Prison, we felt vastly relieved, and began to breathe much easier.

In contrast to Bailov Prison, which was situated in the outskirts, this prison was in the center of the city, surrounded by residential buildings. The prison itself was in a mill building constructed by the Azerbaijani capitalist, Tagiyev, who later made a gift of the building to the Tsarist authorities for use as a prison. Thus it was called Tagiyev Prison.

We were taken up to the fifth floor to the cell block for lifers and prisoners under sentence of death, where I had been previously incarcerated. Here they untied our arms. The prisoners in neighboring cells woke up and looked us over through the slits in their doors, and one of them called out from his cell, "Comrade Mikoyan?" Fortunately, there was no policemen around to hear him. I immediately went up to the cell door and said very quietly, "I'm not Mikoyan, I'm Ter-Israyelian." Apparently whoever it was got the message, because not another sound came from that cell, and the entire incident happened so quickly and quietly that the guards didn't notice anything. They led us down the hall and put us in a cell at the end of the corridor.

A few days later, on passing that cell on my way to the toilet I asked who was in there, and found out that it was one of the former Red Army men in the brigade of which I had been commissar. I didn't know him personally, but he remembered me. He was in prison for committing a terrorist act—he had shot and killed a former Communist named Gelovani, who had turned out to be a provocateur. He had done this in order to carry out a party resolution and at his commander's order, but more about that later.

In this cell block the prisoners were kept isolated in individual cells, but we five were put in one cell. We slept right on the cement floor, but soon we were supplied with sheets and pillows from the outside, and the sending of food parcels was arranged—again in Gogoberidze's name. In addition to newspapers we even began to receive books. I remember with what great pleasure I read Chekhov's stories.

The flow of foodstuffs, books, and newspapers came to us through a group of upper-form schoolgirls, members of the Communist youth organization. They did this with great enthusiasm, very punctiliously, and it should be noted that they somehow managed very successfully, unbeknownst to the guards, to slip notes with crucial information into the parcels.

From the newspapers we learned that the British command had begun to remove their troops from Azerbaijan, leaving only a representative in Baku. Evidently they had decided that their influence and general position here was secure and that there was no longer anything here that threatened their authority. We also learned from a note from outside that our comrades in the Baku Party Committee were actively discussing this issue, arguing about the meaning of such an event for our revolutionary struggle, trying to understand the real reasons behind their departure from Baku, since there were serious fears (which later, it is true, proved ungrounded) that in leaving, the British might give the city up to the Denikinites who were continuing their offensive in the south of Russia.[24]

Up until the time of this arrest I didn't smoke. My cell-mates, particularly Gogoberidze, were smokers—and they smoked a lot. Gogoberidze began to try to persuade me to take up smoking, too. I stubbornly refused, saying there was no point to it, that I had no interest in smoking whatsoever. Naturally, I wasn't thinking about my health at the time. Levan persistently tried to convince me of how much pleasure a person gets from smoking. He was a first-rate agitator—all one had to do was recall the swimming and the dinner in the restaurant! And this time, too, I gave in to his powers of persuasion and began to have a smoke from time to time. At first I didn't really enjoy it, but gradually I got hooked, and I smoked for the next twelve years. It had a bad effect on my health, however, so I quit smoking once and for all and never smoked again.

Being in the full bloom of youth and possessed of a particularly cheerful character, Gogoberidze was prone to almost childish pranks. In our cell, on the fifth floor there was a small window placed very high up on the wall. By standing on a stool one could see the flat roof of the one-story building on which the sentry marched back and forth keeping watch over the windows of the prison. Gogoberidze would spend a long time standing on the stool staring out the window at the sentry. At the precise moment when the sentry was moving away from him, Gogoberidze would loudly cry out to him: "*Asker*, ay, *asker!*" (in Azerbaijani, *asker* means soldier). The sentry would whirl around, raising his rifle, and Gogoberidze would quickly duck down out of sight. He would repeat this routine several times until the soldier almost went crazy trying to find the right window to shoot at, leaving Gogoberidze chuckling with glee. I began to try to dissuade him to stop these pranks: "Levan, don't you understand your cries are fraying this soldier's nerves. What is he to blame for? He's a simple fellow, who is carrying out military orders, doing his duty, and not bothering you in any way. But you tease him, and after all, he could really shoot. He's hardly likely to hit you, but even if the bullet goes through the window and hits the ceiling of the cell, the ricochet could hit one of us. It's not the right thing to do, and besides, it's dangerous." Only after these arguments did Gogoberidze stop teasing the soldier.

We gave a lot of thought to what could be done to bring about our release. Our comrades on the outside were trying to accomplish this, but so far nothing had come of their efforts. Once, during a

[24]See Appendix.

discussion of this question the always serious Yuri Figatner said, knitting his brows and gesturing with his arms like a speaker on the rostrum, "We have to bombard the workers with slogans, bombard them with slogans!" Gogoberidze immediately took up the refrain, humorously mimicking Yuri, "Bombard them with slogans, bombard them with slogans." We laughed, not only because Gogoberidze's mimicry was very funny, but also at Yuri's suggestion in itself, which was simply senseless, under the circumstances. We had no doubts that Gogoberidze, as chairman of the Worker's Conference, had to be released anyway. Our comrades were doing everything possible to bring this about. Consequently, there was no sense in calling the workers out to a demonstration and "bombard them with slogans" on Gogoberidze's account. As for the rest of us, the suggestion was also absurd since we were all in prison under assumed names which would mean nothing to the Baku proletariat, while to reveal our real names would mean creating truly great danger for ourselves. Other means of obtaining our release were needed, and we did not doubt that our comrades would do everything necessary to find them. Besides, when Gogoberidze was released he would join the efforts to get us out.

Both in the Bailov Prison and here, neither any of the other prisoners nor we were let out for walks in the courtyard; we were kept in our cells all day long. Once a day, at a scheduled time, all the prisoners were allowed out to go to the toilet. It was then that we had a chance to see the other prisoners, the criminals, who treated us "politicals" with great respect; two Georgians were particularly attentive to us. Gogoberidze, moreover, was known to the whole city and enjoyed widespread respect, and we were his comrades. Our guard was a Russian, a middle-aged man with a rather pleasant disposition, who aroused our pity when the prisoners would ignore his order to return to their cells after going to the toilet. They paid no attention to him whatsoever, and continued to talk among themselves.

The following scene took place on the first day of our stay in the prison. We were very surprised at the strange relationship which developed there between the prisoners and the guard: he was almost reduced to tears by his lack of authority with the prisoners in his charge. One of the Georgians, a thickset, interesting-looking man around thirty with a typically strong-willed countenance, who obviously had the respect of the other prisoners, and wanting to show his authority, unexpectedly shouted: "What kind of tomfoolery is this, you devils! Get back to your cells, on the double!" He said this in such an imperious tone that the prisoners suddenly calmed down and began to go back to their separate cells in silence. We were in our cell at the time and observed the scene through the open door. The Georgian came up to us and said: "You know, they're such hooligans that if someone doesn't interfere, there wouldn't be any order. The guard can't do anything with them." He came into our cell, then, to ask if one of us had a smoke, and after Gogoberidze offered him a cigarette, we got to talking. He conducted himself throughout with great dignity, and when he left we spent a long time discussing what were the special merits of this criminal which made it possible for him to wield such power over the other prisoners, and indeed, even over the guard who treated him with a particular respect.

One day the door to our cell opened and the guard informed us that he had orders to take the five of us out into the prison courtyard to be photographed. This wasn't exactly good news to us, since I could be recognized from a photograph, and after we got out it would be easier for the police detectives to keep an eye on us.

Gogoberidze agreed with me and we tried to think of some way to circumvent the situation. When we were led out of the prison block, the same Georgian about whom I spoke above kept hovering around us, and went out into the courtyard with us. As we were going down the stairs Gogoberidze conversed with him in Georgian, and then he whispered to me that he had made an agreement with his countryman to have him overexpose (and ruin) the photographic plates on which our pictures were taken. I was overjoyed to hear this.

There being nothing for us to sit on in the courtyard, we five all stood, but for some reason this grouping didn't suit the photographer, so one of us took two rusty buckets which happened to be in the yard, and turned them upside down. Figatner and I, as the oldest ones present, sat on these "chairs," and the others stood up behind us. After the picture had been taken we began a lively conversation on the most diverse subjects with the photographer who, fortunately, turned out to be very talkative. We surrounded him in a closely knit, wall-like circle while, out of the corner of my eye, I noticed that our Georgian was hanging around the camera. When he had obviously finished his "work" and moved off to the side, we ended our conversation and said good-bye to the photographer. The Georgian told us that "everything is in order, the plates are ruined," and we were pleased that everything had turned out so well. The conviction that our group photograph had not come out stayed with us until 1920 when, after the proclamation of Soviet power in Baku, some comrades in checking the files of the gendarmerie found it in the archives, and showed us the photograph which we all thought "hadn't come out." But by then we were glad that it had, in fact, "come out," since it recalled to us very vividly a period of our lives and of the past struggle. I still have the photograph in my possession.

No comforting news had come from the outside concerning our release, and we on our part couldn't figure out what the gendarmes intended to do with us. The only thing that was clear was that sooner or later Gogoberidze would be released, but we could not understand why they had attached Boris Sheboldaev to our group. We didn't even consider the idea that they had some sort of evidence that might tie

Sheboldaev in with one of us. It was also surprising that after the interrogation at the police station none of us had been questioned again: they were conducting an investigation, but they weren't interrogating us!

I couldn't get out of my mind those documents which I had left under the tablecloth in the restaurant. There was no doubt that the waiters discovered them, and if they had handed them over to the police, then our situation would have become significantly more complicated. We found out later that the waiters had indeed found the documents, but they hadn't given them either to the administration or to the police. One of the waiters had taken them to the Workers' Club and given them to the secretary of the Workers' Conference to be returned to the Baku Party Committee. I would like to note that the waiter was not a Communist, and as a matter of fact, there wasn't a single Communist among the service personnel in the restaurant.

Soon we were moved to another section of the prison where common criminals were kept. The corridor there was much longer and the cells larger and contained more inmates, and instead of a wall dividing the cell from the corridor there was a continuous set of iron bars with a gate that was kept locked by the guard. We were installed in a cell that was specially prepared for us, where there were no other prisoners. The guard walked back and forth along the corridor, keeping a constant watch over what was going on in the cells, so that one was under surveillance most of the time and it was difficult to hide even our conversations from him.

It's true that there was a constant and incredible din in the corridor from the uninterrupted noise of quarrels, arguments, and even fistfights between the prisoners, so that our heads ached from the noise and commotion. It was only a little quieter at night. We slept right on the stone floor in this cell as well.

The day after we had been moved to our new cell the senior guard came to our section with some other fellow, and addressing me he asked: "Are you Mikoyan?" Without blinking an eye I replied, "How did you get that into your head? I'm Ter-Israyelian. And I don't even know Mikoyan." At that point Gogoberidze intruded into our conversation: "Don't you know that Mikoyan has been in Tiflis for ages?" he said. "He escaped from this prison and is in Tiflis now." With this the conversation ended.

After they left, we held a council of war. It was clear that someone had recognized me, but who? Had we managed to dispel the guard's doubts, and did he believe that Mikoyan was in Tiflis? If he didn't believe us, it would not be difficult for him to uncover my true identity once and for all, for there were people in the police who knew me by sight. Gogoberidze was particularly worried. He suggested sending a note to the outside requesting that some person, fully empowered by our comrades, immediately meet with the guard and threaten him that if the position of the political prisoners worsened as a result of any denunciations he made, he would answer with his head, but if he promised not to denounce any of us he would receive money. This was in fact done. It was later revealed that he was the one who had recognized me, but he promised to keep quiet.

About two or three days later the senior guard came and informed us that Gogoberidze was being released from the prison. We were all very happy about that. In saying good-bye to us, Gogoberidze told us not to be depressed: he would take measures to insure our speedy release.

A week passed. Just then the guard informed us that three of us were being released; only Boris Sheboldaev whose case had still not been decided, would remain in prison. The news concerning Boris was not at all good, particularly since his case was directly connected to the operations of our military intelligence. We were extremely concerned about him and realized what difficulties we would have to overcome to get him out of prison. However, we didn't let our anxiety show since we had been keeping up the pretense of not even knowing him.

At our release a further blow awaited us when we were informed by the prison superintendent that, "The Azerbaijani government does not want you to remain in Azerbaijan." They were, in effect, exiling us from Azerbaijani territory, and gave us three days to get our things together. Since at our arrest we had represented ourselves as people who had recently come to Azerbaijan in search of work, we weren't even able to make a protest against exile.

On the day of our release we met with our comrades at the Kasparovs' secret apartment where it was decided that, for security reasons, the best place for us to go was either to Georgia or Armenia. Armenia refused to accept us because we were Bolsheviks, however, so we were glad to learn that the Georgian government would grant entry visas to all three of us. It was decided to travel separately so as not to draw attention to ourselves, and since they feared that the Azerbaijani government knew that Ter-Israyelian was really Mikoyan and would want to settle accounts with me en route, the comrades decided that Karayev, a member of parliament, would accompany me to the border. Karayev was to be traveling as a deputy on official business in the border region.

We traveled in the same compartment, pretending to be strangers. An Azerbaijani policeman accompanied me to the border, but seeing a member of parliament sitting next to me in the compartment he stood outside in the corridor. Eventually Karayev went outside in order to draw him into animated conversation in Azerbaijani and get on his good side, and later, when Karayev and I, having "introduced" ourselves, ordered tea, we asked the policeman to join us. Everything was quite nicely arranged.

After that, alone together in the compartment, Karayev and I conversed warmly the whole way. He was a sympathetic, intelligent, straightforward person and a pleasant conversationalist. He spoke about the past, about how he had worked in Tiflis, and about his move to Baku. I had known him before, but now I particularly appreciated his high degree of sincerity. I should say, in general, that Karayev had had his change of heart and came over to the Bolshevik side literally in front of my eyes; in difficult circumstances he had shown himself an utterly dependable and self-possessed political activist.

At the Georgian border we had to change cars for the train to Georgia. According to his instructions, the policeman accompanying me was to hand me over officially to the Georgian police authorities. Of course, this was very undesirable for me, it would have been much better if I could have boarded the Georgian train like any ordinary citizen so that the Georgian police would not know from the start that I had been exiled to them. Therefore, Karayev once again invited my policeman into the compartment to have some tea and struck up a friendly conversation with him. The policeman was very flattered at the attention paid him by a deputy of parliament. When the policeman was completely softened up, Karayev explained to him that, speaking in general terms, his job was to make sure that the exile actually did leave Azerbaijani territory: in this way he was carrying out his official duty, but that once he had seen me onto the Georgian train it was completely unnecessary for him to pass me "from hand to hand" to the Georgian police. My policeman had no problem agreeing with this. Therefore, when we arrived at the border station, Karayev and the policeman simply escorted me to the Georgian car, where I said goodbye to them as the train started up. I was now officially "exiled."

After spending a few days in Tiflis, I once again, but now under a different name, illegally returned to Baku and joined in underground party operations.

And at that time Gogoberidze and Karayev, acting on orders from the Party Regional Committee were working on the problem of getting Boris Sheboldaev released from prison. Various means were set in motion to bring pressure upon both the prosecutor and those other people in authority upon whom the whole case depended. Everything indicated that the authorities had no seriously compromising evidence against Sheboldaev. And, in fact, he had done nothing against the Azerbaijani bourgeois government, since the intelligence network he headed was operating in the rear of Denikin's army; therefore the Azerbaijani authorities had no particular basis on which to inflate Sheboldaev's case out of proportion.

One day, however, the Baku Bureau of the Party Regional Committee was particularly disturbed by a note from Sheboldaev containing the news that the criminals in jail had begun to persecute him. They were demanding money from him, and if he didn't pay they threatened violence. (Sheboldaev had been receiving food parcels regularly from the outside through our young comrades, and thinking that Boris was a rich man, the criminals had decided to try extortion.) Even if the criminals were given a certain sum in the form of a ransom, Boris wrote, it wouldn't alleviate the situation: blackmail and extortion would continue. Therefore, he requested not money, but a more radical form of interference. The letter made us terribly upset. Gogoberidze thereupon took it upon himself to help him.

Recalling certain habits and "traditions" of the Baku criminal world, he arranged through his restaurant-owner friend a meeting with a certain Georgian, whom the criminals were rather afraid of. The meeting with him had the desired effect, for a day of so later we received a letter from Sheboldaev in which he said that a "miracle" had occurred. Not only were there no more threats, but the criminals were treating him with respect, to boot. He just couldn't understand it: what, he asked, had happened? When we managed a short time later to get Sheboldaev released from prison he had a good laugh when he found out how the "miracle" had been achieved.

Chapter IX

The Unification of the Communist Organizations of Azerbaijan

Gummet and Adalet—Political work among the British soldiers—We try to get the cooperatives on our side—The terrorist acts of the bourgeois government—The Transcaucasian Conference of Communist Youth—Mikha Tskhakaya.

The economic situation of the Baku workers had, in the meantime, kept on deteriorating. As a result of the general inflation the workers' real wages had fallen; there was no one to buy the oil which had been extracted in such large quantities, and now the oil output itself was gradually declining. The capitalists, of course, incurred the expenses for its production, but to all intents and purposes they did not receive the necessary revenues. Some of them managed to hold on somehow with the help of the banks, while others were ruined. The promises given by the bourgeois government of Azerbaijan during the May strike to improve the workers' economic situation turned out to be hollow, as was to be expected. Those Azerbaijani workers who had believed the government and had not supported the strike were now totally disillusioned, for by this time the deception was obvious to everyone.

In these conditions the work of Gummet and Adalet yielded especially positive results; these organizations increased their number of active workers and exerted more and more influence on the Azerbaijani segment of the Baku proletariat.

Throughout all the district organizations of the RKP(B) Workers' Communist party (Bolsheviks) intensive work was carried out to educate the Communists, and their positions in the trade unions became stronger. The workers' clubs were also greatly revitalized as centers of party propaganda work, as were the activities in the worker's consumer cooperative, which were new for us.

Preparations for an armed rebellion went on successfully; plans for carrying it out were drawn up, cadres were trained, and arms were laid by. After the slogan for an independent Soviet Azerbaijan, closely tied with the other Transcaucasian Republics and with Soviet Russia was proclaimed as the rallying cry of the struggle, we, and I in particular, became most concerned about an organizational issue within the party.

It seemed impossible to start a rebellion within the existing conditions of organizational parallelism which we then had in Azerbaijan. As I have already stated, the Moslem-Communists were united into two organizations—Gummet and Adalet. Both were under the auspices of the Baku Party Committee. Adalet operated primarily in Baku and Mugan, and Gummet took in all of Azerbaijan. Our Baku Party Committee was responsible for the leadership of all party work in general, and it also provided aid and assistance to the Dagestani Communists.

As far back as the Sixth Party Congress Alyosha Dzhaparidze had given a very high evaluation of Gummet. A delegate of Gummet at the Congress, Yusifzade, had made a request that it be given material aid, and V. P. Nogin had supported this request. Speaking after Nogin, Dzhaparidze had said, "This organization has its own name, but don't think that a second 'Bund' has appeared in our organization . . ." for "Now, as from the very beginning, Gummet is merged with our organization." And further on Dzhaparidze assured those present: ". . . Gummet is worthy of our confidence and will selflessly serve the cause of bringing class consciousness to the Moslem workers and uniting them with the all-Russian revolutionary proletariat." And, in actual fact, Gummet and Adalet did play a major role in drawing the Moslem workers into the revolutionary movement.

Now, however, the situation had changed. New problems had sprung up whose solution required that we convert to the standard organizational norms of the party. This was felt particularly sharply on the eve of the conquest of power. Thought had to be given to what form the party organization should have when it assumed the leadership of a Soviet Azerbaijan.

In particular the question arose: how should an appeal to the workers and peasants concerning the rebellion best be signed? If the appeal were to emanate from only one of the organizations, then it was hardly possible to count on its success. But if it were sponsored by all three, the Baku Party Committee, Adalet, and Gummet, the broad masses might not understand why there were three separate sponsors.

So the idea gradually took shape of unifying all the Communist organizations into one single organization. It was our conviction that such an organization should become the sole Communist party of Azerbaijan, uniting all the Communists of Azerbaijan and forming a component part of the RKP(B) and its All-Caucasian territorial organization. The majority of Communists supported the idea of organizing a single Communist party, and the issue came up for discussion at a meeting which included a large number of comrades, and the leaders of Gummet, Adalet, and our own Baku Party Committee of Bolsheviks.

At first opinions were divided. Karayev—one of the most influential leaders of Gummet—spoke in favor of keeping Gummet as an All-Azerbaijani organization (with which he suggested Adalet be merged) and as an independent organization of the RKP(B) existing side by side with the united Gummet. Another no less influential leader of Gummet, Mirza Davud Guseynov, proposed that a Turkish Communist party be formed in place of Gummet and Adalet which would exist side by side with the organization of the RKP(B).

Some of the members of our Baku Party Committee who spoke at this meeting did not agree with either Karayev or Guseinov, feeling it was necessary to get rid of Gummet and Adalet and to include the members of these organizations in the RKP(B).

We found out somewhat later that one of the prominent leaders of Gummet, S. M. Efendiyev, who had been evacuated to Moscow as a result of the temporary collapse of Soviet power in Baku, had published an article in July 27, 1919, in the newspaper *The Life of the Nationalities* in which he had written:

> During the present time a need has arisen to expand the field of activities of the Gummet organization, since in socialist Russia it is necessary to draw into Gummet's sphere of influence all the millions of peoples of the East. Not only the Moslems of the former Russian empire are needed in organizational and party work, but also the foreign Moslems of Persia, Turkey, Afghanistan and other countries. The task now falls to Gummet to be the alarm bell of communism in the East.

When we read this article, we were utterly amazed: how was it possible to use religion to bring together in Gummet all the Moslems of different nationalities, and, moreover, not only in Russia, but also in the foreign countries of the East?

After long debates and discussions we finally reached a common conclusion: all the Communists of Azerbaijan should join a single Communist party, irrespective of their nationality.

When we brought this issue up for discussion in the Tiflis Bureau of the Party Regional Committee, our position was not supported there, the Tiflis comrades didn't want to change the existing situation in any way. Georgia preserved the old system of the direct subordination of all the local organizations to the Caucasian Party Regional Committee without the formation of an All-Georgian party center.

The Communists of Armenia found themselves in a rather peculiar situation. The fact is that in mid-1918 a group of Armenian Communists headed by the Bolshevik poet Aikuni had formed a Communist party to do work among the western Armenians, who had left Turkey when the Russian troops withdrew from there. They published their own newspaper in Tiflis. When the Georgian Menshevik government's campaign of terror against the Communists intensified, this group moved to the northern Caucasus, where there was Soviet power, later leaving there with units of the Red Army for Moscow, where they set themselves up as the Central Committee of the Communist Party of Armenia. At the First Congress of the Comintern they came forward as an independent party. However, although it called itself the Central Committee of the Communist Party of Armenia, this group had no ties with Transcaucasia in general, or Armenia in particular, doing its work among the groups of Armenian Communists then living in Moscow, Saratov, and other cities of Central Russia. It should be noted that they enjoyed the recognition and support of the Central Committee of the RKP(B).

We in the Caucasus didn't know anything about their activities, and since they in turn didn't inform us about what they were doing, it was only in the fall of 1919 that we learned this Central Committee (which considered itself the ruling organ of the Communist organization of Armenia) had chosen a group of Communists to send via Astrakhan and Baku to Armenia as leaders of the local party organizations.

The Communists who were working in Armenia, Georgia, and Azerbaijan were unanimous in their condemnation of this policy, which had not been given any thought to the Transcaucasian Party Regional Committee—the real leading center of the Communists of all the Transcaucasian Republics, with whom they had no desire to work.

By that time the Communists of Armenia had decided to create their own center to provide leadership for the Communist organizations of Armenia—Armenkom, which would be set up as a regional committee of the RKP(B), and would work under the supervision of the Party Regional Committee. I recall that we, the Communists of Baku, spoke out in favor at that time of the creation of a single Communist party of Armenia along the lines of Azerbaijan. We also joined with the Communists of Armenia in refusing to recognize Aikuni's group as the Central Committee of the Communist party of Armenia, for it was our opinion that the Central Committee should be elected legitimately at a constituent assembly of the Communist party of Armenia, or at a conference of the party organizations of Armenia.

The issue of creating a single Communist party of Armenia (as of Azerbaijan and Georgia) ultimately had to be resolved by the Central Committee of the RKP(B). We knew that there would still be many difficulties and debates on the road ahead of us, but we firmly believed in the rightness of our position.

* * *

The presence of the English occupation forces in Baku behooved us to establish contacts with the British soldiers and explain our views and ideas to them in order to counter the false propaganda preached against us by the British military command. The need for such a measure had been brought home to us during preparations for the May general strike when it was not ruled out that the British command might use its troops against us.

Since we had only two young translators whose command of English, moreover, was quite inadequate, we sent repeated appeals through Astrakhan to the Central Committee of the party to send us two or three British Communists for use in practical work. Finally, in June I think, one such comrade arrived. Jack wasn't very young, he always had a pipe in his mouth, and was calm and composed to the point of being phlegmatic, but he spoke Russian with ease and fluency. I met with him several times to acquaint him with our situation and explain the nature of the propaganda work needed for the British garrison, and we never had any disagreements on that score.

I advised him to make use of the fact that the British soldiers were sick and tired of military service and yearned for their country, their homes, and their families. But the British military command was also aware of how demoralized their soldiers were in the atmosphere of Baku and they accordingly replaced large groups of soldiers at frequent intervals, which meant that the Communist cells we had organized among the soldiers would disintegrate and have to be set up all over again.

Once I asked Jack to bring one of the soldiers, a British Communist, to a secret apartment to take part in a discussion. I wanted to meet a "real life" British soldier Communist. Jack arrived with a young British soldier, a tall, smart-looking fellow with a severe expression on his face. He was dressed in the summer white uniform of the colonial forces, which featured shorts instead of the usual trousers. At the time such a uniform struck us as strange, to say the least, but a uniform is a uniform. He sat opposite us, looking very self-contained and assuming a kind of "constrained" pose which seemed to say that he wasn't ready for a comradely chat, but would rather be standing in front of his commanding office awaiting an order.

I asked him questions which Jack translated. I asked him where he was born, what he had done before joining the army, and whether he had a family. Then I asked him what had led him to embrace communism. What had been the biggest influence on him in this regard? I also asked him about the mood of the other soldiers in his unit. His answers were precise, brief, and reserved, without one unnecessary word, and no details. It was hardly a conversation with much substance to it, and I was completely mystified as to what the problem was. I understood only that our conversation hadn't worked out, and I made no effort to detain him. Evidently, what had been needed was some kind of preliminary icebreaker, for, in general, he was a nice lad and obviously quite intelligent. I thanked him, and we parted.

Jack did a lot of work for us among the British soldiers until the latter half of August 1919 when the British troops were pulled out. The withdrawal was peaceful, and we read an announcement in the Baku newspaper *Azerbaijan* that the Azerbaijani government had given a farewell banquet in honor of the British military command at the Metropol restaurant on August 24. Foreign representatives had attended, as well as all the ministers and deputies; and Usubekov, the head of the government, and the British General Shuttleworth had exchanged cordial toasts.

I had mentioned earlier that at the end of July 1919 the Baku party workers—Victor Naneishvili, Gamid Sultanov, and Dadash Buniatzade had come back to us from Astrakhan. Victor Neneishvili had been an active member of the Baku Party Committee; when the counterrevolution began in Dagestan in May 1918 he was sent there as a Special Commissar of the region, with a detachment of Red Guardsmen under his command to reinforce Soviet power in Dagestan. After the temporary collapse of Soviet power, Naneishvili and a group of comrades were forced to go to Astrakhan. Sultanov had done a lot of organizational work to strengthen the Red Guard in Baku, and Buniatzade had contributed actively to the creation and strengthening of the Peasants' Soviets in the Baku district, heading the struggle against the counterrevolution.

We were very pleased to have Naneishvili, Sultanov, and Buniatzade return to Baku. This was the first group of prominent Baku Communists to return from Astrakhan for work in Azerbaijan.

One evening those of us who represented the Baku party leadership met in a limited body with these comrades at a workers' apartment in Chernyi Gorod. At this meeting I was given the job of informing our newly arrived comrades of the general situation in Baku and of the tasks we had set for ourselves. We wanted to brief them about what was happening and put them to active work immediately. Their arrival enabled us to intensify our work in the districts of Azerbaijan in the days that followed when the movement against the landowners was gathering steam. The party organizations tried to stand at the head of this movement.

Sultanov and Buniatzade were sent into the wilds of Azerbaijan where with their arrival the peasant movement acquired a more organized coloring. Buniatzade, Karayev, Guseynov, Naneishvili, and Yusifzade spoke with great success at the peasant congress in Kazakh, managing to get a resolution passed which had a Bolshevik organization. This made a great impression on the Muscavatists, for although their representatives had spoken at this congress, they had ended up in the minority. The success of the congress was symptomatic of the times, reflecting the mood of the entire Azerbaijani peasantry.

Naneishvili, Sultanov, and Buniatzade were, without doubt, exceptional party workers. Their subsequent work brilliantly confirmed this. In 1920 after Soviet power had been restored, Sultanov and Buniadzade became part of the first Revolutionary Committee of Azerbaijan, and in later years they worked in the highest posts in the Soviet government of Azerbaijan. Naneishvili was elected to the Central Committee of the Communist party of Azerbaijan and was Secretary of the Central Committee.

* * *

In June or July of 1919 Shiga Ionesian—a member of the board of the Caspian Cooperative Association—requested the secretary of the Workers' Conference to arrange a meeting with me in order to discuss an important matter, one that I found of interest.

"Soon," Ionesian said, "there'll be new elections to the board of the Cooperative Association. On the current board the Bolsheviks have only one representative, and that's me—the majority of members are Mensheviks and SRs. But there's a real possibility of the Bolsheviks' getting a majority of votes at the board of elections if they'll do the necessary preparatory work.

"The membership of the union numbers several thousand," he continued. "Many of them are inactive and don't even attend the meetings. At the reelection meeting, for example, no more than 10 or 15 percent of them will be there."

Ionesian's plan was to have some 800 and 900 Bolsheviks join the Cooperative before the elections, and have them all attend the election meeting to insure a majority of votes for the Communists, thus guaranteeing the Bolsheviks taking control of the union leadership.

I should say honestly that at that time we had only the vaguest idea of the work of the cooperatives, but after a lengthy discussion with Ionesian I realized what a big mistake we had made in underestimating their work. The cooperatives included vast numbers of workers employed in the bakeries, warehouses, stores, and purchasing stations, and moreover, the cooperatives made it possible to send people into the different districts legally. What an extraordinary benefit this could be to our illegal party work! And what splendid use we could make of those bakeries and warehouses for storing arms, and of the stores for depositing and disseminating party literature! How fortunate it would be to place in the cooperatives our party organizers who could then, in the course of their travels, carry out party assignments, establish contacts, secret meetings, etc., at the same time that they were fulfilling their official duties!

The results of our conversation were immediately reported to the Baku Party Committee where they provoked great interest. The comrades were amazed that they had never given a thought to this before. A resolution was promptly passed to recruit into the cooperatives up to a thousand Communists and members of the Young Communists' organization—the International Union of Working Youth of the city of Baku and its districts. The work of organizing the enlistment drive into the cooperatives and of carrying out work among the members of the Cooperative Union was entrusted to our most experienced organizers—Sarkis, Vasia Egorov, and also Isai Dovlatov, who was charged, in particular, with responsibility for utilizing the cooperation for party purposes. This resolution was carried out successfully with the hoped-for Communist majority secured at the reelection meeting of the Caspian Union, and a new slate passed which had been proposed by us.

All this came as a complete surprise to the Mensheviks. They assumed, however, that our success had come about by pure chance, and declaring the elections to be unlawful, they set a time for a new reelection meeting and took all the measures necessary to mobilize their members. However, they still didn't succeed. At the second meeting the previous resolution was confirmed by a majority of votes despite all the efforts of the Mensheviks to the contrary, and they had no other recourse: they had to turn the leadership over to the new board.

* * *

In the beginning of September, on a day when we were planning to go to Tiflis to the regular session of the Party Regional Committee, we learned that a boat which had arrived from Astrakhan with people, party literature, and arms had fallen into the hands of the Azerbaijani police. We had to rescue them at all costs. This could be done most successfully with the help of Musevi—the man who was then head of counterespionage in the Azerbaijani government, with whom we had contacts.

Musevi had joined Gummet in the latter half of 1918 but, like Karayev, he had soon expressed his desire to join the Communist party. We accepted him, but obliged him never to say anything about this to anyone and to conduct himself as before—we had our own reasons for this. We assigned Musevi three basic duties: 1) in the event of a breakdown of our communications with Astrakhan, to aid us in the rescue of men and property; 2) to provide us with information concerning all measures the government was contemplating against the Communists and Soviet power; and 3) to keep us informed on the activities of Denikin's agents in Azerbaijan, and on Denikin's interrelations with the Azerbaijani bourgeois government.

We had arranged with Musevi that Gogoberidze, as the legal worker for the Workers' Conference, would be the one to meet with him for the exchange of information. And that's why we gave Gogoberidze the task, as we were leaving for Tiflis, of staying behind in Baku, and of rescuing from the police the people and property that had been detained. He understood what had to be done, and readily agreed to stay.

Two days later in Tiflis we learned by telephone from Baku that something extraordinary had happened: at midnight on September 5 in the New World restaurant, Musevi and Ashum Aliyev had been killed, and Levan Gogoberidze seriously wounded. The murderer was the second cousin of the former vice-governor, Seidbekov.

This was a heavy blow for us. We left for Baku immediately and got there the day after Musevi's and Aliyev's funeral.

Gogoberidze had received two deep bullet wounds, one in the stomach which exited through his liver, and another in the shoulder. Luckily, despite the seriousness of his wounds, there was no danger of his losing his life. This terrorist act, which had been carried out by a direct agent of the Azerbaijani bourgeois government, stirred up the workers of Baku. They organized a huge demonstration at the funeral. On that day the workers started gathering at the Workers' Club in the morning, and from there they proceeded in a multi-thousand throng to the main Moslem mosque, where the bodies of Musevi and Aliyev were, and held a mass meeting in the yard of the mosque. Abilov and Lominadze addressed the crowds, and after the meeting the workers' demonstration moved down Nikolaev Street, past the parliament, where angry speeches were also being delivered against the executioners. The burial took place at the Moslem cemetery; there another large rally was held.

The murder of Musevi and Aliyev had a very strong effect on the Moslem population of Baku. Dissatisfaction with the bourgeois government intensified even more among the Azerbaijani workers and employees, and the authority of the Bolshevik Gummet grew much stronger. Musevi's death, of course, introduced huge complications into our work. But, again thanks to him, we had kept our own people doing counterespionage work in the government, and we relied on their services for a long time to come in order to secure and maintain our contacts with Astrakhan.

Gogoberidze soon recovered, and after two weeks in the hospital he returned to work. He gave us a detailed account of everything that had happened. After our comrades from Astrakhan had been rescued from the police with Musevi's help, Gogoberidze had needed to talk over some additional matters with Musevi, and had arranged to meet him and Aliyev at the New World restaurant. There he sat down at their table as if meeting them by chance. The business discussion with Musevi was brief, and Gogoberidze had already concluded it when a certain officer of the Azerbaijani government appeared along with another man and seated themselves at a neighboring table. At that point Musevi started a discussion in jest about the Bolsheviks and the Denikinites, talking so loudly that his neighbors could hear him.

The officer thereupon made some remarks of an insolent nature, evidently wanting to start an argument. Then Aliyev, who saw that the argument was beginning to turn into a quarrel with this fellow whom he knew and hated, appealed to Musevi to continue their conversation at another time and place. The officer seized on these words as a pretext to be insulted, and gave Aliyev a slap in the face. Gogoberidze tried to separate them and was pulling the officer aside when the latter unexpectedly pulled out a gun. Gogoberidze jumped him and seized the hand holding the gun, managing to push his finger under the trigger so that the officer couldn't fire the gun. The latter succeeded in tearing himself free, however, and after pushing Gogoberidze to the side, shot Aliyev, and then he turned and shot Gogoberidze. Musevi got up from the table, but he didn't even manage to get out his gun before he was mortally felled by a bullet. The officer shot Gogoberidze once more, and leaving him for dead, fled the restaurant.

Gogoberidze began feebly to call for help, but no one responded, until finally, as Gogoberidze told it, a police officer arrived whom he could see as if in a haze. Instead of questioning those present as to what had happened, the police officer began to set forth his own version of what had happened and did everything he could to foist it on the witnesses. He said: "I know what happened better than anyone. A Turkish soldier came in and wanted to arrest Ashum Aliyev as a Lenkoran Bolshevik. Musevi protested and began to fire at him. Then he, attempting to defend himself, began to shoot back and killed all three."

Gogoberidze, who had heard all this, cried out indignantly, "You're lying! You ought to be ashamed of yourself for making all that up!" The police officer turned around, and when he saw that Gogoberidze was alive, he taunted: "Ah, so you're still alive, Mr. Gogoberidze!"

The town governor, Gudiyev, who arrived soon after, hissed through his teeth when he saw the wounded Gogoberidze, "Well, at long last, it's been long overdue!" It was more than three hours later that the policemen dragged Gogoberidze to the police carriage and took him to the hospital. But even here he was not given medical assistance until six o'clock in the morning, at which time a surgeon appeared at the hospital who did everything possible to save him.

The workers held mass meetings in their local districts to protest against the White terror. You can judge, for example, the character of these meetings by a resolution which was passed at a workers' meeting in the Chernyi Gorod district of Baku:

> We, the workers of Chernyi Gorod, have discussed the report of the presidium of the Workers' Conference concerning the murder of our comrades Musevi and Ashum Aliyev, and the mortal wounding of the President of the Workers' Conference, Comrade Gogoberidze, and we protest the terror being carried out against the leaders of the working class and we declare that we shall not tolerate other such brutal acts and will fight the butchers of the bourgeoisie with every means possible. We demand that the murderer be punished according to the full severity of the law. We demand that all the known and secret agents of Denikin be banished from the boundaries of the Azerbaijani Republic. We declare that on the day of the funeral we shall suspend work at all the factories of Chernyi Gorod. We call upon the workers of the city of Baku to repulse the butchers and the murderers of the working class. Death to all the butchers and murderers. Down with all Denikin's known and secret agents!

As a result of this tragic event, which had made such a lasting impression on them, some of our comrades, in their speeches and newspapers, called for a Red terror to combat the White terror. When we discussed this issue at the Bureau of the Regional Committee, we explained to these comrades that although their demands were understandable from a human point of view, it was politically unfeasible for us to engage in retaliatory acts of terror against the bourgeois government. "Our party's principle of a negative attitude toward terror as a means of political struggle is well known to you," we said to them. "Let's adhere to it in this instance as well." The comrades agreed to this.

I should say in connection with this, however, that in accordance with a party tribunal's decision we were forced in two instances to liquidate traitors. The first incident occurred in the spring of 1919, some time after our return from prison, when Sturua raised the question of Gelovani in the presence of Gogoberidze.

I had known Gelovani from my first days in Baku when I had had to camp out in the office of the Baku Party Committee. Gelovani, an old-guard Social Democrat who conducted himself modestly and well, didn't have a place to live either so he and I both slept in the committee office. Later the committee found out by chance that, after the collapse of Soviet power in Baku and the rise of the bourgeois Azerbaijani government during the Turkish and then the British occupations, Gelovani had entered into the counterespionage service of the bourgeois Azerbaijani government. This fact had been established by Sturua, who also knew Gelovani.

We were stunned by this revelation, which was fraught with danger for us. Gelovani knew every one of us well, and as a counterespionage agent he could betray us all. We had so few experienced party workers at that time that if Gelovani continued his "activities" we could incur irretrievable losses. For that reason Sturua suggested that he be given permission to do away with this dangerous traitor. He wanted to meet Gelovani, invite him to go to a restaurant for a cup of coffee, and then poison him.

We agreed that we had to get rid of Gelovani, and gave Sturua and two other comrades the task of organizing his trial, but as far as Sturua's suggestion was concerned, we categorically turned it down since Sturua was a member of the Baku party leadership, and we could not afford to lose him, should he fail. We met once again to discuss the best way of carrying out the act of retribution.

I remembered that Safarov was in Baku at that time. I had known him well at the front and during the struggle against the Turks, and I knew him to be an extremely courageous and self-sacrificing man. When we met I asked him if he could find a man who could carry out the party assignment of liquidating a certain traitor. Safarov said that he would undertake the task readily; he had comrades of proved worth who had fought with him at the front.

Then we instructed Maro, a Communist woman whom Gelovani had recently taken up with, to invite Gelovani to her apartment on Stanislavsky Street, and we gave the address to Safarov. Some days later our mission was carried out—by, it turned out, the man who later was in jail with us when we were arrested at the Tilipuchur restaurant.

The second case of dispatching a traitor concerned a sailor in the Caspian Fleet. After the British command had carried out its purge of the staff of the Fleet there remained very few commanders and sailors who were devoted to the cause of revolution, but there were many free posts, and we had the opportunity to find places on the ships for our own people. We wrote a letter to Kirov in Astrakhan in this connection, expressing the thought that it might prove possible to seize control of some of these warships if they would send us some military-naval specialists and sailors from Astrakhan, whom we would place on board.

Our letter elicited a fairly speedy response. In June, around thirty specialists and sailors arrived on the boats which transported gasoline to Astrakhan. We quickly organized reception and resettlement centers for these men. Those who had come were for the most part nonparty sailors who were, however, inclined toward a revolutionary outlook.

There were no dry laws in Baku at that time, in contrast to Central Russia, and vodka and wine were sold openly. Some of the sailors who had arrived, but were still not posted on ships, began to drink heavily. This was very dangerous. In an attempt to protect the rest of the sailors from drunkenness and demoralization, we set up rules, but there was one young sailor who would not submit to any conditions and demanded that his allowance be increased since he, naturally, didn't have enough money to go out on a binge. When we turned him down he threatened to go to the Azerbaijani police and tell them everything.

This began to take on ominous overtones. The matter came up for a special discussion at a session of the Baku Party Committee. We were alarmed by the situation and decided to increase this sailor's allowance for a few days in order to calm him down somehow, while we instructed Barkhasho, the secretary of the Baku Komsomol Committee, to look into the matter. A short time later Barkhasho reported that he hadn't had any success, the sailor was still using blackmail and threatening to tell the police everything that he knew. He had turned out to be a dangerous, completely demoralized person. We then instructed the Komsomol Committee to organize his trial. Beso Lominadze assisted the Komosomol in this.

The Komsomol court sentenced the traitor to death. The sentence was carried out by one of the members of the Baku Komsomol Committee.

I should, perhaps, say something about yet a third incident I recall when we decided to resort to the physical annihilation of a prominent representative of Denikin's command, judging this to be a natural act of retribution against the White Guards. This incident concerns General Baratov, who was then Denikin's chief representative in Tiflis. The decision to eliminate him was made by the Party Regional Committee. Arkadii Elbakidze (Agordiya) and one other comrade (I can't remember his name) were given the task of implementing the decision.

However, more than a month passed and it still hadn't been done.

I remember that on one of my visits to Tiflis I spoke with Makharadze and expressed my dissatisfaction at the comrades' delay in doing away with General Baratov. Shortly afterward I happened to meet Safarov on the street. "I know that you decided to kill Baratov, and it's a fair decision," he said to me. "I also know that your men are following Baratov, trailing him, observing him, but obviously putting off the execution of the sentence—they won't have any success. Leave it up to me; I'll kill Baratov!"

I replied that we had already assigned comrades to carry out the job. But Safarov kept on insisting: "I'll do it quickly and accurately. I've been watching Baratov for a couple of weeks now. I know that he lives on Mikhailovsky Street, I know when he leaves the house, I know his car, I've even picked a tree that I could hide behind. I'm an excellent shot, I won't miss."

I told Safarov that I couldn't assume the responsibility for such a decision, and that I'd have to consult with the other comrades. During our discussion of this matter at a session of the Regional Committee I proposed that Safarov be given the job of carrying out the sentence against Baratov.

I remember that Elbakidze, who was present at the session, resented me greatly for this, and took my proposal as a sign of my lack of confidence in him. He claimed that they were all set to eliminate Baratov. I then withdrew my original suggestion. Two days later Elbakidze and his comrade came to our secret apartment and announced that they would get rid of Baratov the next day and would come to see us immediately after.

And, indeed, the next day when I was walking down the street near Tsitsianov Rise I heard a powerful explosion. I realized that it was our comrades who had thrown a bomb at Baratov. I returned home in a hurry because it was obvious that a general roundup of Communists would begin, followed by arrests. We soon learned that Elbakidze had indeed thrown a bomb at Baratov's open car when he was driving up Veriisky hill, killing the general's aide and severely wounding General Baratov. Elbakidze had run from the scene, the object of a chase. A bullet in the back struck him down on the bank of the River Kura.

* * *

A task we were especially concerned with at that time was the organizing of youth. The Communist youth organizations of Transcaucasia were in a disorganized state, while the Communist youth organization of Georgia and Armenia, which had been in existence since 1917 under the name Spartak, worked under difficult illegal conditions. Azerbaijan had its own Communist youth organization, which was called the International Union of Young Workers of the city of Baku and its various districts.

We were aware of the weakness of our work among the youth and we proposed that the leaders of the Baku youth get in touch with the leading comrades from Spartak and take the necessary steps to convene a Conference of Transcaucasian Communist Youth. Such a conference should, in our opinion, be held in Baku because as a sufficiently strong organization of the working class, we were able to ensure that the conference would take place under optimal conditions, and also, holding the conference in Baku would help to raise the level of party work.

Our proposal was adopted, and on September 22, 1919, an illegal Conference of the Communist Youth of Transcaucasia opened in the Workers' Club. Thirteen delegates attended, representing 1300 members of the Communist Youth Unions of Azerbaijan, Georgia, and Armenia. Of course, these numbers may seem very insignificant now, but for those days it represented a considerable political force. And the conference demonstrated this with dignity and honor, for despite its small numbers the conference began and ended on a note of high political enthusiasm.

With great fervor the delegates chose Vladimir Ilyich Lenin as the Honorary Chairman of the conference, and they chose Lunacharsky and Mikha Tskhakaya as Honorary Members of the presidium. Karayev, Lominadze, Agayev, and Suren Agamirov from our regional party organization took part in the work of the conference, and I welcomed the delegates on behalf of the Party Regional Committee and gave a report on the tasks facing working class and peasant youth.

The primary focus of the conference was on the reports from the provinces. This was very important, since the delegates were poorly informed about the situation in the various districts.

Of those delegates attending the conference from Georgia, I was well acquainted with Dzneladze and Garegin Gardashyan. We had worked with Dzneladze in Marxist youth circles even before the Revolution. Together with him we had fought for the Bolshevik line at the conference of these circles in March 1917 in Tiflis. He was a capable, well-principled Bolshevik, who had devoted himself completely to the revolutionary struggle. Our meeting was a great joy for both of us in view of our long separation.

I was also happy about meeting with my school friend Gardashyan. He was doing active political work even in those days and had played a rather prominent role in the Tiflis organization Spartak. After the victory of Soviet power Gardashyan went on to study medicine and subsequently became a talented surgeon and a close associate of the surgeon Rozanov, who was well known throughout the entire country. He later became People's Commissar of Health for Armenia.

It was at the conference that I first met and became friends with Gukas Gukasian, who represented the Armenian Spartak. In addition to his official report at the conference, he told me in detail in a personal conversation of those oppressive conditions under which the Communists and the members of Spartak had to work in Dashnak Armenia.

Gukasian was a thoughtful and already rather mature Communist. He had traveled widely around the districts of Armenia, setting up groups of Spartak, at the same time that he carried a heavy load of party work. He subsequently played an outstanding role in the struggle for the victory of Soviet power in Armenia and died the death of a hero. The Communists and the Komsomol members of Armenia honor his memory even today, and one of the districts of the Republic bears his name.

Karayev welcomed the Conference of Youth on behalf of Gummet. In his speech he spoke in detail of the work done among the young workers and peasants of Azerbaijan, and he pointed out the unsatisfactory state of this work. This issue was also touched on by Agayev, who gave a welcoming speech on behalf of Adalet.

In my report I threw some light on the international situation of Soviet Russia, and also the internal situation and the general conditions at the different fronts. This was the time when Denikin was gaining victories over the Red Army in the south of Russia. In the east the Red Army was advancing into the Urals and Siberia, routing the Kolchak hordes. To the approval of all those present, I expressed my complete confidence in the victory of the Red Army over Denikin, declaring that the same fate that befell Kolchak awaited Denikin.

When I spoke of the political situation in Transcaucasia I lingered over the fundamental tasks which the Party Regional Committee, in preparing for an armed rebellion, had put before the Bolshevik organizations, the Komsomol, and all the working youth of Transcaucasia. I called upon the Komsomol members to intensify the struggle against the nationalist blight, which had then infected a fairly considerable segment of the youth of the Transcaucasian nationalities.[25]

I also spoke of the need for special caution when accepting young workers, peasants, and students into the Komsomol. "Don't forget that for the present we are working in illegal conditions. Don't forget that we still have before us a very difficult struggle with all manner of anarchistic tendencies and nationalist deviations."

In the course of the conference certain disagreements turned up which reflected the different economic and political conditions under which the delegates worked. On one side there were the members of Spartak, and on the other there were the members of the Baku Council of Youth. These disagreements were, to some extent, a result of the fact that they worked independently, in isolation from each other. "There is no doubt," I maintained, "that after you have been at this forum of Transcaucasian Communist youth you will leave here closer in your views and with your ranks strengthened, both in organizational terms and in a political-ideological sense as well."[26]

"The lack of unity among the national revolutionary organizations," I concluded, "is harmful to the proletarian movement, it splits its ranks and opposes the workers of one nation to another, which plays into the hands of the bourgeoisie. We respect the national languages, but for us language is primarily a means of communication. You can speak in different languages, but the organization must be united and international. For this reason your primary task is to impart to youth an international, proletarian consciousness, a readiness and an ability to struggle with nationalistic enmity and to strengthen fraternal friendship between the youth of different nationalities."

Lively debates arose at the conference. The delegates spoke several times. This was useful in clarifying the different points of view. Everyone was conscious of the fruitfulness of the discussions, and despite

[25]See Appendix.
[26]See Appendix.

the differences of opinion the conference unanimously passed a resolution to unite all the Communist youth organizations of Transcaucasia into a single Transcaucasian organization of the All-Russian Communist Union of Youth (Komsomol). The Regional Committee of the Transcaucasian Komsomol was chosen in the same spirit of unanimity.

The work of the conference concluded with the issuing of a declaration to all the working class and peasant youth of Transcaucasia, appealing to them to close ranks in the struggle for Soviet victory and for the triumph of the socialist revolution. This declaration stated, in particular:

> The Transcaucasian Conference of the Unions of Communist Working Class and Peasant Youth has taken into account the existence of extremely intricate national interrelations and the deeply rooted chauvinistic nationalism of the local population, which is always assiduously being kindled by the cries of the Transcaucasian nationalists from bourgeois-landowner circles, and it finds it necessary both in the interests of the union itself (i.e., the Komsomol) and in the interests of the broad strata of Transcaucasian youth, to wage a merciless battle with chauvinistic passions which eat away at the consciousness and emotions of man; it obligates every member of the union to carry on a never-ending crusade for the ideas of internationalism and the brotherhood of all nations. Considering itself to be part of the working class-peasant youth, the Transcaucasian Regional Conference recognizes the ideological authority of the Communist party alone as the brilliant mouthpiece of the workers' and peasants' interests.

The conference received greetings from the Third Communist International, the Central Committee of the Komsomol, and Mikha Tskhakaya. While the Youth Conference was going on, Mikha Tskhakaya was in Kutaisi prison, where the Menshevik government of Georgia had interned him along with a large group of Communists, because it feared the tremendous authority and influence enjoyed by him and his friends among the population.

It said in its own message of greetings: "The Conference of Proletarian Youth sends greetings through you to all the Communist proletariat of the Caucasus, and it expresses its conviction that those oppressors and traitors of the Caucasus, who have dared to insult the Caucasian proletariat by arresting the pioneer of the proletarian revolution in the Caucasus, Mikha Tskhakaya, will not longer be masters of the situation, for in the very near future there will be flying over the Caucasus the Red banner of the proletarian dictatorship and the Communist International, for the ideas of which you are the fighter. Long live the Soviet Caucasus!"

Chapter X

Moscow

On a mission for the Regional Committee—My first meeting with Kirov—Reporting to Lenin—At the Second All-Russian Congress of Communist Organizations of the East—E. D. Stasova—Some conclusions.

By the middle of 1919 the party organizations and the revolutionary movement in Baku had recouped their strength to a considerable extent after their defeat in 1918.

As had been correctly noted in our letter to Lenin, the formation of "independent" national bourgeois states, which appeared to be a disaster for the proletarian movement in the Caucasus and in Soviet Russia ". . . had beneficial effects in terms of clarifying and crystallizing the class consciousness of the masses . . . The process of delineation on the basis of different nationalities and uniting all a nation's forces around the idea of an independent national republic gave way abruptly to a thoroughgoing process of unification of the worker-peasant masses under the aegis of the Communist party. The shift away from Soviet Russia toward one's own bourgeoisie gave way to a break with one's own bourgeoisie and a surge toward Soviet power and Soviet Russia."*

The Caucasian Party Regional Committee met in Baku at the end of May on behalf of the All-Caucasian Party Conference for a plenary session of its newly constituted membership, and after a detailed discussion, resolved not to come out against the national republics but, rather, to consider the next order of the day to be the overthrow of the governments of the existing republics and the establishment of Soviet power within their boundaries; the creation of the Soviet Republics of Azerbaijan, Georgia, and Armenia, and their voluntary unification with Soviet Russia in regard to military and economic matters.**

*TsPA (Central Party Archive) IML (Institute of Marxism-Leninism) f. 5, op. 1, ed. xr 745, line 2.
**TsPA IML, f. 5, op. 1, ed. xr 745, line 4.

First of all, it was necessary in this connection to make preparations for an armed rebellion in Azerbaijan, Georgia, and after that in Armenia, to set the timing for these rebellions and to coordinate them with the advance of the Red Army. Issues relating to party organization came to the fore and acquired even greater importance. There was no unified opinion regarding these issues, which complicated the situation even further.

Since all of this combined, together with the existing situation, demanded new guidelines on the part of the Caucasian regional organizations, we needed to make a trip to Moscow where we could report on the current state of affairs and receive the necessary instructions. Moreover, for more than a year after the defeat of the Bolsheviks in the Caucasus not one member of the Regional Committee had been in the party Central Committee. The Caucasian Party Regional Committee resolved that I should go to Moscow to report to the Central Committee on its behalf. I looked forward with relish to a personal meeting with Lenin and I have to confess that I was delighted when my comrades put my name forward as the one to go. It was known in Baku that the Seventh All-Russian Congress of Soviets was to convene in November (it, in fact, took place in December 1919). On September 26 the presidium of the Workers' Conference of Baku and its districts issued the following mandate:

> This mandate is issued by the presidium of the Central Workers' Conference of the city of Baku to Anastas Mikoyan, who is delegated to attend the All-Russian Congress of Soviets, scheduled to take place in the month of November.*

There was only one route to Moscow: a five- or six-day voyage on the Caspian Sea north to Astrakhan on one of the fishing boats in the fleet the Bolsheviks had organized to transport airplane fuel for Soviet Russia from Baku. Our boat was to leave from a pier on an avenue in the center of the city. We had managed to procure official documents for five passengers allegedly en route to Persia, but the first attempt, on October 8, fell through when the police got wind of our trip.

This is what Shura Bertsinskaya recalled of the incident:

> Olga Shatunovskaya, Skachko, Tigran, and one other comrade were going with Anastas. The boat was to leave from a pier right in the middle of the city. It was evening. Four of our group were already on the boat and were waiting for Anastas, who was to arrive at the last minute for security reasons.
>
> The police showed up unexpectedly and everyone was arrested, but they weren't taken away. It was obvious that the police knew

---

*TsGAOR, USSR, f. 1235, op. 6, ed. xr. 81, line 77.

about Anastas and were waiting for his arrival. Someone had betrayed us.

Olga and Tigran were preoccupied with only one thought: how to warn Anastas? The most important thing was to prevent his boarding the boat. When they were taking Olga down the street along the quay to the police station, Anastas, who was heading in their direction, didn't catch sight of the policeman immediately, and was about to shout out to Olga but she walked quietly past. Tigran asked that he be taken out off the boat to urinate. When he disembarked he looked around for Anastas, moving from place to place in order to gain time. When Anastas showed up a minute or so later he caught on at once to what was happening, and addressed Tigran with a cigarette in his teeth, "Have you got a match?" "No, I don't, get going," Tigran replied. The police urged Tigran to finish his business quickly. He showed no interest whatsoever in the passerby who wanted a light for his cigarette, beyond muttering, "Get going, get going!" Thus, without waiting for the one remaining passenger who hadn't shown up yet, the police took the four who had been arrested off to the police station.

It was nighttime by now and they had to wait for the chief of police who was detained, it turns out, at a party at the governor's. In the meantime a police officer carried out the interrogation. Olga made herself out to be the widow of a merchant who had suddenly died in Persia. The faithful wife was traveling to Persia to recover the body of her beloved husband. She wept copiously and bemoaned the fact that she had been deprived of her only consolation in life. Skachko was a merchant trading in tobacco who demanded that he and his assistant (Tigran was cast in this role) be released, and threatened that all the losses he incurred because of being detained would be charged to the police.

The tears, cries, threats, and apparent wealth of those arrested (they were all very well dressed) caused the policeman some consternation, but he couldn't do anything. "Ladies and gentlemen," he said, "the chief of police will come and release all of you, but I cannot, I haven't the authority. The chief said we had to wait for him." After managing to find a place for themselves some way or other, the detainees finally dozed off. Suddenly in the middle of the night an order was given: "You're all free, you can go." Evidently Anastas had arranged for a sizeable bribe to be given to the policeman.

They left the police station and immediately scattered in different directions down the narrow and confusing alleys of the Baku fortress. Almost at once there was the sound of shouts and the tramping of feet—the chase had begun. But it wasn't difficult to hide in the labyrinth of the fortress. We found out later that as soon as the prisoners had been released, the chief of police had showed up and sent the policemen rushing out to look for them, but no trace was ever found of this motley group—the angry merchant and his assistant, and the weeping widow.

After this aborted trip a new group was given the task of planning a second attempt. This time Fedya Gubanov, the president of the Water Transport Workers' Union, was to make the trip along with me to Astrakhan, and we decided to leave in separate boats, a day apart from each other, from another pier. We figured that if one boat didn't make it, then perhaps the other might get to Astrakhan.

Several days later, as arranged, I left Baku illegally, from a different pier, in a fishing sailboat which also had a motor. I posed as an ordinary tradesman taking tobacco to Enzeli (Persia). I arrived just before the boat was to leave, carrying a huge watermelon in my arms for cover. With me were two Baku party workers who were also dressed as tradesmen, of whom there were many around the pier. There was also a representative of the mountain peoples who was going to see Kirov in order to establish contact with the Eleventh Army and to enlist support for the rebel detachments then operating in the Caucasian mountains.

Our departure took place on a bright, sunny morning. This time we embarked without attracting any attention and managed to pull away from the pier without complications. After clearing the coastal waters and getting out into the open sea, we plotted a course, naturally, not for Persia, but for Astrakhan. Usually everyone who goes to sea hopes for calm weather and quiet seas, but on this trip we, on the contrary, dreamed of stormy weather and gale-force winds.

The trip ahead was long and dangerous. Since the motor was very weak, all our hopes lay in the sails, but we couldn't get anywhere in calm weather. Moreover, we knew that Denikin's warships which dominated the Caspian were patrolling the sea lanes night and day in good weather, whereas they stayed in port as a rule in bad weather. Luckily for us, a gale blew up shortly after we left the port. It was so cramped and stuffy in the one small cabin, however, that we went up on deck and braved the showers of spray to breathe in the fresh sea air.

The boat was small and the waves huge. When you sat in the stern and looked behind you it seemed as it a four-story wall of foamy water were about to fall upon you and swamp you. But our boat, as if obeying some mysterious force, rushed forward, climbing swiftly up the crest of the wave and then sinking from it just as swiftly, only to repeat the whole process all over again. At first it was terrifying to feel we were about to be engulfed by the watery abyss, but then we gradually got used to it, and even began to feel proud that man could successfully combat the raging elements.

The first serious danger we would confront was the stretch of the journey from Fort Alexandrovsk (now Shevchenko) to the Kizel coast.

This was the narrowest mouth of the Caspian where any passing craft could not fail to notice our boat. It was either on the third or fourth day when we approached Alexandrovsk. By now the weather had improved considerably so we decided to head for the nearest empty cove in the Kara-Bogas Gulf, wait for the appropriate (rougher) weather, and continue our journey at night with a favorable wind, passing through the dangerous strip and entering the Volga delta.

In the cove we met an old Kazakh and a boy of about twelve who were the only inhabitants there. The old man earned his living as a fisherman, and it turned out that he hadn't seen a crust of bread for about five months, so we gave him a few loaves from our supply. The fisherman treated us hospitably. He slaughtered a sheep, grilled the meat on an open fire, and cooked us a delicious fish soup. We had a good rest, and when it was dark we resumed our voyage. The old man insisted that we take some fish with us for the trip, which we were glad to have.

We slipped past Aleksandrovsk without being noticed and sailed on for almost an entire day. As we got closer to the Volga we noticed that the color of the sea water gradually changed from dark blue to a yellowish cast—the waters of the Volga were already showing through. We were now near the mouth of the Volga, but we knew that Denikin's warships were especially diligent in their patrol of the entrance to the Volga delta. However, as the depth of the water began to decrease the dread chance of meeting up with a large enemy ship became less and less.

When we had begun our journey we had prudently hidden three combat rifles and several Mausers and grenades in the boat in the event that if the Denikinites seized us, we would prove ourselves worthy opponents and would not sell our lives cheaply. Thus we approached the Volga delta in a state of combat readiness, scanning the horizon vigilantly for Denikin's ships. The horizon was clear, no ships could be seen. Then, just before the sun set we saw a ship in the distance. It was moving toward us very quickly and there was no way for us to tell if it were a White ship or one of ours, a Red. A warning shot sounded across the waters and an order was given for us to raise our flag since we had been traveling without one.

I thought for a minute and then ordered that a white flag be raised. My comrades protested, "Why should we raise a white flag when we're Communists?" I explained that a white flag was good because it signified a formal refusal to put up any resistance. The ship heading toward us would not commence firing when it saw the white flag. It the ship turned out to be ours, a Red, then everything would turn out all right. If, however, it turned out to be one of Denikin's ships, then we could at least take up arms, destroy as many of the enemy as possible, and not be taken alive.

The white flag was raised. We stopped, lowered our sails, and turned off the engine, waiting nervously for the ship to approach. This was one of the most tense moments, with everything at stake. As the ship drew near we scrutinized it to see whose it was, trying to ascertain if the officers had shoulder straps. We didn't see any, which meant that they were Reds! We put down our arms and when the ship drew alongside, three sailors without any badges of rank jumped on board our boat; one of them was the captain.

I introduced myself and said that I was going from Baku to Astrakhan on a mission to Kirov, but that I couldn't go into any details. "Take us to Astrakhan, to Kirov, and you can have everything in our boat as trophies, including our large stash of tobacco." I said. The sailors were very pleased with this offer; they hadn't, as I'd suspected, had any tobacco for a long time and had had to make do with all manner of substitutes. We changed over to their ship, and the next day, October 16, we arrived in Astrakhan.

* * *

The first time I was taken to Sergei Mironovich Kirov's apartment he wasn't home. When I asked where he was, an old woman, evidently the landlady, replied, "He's been at the Soviet since morning, giving speeches." Kirov was, in fact, in the middle of a difficult situation at that time. Astrakhan was a hungry city, with not enough to eat and problems with rationing. Kirov had to rush everywhere, explain things to everyone, clear things up, and most important of all, raise the spirit of a hungry people. Later on, walking the streets of the city I saw people here and there trading in dried apples and occasionally dried fish. There was nothing else to be had.

It was several hours later before Kirov showed up for what was to be my first meeting with him. Until now we had known each other only through a spirited correspondence which had started about a half-year earlier, but we greeted each other like longtime friends. I told him in detail everything I intended to say at the Party Central Committee. Our discussion dwelled in particular on concrete ways of helping the Caucasian mountain rebels, and we also planned ways to increase the gas shipments from Baku to Astrakhan.

A lively, inquisitive, and intelligent man, with a clear, precise mind, Kirov could grasp the subtleties of an issue in an instant. His positive attitude toward our policy on issues was encouraging and convinced me that questions of concern to us would be successfully examined in the Party Central Committee. He astonished me with his efficiency and drive, and his ability to make decisions on the spot. It was clear that all channels of military, governmental, and party work led directly to him, and that he carried his leadership role skillfully, relying on the trust of his comrades and enjoying authority among them. We got to know each other well while I was in Astrakhan, and became lifelong friends.

As I remember him, the Kirov of those days was an exceptionally self-disciplined and reserved person with a very strong character. He had an extraordinary ability to win people over, even with his appearance; he was of a rather short, husky build, with a very likable personality, a distinctive voice, and an unusual gift with words. When he spoke from a podium he could captivate the great masses of his audience. In personal conversations he spoke sparingly, but expressed his ideas very clearly and was a good listener when others were speaking. He loved a pointed remark and was himself an excellent storyteller.

It turned out the Sergei Mironovich Kirov was in constant touch with Lenin over the telegraph, informing him regularly about what was happening, asking for instructions, and passing on to the Moscow center the news which came into Astrakhan from the Caucasus. On my first day in town he sent Lenin a long telegram about my arrival in Astrakhan and my forthcoming trip to Moscow, informing him of all I had told him concerning the scale of the uprising of the highlanders in Chechnya under the leadership of N. F. Gikalo, and of the measures which Denikin was planning to take in connection with this.[27]

Kirov and I decided that I should stay in Astrakhan for a few days; I wanted to wait for Fedya Gubanov so that we could go on to Moscow together, as agreed. While in Astrakhan I found out unexpectedly that there was a large group of Armenian Communists there, headed by Aikuni, who planned to leave for the Caucasus. A meeting of these Armenian comrades was arranged at which Aikuni was to give a report and I was to give a counterreport.

In my address to the group I declared that the Communists of Armenia and the Transcaucasian Party Regional Committee did not recognize the Central Committee of the Communist Party of Armenia, headed by Aikuni. He and his group had not been elected by the Communists of Armenia and had no contact with the local party organizations there, whose work was now led by the recently formed *Armenkom*, which didn't recognize Aikuni's group. The Communists of Armenia were part of the Transcaucasian party organization and recognized the leadership of the Party Regional Committee, with which Aikuni and his group had nothing in common. This, I said, is one of the manifestations of the nationalistic tendencies of the group which has split the ranks of Transcaucasian Communists, and undermines their unity.

I claimed that the comrades who planned to go to the Caucasus could count fully on the friendly welcome and support of the Communists of Armenia and the Transcaucasian Regional Committee, if

[27]See Appendix.

they would join the ranks of the local party organizations in a calm and organized fashion, instead of following Aikuni. To my surprise the overwhelming majority of Communists present at the meeting supported me.

During the following days I waited anxiously for Fedya Gubanov to appear, fearing the he might have fallen into the clutches of the Denikinites. Our fears were confirmed, for we soon learned that the boat had been seized by the White Guardsmen, and that Gubanov had been arrested and soon after had perished. I had no alternative now except to make the trip alone. It was not an easy matter to get to Moscow, however, for train service was irregular and sporadic, not more than once a week, and unless the opportunity presented itself, one could not manage it.

"Such an opportunity has arisen," Kirov told me one day. "A member of the Revolutionary Military Council of the Republic, named Smilga, is arriving here in a few days on his own train with a group of military workers. He'll spend two days in Astrakhan, and then you'll be able to leave with him for Moscow."

And that's what happened. On October 26 I left for Moscow in the same train with Smilga. Kirov notified Elena Stasova of my departure for Moscow in a telegram on October 24, and in telegrams to Lenin and Stasova on October 27.

We reached Moscow in about two weeks' time. Railway transport was then in catastrophic shape. There was not enough fuel, the rolling stock was damaged, and there was no proper order on the railways, with unscheduled stops all along the way. During the endless journey I continued to mull over my report to the Central Committee of the party. I thought a great deal about my forthcoming meeting with Lenin, which filled me with excitement.

My arrival in Moscow came at a time when the situation was very difficult for Soviet Russia. The civil war was in progress, with counterrevolutionary revolts flaring up all over the country and famine and epidemics raging everywhere. The capital city was living a very strained existence.

The Central Committee of the party was then housed in a building on the Vodzdvizhenka (now Kalinin Prospect). I was sent to the room where Elena Dmitriyevna Stasova worked. When I entered the small and, as I recall, rather dark office, a tall blond woman of aristocratic appearance was standing at a table talking with some comrades. She turned when she heard the door close, and motioning to her colleague to stop talking she inquired whom I was looking for and in regard to what. I introduced myself and said that I had to see the Secretary of the Central Committee, Comrade Stasova, to which she replied with a brief: "That is I."

Elena Dmitriyevna already knew of my impending arrival from Kirov's telegram. She smiled cordially, and asked me to sit down and wait while she finished her conversation with the comrade. A few minutes later I was answering Elena Dmitriyevna's questions.

* * *

After she had inquired about my journey to Moscow she sent me to Vladmir Ilyich in the Kremlin. Lenin, she said, had given instructions that I was to be sent to him as soon as I arrived. That same evening he received me in his office, alone. When I opened the door to Lenin's office, he got up from his desk, smiling cordially and screwing up his eyes, and walked over to meet me. He pressed my hand in a friendly grip and asked me how I had come. Then he offered me a chair next to his desk and returned to his seat. "Tell me, tell me . . ." he said. When I began to speak he was immediately transfigured, he became totally absorbed in what I was saying, the smile left his face, and his eyes became serious and intent.

I had heard that Lenin was a simple man, but I had not imagined him to be the way he was. He immediately created an atmosphere of a relaxed, businesslike conversation. At first I was extremely nervous, but soon I pulled myself together and began to report to him boldly and without embarrassment.

I told him of the great successes of the Bolsheviks of Azerbaijan during the past six months since the spring of 1919; how the Baku proletariat had rallied around our party, and how underground party work had been combined with legal forms of activity; how there was a permanently functioning Workers' Conference in Baku consisting of delegates from the different plants—something like the Council of Deputies—that was the strongest organ and the virtual leader of the workers' movement in the city. In March of the current year we, the Communists, had managed to oust the SR—Menshevik majority from the presidium of the conference, and occupy the leading position in it and in the presidium.

* * *

I spoke of how the Baku unions had also been in the hands of the Mensheviks in the beginning of 1919, how they had been very weak organizationally and ideologically, and even helpless at times, but we had managed nonetheless to occupy the leading position in the unions, and also in the workers' clubs of the Baku districts, turning them into bases for the expansion of mass political work among the workers, and into communication centers and secret meeting places for party organizations. I told about the All-Caucasian Congress of Trade Unions and how it had passed a resolution which had been proposed by the Communists, noting that at this congress there had been representatives from only a few unions in Georgia because the unions, which were under the leadership of the Mensheviks, had seen that they would be in the minority at the congress, and so they had not sent their representatives but had, instead, organized their own all-Caucasian union organization.

I deemed it necessary to tell Lenin how we had managed, despite the efforts of the Mensheviks, to receive a majority at the reelections of the board of the Caspian Cooperative Association, and to transfer control of it to our hands so that this Cooperative Association had become one of the mainstays of our party work. Its enterprises were used to organize unemployed Communists, to send people out on party assignments under the guise of doing work for the Cooperative, to organize contacts, set up secret meeting places, etc.

While I was still en route to Moscow and thinking about how best to make my report to Lenin, I had decided that I would first set forth the facts and then analyze them, making generalizations and drawing conclusions, and that's exactly what I did. Vladimir Ilyich listened to me avidly when I dealt with facts, but as soon as I tried to draw any deductions, he interrupted me politely, got up, walked around the room, and asked me for additional facts. He would go over to the map: "Now, let's see, where's Dagestan and Chechnya?" We would look. "How many detachments are there there?" I would answer and again begin to generalize, and again Lenin would ask questions aimed at defining the issue more precisely. It was only then that I realized that I should state the facts in more detail and not try to draw any conclusions from there: Lenin could do that himself, and far better than I.

I gave Lenin a report about the political situation in the Transcaucasian Republics, in Dagestan, Chechnya, Ingushetiya, and Kabarda. I noted that a revolutionary situation was shaping up in Azerbaijan, where the struggle against the bourgeois government and the landowners for seizure of the lands as well. The Communists had managed to seize control of this movement for themselves.

After the tragic demise of the Lenkoran Soviet Republic, the strongest antigovernment movement had developed in the district of Kazakh and in Karabakh, whereas in Dagestan, Chechnya, and Kabarda a partisan struggle was being waged against Denikin's army. The mountain peoples of the Caucasus had formed a government, through the efforts of the Georgian Mensheviks, and with the support of the Muscavatist government, but it had no authority in actual fact, and the decisive power there was the Red partisan detachments. In the Kizlyar region and especially in the districts of Novorossiisk and Tuapse detachments of so-called Green Guardsmen were operating, which consisted of dissatisfied peasants, former northern Caucasian Red Army men, and deserters from Denikin's forces. They were carrying out military operations against Denikin. The Party Regional Committee was taking strenuous measures to establish contact with them in order to take advantage of their services in the fight with Denikin's army, and to destroy the enemy's rear.

"What is the economic situation in the bourgeois republics of Transcaucasia?" Lenin asked.

I replied that the situation in Georgia was relatively good. Azerbaijan possessed large reserves of already extracted oil, but was deprived of a Russian market and was going through a depression because of the impossibility of exporting and selling oil. The most serious situation was in Dashnak Armenia. Given the general poverty of the population, the absence of industry, and the insufficient amount of arable land, the economic situation was catastrophic. Around 300,000 Armenians—refugees who had come from western Armenia along with the retreating Russian troops—found themselves in terrible straits. The widely publicized aid of Britain and America had been, in fact, insignificant, and the people were starving.

Lenin asked what were the mutual relations between the various national governments of Transcaucasia. I answered that they bickered among themselves and were torn by territorial disputes. The attempts made over the summer at the meeting of the different governments to pave the way for cooperation had not yielded any results.

In Georgia, despite the Menshevik government, the combination of internal social conditions was developing to our advantage. The majority of the active workers and peasants, and individual subunits of the regular army were not only well-disposed to us, but were even ready to revolt at a sign from our party. In certain districts and military units it wasn't even always possible to stem premature, uncontrolled outbursts. That's what had happened in places like Poti, Mikhailov, Lanchkhuty, Sukhum, etc.

In Armenia both the economic and political situations were difficult, and the Communists were subjected to severe persecutions and were forced to engage only in underground party work, since there were no opportunities for legal work. The inclination toward Russia was very strong among the popular masses and those dissatisfied with the existing situation. In the event of a victory over Denikin the laboring population of Armenia, including the Moslems, would join the struggle for Soviet power.

Suddenly Lenin, grinning slyly, asked: "And how do the Georgian Mensheviks implement democracy?"

I replied that there was no democracy in Menshevik Georgia, not even a whiff of it. On the contrary, there they savagely suppressed the peasant revolts which had broken out in some districts. A large number of our party comrades had been arrested and were being held in prisons without investigation or trial. Mikha Tskhakaya, who was well known to Lenin from exile in Switzerland, had been arrested and was in Kutaisi prison. Naturally, there was no question of freedom of the press. The Bolsheviks could publish neither newspapers nor pamphlets, and worked underground.

Lenin reacted very heatedly to my account. "I'm well acquainted with these Mensheviks," he said. "You can't expect anything else from them!"

I said that in Georgia one newspaper with a Bolshevik orientation had lasted only two weeks before the Mensheviks closed it down. That happened in the days when we had appealed to the petit bourgeois parties and the national bourgeois governments with a proposal to form a united front against the Denikin threat, promising them all possible assistance in this regard. Our appeal found a response in the broad masses of workers, under whose pressure the Georgian Mensheviks agreed to call an all-Caucasian workers' congress to organize the struggle against an invasion by Denikin. We took advantage of this period to make legal speeches among the workers of Tiflis, but when the Mensheviks saw that this campaign was adding fuel to the Bolshevik fire, they managed to get guarantees from the British that they would not permit Denikin to invade Transcaucasia, and after this they cut off all negotiations with us. Thus began a new wave of repressions against the Communists and an onslaught of anti-Bolshevik propaganda.

I told Lenin how the party organizations of Transcaucasia were successfully preparing for armed rebellion and were awaiting the instructions of the Central Committee as to when it should take place, since we wanted to make the rebellion coincide with the arrival of the Red Army in the Caucasus. I went on to say that in addition to the RKP(B) there were two other Communist organizations in Azerbaijan: Gummet, which united the local Azerbaijani workers, and Adalet, which united the refugees from Persia, but while Gummet and Adalet had played an important role in attracting the Moslem workers into the Communist movement, it was necessary now to have a united Communist organization for everyone, irrespective of their national affiliation.

Insofar as we had advanced the slogan of creating an independent Soviet Azerbaijan, closely aligned with Soviet Russia, it was the Communist party of Azerbaijan, uniting all the Communists of the Republic and forming part of the RKP(B), that had to become such a united organization.

I told Lenin that this unanimous opinion of the Communists of Azerbaijan was in contradiction to the resolution of the Central Committee that had been passed in July, to the effect that Gummet was recognized as an independent Communist party of Azerbaijan, with the rights of a party regional committee. I said further that the Communists of Armenia and the Caucasian Regional Committee of the RKP(B) did not recognize the Central Committee of the Communist Party of Armenia, which was headed by G. Aikuni and was then based in Moscow. He and his group had no contacts with the local party organizations of Armenia.

The Communist organizations of Armenia had elected a committee under underground conditions which had been given the name of Armenkom, and which was to lead the work of all the Communists of Armenia. The Armenkom formed part of the Caucasian regional party organization and thus part of the RKP(B). It recognized the leadership of the Regional Committee and did not recognize Aikuni's group. This group, in its turn, recognized only the direct authority of the Central Committee of the RKP(B), and rejected the leadership of the Caucasian Party Regional Committee, did not wish to join the Caucasion regional organization of Communists, and thus disrupted the ranks of Caucasian Communists.

I told Lenin about my meeting in Astrakhan with the group of Armenian Communists, including Aikuni, who planned to go to the Caucasus. I told him I had spoken against Aikuni's policies, and that everyone present had adopted a resolution proposed by me, and that Aikuni had remained in isolation and had decided to return to Moscow.

"The Caucasian Communists cannot think of victory without the unity of the Communists of the Caucasus," I said.

Out of my naiveté I thought that in the course of the conversation or immediately upon its conclusion Lenin would give me specific answers to the questions I had raised, and would express his opinion and thoughts concerning them. Lenin said that all these issues concerning the Communist parties of Azerbaijan and Armenia, and their relations with the Regional Committee, and also all the other issues I had raised, had to be studied and discussed in the Central Committee, and then resolutions would be passed regarding them. In order to make this possible, I had to set forth everything in written form, which I later did.

Lenin said that the unification of the Communists in one organization according to the territorial-production principle, irrespective of their national affiliation, was completely correct. It was also correct that in the independent states which had been organized on the borders of Russia the Communist organizations should work as independent Communist parties forming part of the RKP(B).

I have never forgotten this first meeting and conversation with Vladimir Ilyich Lenin.

* * *

After my conversation with Lenin I returned to Vozdvizhenka Street, where I was sent to the Third House of the Soviets on the former Bozhedomsky (House of God) Lane, now Delegate Street. (Afterwards for many years the Presidium of the Supreme Soviet and the Council of Ministers of the RSFSr were in this building.) It was my first time in Moscow, and since neither the tram cars nor any other form of city transport was in operation, I was given a car to take me about.

The next morning I managed to find out where Shaumian's family was living. It turned out that they were occupying an apartment in the courtyard of that same building, and I went to see them. Ekaterina Sergeevna Shaumian insisted that I move in with them. So, once again, as in Baku, Shaumian's home became my own.

I had another meeting that morning with Stasova at which I spoke in detail of the state of affairs in the Caucasus and of those issues which the Regional Committee was going to put to the Central Committee for consideration. Then, on Stasova's recommendation, I went to see K. T. Novgorodtseva, who was currently in charge of the Central Committee Organizational Department, and gave her all the information as well. She impressed me very favorably as a serious, thoughtful party worker, who had a simple manner and was able to win over those she talked to. I found out later that she was the wife of Ya. M. Sverdlov, the secretary of the Central Committee and the president of the All-Russian Central Executive Committee (VTsIK), who had passed away at the beginning of the year.

During the following days I relived and reflected upon my conversation with Lenin. I had been struck by the fact that Vladimir Ilyich had been interested in the most minute details of all the issues relevant to the Caucasus. He had asked innumerable questions about what was happening in Baku and in Azerbaijan as a whole, and in Georgia, Armenia, Dagestan, Chechnya, and among the other peoples of the Caucasus. You wouldn't think he had time for such things!

For this was a critical time when both the internal and external situation of the Russian republic continued to be in jeopardy. At the end of the year, according to facts presented by the Central Statistical Board to Lenin (TsSU), Soviet power had spread to seventy-five million people, that is roughly half the population of Tsarist Russia, in thirty-three provinces of the Republic. These were principally the consumer provinces.

The British troops were advancing to the south, to Vologda from Arkhangelsk and Murmansk, while Yudenich's White Army was outside Petrograd, and after taking the offensive at the end of September it was closing in on the city by the middle of October. A turning point occurred at the end of October on Denikin's primary front, however, checking the movement of Denikin's hordes to the north and throwing them back from Oryol, thus relieving the direct threat of an invasion of Moscow by Denikin's troops. However, at the end of the year the situation was still critical and complicated. In the middle of October the Soviet troops had been forced by Kolchak's troops to retreat to Tobol. The Soviet Republic was virtually deprived of their sources of raw materials, fuel, and grain, and its defense factories had, with few exceptions, come to a standstill. Metal production was down to a minimum. Communication with Soviet Turkestan had been severed by Dutov's Cossack troops, as a result of which Russia's cotton supply was curtailed.

How great was Lenin's faith in victory if at such a difficult time he found it possible to spend more than two hours questioning me about what was happening in the far-off Caucasus, about what issues were of concern to the local Communists, and how we planned to resolve them!

* * *

In March 1919 the Communist International was created. Lenin had spoken of the necessity for this as far back as his April Theses. The First Congress of the International which took place heralded the beginning of the ideological and organizational unification of all Communists of all countries under the slogan of the dictatorship of the proletariat. Herein constituted the fundamental distinction between the Comintern and the international association of Social Democrats. But there was one more essential difference: at the Comintern Congress there were united for the first time in the history of the revolutionary movement Communists from a country which had been triumphant in a socialist revolution, and also the proletarians of Europe and America, along with the toilers of the colonial and dependent countries. The Congresses of Communist Organizations of the East which were being held at that time also played an important role in this scheme.

On November 22, 1919, the Second All-Russian Congress of Communist Organizations of the East opened in Moscow. I wanted to attend the congress as a guest in order to gather first-hand information on the opinion of these organizations concerning the nationality issue. The congress convened in the Kremlin, not in Sverdlov Hall, but in Mitrofanya Hall, which held around eighty delegates.

Lenin's appearance on the opening day of the congress came as an unexpected and delightful surprise to everyone. On welcoming the delegates, Lenin said immediately that the present most vital issue was the revolutionary movement of the peoples of the East against imperialism. He declared that this movement "could now develop successfully, and reach its resolution only if it were in direct association with the revolutionary struggle of our Soviet republic against international imperialism."*

While speaking in detail of the victories of the Red Army and of

*V. I. Lenin. Complete Works, Vol. 39, p. 318.
**Ibid. p. 319, 321.

the turning point in the civil war, he added that "these victories which we are now winning over Kolchak, Yudenich, and Denikin represent the onset of a new phase in the history of . . . imperialism against the countries and the nations which have begun a struggle for their own liberation." And further on: "I think that everything the Red Army has accomplished, its struggle and the history of its victories will have utmost significance for all the peoples of the East."**

After he had lingered over the causes which led to these victories, and to the significance of the internal disintegration of imperialism, Lenin concentrated his primary attention on defining the character of the struggle and of the tasks facing the peoples of the East, the solution for which, as he expressed it, "you will find in the overall general struggle which Russia had begun."* In Russia it had been possible to join the struggle of the workers with that of the peasants; the same thing, although in different forms, had to be accomplished on an international scale.

The most important theoretical conclusion which Lenin came to was that "the socialist revolution will not be solely and primarily a struggle of the revolutionary proletarians in each country against their own bourgeoisie—on the contrary, it will be a struggle of all colonies and countries oppressed by imperialism, and of all dependent countries against international imperialism."**

This is why, said Lenin, appealing to the representatives of the peoples of the East, "your task in the history of the development of world revolution—which will, judging by its beginning, continue for many years and demand many exertions—is to play a major role in the revolutionary struggle, and to merge this struggle with our struggle against international imperialism . . . to join in the general struggle with the proletarians of all countries."***

The laboring masses of the countries of the East must know "that the international proletariat is the sole ally of all the workers and the exploited hundred millions of the peoples of the East."****

Lenin put special emphasis on the search for the distinctive forms to be assumed by this union of worldwide vanguard proletarians with the working masses of the East who often lived in medieval conditions. He stressed: "The task which lies before you, and which has not confronted the Communists of the entire world before this time is as follows: while seeking support for general Communist theory and practice, you must adjust to the conditions which are nonexistent in European countries and make this theory and practice conform to a

---

*Ibid., p. 330.
**V. I. Lenin. Complete Works, Vol. 38, p. 327.
***V. I. Lenin. Complete Works, Vol. 39, pp. 328, 330.
****V. I. Lenin. Complete Works, Vol. 39, p. 330.

situation in which the peasants constitute the primary mass of the population . . ."*

These thoughts of Lenin's concerning the organic link between the proletarian revolution and the struggle of the oppressed peoples against imperialism have proved to be of fundamental and ineradicable significance. When speaking at the Congress of Communists representing the Moslem peoples of Russia, Lenin turned directly to all the peoples of the East, showing them the prospects for their revolutionary struggle given the support of the Land of the Soviets and of the World Proletariat.

Today we can see how the victory of the first socialist state over fascism has contributed to the growth of revolutionary forces in the countries of the East and West. A worldwide socialist system has been established. The demise of the old colonialist system and the victory of formerly oppressed peoples in the struggle for national-liberation have occurred precisely because of the unified actions and the mutual assistance provided by the worldwide system of socialism, the proletariat of the capitalist countries, and the oppressed peoples in the difficult battles which they have fought for their independence, thanks to the union of these three of great revolutionary forces of present-day socialist development.

The freshness and the immediacy of Lenin's ideas half a century later, in a new international situation, can be seen in the fact that they lie at the base of the documents of the International Conference of Communist and Workers' Parties, which was convened in Moscow in 1969. And how removed from Lenin's ideas are the vain attempts of those who, calling themselves Marxist-Leninists, advance a newfangled theory of world revolution resulting from "the encirclement of the world city by the world country."

Lenin stressed the growing importance of the peoples of the East in the world revolution. He not only said that Soviet Russia would assist the peoples of the East in the revolutionary movement, but that the developing struggle of the peoples of the East against imperialism was a great help to the proletariat of the West in its struggle for the victory of revolution in their countries.

Lenin's speech made a tremendous impression on the audience. And a special impression on me because it was the first time that I had heard Lenin give a public address. I noted to myself that Lenin had not touched upon the concrete issues which were of concern to the delegates and which they were going to have to discuss and then adopt specific resolutions. The day before he had held a special meeting with the delegates and knew of the issues which caused them concern. With his speech it was as if Lenin made it known that all the concrete problems could be solved only after a thorough discussion had intro-

*Ibid., p. 329.

duced clarity into the new issues of the Revolution raised by the practical problems of life.

I attended several other of the sessions of this congress, striving to learn from the comrades who spoke there, and to glean everything valuable from their experience in resolving the nationality issue among the Moslem peoples of the Povolzhe area, and the Urals. I knew almost nothing about their life.

The Russian Communists were more knowledgeable than any other party in the theory of the nationality issue. On the eve of and during the First World War Lenin had worked in detail on the nationality question which had acquired especially pressing significance just at that time. In particular, he had eluded the position of the Party Programme of 1903 concerning the right of nations to self-determination, which included the prerogative of self-determination to the point of forming separate states. This, of course, as Lenin himself emphasized, didn't mean that the Bolsheviks recommended or advocated separation. Lenin wrote that the proletariat, "recognizes equality of rights and equal rights to form a national state . . . but it values and puts higher than anything else the union of the proletarians of all nations, and evaluates every national claim and every national form of separatism from the point of view of the workers' class struggle,"* But Lenin firmly insisted that the various nations which inhabited Russia be insured the right of self-determination and even separation from Russia and that they themselves should be able to decide freely whether to remain within the framework of a single state or to split off and form an independent state.

Lenin firmly insisted that the colonial peoples should have the right to self-determination and separation from the seat of empire. This was the slogan of freedom and independence for the colonies, and it was considerably more profound than many of the sentimental but abstract speeches about "the equalization of rights" and "the restoration of humanism" in regard to the oppressed peoples.

European right-wing Social Democrats certainly didn't come out for the freedom of the colonies and their right to separate from the seat of empire—for they were bound to the imperialist policies of their own bourgeois countries. The opportunists preferred to hold forth on "cultural autonomy," as the Austrian Social Democrats did. Lenin's labors in this area struck a strong blow against right-wing social democracy as well.

The activities of the Central Bureau of the Communist Organizations of the Peoples of the East, which was attached to the Central Committee of the RKP(B), and which had worked unsatisfactorily in all respects, was subjected to harsh, even merciless criticism during the ensuing debates. It was obvious that the conditions under which

*V.I. Lenin, Complete Works, Vol. 25, pp. 274-75.

work had to be carried out were difficult; indeed, in a number of provinces of the Povolzhe region and the Urals where as a result of Kolchak's invasion, the Czechoslovakian Corps, Ataman Dutov, and others, the ruling authority was constantly changing.

There were many sharp arguments and clashes of opinion between those who spoke at the congress. There was a sense that there was a great divergence of opinion as to how to resolve the concrete issues of the nationality policy, and it even got to the point where separate factions were formed which held irreconcilable views.

I went to the congress with a more optimistic conception of how things stood. Much of what I saw and heard came as a complete surprise to me. My first days at the congress made a painful impression on me, but I was pleased to have learned a great deal about the concrete state of affairs in these districts.

Included among the other speakers was a representative of the Moscow group of the Azerbaijani Gummet, Israfilbekov. He, by the way, spoke of my coming to Moscow, and also of the reference I had made in my statements to the opinion of the Baku comrades that the opportunity of working successfully in Azerbaijan under the aegis of the Gummet party had not presented itself. I should explain that soon after my arrival in Moscow I had a meeting with Narimanov and the other active workers of Gummet, who were also there at the time, and filled them in on what was happening in Azerbaijan, setting forth in special detail our views in Baku on the necessity for a single Communist party of Azerbaijan. Contrary to my expectations, they agreed with me, which helped to simplify resolution of the issue.

On the whole, the congress turned out to be a fine school for the young national cadres of the peoples of the East who inhabited our country. It had promoted rapprochement, and the subsequent solution of the problems of national reconstruction and cooperation among nations, thanks to the new and profound ideas of Lenin concerning the peculiarities and the progress of the national liberation movement in the undeveloped countries.

* * *

I want to say a few more words about E. D. Stasova—a great toiler of our revolution, and one of those who, under Lenin's leadership took part in the creation of our party. On the day after my conversation with Lenin, Elena Dmitriyevna received me and asked me to inform her about the work of our organization, the political situation in the Caucasus, and the issues which needed to be resolved in the Central Committee. What has remained in my memory from this conversation is the businesslike spirit, the concreteness, and the reserve she manifested in our first long discussion.

During my stay in Moscow I had many other occasions to turn to Elena Dmitriyevna, and discuss with her the most varied aspects of our work. Since I was frequently in the offices of the Central Committee Secretariat I was often a witness to her businesslike conversations with comrades from the various fronts of the civil war and from the local party organizations. I became more and more impressed with her unusual efficiency, single-mindedness, precision, and the energy and thoroughness which she displayed in her work.

In 1920, after the victory of Soviet power in Azerbaijan and prior to my departure from the Caucasus Elena Dmitriyevna came to see us in Baku in her capacity as Secretary of the Caucasian Bureau of the Central Committee of the RKP(B) and I had occasion to work side by side with her as members of the Organizational Bureau, headed by Ordzhonikidze, which was empowered to convene in Baku a Congress of the Peoples of the East. I remember her high degree of concentration and her ability to stand up firmly for matters of principle. You could sense great culture and nobility behind her modesty. When it concerned carrying out a party decision, she was mercilessly demanding both of herself and her comrades, and yet at the same time Elena Dmitriyevna was, I repeat, very warm and solicitous in her dealings with young party workers, despite the severity and extreme reserve of her manner. You could sense both her warmth and her skill in working with youth and her desire to share her experience with them. Moreover, she displayed almost a maternal concern toward us and was always ready to render assistance at the very moment when help was required.

It's impossible to imagine Elena Dmitriyevna not working. Even in her declining years when her health had deteriorated, she continued to do party work. She devoted a lot of attention to the education of youth, wrote her memoirs, and took up literary work. Everyone who turned to her for help and advice found her to be very cordial. She responded very sharply to all shortcomings in the work of the party and state organs, especially when they encroached upon the rights of citizens.

That was the way Stasova was throughout her life. I remember when I went to visit her three years before her death, when she was in the hospital, she asked me in detail about the work of the presidium of the Supreme Soviet of the USSR. She was particularly concerned about the necessity of guaranteeing some kind of system whereby all the workers of the state apparatus would give consideration to the claims and complaints of the workers. "I myself," she said, "receive many letters from citizens and I devote quite a lot of time to them. You must do everything possible to insure that the letters, claims, and complaints of the citizens always receive a fair and objective hearing. It is very important in a socialist society that every citizen, no matter who he is, be able to assert his lawful interests and rights. Besides, the signals given by the workers help us to learn where violations of legality and neglect of the workers' interests are occurring, and in which regions and districts the Soviets work better or worse, and they enable us to take measures to correct the situation."

I expressed agreement with her ideas. I was well aware of the fact that she spent a lot of time going over claims made by the workers, because I often received notes from her in regard to individual claims, with a request that I consider them, and she often called me on the telephone whenever important issues were raised in the citizens' letters. Even in the hospital she devoted herself to going over the citizens' letters as soon as she began to feel better.

When I was in her apartment I saw neat piles of letters on her desk from all over the country. People wrote to Elena Dmitriyevna from far-off Yakutia, the Republics of Central Asia and the Caucasus, the big cities and the small villages of our country. Not one letter went unanswered. Stasova loved to repeat that correspondence with the workers was a form of political work, a form of communication with the laboring masses. The huge quantity of letters which she received were a clear testament to her indissoluble ties with the people, to whom she devoted her entire life.

To many of the people who came in contact with her only through work, Stasova seemed an extremely reserved, and even stern woman. But that was only a first impression. Her friends and comrades knew Elena Dmitriyevna to be a person of profound sensibility who had deep feelings and responses. Her love and deep understanding of classical music, for example, is well known.

Once in the summer of 1966 Olga Shatunovskaya called me and said that Elena Dmitriyevna wasn't feeling well and wanted me to come see her. Alarmed, I set off for Elena Dmitriyevna's home immediately with Shatunovskaya. She was lying in bed in her small bedroom, and we saw at once that her condition was serious. She spoke with difficulty and said she wanted to listen to Beethoven. She didn't have either a record player, records, or a tape recorder, but I telephoned my son's family who lived in the same building, and my granddaughter brought over a record player and several Beethoven records. It was evident that the music gave her great joy and provided a certain relief.

Some two months later Voroshilov and I were at a mass meeting at the Kremlin wall to celebrate the laying of the Monument to the Unknown Soldier, and after the ceremony I told him that I had been to see Stasova, and suggested that we both pay her a visit. He agreed with pleasure, and we went off to see her. She was delighted with our visit. We told her about the ceremony, and about the solemn procession of the troops and the workers of Moscow. Her good spirits lasted throughout our visit. As we left we felt that her condition was improving and that there was nothing to fear on that score.

On December 31, 1966, I decided to go to see Elena Dmitriyevna and wish her a Happy New Year. To my surprise I found her in a hopeless condition; unconscious, and breathing with difficulty. A young nurse was giving her a massage, trying to stimulate her heart and keep her breathing. Soon after the doctors appeared but their attempts to save her life were in vain, and a short time later she passed away.

In my heart I treasure the bright image of this remarkable revolutionary.

* * *

On more than one occasion during that time we in the Baku organization have discussed everything that our group had experienced during the period of the underground at the time of the British occupation.

Less than a year had passed, yet how much had changed in our lives! During the course of some two months after the British interventionists' arrival in Baku our party organization had literally ceased to exist. Thousands of Communists had been evacuated to Astrakhan or had scattered to other regions of the country. The twenty-six Baku commissars had met with their tragic deaths. A large group of Communists had been thrown into prison. Only isolated Communists remained in Baku.

And then small groups gradually began to form. They began to join forces with one another until from mid-December 1918 through February 1919 the Baku Bolshevik organization—although still very small—was reestablished once again. Thus a group of young Communists had taken upon itself the functions of a leadership center, albeit not entirely in the usual way, without either conferences or elections, and having quickly achieved a sound organization and political influence in underground conditions, it conducted a party conference in the first half of March at which, in complete accordance with the Party Statutes, a Baku Party Committee had been elected.

By fall 1919 the organization already numbered around two thousand persons, whose acceptance into the party was accomplished through an extremely strict process of selection: only those were accepted whose actions had given proof of their devotion, capacity for struggle, fearlessness, and readiness to give themselves wholly to the party cause. As events unfolded with incredible speed the political consciousness of the workers grew with the same velocity. The young cadres gathered organizational and leadership experience and were tempered in direct clashes and face-to-face combat with the numerous enemies of the revolution. By autumn 1919 our organization had won indisputable political influence among the working masses, while the representatives of the treasonous petit bourgeois parties had been completely discredited and toppled from leadership of the workers' organizations.

The kaleidoscope of events had illumined the workers' consciousness like lightning, cleansing it of many illusions. The image of the British as civilized allies had been smashed forever, and confidence in the appeasement-oriented parties had been finally undermined when the treasonous role played by their leaders had been revealed after their participation in the villainous execution of the Baku commissars. All this took place against the backdrop of a most severe civil war foisted upon Soviet Russia by the capitalists and landowners of the fatherland on one hand, and their foreign interventionist-allies on the other.

The workers were inspired by the heroic struggle of both the Red Army and of the Communist party against its numerous enemies. This strengthened even more the masses' faith in the righteousness of the Bolsheviks and inspired them to contribute their mite to the cause of the ultimate victory of Soviet power.

The Baku Communists had been forced to work for many months without any contact with Moscow or the central organs of the party and Soviet authority. In responding to the rapidly changing local conditions of the struggle it was necessary to orient oneself in the situation without any outside help, to work out tactics and to pass appropriate resolutions on one's own.

And so, looking back on the path we had traveled we had every basis for reaching certain conclusions: a Baku underground Communist organization had been created; leadership had been won in the Baku Workers' Conference—that distinctive workers' parliament which had thrived in our area during the British occupation; the influence of the Communists in the districts Workers' Conferences, workers' clubs, trade unions and cooperatives had been firmly secured; contacts had been reestablished with the Communists of Georgia and Armenia, the Transcaucasian Party Regional Committee, the Dagestani and northern Caucasian Communists, and with the Bolsheviks of the Transcaspian area; the leadership of the Communists' operations in the districts of Azerbaijan had been strengthened.

The most remarkable thing of all was the fact that in a very short time the Communists had succeeded in undermining the confidence of the backward segments of the Azerbaijani laborers in "their own" national government, and in dispelling the illusions that "their" bourgeoisie was better than any other, for example, the Russian bourgeoisie. The idea that the Azerbaijani workers and peasants would soon win from "their" government an improved standard of living, political rights, and the proper conditions for spiritual development —such thoughts burst like a bubble.

The influential counterrevolutionary Muscavatist party, which had stepped forward before the workers with loud-mouthed, demagogic declarations and pronouncements, while unconditionally supporting the bourgeois government, had been isolated. Bolshevik propaganda fell on fertile soil, finding support among the workers and raising the political consciousness of the Azerbaijani workers. The same thing happened in the countryside. The peasants began to realize that the landowners did not want to yield a single inch of their land to them, and if some peasants had once had any illusions that "their" national government would give them land, their political understanding was rapidly clarified as soon as they began to get bullets instead of land.

The Bolsheviks used all legal means for intensifying operations in the organizations. To this end we did not ignore even so feeble, yet nonetheless representative an organ as the Azerbaijani parliament, where the Muscavatists were playing first violin at that time.

Finally, when the Denikinite danger had hung directly over Transcaucasia we had advanced the idea of a united front against it. And although in the end the Tiflis Mensheviks wrecked the creation of the united front (and thereby suffering a major political loss) the very process of the negotiations and our appearances on behalf of a united front in Baku and Tiflis had yielded their own important political gains, allowing us to expose the treasonous essence of the Menshevik, Muscavatist, and Dashnak parties, and to strengthen the influence of the Bolsheviks on, among others, that portion of the workers who had previously supported the other parties.

Of course, it wasn't all smooth sailing for us. In defining tactics and our general line of conduct, we experienced differences of opinion, and hot-tempered debates had ensued as different points of view were expressed. This was natural. In the complex, constantly changing circumstances of acute class struggle, not everything lent itself immediately to profound analysis and a precise assessment. But usually, in the course of discussion and debate, the situation became clear, and we as a rule came to harmonious, unanimous decisions.

There were those issues, of course, on which debate did not immediately lead to a unanimous decision. For example, the majority of us thought that on the position of a united front in the struggle against Denikin that hardly anyone could question the correctness of this tactic. There were comrades among us, however, who found intolerable the very idea of united action with the appeasement-oriented parties, and even more so, with the bourgeois government. Lominadze, for example, was of this opinion. He expressed his own particular point of view on certain other issues as well. But he did so in all sincerity, and it was clear that he was simply mistaken. Subsequently, in the course of the discussion on the question being argued he would frequently renounce his position once he had become convinced that it was wrong. Sometimes, however, and in particular on the question of the united front, he persisted stubbornly in his own view.

Since we had not reached consensus on this important issue, it was put before the Baku Party Conference for discussion where Lominadze, Agamirov, and Shatunovskaya spoke out against our position hoping that the conference would not support us. But they were wrong. I was the one who read the report on this question. Other comrades, including Sarkis, also spoke out at the conference in support of a united front, and when the vote was taken, only two people —Lominadze and Agamirov—had voted against it. They had not only failed to get new "allies," but had even lost one—Shatunovskaya, who had become convinced of her mistake. And even Lominadze himself, eventually became convinced of our view and took our side. Later, in one of his speeches at the Baku Party Conference he said, "I was an opponent of the united front simply out of stupidity. I realize this now, and it's good that the Baku organization put these tactics into effect in the best possible way at the time." Lominadze's openness and sincerity impressed all of us most favorably.

I recall that one other question which called forth much debate was that of the participation of the workers' organizations in the Azerbaijani bourgeois government's organs for the protection of labor. At that time the Ministry of Labor had organized a governing board to direct measures for the protection of labor and had invited representatives of the trade unions and other workers' organizations to take part in its work. The Mensheviks were very active in "pushing" this question (they even managed to camouflage from us the fact of their pushing through a relevant decision on this question amidst a large number of diverse resolutions at the April Congress of Revolutionary Trade Unions from Transcaucasia, Dagestan, and the Transcaspian Area).

The situation of the workers in Baku was oppressive. Arbitrariness and caprice were on the increase, and the workers could not find protection anywhere. Some comrades—Sarkis, Lominadze, and others—were in favor of our participating in the government organs for the protection of labor. They were moved not by political considerations, but merely by the hope that such participation would relieve even somewhat the workers' situation.

Our discussions of this issue were stormy. I was decisively against the proposal, considering our participation in the organs of the bourgeois government inadmissable on principle. I was convinced that such participation would strengthen the false impression among backward workers that the bourgeois government was in a position to improve their lot. The Baku committee rejected the proposal by a huge majority. At the next meeting, however, Lominadze and Sarkis tried to bring this question up again for discussion since they sincerely believed that the improvement of the workers' position depended on it, but the Baku committee refused to reexamine its previous decision. Later, the question surfaced more than once at the meeting of the Regional Committee, now at the party conference, but the original decision of the Baku Party Committee remained in force.

After our release from prison, when autumn was fast approaching, we would ask ourselves more and more frequently: what is to be done? The workers' situation continued to worsen; the oil depositories were filled to overflowing, the entrepreneurs began decreasing oil production, closing the oil fields and laying off the workers, stopping their wages, etc. Capital's economic offensive had begun. The workers were spontaneously straining at the bit, moving in the direction of uprisings and struggle. At meetings separate groups of politically active workers introduced resolutions calling for strikes. It was clear to us from the May strike in Baku that the economic dilemma turned on the question of political power; it was impossible to better the situation of the workers without overthrowing the old regime.

We protested separate outbreaks because in our opinion they only weakened the workers, calling instead for an organized buildup of forces, and for holding them in reserve until spring when the possibility of exporting oil from Baku to Astrakhan might open up. And besides, we were counting on help from the Red Fleet in the spring. Not everyone, of course, including some in our own party, believed that we would succeed in preventing isolated segmented outbreaks and in convincing the workers to wait until spring.

In this regard the story of the workers' strike at the former Eisenschmidt Central Factory Shops is worth relating. We had a strong party cell there which had great influence on the workers, but when their situation became particularly oppressive, the factory committee declared a strike without first obtaining the Baku committee's consent. We invited the strike leaders to the Baku committee where one of them, Mir-Bashir Kasumov, stepped forward and tried to convince us of the correctness of their decision, saying that it had been impossible for them to withstand the pressure put on them by the workers, and that the strike would have begun spontaneously in any case, and so forth.

The Baku committee recognized the strike as inexpedient, but in view of the fact that it had already begun we decided to support it out of a feeling of solidarity, and our press and the Workers' Conference duly came out in its support. However, as might have been expected, the shop workers won only insignificant concessions from the bosses, and I think that the strike was called off within a week.

I remember that on another occasion debate flared up in our midst concerning the work of the trade unions. Certain leading workers such as Mirzoyan and Anashkin, who were working in the trade unions at the time, began to have doubts about our policy of excluding the Mensheviks and SRs from trade unions, citing the fact that there were very experienced and efficient workers among them. Our primary interest was in securing firm Bolshevik leadership of the trade unions;

everything else was of secondary importance to us. However, we respected the desires of our comrades and kept in trade union work some of the most experienced SRs and Mensheviks who had no designs on leading the unions and thus did not pose a threat to us.

All the differences of opinion which arose in our midst were smoothed out and dispensed with rather peaceably and, what was most important, did not lead to the basic dissensions which result in the formation of interparty cliques or factions. In other words, they did not threaten our unity.

I simply cannot recall any instances among us of political intrigue or cliquishness; in everything we did there reigned supreme the spirit of honest revolutionary adherence to principles, of dedication to Leninist ideas, of political awareness and orderliness. We owed this first and foremost to the glorious traditions of the Baku party organization which had been nurtured by our elder comrades, and most of all, of course, by Stepan Shaumian and his comrades-in-arms.

We, the representatives of an already younger generation of Baku leaders, preserved the sanctity of those traditions in the complicated conditions of the oppressive underground during the British occupation of 1919. This underground was the second in the history of the Baku Bolshevik organization; the first underground had existed during the years of the struggle against tsarism. The second underground was not a simple repetition of the previous one—a struggle was being waged for the final victory of Soviet power in Azerbaijan, for the transformation of Baku into an indestructible bastion of bolshevism in the Caucasus.

As we looked back over the path we had travelled, we tried by means of self-criticism to come to grips with and understand whether or not we had permitted any major, fundamental political mistakes to be made in the conditions of the underground.

And with a feeling of great pride in the Baku Bolshevik organization which had always stood firmly on Leninist positions, one can say that even in the difficult conditions of the underground during the British occupation and the Muscavatist bourgeois government, after suffering the terrible loss of its enormously experienced leaders and filling its ranks with the revolutionary, proletarian youth, the Baku organization had proved worthy of its tasks and had kept faith with the great banner of Lenin's party.

PART V

THE BIRTH OF SOVIET TRANSCAUCASIA

Chapter I

Return from Moscow

Meetings with M. V. Frunze and V. V. Kuibyshev—The Eleventh Army marches to Baku.

I spent about two months in Moscow on that first trip in 1919, having arrived there early in November, and leaving at the beginning of January 1920 in company with a group of comrades—Mikhail Kakhiani, Vladimir Ivanov-Kavkazsky, and Olga Shatunovskaya. We returned to Baku via a different route, by way of Tashkent and Krasnovodsk, which had already been liberated by the Red Army, since the route through Astrakhan was closed because of ice.

It took us more than a month to go from Samara to Tashkent, traveling by train with M. V. Frunze who was in command of the Red Army troops in Central Asia. In Tashkent I met still another remarkable Bolshevik-Leninist for the first time—V. V. Kuibyshev. He had arrived in Turkestan at the end of 1919 as part of a special commission which the Party Central Committee had sent there to consolidate Soviet power in an area where Basmach bands were still on the rampage.

I remember how struck I was by Kuibyshev's untiring efforts to gain a thorough knowledge of the situation in the area and to get acquainted with the people. His sober-minded approach to the solution of any question impressed me, and one felt that he had the enormous respect of his fellow comrades in the party organization of Turkestan. They knew him well and loved him; he was greeted everywhere as one of their own people. In everyday life, at work, and in dealings with his comrades he was unusually modest and straightforward. A man of uncommon personal charm who had a certain special sensitivity to those around him, he possessed at the same time enormous inner strength and the ability to captivate people. Later on I met with Valerian Vladimirovich Kuibyshev many times on a regular basis, and thus our first acquaintance grew into a solid friendship which lasted until the time of his death.

While I was in Tashkent, I had a rather detailed conversation with

Kirov on a direct line to Astrakhan. At the time, March 6, 1920, Astrakhan was still cut off from Baku by ice, but a group of comrades had just arrived in the Transcaspian area from Baku and they gave us the most detailed information about what was happening there. I passed on to Kirov their news about the suppression of the rebellion in Georgia and about the mass arrests and shootings, as a result of which few Georgian party activists remained at liberty.

"Because of this," I said, "the center of the movement is now being transferred to Azerbaijan," where our influence among the masses was sharply on the increase in spite of the repressive measures to which the Communists there had also been subjected, particularly after the events in Georgia.

I informed Sergei Mironovich Kirov about the decision of the Azerbaijani Bolsheviks to prepare for a revolutionary coup, and about the steps we were taking in that direction; and also about the aid in arms and money which I had managed to arrange for there, in Turkestan. I asked him, too, to do everything necessary in order to improve, first and foremost, Baku and Dagestan's supply of arms and uniforms which they sorely needed.

Having thanked me for the information and at the same time asking that I be in contact with him more frequently in the future, Sergei Mironovich Kirov promised to give immediate attention to my requests and suggestions, and assured me that he would give the people of Baku the most effective assistance possible. He was, in fact, very disturbed by the situation in Baku, fearing that because of the repressive measures instituted against the Communists there "our comrades might give in to panic," and in order to boost their fighting spirit he asked particularly that I inform them in detail "of our brilliant situation here on the Caucasian front." He spoke of the critical shortage in Astrakhan (and in all of Soviet Russia, for that matter) of oil, gasoline, machine oil, and requested that these products be sent to them as soon, and in as large quantities, as possible.

I assured him that the mood of our Baku, and in general, of the Azerbaijani Communists was cheerful and revolutionary, that they were not being broken by the repressions, and that recently they had begun to receive more detailed information about the state of affairs in Soviet Russia and were inspired by its victories. I also gave Kirov a firm promise to send petroleum products to Astrakhan immediately upon the opening of navigation.

The purely official notes of my conversation with Kirov, which were published many years later,* do not convey the warmth and sincerity which always permeated my long friendship with the unforgettable Sergei Mironovich.

From Tashkent I went to Krasnovodsk via Ashkhabad. A congress

---

*S. M. Kirov, Articles, Speeches, Documents, Vol. 1, pp. 218–220.

of workers' representatives was taking place in Ashkhabad, and at Frunze's suggestion I greeted the congress on behalf of the Baku proletariat and congratulated the Turkmeni workers on their victory.

In Krasnovodsk I boarded a launch, which our comrades had sailed up from Baku, and headed home on the last lap of my journey. Since Denikinite naval forces were still active on the Caspian Sea, the launch had been armed with cannon and machine guns, and we set out for the Azerbaijani coast in order to make our way illegally to Baku for the purpose of preparing an armed rebellion against the Muscavatist government. We were a day out to sea when our compass broke; we soon found ourselves off course and were unable to determine our location. After a long time shore lights appeared in the distance. Some thought they were the lights from the beacon in the Iranian port of Enzeli (Pekhlevi); others thought they were the lights of Dagestan. An experienced sailor by the name of Semyonov who was traveling with us declared that from the shape of the cliffs he determined we had reached the shores of Derbent. However, in his opinion it was impossible to disembark at this point, particularly since the sea was rough.

We decided to go in the direction of Petrovsk, since we knew that the Red Army was already approaching Dagestan but we didn't know if it had reached it yet. We began our approach to Petrovsk in the early morning, reasoning that if there were no vessels in the harbor, the port had been liberated and the Whites had gone. As we drew closer we soon saw that there were no trucks on the road or ships in the harbor, which meant clear sailing. But another question arose to worry us: might not the harbor be mined? Therefore we approached our moorings with extreme care and caution.

We moored the launch safely, however, and as we disembarked Red Army men came down to the dock to greet us. Learning from them that the headquarters of the Eleventh Army was located in a special train at the station, we went at once to the headquarters' cars and there I met once again with Sergo Ordzhonikidze and Sergei Kirov, who introduced me to the commander of the Eleventh Army, M. K. Levandovsky, and to a member of the Military Council of the Army by the name of Mekhanoshin.

I remember how wildly happy Sergo was about the successful campaign of the Eleventh Army. But we were even happier about the approaching liberation of Azerbaijan, on behalf of which the Baku proletariat had for so many years and under such difficult conditions been carrying out self-sacrificing, fighting work, and for which it had suffered many casualties. We dreamed that the liberation of Georgia and Armenia would soon follow. And this was more than a dream now, it was an utterly real possibility, for now at last Soviet power was winning victories on all fronts. The Red Army had already smashed the Denikinite units in the northern Caucasus (the White

Guards had succeeded only temporarily in holding their ground on the Crimean peninsula). After liberating the northern Caucasus, the Eleventh Army had reached the very borders of the Azerbaijani bourgeois state, and Levandovsky was preparing the army for further military operations, having stationed forward units along the left bank of the Samur River and having concentrated a detachment of armored trains here in Petrovsk.

The day after our arrival in Petrovsk, Lominadze, who was a member of the Baku Party Committee, arrived there from Baku illegally and assured us of the Baku Bolsheviks' readiness for armed rebellion. "Although many comrades from the Baku party leadership are in prison," he told us, "there is nevertheless a large group of Communists in the city who are capable of taking charge of the rebellion." Great hopes were also being put, of course, in help from the Red Army.

Levandovsky acquainted all of us with the order issued by Tukhachevsky, commander of the Caucasian front, and Ordzhonikidze, member of the *Revvoensovet* at the front, which order commanded our troops to cross the Azerbaijani border on April 27 and come to the aid of the Baku workers. Because of this we ordered Lominadze to return to Baku immediately (again illegally), to inform the Baku comrades of the date when the Red Army would be there.

I remember how anxious Sergo was about making sure that the Muscavatists did not set fire to the Baku oil fields With defeat so certain they were capable of taking such a desperate step, and Ordzhonikidze told us that he had special instructions from Lenin on this matter. Therefore, we made Lominadze promise that the prevention of any explosion at the oil fields would be regarded by the Baku Party organization as one of its highest priorities.

Levandovsky briefed us on the military situation that had taken shape. According to intelligence reports, the Muscavatist government of Azerbaijan had an army of 30,000 at its disposal; 20,000 were stationed at the borders of Armenia, and 10,000 were dispersed around garrisons in several built-up areas. In Baku itself there were about 2,000 infantry men, the Junker (Training) School, and artillery. There were a little more than 3,000 others directly on a line of contact with our army. Thus, the Muscavatist government did not have a significant enough troop-capability to enable it to offer serious resistance to the Eleventh Army on its march into Baku. This made it possible for us to carry out the will of the Baku workers without stubborn fighting, after aiding them in the establishment of Soviet rule. Of course, in the face of a further advance on the part of the Red Army, the Muscavatist troops, which were mainly concentrated on the Armenian border, could create difficulties for us, which was in fact what happened.

I remember that we worked out a detailed plan for concerted actions by our troops: four armored trains under the general command of Yefremov* were to push their way through to Baku; the cavalry was to operate on the right flank; rifle units would follow the trains on foot. Gogoberidze who had arrived in Baku illegally a short time before, was named commissar of the cavalry division.

I wanted to get to Baku as quickly as possible so I asked Levandovsky to help me enter the city along with the first military units. He said that the armored trains would enter Baku first. They were the vanguard assigned to push their way through to the oil fields to help safeguard this vitally important objective. "Therefore," Levandovsky said, "if you want to reach Baku before the rest, go with that detachment." I requested permission from Kirov, as a member of the *Revvoensovet* of the Eleventh Army, to do this, and he supported me and suggested I go as a fully empowered political representative of the Army *Revvoensovet*.

Before our departure from Moscow, when we still expected to reach Baku before the Red Army, Elena Dmitrievna Stasova had given me the idea, upon my arrival in Baku, of organizing intelligence operations in order to transmit information to the command of the Red Army. I had been instructed in the use of and supplied with materials necessary to carrying out intelligence operations: codes, money, and valuables. I was bringing all of it with me to Baku, but none of the comrades who were traveling with me knew about it. To avoid the risk of taking such property onto an armored train about to become involved in a military operation, Ordzhonikidze and I decided to apprise Olga Shatunovskaya of what I was doing and to entrust her with getting all the valuables to Baku. Disguised as a nurse, Shatunovskaya brought them to Baku safely and handed them over to I. I. Dovlatov, the treasurer of the Caucasian Regional Party Committee.

Comrades Kakhiani and Ivanov-Kavkazsky traveled to Baku with other units of the Eleventh Army. Up until then they had been doing party work in the Zamoskvorechiye district of Moscow and were now on their way to Georgia to do illegal work. I became friends with them en route. The Communist Zavyalova, who had been recommended to do illegal work among the women of Baku, was also traveling with us.

Comrades Musavekov and Dzhabiyev traveled from Astrakhan with the Eleventh Army. They were prominent Azerbaijani Communists who had temporarily been evacuated to Astrakhan after the fall of Soviet power in Baku in 1918. They had already undergone some schooling in practical work under Soviet rule in Russia, and could not wait to get back to their native land. They were dedicated party activ-

---

*Later on, Mikhail Grigorievich Yefremov became the commander of the Orel military district. With the beginning of the Great Fatherland War he commanded the army. He died in the midst of organizing a breakthrough of the enemy front to the west of Moscow. There is a monument dedicated to him in Vyazma.

ists, capable party workers who, immediately upon arrival in Baku occupied leadership positions inn Soviet Azerbaijan.

We were also accompanied on our journey from Astrakhan to Baku by the Communist Museib Shakhbazov and his wife, Lyuba, also a Communist. We left them temporarily in Derbent for the purpose of smoothing out party operations there. Later, Shakhbazov became the People's Commissar of Education of the Azerbaijani Republic.

After leaving Petrovsk for the place where the armored trains were stationed I met in the Derbent district with Mikhail Grigorievich Yefremov, the commander of the armored train contingent, and Dudin, the military commissar attached to this group. The armored train, the "Third Internationale," in which Yefremov's command center was located, was standing some 200 to 300 meters from a bridge over the Samur River, constituting the border of Azerbaijan. The train was hidden in the woods and could not be seen from the opposite shore.

It was a warm spring day; the trees were already covered with leaves. Yefremov read the command's order, signed by Levandovsky. It was very brief and unusual, ordering him to begin the advance on April 27: to break into Baku by means of a swift surprise attack, to consolidate our hold on the Baku oil fields, and, relying for support on the workers' detachments, to guard the oil fields from possible attempts at arson until the arrival of the main forces of the Eleventh Army.

"What time will we start?" I asked Yefremov.

"Only the day of the advance is designated in the order," he replied, "but the hour isn't indicated, and since it isn't, that means that we have the right to decide that ourselves."

"Well, when will we get going, then?"

"At night, about five or ten past twelve. The earlier the better."

This was the morning of the twenty-sixth of April; in less than twenty-four hours we would be on the move. We spent the morning familiarizing ourselves with the state of readiness of our armored trains and chatting with the fighting men. Our intelligence informed us that a detachment of Azerbaijani troops was stationed on the opposite shore, and that they had an armored train under the command of a Georgian officer. After discussing the situation with Yefremov we decided to try to enter into negotiations with the commander of that train and also to determine whether or not the bridge was mined, and where the telephone and telegraph lines connecting the sentry post at the bridge with the adjacent station were located.

The bridge was no-man's-land. Our sentry stood at one end of it and an Azerbaijani sentry at the other end. We crossed the bridge to the Azerbaijani sentry and asked him to convey our request to his commander to come to the bridge for talks with us. The sentry summoned the chief of the sentry detachment who came over and said that the train had left for another station and that therefore he could not transmit our request to the commander. We asked him to do so when the train returned, and he promised that he would.

Since the chief of the guardpost gave the impression of a man who was well disposed to us, we decided to take a bigger risk—try going further into the territory beyond the bridge. We strolled casually forward and Azerbaijani soldiers came up to us and greeted us. In order to conceal the aim of our visit and to weaken their vigilance, we struck up the following conversations:

"We've come to you as guests. The only thing is, we don't have any vodka—we're forbidden to have it. Perhaps you've got some cognac?"

"We haven't any. But it can be bought, all you need is money," the chief of the guard replied.

"So, if we give you the money and come back tomorrow at the same time, we can have a few drinks together?" I gave them a packet of money which we had brought with us for this purpose (it was Tsarist money, worthless to us, but worth more in Azerbaijan than the local currency), and we began telling the soldiers that there wouldn't be a war, that an Azerbaijani delegation, headed by Pepinov, a member of parliament, had gone to Moscow for talks, and that a peace treaty would be signed. We asked them how things were going for them, and they said, "Take a look and see, if you want to."

We strolled over to the guard house which was a few feet beyond the bridge. It contained a small barracks for fifty or sixty men. Along the way Yefremov and I examined everything carefully and concluded that the bridge wasn't mined and that the communications lines with the rear were laid along the railway embankment. We chatted with the Azerbaijani soldiers as we strolled along, behaving like comrades toward them and they responded to us in the same way. Everything indicated that a very calm atmosphere prevailed on the Azerbaijani side.

It was time for the changing of the guard, however, and all of a sudden we heard music and then the new guard shift, stepping smartly to the sound of the *zurna* and the beat of a drum, came into view. Our little sally could turn out badly, we thought then. If the chief of the new guard is just ordinary folk like the soldiers, it will be all right, but if he's a reactionary officer he might suddenly take it into his head to detain us. In order to allay suspicion we joked in a relaxed way and we even began to dance to the music. The soldiers began to be more and more well disposed to us, and luckily for us, the new chief of the guard turned out to be a talkative and obliging person. Nevertheless, we felt it was time to leave before they realized what was going on and decided to hold us for questioning. We told them that we'd come back for the cognac the next day and made haste to return to our side of the bridge.

And so our risky excursion ended in complete success. We had learned quite a bit. But what if we had suddenly been arrested just a few hours before our advance. But all's well that ends well. We were informed that the commander of the Azerbaijanis' train had arrived at the bridge to meet with us.

We walked out to the middle of the bridge where the commander of their train awaited us. He was tall, slightly older than Yefremov, a little bit on the heavy side, but manly, with a moustache and a deep scar on his face. He conducted himself in a stern and officious manner, introducing himself as a staff captain in the Tsarist army, Lordkipanidze. Wanting to win him over, and also to find out what kind of a person he was, we said to him in the course of the conversation, "There won't be a war. You were a staff captain in the Tsarist army, you're an experienced commander, and if you were in our service you would receive command of a division, since we have few officers. Come over to our side. We will receive you gladly."

Until this moment, the officer had been calm, and had spoken in a restrained way, but upon hearing our proposal, he raised his voice. "How can you suggest such a thing to me!" he said. "That I, a nobleman from Kutaisi province, and officer in the Russian army who was wounded in the war," he pointed to the scar on his face, "that I should go over to the side of the Bolsheviks? Inconceivable!"

"What has made you so upset, staff captain? We have many officers who were once in the Tsarist army and are now in our service."

"No, and no again," he replied.

Nevertheless, we parted peaceably, and felt we had achieved our goal; we had managed to weaken the vigilance of the Azerbaijani officers and soldiers and to mislead them into thinking that we weren't preparing to attack.

On the night of April 26 the weather was warm, calm, and the moon shone brightly. Yefremov held a meeting with the commanders of the armored trains and gave them the necessary instructions. A meeting of the armored train crews and the landing forces was set later, for 11:30 p.m., at which Yefremov explained the operational task which the Army Command had assigned to the armored trains. His speech did not take more than five or six minutes. In just as short a time I explained the political significance of our attack, the significance of Azerbaijan as a doorway in the East to the victory of worldwide revolution, and the significance of Baku oil for the whole economy of our country. We had to be prepared for any sacrifices in order to carry out this historic assignment of breaking into Baku by means of a swift surprise attack.

The Red Army men took their positions in the cars. Yefremov and I were stationed in the weapons' area of the armored train. A platform had been hitched to the front of the train, on it stood two Red Army men who had been briefed in advance. Yefremov held a watch in his hands and at exactly 12:10 a.m. he gave the order to advance. We moved quietly and cautiously in order to avoid being spotted too soon, rolling onto the bridge toward the Azerbaijani positions. They had no idea there what was going on. "What's up?" they asked as we approached. "We're going to Baku. If you want to, come along with us," we said. While we were talking to the Azerbaijani soldiers the two Red Army men jumped off the forward platform and cut the telephone and telegraph wires which went from the guardhouse to Yaloma station.

As soon as the Red Army men had carried out their assignment Yefremov commanded in a loud voice: "Landing forces out!"

The Azerbaijani soldiers scattered. One of them opened fire, but the shooting ceased as soon as it had begun. Fifteen minutes later our armored train had resumed its forward motion.

According to intelligence reports a detachment of police was stationed in the woods, not far from Yaloma station, to the left of the train tracks. The landing force disembarked for the purpose of getting to the barracks and surrounding the police. Our artillery fired a few shots at the barracks to cover their attacks. In the next few minutes some brief skirmishes took place during which the police scattered, and then it was all over. The landing force, made up of Lettish riflemen, lost three men killed in the skirmishes, and a few were wounded.

But a new danger had arisen in the form of a locomotive under steam on the tracks at Yaloma station. Yefremov was afraid that they might drive it straight at our armored train and wreck us, so he gave the command to fire at the front wheels of the locomotive. We had good artillery men: the third missile was right on target and the locomotive crashed to the ground, falling over on its side. We rolled on into the station. There was no one around. Yefremov and I got off the train with our carbines slung on our shoulders, but as I started to enter the station depot he stopped me.

"No, that's not the way it's done," he said, and pulling out a grenade as a precautionary measure, he opened the door and we went in. There was no one in the hall. We opened the door into the next room and saw about fifteen railway workers and several passengers hiding behind tables and chairs. The majority of the railway workers there were Russian. We turned to them and said: "Why are you hiding? What's the matter with you, are you against Soviet power?"

"No," they said.

"The Red Army is on the move, railway workers should be at their posts," we said. They relaxed, then, and began to go about their duties. Among those in the room was a fat Azerbaijani wearing the uniform of a clerk in the forestry department. He was very upset and asked if he could leave and go to his official carriage. Naturally, we let him go, for we were trying to create an atmosphere of a peaceful campaign rather than of a wartime situation.

After talking with the railway workers we left them in charge of maintaining order at Yaloma station. We left two Red Army men there as well, and then we boarded our armored train and continued on our way toward Baku.

When it grew light and the new day dawned we saw from the moving train an unfolding panorama of peasants working, ploughing and sowing their fields. As our train moved along through the countryside with its red banner raised aloft, the peasants greeted us with exclamations and raised arms, and we called back in reply: "Long live Soviet Azerbaijan!"; "Long live the friendship between the Azerbaijani and Russian peoples!" As we neared Khudat station, however, the Muscavatist armored train made its presence felt by opening fire on our armored train from behind a hill, but its aim was bad and it did not score a single hit. We did not fire back since the target couldn't be seen, and we didn't want to waste shells for no reason. We proceeded cautiously ahead, fearful that the tracks might be mined, while the Muscavatist train kept up its fire, but it kept moving further and further away, and soon it disappeared entirely. At Khudat station I called the next station on the phone and asked whether the Azerbaijani armored train and its commander, Lordkipanidze, were there. In a few minutes Lordkipanidze himself came to the phone.

Once again we asked him to come over to the side of the Red Army, and once again he replied with a categorical refusal.

As we were approaching Khachmas station we noticed that a large group of workers were repairing the damaged roadbed and track. We were deeply touched and delighted by the workers' concern for the forward movement of the Red Army, and were able to continue our journey without delays. When our train arrived at Khachmas station we organized a political meeting on the spot in which everyone who was there took part. We spoke about the significance of the establishment of Soviet power, about the friendship between peoples, and about the necessity for maintaining order.

Suddenly someone threw himself around my neck with tears of joy in his eyes. I recognized him as the Baku Communist, Pereverdiev, with whom I had worked in the underground, and we embraced with emotion, but I recall saying to him that now was not a time for tears, that there was work to be done. Right there at the meeting he was proclaimed chairman of the Khachmas Revolutionary Committee. When the meeting was over we boarded our train again and continued our journey and soon we were approaching Sumgait station. All along the way Yefremov had been sending regular reports on our forward progress to Levandovsky. The commander knew where we were and was evidently quite pleased with our actions. To our surprise, however, we received a telegram from Levandovsky at Sumgait station saying that our armored train, the "Third Internationale," had got too far ahead of the main units for its own safety and that it was therefore necessary to fall back somewhat and keep closer to our troops.

The telegram upset us. We didn't understand what had caused

the change in the original order or what information could have served as a basis for the new one. The only reason we could think of was caution—or as we thought of it—as undue caution, for if we moved back, the enemy would think that we were retreating, and they could destroy the roadbed and track. We replied to Levandovsky that we were in Sumgait station and therefore had already traversed the coastal segment of the railroad. If we were to go back, the Caspian flotilla of the Muscavatist government could fire at us from the sea. In other words, we had already passed that danger. In our opinion, we replied, it was no longer possible to move backward; rather, we should continue to carry out our orders and make our way through to Baku, which was, in fact, what we were doing. And since the instructions had not been phrased by Levandovsky in a particularly strict way, we decided to move further on without waiting for an answer to our telegram.

Back at the Khachmas station we had got hold of a recent Baku newspaper from which we learned that on the evening of April 27 the Azerbaijani government was holding a reception in Baku for foreign representatives.

"It would be good if we could occupy Baladzhary,* cut off the diplomats' train route out of Baku, and prevent them from escaping to Tiflis," I said to Yefremov.

He looked at the clock, thought a moment, and said, "We'll most likely have time to do it if we start at once."

"Do you know why I think it important to detain the foreign diplomats in Baku?" I said, "We'll then be able to exchange them as hostages for the revolutionaries who have been arrested by the British."

Yefremov gave me his support on the entire project, and we pulled it off; the diplomats were arrested by us without even managing to get out of Baku. However, a short time later Chicherin sent a telegram demanding their release, and we were forced to comply with the demand of the People's Commissar of Foreign Affairs.

The case of the diplomats reminded me of the incident with the British officer Captain Walton, which I have already related. In reply to Walton's spiteful remark that the leaflets we had put out for the British soldiers were written in bad English and would not achieve their goal, I had promised Walton, not without irony, that as soon as Soviet power was victorious in Baku we would be pleased to use him himself as our most qualified translator. My remark had drawn a rather sour smile out of Walton. When I now related the incident to Yefremov, we both had a good laugh imagining how funny Walton would look if he happened to be one of the diplomats we detain and again heard the suggestion that he become our translator!

*Baladzhary was the railway junction from which a side track ran to Baku.

It was dark when we made our approach to Baladzhary. Along the way our train was bombarded by artillery fire. Yefremov ordered the landing forces to disembark and to advance along the railroad track on both sides of the armored train while a few Red Army men went ahead to check whether the tracks were mined. The train moved slowly and arrived in Baladzhary at midnight.

Here we held a brief political meeting which was attended by many workers. After the meeting we gathered the Communists from the local cell—there were about thirty of them—and distributed rifles to them. We set up a Revolutionary Committee at Baladzhary station selected from among the local workers.

Yefremov received information by telegraph to the effect that the Azerbaijani government had requested the Georgian Menshevik government to send an armored train. The Georgian government had given instructions to detach an armored train and some other forces as well, and consequently Yefremov made the decision to have one of our armored trains move 150 to 200 kilometers further on, deep into the area of Elizavetpol (the name of the ancient Azerbaijani city of Gyandzha during the Tsarist period). If the Georgians' armored train appeared, we were to avoid a skirmish with it, but the bridge lying in its path should be blown up in order to delay its advance for forty-eight hours until our main forces had reached their destination. The rest of the armored trains were ordered to insure the advance of the units of the Eleventh Army.

That night Kamo telephoned me from Baku and reported that our comrades had proposed to the Azerbaijani government, under threat of rebellion, that it peacefully relinquish power to the Communists. In view of the fighting mood of the Baku workers, and of the fact that we had occupied Baladzhary with an armored train and that Red Army units had mounted a general campaign, the Azerbaijani government decided to comply with the demand. On the night of April 28 they released all the arrested Bolsheviks from prison and relinquished power. To put it simply, the Muscavatist ministers ran scattering in all directions.

Overjoyed by the news I told Kamo that we would leave immediately and be in Baku by early morning. The train moved forward. It went slowly as we were afraid that the roadbed and track had been mined.

Around six o'clock in the morning on April 28 our armored train arrived safely at the Baku station. Kamo met us. We went with him by car to the building that housed the Azerbaijani parliament where members of the Military Revolutionary Committee and the Central Committee of the Communist party of Azerbaijan had already been meeting for several hours.

Chapter II
Farewell, Baku!

May Day in freedom—The Congress of Peoples of the East—Saying farewell to Baku.

When we arrived in Baku early on the morning of April 28, 1920, the old bourgeois regime no longer existed, but a new Soviet rule had not yet been established, and as a consequence we came across policemen on the streets who were continuing to carry out their duties, still unaware that there had been a change of power in the city. The shops had opened at dawn and at first life appeared to be flowing along its normal course. The local Red Guard had assumed responsibility for guarding the oil fields and oil depots and for maintaining necessary order in the city. The Red Guard was armed from the weapons which we had been secretly stockpiling in the underground and also by weapons sent to Baku from the Turkestani front on Frunze's orders. These arms had been delivered directly to Baku by one of our daredevils, the Communist sailor, Storozhuk. Late at night on April 28, 1920 a proclamation of Soviet rule in Azerbaijan had been issued by a provisional government headed by N. Narimanov, and as we rode through the city in our car, we saw that this proclamation had already been posted on the walls of the buildings:

TO EVERYONE, EVERYONE, EVERYONE!
Moscow. To Lenin.
The provisional Military Revolutionary Committee of the Independent Azerbaijani Soviet Republic, which has come to power at the will of the revolutionary proletariat of the city of Baku and the laboring peasantry of Azerbaijan, declares the former Muscavatist government to be traitors to the people and enemies of the country's independence, and is breaking off all relations with the Entente and with other enemies of Soviet Russia.

Unable with its own forces to hold back the onslaught of the united bands of the external and internal counterrevolution, the Military Revolutionary Committee of Azerbaijan invites the government of the All-Russian Soviet Republic to enter into a fraternal alliance for the purpose of waging a joint struggle against worldwide imperialism and requests immediate practical assistance in the form of a dispatch of Red Army detachments.

The Military Revolutionary Committee of the Azerbaijani Republic:
Nariman Narimanov, Mirza Davud Guseynov,
Gamid Sultanov, Alimov, and Ali Geydar Karayev.

We drove up to the parliament building and on entering it we were seized by an acute feeling of joy. I recalled in the most minute detail the events had unfolded in this very place exactly a year ago at the May Day demonstration . . . I recalled the political rally and the speech I delivered from the truck. I also recalled the concluding words of that speech: "Gentlemen, landowners and capitalists! Know, that by next year's international holiday of the Soviet, Workers', Peasants', and Soldiers' Deputies of Azerbaijan will be sitting in this building!"

It gave me joy to recall all these things now, a year later, on the eve of the May Day celebration, when the workers' and peasants' government of Soviet Azerbaijan was in fact already in session in this building The dream of the Baku revolutionaries had come true. And once again the unforgettable images of our friends, the Baku commissars, rose up in my mind's eye . . . What a sorrow it was that they weren't with us now!

We had not closed our eyes for two nights: there had been no time nor any desire to sleep. The joy of victory and an insatiable desire for activity filled our hearts to overflowing.

Units of the Eleventh Army began to enter Baku on April 30. Up until then, over a period of two days, our armored train the "Third Internationale" had been the only military unit in Baku. The rest of the troops were still on the move.

Ordzhonikidze, Kirov, Levandovsky, and Mekhanoshin arrived in the city, stopping off to pay a visit to our armored train in order to congratulate the Red Army men on a successful raid. Ordzhonikidze awarded Yefremov the military order of the Red Banner, and we were all just as happy about Yefremov's award as he was himself. It was rare at that time for anyone to receive the order of the Red Banner, the only military decoration then in existence. It was awarded in recognition of the great service Yefremov had rendered as a Red commander.

The whole two or three days I had spent with Yefremov left a deep trace in my consciousness. There are times in the struggle, in life, when in the course of a few days one sometimes comes to know a man better than after many years of collaborative work. I had been penetrated through and through with profound respect and comradely love for this man whom I hadn't known before, a man of great charm in interpersonal relations who was fearless in battle, calm and decisive, and who inspired the confidence of his comrades and subordinates. That is the image of this commander-hero which I shall always carry with me in my memory.

I in turn requested that Ordzhonikidze be decorated with the order of the Red Banner for the successful secret delivery of arms from the Turkestani front, and that our Storozhut also be awarded the order for the important role he played in the illegal transporting of gasoline to Astrakhan. The order was presented to him by me, to the applause of the workers who filled the municipal theater.

May Day had arrived. The major segment of the infantry was entering the city at that time, and from early morning. Columns of exhausted, dusty, but joyful Red Army men marched through the streets. From every district, industrial plant, and factory the workers and their families came streaming into the city and filled the streets. A genuine fraternization between the population and Red Army men ensued amid embraces and general rejoicing. It was a remarkable, joyful May Day in Baku. Political rallies sprang up everywhere—at crossroads and in the square.

A workers' conference took place, attended by delegates from every industrial plant and place of business. Warm and excited congratulations could be heard in honor of Lenin, the Red Army, and Soviet Russia. Sergo Ordzhonikidze delivered a speech of welcome on behalf of the government of the All-Russia Federation and the Red Army. He congratulated the Baku proletariat and the entire Azerbaijani nation on the victory of Soviet power and on the raising of the banner of socialism on Azerbaijani soil.

In the course of several days Red Army units had been billeted in all the centers of Azerbaijan, and the Red Army now reached the borders of Menshevik Georgia and Dashnak Armenia.

* * *

The state of calm in Azerbaijan did not last long, however. By the end of May the Azerbaijani feudal lords, who had gathered in the provincial city of Gyandzha (now Kirovabad) along with the ministers, counterrevolutionary officers, and the civil servants who had fled Baku, rose up in armed rebellion. Street fighting ensued, in the course of which the mutiny was suppressed. But hardly two weeks had passed when the defeated units of White rebels made their way to Karabakh and started another rebellion there, which was also suppressed. A short time later counterrevolutionary disturbances began in a few other areas, but they were soon put down. The military operations instituted for routing these White revolts, in which Muscavatist troops stationed here also participated, were carried out by units of the Eleventh Army under the command of Levandovsky and the Military Revolutionary Committee of the Azerbaijani Republic.

I would like to share my impressions of Levandovsky, an exceptional man. I remember Mikhail Karlovich Levandovsky as a decisive and strong-willed commander. Thoughtful and sparing of words, he usually spoke little, but always in a well thought-out and meaningful way. The main character traits of this remarkable man were his great organizational abilities, his outstanding talent as a military leader, and his boundless devotion to the Communist party and the Soviet people.

* * *

As we later found out after the Red Army's arrival in Azerbaijan, the rebellious mood in Aleksandropol (now Leninakan)—the most important party center in Armenia—began to intensify. Under pressure from the masses, the Aleksandropol Party Municipal Committee requested permission from the Armenian Committee (Armenkom) to begin an organized rebellion for the purpose of overthrowing the Dashnak government. On May 10 the Aleksandropol Military Revolutionary Committee proclaimed Soviet rule in Armenia.

On that very day the flag of Red rebellion was raised in Kars. Sarykamysh, Kavtarly, Nor-Bayazet, Shamshadin, and Idzhivan followed suit. Later, it spread to Zangezur as well. In all these places Red revolutionary committees proclaimed the overthrow of the Dashnak government and announced the establishment of Soviet rule in Armenia. The Armenian Communists were very late in reporting all these things to Baku and the Caucasian Bureau of the Central Committee. Therefore, we didn't discuss this issue in Baku and, naturally enough, had no plan for aiding the Bolshevik rebels, since at the time the Red Army was occupied with suppressing the rebel Muscavatist units. Sergo Ordzhonikidze and field headquarters were stationed in those locales. Soon the sad news reached us that after three days of fighting in Aleksandropol, and then, in the other regions, the Bolshevik rebels had been defeated. Many participants in the rebellion had been arrested, and, as we learned later, eleven of the leaders, including Alaverdian, Musayelian, Garibdzhanian, and Gukasian, had been shot.

After receiving a report about this from the Armenian comrades, which came with a request for aid for the Armenian Bolshevik rebels in the Kazakh area, I immediately sent a telegram to Sergo Ordzhonikidze, addressing it to the city of Shusha. It turned out that he was in another region, so the telegram was forwarded to him. It had been preserved in the Ordzhonikidze archive, and here is what it said:

To the chairman of the Caucasian Military Council of the Caucasian front

Comrade Ordzhonikidze

By the morning of the twentieth a delegation from the Armenian Communist rebels will be here. They have telegraphed news of their rout in Aleksandropol and request that armed assistance be given to their rebel forces in the Armenian Kazakh area. The telegram from Karakhan says that Russia has assumed the role of mediator between Armenia and Azerbaijan. All contested territories are temporarily occupied by the Russian Red Army. This circumstance is propitious for bringing the Red Army into Armenia in accord with the request of the insurgent nation. Please advise immediately as to what steps you consider it possible to take.

Mikoyan.

At the time the Red Army units were still busy suppressing the Muscavatist rebels and no one on the local scene could make a decision on such an important political question without the consent of the center. Moreover, the question had been put before us after the rebellion in the main centers had already been suppressed.

Although it ended in defeat, the May rebellion of the Communists in Armenia was an historic event in the struggle for the overthrow of the Dashnaks' antipopulist government, for the establishment of Soviet rule in Armenia, and for its unification with Soviet Russia. The Communist rebellion bore a mass character, which indicated the intensification of the socialist revolution in Armenia. It had a great influence on the working people of Armenia. The defeat suffered by that rebellion paved the way for the later victorious rebellion in all of Armenia which took place in November of that year.

On November 29 Armenia was proclaimed a Soviet Socialist Republic.

On February 28 the Georgian Soviet Socialist Republic was proclaimed. All of Transcaucasia had become Soviet.

* * *

After the liquidation of all centers of White uprisings in Azerbaijan, Sergo Ordzhonikidze, who headed the Caucasian Bureau of the Central Committee, initiated jointly with the Central Committee of the Communist party of Azerbaijan the execution of political and economic measures designed to pull the rug out from under the feet of the landowners and capitalists. Soon a decree was issued from the Azerbaijani Revolutionary Committee concerning the confiscation of gentry lands, their nationalization, and their transfer to the use of the laboring peasantry. This undermined the economic base of the landowners and disposed the peasant favorably to Soviet rule.

At the end of May 1920 the oil industry—more than 250 plants belonging to private companies—was nationalized for the second time (the first nationalization had taken place back in 1918 before the defeat of Soviet power in Baku). The 250 nationalized plants provided the base on which the state association "Azneft" (Azerbaijani Oil) was organized. It was necessary to arrange as quickly as possible the export by sea of the enormous quantity of stockpiled oil to Russia, where there was a critical fuel shortage. The route through Astrakhan had already been opened and navigation had begun. At the same time it was necessary to increase oil output.

Ordzhonikidze organized the local comrades to carry out these important tasks. A. P. Serebrovsky, an old-time Bolshevik who had arrived from Moscow, gave a great deal of assistance. An experienced engineer and a talented organizer, Serebrovsky became the head of Azneft. Relying on the trade union of oil industry workers, the Sailors' Union, and on that segment of the technically skilled intelligentsia which even before had maintained professional contact with us, he got oil drilling operations going and arranged for the export of oil to Astrakhan.

On June 13, 1920, the first triumphant meeting of the Baku Soviet of Workers', Red Army, and Sailors' Deputies took place. Ordzhonikidze delivered the welcoming speech and was elected permanent chairman of the Baku Soviet. At the meeting Sergo was directed on behalf of the Baku proletariat to convey fraternal greetings to the First Cavalry Army in conjunction with its first stunning victory over the White Poles, and to present Army Commander Budyonny and Voroshilov, a member of the Revolutionary Military Council of the First Cavalry, with gifts from the Baku workers—two gold daggers.

Sergo was up to his neck in work; it was hard for him to cope with everything. But at thirty-four he was strong and healthy, always cheerful and on top of things, and didn't seem to know the meaning of exhaustion. We worked without rest, without even thinking of any days off. Sergo was very fond of horseback riding, but we managed only twice to go for a ride together in the environs of Baku, glad of the chance to talk in a tranquil setting.

Concern for the Eleventh Army and Azerbaijan were not Sergo's only responsibilities. As head of the Caucasian Bureau of the Central Committee of the Workers' Communist party (Bolsheviks), he was in charge of the party organizations in all of Transcaucasia and the northern Caucasus, often made trips to Stravropol, Rostov, Grozny, Krasnodar, Vladikavkaz (now Ordzhonikidze), and spoke there at congresses, gatherings, and meetings with the most active and prominent party members. Work on strengthening Soviet power at the local level was in full swing. Complicated issues had to be resolved among the mountain peoples, who had for the first time received the opportunity for free national development. Economic life had to be arranged and regulated. The delivery of grain and other goods to Moscow had to be speeded up: there were large supplies of grain in the northern Caucasus, but the cities of central Russia were suffering severe food shortages.

The Red Army was winning major victories, and the Civil War was nearing its end. On his travels through the cities and villages, Sergo set before the Soviet and party organizations the tasks of reviving the economy and shifting operations to peacetime construction. In August 1920 however, it became clear that the Entente had decided to try one more time to salvage the defeated counterrevolution by inciting bourgeois-gentry Poland to attack Soviet Russia. In spite of several attempts on the part of Lenin and the Soviet government to regulate relations with Poland by peaceful means, the White Poles stirred up war while at the same time the Entente was helping to arm Wrangel who, after consolidating his hold on the Crimea, led Ulagay's landing force to the Kuban and raised a rebellion there.

On August 20, Sergo received a telegram from Lenin which contained the order of the Politburo of the Central Committee of the RKP(B) to leave immediately in order to take part in the destruction of Wrangel's landing force. He left at once and by September 4 telegraphed Lenin, saying that the landing force had been routed in the region of Akhtyrka. Wrangel, however, was still strong, and aimed an attack at the north from the Crimea. His troops were well armed, with England, France and America providing them with all the necessities.

Once again the Central Committee of our party was forced to summon the country to the defense. Following the example of the other party organizations, the Central Committee of the Communist party of Azerbaijan decided to mobilize 10 percent of the total number of Communists for action at the front. The Council of Trade Unions of the Republic established a mobilization order for 2 percent of the union membership for the purpose of filling out the ranks of the Red Army in the struggle with the White Poles and Wrangelites. The Baku Soviet of Workers', Red Army, and Sailors' Deputies passed a mobilization order for 5 percent of its deputies to be sent to the front.

* * *

In September 1920 less than a year after the Second All-Russian Congress of the Communist Organizations of Peoples of the East, the first congress of Eastern peoples was convened in which the representatives of the Communist parties of several foreign countries took part.

The organizers of this congress were the Executive Committee of the Comintern, along with some of the delegates who had arrived in Moscow at the end of June 1920 for the Second Worldwide Congress of the Comintern.

The Central Committee of the Workers' Communist party (Bolsheviks) and the Executive Committee of the Comintern created the Organizational Bureau which, on behalf of several countries, appealed to the peoples of the East with the proposal to prepare for a congress of representatives of the working peoples of the entire East.

Sergo Ordzhonikidze and Elena Stasova conducted the basic organization work involved in preparing for the congress. Since I was a member of the Organizational Bureau, I was also knowledgeable about all issues.

Delegates were expected at the congress from the nationalities of the former Russian Empire, including the Transcaucasian Republics and Dagestan; from the other northern Caucasian nationalities, from the peoples who lived in the Turkestan Republic; from Khiva, Bukhara, and Bashkiria; from the Tatars and Kalmyks; from China, India, Afghanistan; from Persia, Turkey, Japan; from the Arabs and the Kurds.

It was necessary to get in contact with all the delegates, but it was difficult to arrange contacts with the Persians and Turks. Since, however, a Bureau of Turkish Communists and a Persian Bureau already existed in Baku, the work of attracting delegates from their countries was conducted through them. They also helped us to establish contact with the Kurds and the Arabs, and with India. It was decided to draw not only Communists into the congress, but also representatives of national revolutionary organizations, and nonparty anti-imperialistic prominent figures from the countries of the East.

The city of Baku was unanimously chosen as the place for convening the congress. It was not by chance that this city was selected as the meeting place of experienced proletarian fighters from the West, and vanguard strugglers for the liberation of the peoples of the East. This working-class city, which had undergone the oppression of both native and foreign capital, was a center of renascent socialist consciousness in the East, the arena of the political and economic struggle for the liberation of all, not only the Azerbaijani people. Now a great honor had been bestowed upon Baku—the honor of being the place where representatives of thirty-seven nationalities would gather with the aim of joining forces in the common struggle against imperialist oppression for the liberation of the peoples of the East.

One thousand, eight hundred and ninety-one delegates attended the congress; one thousand two hundred and seventy-three of them were Communists. Thus, a third of the delegates had no party affiliation. Fifty-five were women.

Representatives of several Communist parties in Europe and the USA came as guests: Bela Kun from Hungary; Welch from England; Rossmer from France; John Reed from America, and others. China was represented at the congress by eight delegates, one of whom, Van, was elected to membership in the Council for Action and Propaganda in the East, which was created at the last session.

The congress took place at an historic moment. The country had just survived an enormous wave of foreign intervention. The venal mercenaries of imperialism—Kolchak, Denikin, and Yudenich—had been routed, and England had begun to initiate peace talks with us, while at the same time inciting White Poland against us. It was just at this time that the Red Army had suffered a defeat after beginning to chase out the White Poles. This defeat had encouraged the capitalists once again, but now they were confronted by the fact that the wave of anti-imperialism had spread from Russia to the peoples of the East.

The congress began its work on September 1, 1920. The Baku Soviet, on meeting the day before to wind up its preparations for the congress, decided to organize a welcome for the representatives of the Comintern and the delegates from the Communist parties of several countries who were arriving that night by train. The Soviet deputies and numerous workers went to meet the train and escorted the guests straight from the station to the theater where the session of the Soviet reconvened. It ended at five the next morning. The Comintern representatives held forth with fiery orations as one after another they gave their opening speeches.

In his brilliant speech John Reed said:

> I represent the revolutionary workers of one of the great imperialist powers, the United States of America, which is exploiting and oppressing the colonial peoples.

"You peoples of the East and of Asia have still not experienced the power of America," he said.

> You know and detest the British, French, and Italian imperialists and probably think that "freedom-loving America" will do better, that she'll feed and protect and liberate the colonial peoples. No—the workers and peasants of the Philippines, Central America, and the Caribbean islands know what it means to live under the rule of "freedom-loving America."
>
> The Filipinos, for example, in 1898, with the help of the Americans, rebelled against the harsh colonial regime of the Spanish government, but after the Spanish had been chased out, the Americans did not want to leave. Then the Filipinos rose up against the Americans and this time the "liberators" began killing them and their wives and children; they tortured them and finally conquered them, seizing their lands, and forcing them to work and earn profits for the American capitalists.
>
> The Americans promised the Filipinos independence. An inde-

pendent Philippine Republic will soon be declared, but that doesn't mean that the American capitalists will leave the Philippines, or that the Filipinos will not continue working to create profits for them. For the American capitalists have given the Filipino rulers a part of the profits—granting them government positions, lands, and money, and creating a Filipino capitalist class which also lives off the profits generated by the workers. And in whose interest is it to maintain the Filipinos in slavery?

It has already happened in Cuba, which was liberated from Spanish control with the help of the Americans. It is now an independent republic, but American millionaires own all the sugar plantations, with the exception of small portions of land which they left to the Cuban capitalists who also rule the country. And as soon as the workers of Cuba try to elect a government which is not in the interests of the American capitalists, the United States of America sends soldiers to Cuba to force the people to vote for their oppressors.

The farsighted young revolutionary's speech, delivered a half-century ago, gives a picture of the American imperialists' contemporary methods, which they use for oppressing the peoples of Asia, Africa, and Latin America. John Reed's speech was the last he made at Communist forums. Having contracted typhus, he died in that very year, 1920, in Russia. Paying homage to the American revolutionary, the Soviet people buried him in Moscow, in Red Square near the Kremlin Wall. His book, *Ten Days That Shook the World*, has been printed in millions of copies and will attract the interest of future generations. Lenin had a high opinion of John Reed, one of the founders of the Communist party of the USA.

The speech of the Hungarian delegate Bela Kun rings just as true in terms of the present day as does John Reed's. "For the purpose of pacifying the colonial peoples, the imperialist-exploiters are mobilizing the European workers," Bela Kun said,

> whom they are trying to lure to their side with a bribe, a tiny bit of the excess profits which they have wrung out of the colonial peoples; this has always been the case in England and in Germany. In this way they want to deflect the workers from the revolutionary path. On the other hand, and particularly of late, the imperialistic bourgeoisie is seriously considering sending recruited colonial troops against the European workers by taking advantage of the colonials' political backwardness; in this way they hope to protect their tottering ruling power from the pressure of the working class.
>
> I, comrades, personally witnessed such an attempt. When we, the Hungarian workers and the poorest peasants, seized power into our own hands, the French bourgeoisie made an attempt to crush our revolution by using Moslem colonial troops. But in spite of the dif-

ficulty we had communicating with those soldiers because of our different languages, we nevertheless managed to find a way to the hearts and minds of the colonial troops. Realizing that the imperialists wanted to force them to drown the revolution in blood, they threw down their weapons.

The imperialistic bourgeoisie usually manages to find in colonial countries that stratum of the local population and in semicolonized countries that ruling class with whose help it can insure that its exploitative policies become less difficult, cheaper, and less costly in terms of human life . . .

The delegate from France, Rossmer, also spoke about the self-seeking policies of imperialism:

When it was necessary to struggle against the Germans, when it was necessary to mobilize some hundred thousand Algerians, Tunisians, and Moroccans, then they were promised various freedoms, but the day after the defeat of Germany all these pitiful freedoms were taken away.

The delegate at the congress from the British Communist party, Welch, made a brilliant speech. He said that "the enemy of the British working class—the British imperialists—is at the same time the enemy of the peoples of the East, of the oppressed East." Therefore, the struggle of the British working class against British capitalism is also the struggle of the oppressed peoples of the East against it.

The meeting was permeated with a great upsurge of enthusiasum, people spoke in inspired tones. Everyone's spirits were lifted, the reigning mood was one of militant readiness to fight for Soviet power against the White Poles, against Wrangel. The speeches and the audience itself also made an unforgettable impression. The delegates, seized upon by a single, common spirit of enthusiasm, rose, and some, shaking the weapons they had with them, vowed to struggle hand in hand with the European workers against the oppressors.

The same enthusiasm and awakened will to struggle for their own liberation accompanied the speeches of the delegates who represented the vanguard strata of the Asian nationalities. The delegate from India, Fazma Kadyr, read a declaration written by a certain Indian revolutionary organization and signed by its president, Mahomed-Abdur-Rabe-Berk. The declaration said that on behalf of the 315 million oppressed people of India, he was appealing to the congress delegates and the representatives of Soviet Russia with a request for aid to India. "Everyone hopes that the aid will be given without any attempt to interfere in internal and religious life."

The speech of the Kazakh delegate Ryskulov also rang out in very impassioned tones. He said that the major task in the unification of working people is the unification of the uncoordinated revolutionary movement in the East with the movement in the West. It is for this question that we have come together here, and this is the question we are resolving here.

The appearance on the rostrum of the Turkish woman Nadzhia Khanum was greeted by wild applause from the audience. Addressing the male delegates she said:

> Hear our demands and give us some real help and cooperation: 1) full equal rights (for women); 2) unconditional admittance for women to male educational institutions and trade schools; 3) equal rights for both partners in marriage; 4) unconditional entry of women into the work force of legislative and administrative insitutions; 5) the universal establishment, in cities, towns, and villages, of committees for the rights and protection of women.
>
> The Communists, who have recognized equal rights for us, have extended us their hand, and we women shall be their most loyal companions.

And Bibinur, the representative of the women of Turkestan, who greeted the congress on behalf of the women proletarian workers of the city of Auela-Ata (now Dzhambul), assured the delegates: "We women have awakened from oppression, from a nightmare, and with each and every day we are pouring our best forces into your ranks. We also must work untiringly for the liberation of all the oppressed peoples of the East."

The congress applauded the proposal to seat the following women on the presidium: Bulach from Dagestan, Nadzhiya Khanum from Turkey, and Shabanova from Azerbaijan.

Unfortunately, Sergo Ordzhonikidze—one of the chief organizers of the congress—could not be present since he was taking part in the destruction of the Wrangel landing force in the Kuban, but he sent a telegram of warm greetings to the Congress of the Peoples of the East.[28]

Many important questions were discussed at the congress: the international situation and the tasks confronting the laboring peoples of the East; the Soviets in the countries of the East; agrarian, national-colonial, and other issues. The entire work of the congress was done under the signpost of unity with the resolutions of the Second Congress of the Communist Internationale which had ended not long before in Moscow.

Vladimir Ilyich Lenin kept a close eye on developments in the work of these historic forums. Highly valuing the significance of con-

[28]See Appendix.

gresses, he said in a speech of October 15, 1920, at a meeting of representatives from the districts, *volost,* and rural executive committees of the Moscow province.

> What has been accomplished by the congress of Communists in Moscow and the congress of Communist representatives of the Eastern peoples in Baku is something that cannot be gauged immediately. It does not lend itself to purely quantitative analysis; but it represents the sort of conquest that means more than military victories, because it shows us that the experience of the Bolsheviks—their activities, their program, and their summons to revolutionary struggle against the capitalists and imperialists—have won recognition for them all over the world. What was accomplished in Moscow in July and in Baku in September will be assimilated and digested by the workers and peasants in all the countries of the world for many months to come.
>
> . . . It is the international congresses which have united the Communists and demonstrated that in all civilized countries and in all Eastern countries the Bolshevik banner, the program of bolshevism, and the image of the Bolsheviks' actions represent for the workers of all civilized countries, and for the peasants of all politically backward colonial countries, the banner of salvation, the banner of struggle. They have shown that during the past three years Soviet Russia has actually not only beaten back those who attacked her in order to crush her, but has also won for her the sympathy of working people all over the world, and that we have not only defeated our enemies, but have acquired and are acquiring, day by day, and hour by hour, allies for ourselves.

In the summer of 1920, when the country was gearing itself towards a peaceful reconstruction, a new military threat to Soviet rule appeared on the horizon; the troops of the White Guard General Wrangel, which had been rearmed by the Entente, went on the offensive. The dangerous hotbed of civil war flared up once again.

Having received the decision of the Party Central Committee to send me to work in the presidium of the Nizhny Novgorod Party Provincial Committee I requested the Central Committee to change its decision and send me to the Wrangel front. The Central Committee, however, left its decision in force.

At the end of September 1920 I moved to Nizhny Novgorod (now Gorky), and after several months there I was elected secretary of the provincial party organization.

The period of my life in Transcaucasia had come to an end.

Appendix

1. Editor's note: When this book was in press, Chief Designer A. I. Mikoyan died. The following announcement appeared in the papers on December 10, 1970:

> From the Central Committee of the Communist Party of the Soviet Union, the Presidium of the Supreme Soviet of the USSR, and the Council of Ministers of the USSR.
>
> The Central Committee of the Communist Party of the Soviet Union, the Presidium of the Supreme Soviet of the USSR, and the Council of Ministers of the USSR announce with deep regret the death of Academician Artem Ivanovich Mikoyan on December 9, 1970, at age 66 after a long and serious illness. Academician Mikoyan was a leading aircraft designer, a Deputy of the Supreme Soviet of the USSR, twice a Hero of Socialist Labor, a Lenin and State Prize recipient, and a Colonel-General of Technical Engineers.
>
> | The Central Committee of the CPSU | The Presidium of the Supreme Soviet of the USSR | The Council of Ministers of the USSR |
> |---|---|---|

2. The order reads as follows:

> A description should be made of all published materials not approved by the censor that are uncovered by the search; of any and all correspondence; of photographs, visiting cards, and addresses, as well as of all objects, the personal possession of which is forbidden, or which could testify to the criminality or criminal intentions of the subject of the search, after which these objects should be affixed with a seal and labeled as to ownership and place of discovery; and, according to the protocol of search and seizure, they are to be dispatched immediately, so labeled, to the District Police Administration.
>
> A special listing should be made of whatever documents and valuables (belonging to the party in question) are taken. These should be sealed in a separate packet, described in detail, and also dispatched to the administration. Seized property will be retained in custody pending the results of the search.
>
> All persons found in the apartment of the subject of the search who are without documents, or unknown to the police, or whose

documents are of doubtful authenticity, shall be unconditionally detained until such time as proof of identity is made known.

3. In 1918, when I was a commissar of a Red Army brigade at the front during the Turkish attack on Baku, I happened to read something about Andranik in the newspaper, *The Baku Worker*. I hadn't heard anything about him since 1915. Two telegrams were published in the paper. One, dated June 14, 1918, was sent by Andranik to Shaumian, the Special Commissar of the Caucasus. Andranik announced that he had proclaimed the Nakhichevani district, where he was stationed with his troops, as an inseparable part of the All-Russian Republic, and he requested that Shaumian make it known to the proper authorities that he and his troops were at the disposal of the All-Russian Central government and would take their orders from it. The second telegram was from Shaumian: "Dzhulfa. To the national leader Andranik. Received your telegram. Communicated full text to Moscow Central. My personal greetings to you, a real hero . . ."

After the temporary capitulation of Soviet power in Transcaucasia, Andranik, finding himself in the newly formed bourgeois state of Armenia, could not reconcile himself to the policies of the Dashnak government which was pro-English and pro-American and in alliance with Denikin. He left Armenian territory, therefore, and emigrated to Bulgaria through Batumi. It's noteworthy that after the establishment of Soviet power in Armenia, Andranik made a gift of his sabre to the Erevan Museum as a token of his admiration for Soviet Armenia.

4. It might perhaps be of some interest if I were to list the most memorable of my comrades from the Tiflis seminary:

Gevorg Alikhanian—later an active party worker in Tiflis, Baku, Armenia, the northern Caucasus, and Leningrad. Subsequently, he became head of one of the departments in the Comintern.

Alikaz Kostanian—an outstanding party organizer. For several years Secretary of the Central Committee of the Armenian Communist party, he later became Secretary of the Profintern.

Vagan Balian—a key party worker, Assistant Chief of Political Administration of the People's Commissariat of Communications of the USSR.

Napoleon Andreassian—after an active party career in the Caucasus, he was elected First Secretary of one of the Moscow Party Raikoms (District Committees), became a member of the Bureau of the Moscow Party Gorkom (City Committee), and after that, worked as a Deputy Peoples' Commissar of the Food Industry of the USSR; a delegate with voting rights to two party congresses.

Suren Akopian—First Secretary of the Vyatka (now Kirov) Party Obkom (Regional Committee), and later an Inspector of the Central

Committee of the All-Union Communist party (Bolsheviks).

Karo Alabian—after the Civil War, he went to architectural school and became one of the outstanding leaders of Soviet architecture, for many years occupying the post of Head of the Architects' Union of the USSR and Vice-President of the Academy of Architecture of the USSR; a street in Moscow is named after him.

Shavarsh Amirkhanian—played a crucial part in the establishment of Soviet power in Armenia. Later, in Moscow, he was one of the leading workers at the Ministry of Communications of the USSR.

Aram Shakhgaldian—did important party work in Armenia, then in Moscow.

Artak Stamboltsian—was actively involved in party work in Transcaucasia, then studied Marxism-Leninism under the auspices of the Central Committee of the All-Union Communist party (Bolsheviks); was the head of the Political Section of the Omsk Railroad, in the organizational division of the Central Committee of the All-Union Communist party, and Secretary of the Lisichan Party Gorkom.

Gevorg Gevondian—an active Communist, he went to medical school and became one of the founding fathers of Soviet public health. For many years he was Peoples' Commissar of Health in the Armenian Soviet Socialist Republic.

Gevorg Abov—poet, scholar and literary researcher, Director of the Institute of Ancient Manuscripts of Armenia.

Vagan Eremian—first Peoples' Commissar of Finance for Armenia, and then for the entire Transcaucasian Federation.

Garegin Gardashian—one of the organizers of the Transcaucasian Komsomol. After the Civil War he went to medical school and became a gifted surgeon. He and Ochkin were the chief assistants to the famous Soviet surgeon, Rozanov, and then he became the Peoples' Commissar of Health in Soviet Armenia.

Ovanes Pogossian—a leading member of the Communist party of Armenia, a member of the Bureau of the Central Committee of the Party of the Republic; rector of the Erevan State University.

Daniel Dznuni—for many years a leader of cinematography in the Armenian Republic.

Suren Avetissian—became an outstanding doctor.

Telemak Arutiunian—received a higher education and went on to become a professor.

Arutiun Grigorian—headed a district party organization. He participated actively in the struggle with the Dashnaks, in which he perished heroically.

Sergei Parsadanian—an old Communist and economic planner, the head of the largest flower-growing combine in our country.

Tigran Galstian—was the editor of a Moscow journal.

Artem Bablumian (Yetum)—an old Communist, who occupied

high positions in leading party organs and in the press.

Mikhail Sarkissian—the former chief radiologist of the Abkhazian Autonomous Soviet Socialist Republic, now living on a retirement pension.

Shamrkhanian—a well-known figure in the Armenian theater.

Drastamat Ter-Simonian—one of the leaders in the May Rebellion in Armenia against the bourgeois government, later a well-known figure in Republic affairs.

Tatevos Mandalian—a participant in the Shanghai Rebellion, one of the leading workers of the Profintern.

Gevorg Perikhanian—an important power-engineering specialist in Armenia.

Gurgen Voskanian—a well-known Soviet and party worker.

Sedrak Markarian—former leading worker in the apparatus of the central security organs.

Ervand Kochar and Mikael Mazmanian—active Communists who achieved success in the fields of sculpture and architecture.

Artavazd Egiazarian—an old Communist who served as Peoples' Commissar of Education in the Armenian Soviet Socialist Republic, and is presently involved in scholarly activities; a Doctor of Sciences.

Semen Sarkisov—Academician of Medicine, Director of the Institute of the Brain.

Ovaness Bannayan—a leading party worker in several districts; at one time First Secretary of the Novorossiisk Party Gorkom.

This is by no means a complete list of those comrades—from both the upper and lower forms—who were classmates of mine. All of them led lives of selfless service to the great cause of our party and the Soviet people. I must note that among them was the outstanding military commander and hero of the Civil War, Gai (Bzhshkian).

I can't help drawing attention to still another fact of note. I have a photograph which was taken in my Kremlin apartment in Moscow. It shows a group of comrades, six of whom were my classmates and members of the Marxist circle. Like myself, Alikhanian (Alikhanov) —Leningrad, Akopian—Moscow, Amirkhanian—Armenia, Artak (Stamboltsian)—Azerbaijan, and Balian—the Don Region—were delegates to the Fifteenth Party Congress. Thus, seven voting delegates to the All-Union Party Congress had been schoolmates at the Nersesyan Armenian Seminary!

Many of my classmates are no longer living. Some of them —Alikhanian, Stamboltsian, Kostanian, Balian, Gardashian, Akopian, Eremian, and Markarian—unfortunately fell victim to the violation of socialist legality, but they were all posthumously rehabilitated, and the honor due them as faultless Communists has been completely restored.

5.   TO THE HIGHLY ESTEEMED BOARD OF DIRECTORS
OF THE ECHMIADZIN GEVORKIAN
ARMENIAN RELIGIOUS ACADEMY

From Anastas I. Mikoyan,
graduate of the Nersesyan School,

A PETITION

Having completed the entire course of study at the Tiflis Nersesyan Armenian Religious School, I hope to continue my education and am therefore petitioning the highly esteemed Board of Directors to accept me as a first-year student at the Academy.

I am enclosing with this petition a copy of the transcript of my academic record. I shall provide a birth certificate in person, but if necessary, reference can be made to the copy of my transcript where my birthdate is noted.

Anastas Mikoyan.

6.   The first issue of the Union's newspaper, *The Proletariat*, published the Union Manifesto, which had been written by Shaumian. In *The Spark* Lenin wrote as follows about the Manifesto: "With all our hearts we hail the Manifesto of the 'Union of Armenian Social Democrats' and we welcome its especially remarkable attempt to offer a correct formulation of the national question."

7.   On December 6, 1913, Vladimir Ilyich wrote:

Dear Friend! I was very happy to receive your letter of November 15. You must know that a person in my position really appreciates the opinions of his comrades in Russia, especially those who are thoughtful and knowledgeable in a given matter. Your quick response was, therefore, particularly pleasing to me. When I receive letters like yours, I feel less cut off from things.

Later on in the letter there was this request: "A popular pamphlet on the nationality question is very much needed. Write one. I await your reply and extend you my warmest regards."

8.   Perhaps it is enough merely to list the titles of only a small part of these articles in order to suggest the kinds of issues that excited us all so keenly at the time: "Fraternization at the Front"; "Will the Bolsheviks Hold on to State Power?"; "Nationalist Peasant Unions and the Situation in the Caucasus"; "The Meaning of Peace Without Annexations and Reprisals"; "Who Are the Bolsheviks and Who Are the

Mensheviks?"; "Concerning the Latest Strike"; "A Collective Agreement"; "The Eight-Hour Working Day and the Present Moment"; "The Agrarian Question and the Landowners' Counterrevolution in Transcaucasia"; "The Georgian Nationalists and Democracy"; "Forms of Bourgeois and Proletarian Power"; "New Tasks and Methods in the Workers' Movement"; "The Peace Policy of the Soviet of Peoples' Commissars" (in defense of the Brest Peace); "Menshevik Hypocrisy"; "Concerning the Purity of the Mensheviks' Principles"; "The Defenders of the People"; "The Left-Wing Dashnaks"; "To the Renegades of Marxism"; "The Nationalization of the Petroleum Industry"; "Poverty and Our Debt"; "The Housing Question and Capitalism"; "On the State of Affairs at the Front and the Gathering-In of the Harvest"; "The Tasks of Cultural Enlightenment Confronting the Workers' Soviet."

9. Radiogram to the Tsentrokaspian Dictatorship

On the night of the fourteenth by the decree of the Dictatorship, and by order of Comrade Dalin, President of the Special Commission, we were released from Bailov prison. The guards who came to get us told us that we were allowed to board the ship *Sevan*, where there was a delegation of the Astrakhan Soviet and naval fleet, which had come regarding the issue of our release. Since the *Sevan* was not to be found at the naval docks, where it was supposed to be to receive us, the guards took us to the city to the Special Commission. There was no one there; so they then took us to the nearest pier Kamvo, to the ship *Turkmen*. There the convoy guard of ten men, in view of the evacuation of the Special Commission and the fact that no other ships were present, asked the head of the detachment to take them on board the *Turkmen* as well—a request which was permitted them. The ship *Turkmen* headed for Petrovsk on the fifteenth, after a temporary stop-over at Nargen and Zhiloy [Island], as part of a convoy of many ships according to the directives of the military judges. On the morning of the sixteenth we found out that the ship was setting course for Krasnovodsk since the north wind which had risen was making movement to the north difficult, and there was little water and fuel on board, whereas there were as many as 800 people on board, primarily women and children, who were completely without provisions. As it turned out, the ship's administration and the head of the detachment decided, given the circumstances, to go straight into the wind and head for Krasnovodsk, quickly disembark the refugees, and after taking on fuel and water in Krasnovodsk, return to Petrovsk again.

In the morning the sailors informed us that Krasnovodsk would not receive the refugees, and that, on the other hand, the weather had improved. The question arose whether it wouldn't be best to take the refugees to Astrakhan, although there was fear that in Astrakhan the Bolsheviks would not receive the refugees, but would impound the ship, etc. It would, of course, have been better for us as well if we could have gone to Astrakhan, in order to avoid the least possible unpleasantness in Krasnovodsk, where a struggle was being waged against the Bolsheviks, and we promised to insist, once we got to Astrakhan, that the refugees be received, and the *Turkmen* returned to Petrovsk. After this, the ship set course for Astrakhan. But in view of the ship committee's protest, and unwillingness to go to Astrakhan, the course was again set for Krasnovodsk, where we arrived on the evening of the sixteenth. Here, due to the provocative statements of some of the passengers, doubts arose among the local leadership as to whether we had escaped from the prison during the panic and had left Baku without the consent of the Dictatorship. This resulted in our arrest, and that of a dozen others, including the head of the detachment himself, Amirov.

We ask the Tsentrokaspian Dictatorship to confirm our release and the fact that we had left on your orders. If you don't find it possible to see to it that we are sent to Astrakhan, then let us be sent to Petrovsk. You know about our situation, there [i.e., in Petrovsk] you can be apprised of all the circumstances surounding our ship to Krasnovodsk from the passengers and the crew of the ship *Turkmen*, who are leaving for Petrovsk. We request prompt attention to our request.

"Signed by all of us:
Dzhaparidze, Fioletov, Shaumian, Korganov, Petrov, L. Zevin."

Out of tactical considerations, we departed somewhat from the facts in the radiogram. I too wrote a telegram in the same spirit.

10. Later we found out that Moscow had also taken steps in arranging an exchange of prisoners which involved us. It is interesting that one of the persons involved in the exchange of prisoners was the outstanding, progressive Armenian writer, the president of the Caucasian Society of Armenian Writers in Tiflis, Hovaness Tumanian. In March of 1919 (at that time we had already been released from prison) Tumanian communicated to the Soviet of People's Commissars in Moscow:

> It has become known to Armenian public figures that the Soviet government has agreed to release Doctor Zavriyev and the writer Nazariyan if the British command releases Varvara Dzhaparidze and Shaumian's two sons. By means of influence exerted on various circles during the past few days, their transfer to Baku and release from prison has been successfully obtained. Although they were let out on bail, they are living in complete freedom. In view of this, we hope that the Soviet government will now release Doctor Zavriyev and the writer Nazariyan.

Upon receiving an answer from Moscow, Tumanian telegraphed Avanesov, the secretary of the All-Russian Central Executive Committee, on April 12, 1919:

> Your radiogram of March 30 has been received. Varvara Dzhaparidze and Shaumian's sons are now free without bail or any other restrictions on their right to travel, a fact which Dzhaparidze herself reported to Stalin in a letter which was sent to Moscow by special courier. For the time being neither Dzhaparidze nor Shaumian's sons have expressed any desire to leave the Caucasus. Dzhaparidze will go to Russia when she learns from Stalin the whereabouts of her children, as she wrote him. Therefore, an exchange of prisoners is no longer necessary, and we request that Zavriyev and Nazariyan be given complete freedom to leave.

It is clear that Tumanian had established contact with Varvara Mikhailovna Dzhaparidze either directly or through a third party, and had sent this radiogram to Moscow after learning the actual state of affairs. Thus, it turned out that Tumanian was the first to report our release to Moscow.

11. The all-Russian proletariat is a fighting detachment of the world-wide proletariat, and is struggling to annihilate the oppressor class with the goal of establishing a just order; the leader in this struggle is the Communist party. The slogans of the Communist party are those of the international proletariat. The proletariat is making its way along the path of class struggle. For a year and a half Soviet power showed the international proletariat that it was possible to establish a just order; it is struggling against the counterrevolutionaries and will continue to do so. The example of the all-Russian proletariat has forced people in the west to do what Soviet power is doing here—for example, the breaking up of the German Constituent Assembly. The cause of revolution is going well, and a world revolution is at hand. If the Great Russian workers are struggling with the British in Murman, then the Baku workers cannot help being the enemies of the British here. The Baku proletariat is a detachment of the international proletariat and cannot prevail with the aid of conversations with the British command. We can triumph only by relying on the tried-and-true method—strikes. The Strike Committee has not been in control of the situation. Rather than struggle actively, the Strike Committee has consented to humiliating negotiations with General Thompson. At the strike committee meeting the Mensheviks behaved like traitors. They write against the strike in their newspapers, thus lowering the dignity of the workers and paving the way for the failure of the strike.

    What should our policy be? Comrade Churayev was right when he said that it is better to be defeated in an unequal struggle than to give up without a fight. The Communists support a strike, but in their opinion the strike should be called only after things have been well-organized.

A life and death struggle confronts us all. One must look at things seriously. It would be better to have fewer soldiers in our camp—soldiers who are devoted to the cause of the workers. With such soldiers we can go boldly forth to battle.

I propose that the existing strike committee should be reconstituted. It should not include people who have compromised themselves through humiliating negotiations with General Thompson.

It is also necessary to reelect the presidium of the conference. Only under such conditions can we win both a real and a moral victory.

12. Mikoyan: Points out that the withholding of vouchers and salaries came about largely as a result of the actions of the British command. Seeing that the Baku proletariat is preparing for a major political battle, it has purposely not paid the workers their salaries from March 7 to the seventeenth with the aim of breaking the political strike and quashing the revolutionary movement in Baku "with the bony hand of hunger," as the millionaire Ryabushinsky said. All this is known to and entirely supported by the Azerbaijani government and the oil-manufacturers. In countering all the intrigues of the united counter-revolution, we should not allow ourselves to be provoked into acting when they want us to. But, nevertheless, we are confronted by the fact of the Nobel workers' spontaneous economic strike. However much one may condemn the separatist action of the Nobel workers, one cannot deny that even without the benefit of close contact with the entire Baku proletariat, the Nobel workers have plenty of revolutionary energy and internal "Nobelian" organization, and that they waited for an answer from the Council of Trade Unions for four whole days, and when it didn't come, they decided not to wait any longer. We suggest that all comrades approach this grave issue seriously and disinterestedly and not simply relegate it to a matter of the honor or pride of one or another group. I am categorically against Ilin's suggestion of declaring an economic and political strike simultaneously. At the given moment, when the worker is hungry, only an economic strike is possible.

But if it is possible, we should avoid this strike since we confront a struggle for a collective agreement. It must be obtained by means of issuing an ultimatum for the immediate payment of wages. The conference should be convened, and a general economic strike should be called only in the event of a continued refusal to pay wages . . .

13. Comrade workers! Today we all join together as one and proclaim to the vile executioners, base traitors, and all the obvious and hidden

enemies of the proletariat that our leaders—great fighters—fell at the filthy hands of the executioners. They have passed away, but their memory is eternally alive in our hearts, and their sacred cause has not died: it has grown apace and acquired broader, more powerful dimensions as the best way of immortalizing their memory. We shall close ranks more tightly, and struggle more energetically for their and our common ideals.

May our protest of today against the international imperialists serve as a laurel wreath upon the fresh graves of the fallen fighters.

Comrades! Let us make a sacred vow that we will self-sacrificingly and devotedly continue the struggle, on the thorny path of which our leaders and comrades heroically lost their lives!

May the memory of the fallen fighters live on forever!

14. "Exactly six months ago," I wrote in this article,

in the desolate Transcaspian steppe, far away from people's eyes, an egregious crime was committed. At the silent signal of the ringleaders of south-Great Russian counterrevolutionary gangs, a handful of counterrevolutionary bandits savagely shot and tore to pieces, without trial or investigation, the great leaders of the Communist party in the Caucasus who had led the movement of the Baku proletariat over a period of decades. During the horrible execution the cowardly criminals and base traitors, acting with inhuman savagery, subjected the unarmed heroes to filthy insults and unimaginable torture. Only the silent steppe and the pale moon, that lighted up the bestial execution of these great harbingers of sacred ideals, bore disinterested witness to that foul, nightmarish tragedy.

Today, in memory of its heroically fallen comrades, the Baku proletariat must put aside all other work and all other thoughts and celebrate with great solemnity the bright memory of its fallen leaders . . . Today the entire Baku proletariat must demonstrate its most magnificent hatred and contempt for the insolent imperialists and their yes-men. It must proudly and boldly declare for all to hear that the words and deeds of its great leaders have not remained "a voice crying out in the wilderness," but have penetrated to the very core of the hearts and minds of the broad masses of the proletariat, revolutionizing and nurturing them in the spirit of communism.

Yes, in savagely shooting the great members of the proletariat, Communist family, the worldwide imperialist aggressors thought they would deprive the working masses of their leaders. But they were mistaken in their calculations, for it is impossible to deprive the proletarian family of its leaders; it is impossible because each and every member of our great commune is in himself, by his very nature, a leader and a stalwart fighter. Our leaders have left us—we will replace them. They struggled to the end of their lives, and we shall struggle to the final destruction of the power of our cowardly enemies. They are dead, long may they live!

15. Iskra *(The Spark)*, No. 93, May 20, 1919.
From the Central Prison.
DECLARATION

> To the socialists in parliament—
> to be read aloud at a session of the Azerbaijani parliament
> and put in print

In all of its speeches, reports, resolutions and actions the Baku Workers' Conference, its presidium and the Central Strike Committee have emphasized in every way possible that the strike is a purely economic one, no matter how much our adversaries insist, for reasons of provocation, on its political nature. And indeed, the demands which we have made for a collective agreement and for the opening of trade relations; the recognition of these demands in every government declaration; and the means we have planned to use for achieving our demands—all these things testify to the economic nature of the strike.

However, the paper *Azerbaijan* and the official representatives of the government, basing themselves on facts and documents known to be false, have advertised and tried to prove in every way possible that the strike is political and has as its goal an armed insurrection against the independence of Azerbaijan, by which they justify their police repressions against the workers, their mass arrests, and their use of armed force to put down the strike.

The representatives of the government and Nicholas II's gendarmerie have gone so far in their provocational zeal that on the barges in the city there have been "discovered" whole stores of arms, cartridges, rifles, bombs and long-range guns allegedly belonging to the Workers' Conference.

All these "facts" and all this "material evidence" are so blatantly fabricated and false, that there's no need to refute them. But since they may acquire a trace of veracity among the masses, we deem it necessary to rebuff the provocational "activities" of the slanderers who have gone too far.

The store of arms discovered on the barge is a legacy from Bicherakhov and the disarmed Caspian Fleet, and in the course of a month, by order and with the knowledge of the British authorities, was under the jurisdiction of the head of the port authority, Denikin's agent, Captain Grigorev, a fact well known both to the Azerbaijani police and the government. Taking advantage of an opportune time in the struggle, the government "discovered" without a touch of shame or a pang of conscience Denikin's store of arms which was long known to them, and, without giving the matter much thought, fortuitously declared it to be the "property" of the Workers' Conference. There were quite a few of these Denikin stockpiles, which were created with the aid and in the full knowledge of the authorities, and, very likely, there are still such in the city, which will now be "discovered" and attributed to the conference. When circumstances didn't allow them to lay the blame on someone else (as was the case with the arrest of Voskresensky, which was carried out only on the insistence of the socialists), the authorities hushed up the fact that they had found the arms' stockpiles of the official representative of the Denikin counterrevolution, Voskresensky, and that they had released him immediately and given him the opportunity to travel to Denikin "with a report."

In all these actions of the government agents we see their secret deals with the Denikinites and their open agreement with the imperialist invaders of Transcaucasis which has the aim of depriving the Baku proletariat of its leaders, subjecting it to blood-letting, so that the united combined reaction of foreign invaders-imperialists, local and all-Russian counterrevolutionaries can set itself up more firmly on the innocent corpses of starving workers.

But we're warning them in advance that they won't succeed in luring the Baku proletariat into bloody adventurism. All authentic and unfabricated facts show that the Baku proletariat was not intending to organize an armed revolt against Azerbaijan, but, rather, was attempting by means of a peaceful strike to realize their economic demands which had reached a head. Let the provocateurs who have overstepped the bounds show that during the arrest of the members of the Central Strike Committee and the district strike committees, or during the search of their apartments, even one pistol or bomb was found, or one word in a piece of paper about an armed aggression, or even some insignificant indication of preparation for military operations. An impossible task for anyone!

Besides, many Moslem workers, and not only socialists, but Muscavatists too, took an active part in all the work of the Workers' Conference, the Central Strike Committee, and the district strike committees. The member of parliament, Comrade Karayev, as the comrade [assistant to] the Chairman of the Workers' Conference and as a member of the Central Strike Committee, was informed about everything that was happening, and nothing was done without him. Comrade Abilov, a member of parliament, was also aware of what we were doing. Can any one of them say with a clear conscience that there was either the slightest hint in our work of armed struggle or any basis for considering it? No one!

If there were any impartial investigators or judges to be found, they would put in the dock all those who declare our peaceful strike to be an armed insurrection, and substantiate their claims with facts known to be false.

In addition to the above, there is the stigmatizing disgrace of the unworthy battle weapon used against us by the Minister of Labor and Justice, Safikiurdsky!

On the evening of May 7 after a long discussion of the workers' demands and the government's proposals the Minister of Labor and Justice informed us officially on behalf of the Azerbaijani government in the presence of the factory inspectors, the representatives of the industrialists and the British Command that complete immunity was guaranteed those on strike for the duration of the negotiations—to insure a peaceful discussion of the government's proposals and the peaceful settlement of the strike. We trusted the minister's word of honor. The following day, without taking any precautionary measures, we openly convened the factory and industry representatives in order to inform them of the government's proposals and to get an idea of their opinion.

Although the meeting was an *open* one, it was treacherously surrounded by a large detachment of police; we were all arrested, along with many chance onlookers and workers from the factories and trades.

In bringing this to the attention of all the citizens of Azerbaijan, we are expressing our loud protest against the unworthy means that have been employed by the Azerbaijani government in their struggle with the Baku proletariat!

The Central Prison, May 11, 1919. Block 15. Cell 112. Signed by: Churayaev, Mikoyan, Anashkin, and Koval.

From the authors. This declaration was sent to the socialist members of parliament via the prosecutor. But due to some formal considerations, he did not pass the declaration on to the addressees.

And so, our declaration didn't reach the parliament. But it is worth recalling, in connection with this, what a repellent role was played in all this by the right-wing socialists—who were members of parliament.

16. In addition to the five of us, the following people were also arrested on May 9: R. Voskanov, A. Meshcheriakov, G. Semenisky, O. Forer, L. Kaplan, Mir-Bashir Kasumov, I. Isakin, N. Merkel, M. Makhmudov, E. Rodionov, V. Ananyev, S. Popoviants, G. Sturua, M. Lelikhin, K. Tsimon, P. Orlov, V. Gorodinskaya, B. Lominadze, Guseinn Kuli Kerbalai Mukhtarogly, Ya. Sherin, Z. Pokonov, M. Tsarev, M. Zalov, F. Dyganenko, F. Sudenko, G. Stepaniants, Ya. Babenko, N. Tiukhtenev, S. Iliukhin, I. Koliubiashkin, N. Liakhovsky, I. Gandiurin, N. Pomerantsev, I. Blagochinny, T. Krivoi, V. Rebrukh, N. Rogov, and F. Gubanov.

By decree of the Azerbaijani government all of them were sentenced to prison "for a period of one month," as stated in the decree of the Chief of Police Captain Gudiyev No. 2223.

17. Evil kites in the garb of "mature and able" politicians have raised their ugly heads . . .

The venerable and tried leaders of the proletariat, as the Menshevik and SR gentlemen like to call themselves, said nothing about the untimeliness or the unpreparedness of the strike before it occurred or at the time when it was declared, but, on the contrary, they were either thoughtfully silent or they insisted impassionedly that it be called immediately. However, *now, after the results of the strike have been revealed, they're using so-called hindsight in their assertions that the strike was not properly prepared for and rash, and they explain its "failure" and "defeat" by just this kind of "philosophizing."*

"Fatal mistakes," committed by the Communist leaders of the Baku proletariat, made the "failure" of the strike inevitable . . .

People who are not close to the work of the Workers' Conference and its presidium, who do not participate in it, who do not know what has happened and what is happening in the working-class districts and organizations, who are not familiar with the material and legal situation of the proletariat that has arisen, with the status of the oil industry—the source of our life, can, on the basis of the numerous, but worthless articles in the appeasement-oriented newspapers, form the incorrect opinion that the strike was declared rashly, at an inopportune time, and that it was possible to postpone it to a more opportune moment which would have provided the possibility for a decisive victory.

The petit bourgeois *intelligent* lives through moments of boundless happiness, infinite sorrow, and . . . childish torpor. His soul is a phosphorescent candle which bursts weakly into flame, and then dies out and disappears in the darkness. Such an *intelligent* has little experience of the objective past and the inevitable future. His intellectual horizon is narrow, limited by his cottage and kitchen garden, and doesn't extend farther than his nose.

The character of this spineless and brainless petit bourgeois *intelligent*—a man who is morbidly revolutionary and at the same time morbidly reactionary—was revealed in all its nakedness during the strike, in the actions of the morbidly-puny SRs and Mensheviks . . .

Could they and can they now, these pale shades of the past, look broadly on the recent strike? Understand it correctly and make a fitting appraisal of it? Mark out future paths of struggle—without looking for any "fatal mistakes," and calmly explain the real reasons and the moving forces behind the struggle? No, of course not: that is beyond the strength of the rotting corpses of the former parties.

The only ones capable of this are the proletariat and the Communist party, who live not for fleeting manifestations of revolutionary outbreaks, alternating with periods of total despair, but who burn with the inextinguishable flame of the revolutionary struggle, who march along not the smooth path of easy victories, but who use a sword to cut out a path for themselves which is illumined by the rays of the rising sun of communism.

A collective agreement is not a new demand on the part of the Baku proletariat. It fought for one even during the reaction of 1907. In 1917 it both declared and conducted successfully a strike for a collective agreement which ended in the complete victory of the Baku proletariat . . .

Neither the bloody victories of Turkish and British imperialism, nor the crushing blows to the workers' organizations could force the workers of Baku to relinquish the collective agreement and give up their struggle for it. The Baku proletariat had barely managed to recover from the blows of the imperialist predators and to stand on their feet once again, when as early as the end of last year it put forward—as a routine battle demand—a collective agreement.

When, however, it found itself without its energetic Communist-leaders, who had either been shot or taken prisoner by the counter-revolution, and feeling that not all of its organizations were reestablished—the Baku proletariat, on the advice of the appeasers, was forced to accept a "crumb" from the capitalists (an increase in pay and a few other items of the collective agreement) and temporarily refuse to continue the immediate struggle. However, even in these conditions it did not roll up its banners but continued in every way possible to prepare for the future battle.

The collective agreement was handed over to an "arbitration commission" which consisted of representatives of the workers and the oil manufacturers, under the chairmanship of a "neutral" person—the president of the government . . .

In the beginning of March the members of the arbitration commission from the workers (and even the most inveterate appeasers among them, as, for example, Slepchenko) came to the conclusion that, in so far as the capitalists would not make any further concessions, it was both pointless and futile to continue the work of the commission. A joint session of the new presidium of the Workers' Conference and the representatives from all the districts and the arbitration commission came to the same conclusion.

However, considering the preparatory organizational work for the conduct of the future struggle to be incomplete, they decided to continue this "game of compromise" in order to win some time to strengthen and accelerate the organization of fighting ranks within the Baku proletariat. Then, at the beginning of March, a political strike almost broke out, the declaration of which was so urgently sought by the representatives of the SR committee.

The Central Strike Committee, adopted on our insistence, the tried and true tactics of the Communists, the tactics of delay and equivocation, of conserving one's forces and closing one's fighting ranks, which made it possible not to capitulate before the enemy (as the Mensheviks Rokhlin and Churayev proposed), or thoughtlessly, without taking forces or circumstances into account, to yield only to emotion, and fall unarmed on the British bayonets, in order to "die a beautiful death" (as the SR "dreamers," Ilin and Belenky, proposed).

Our tactics were victorious: we didn't capitulate, but, rather, achieved, the realization of almost all our demands.

18. The resolution adopted on May 2 at a political meeting of soldiers in the Lenkoran detachments and working people in the city is of interest. It said:

> ... With all (our) forces (we) support Soviet rule as the only power that protects the interests of workers and peasants. We are convinced that the only possibility of freeing labor from the oppression of capital lies in unity with the Great Russian workers under the common banner of the Third Communist Internationale, solidarity between the workers of all countries, and the dictatorship of the proletariat. All of us, as one person, are ready with weapons in hand to answer the call of the Communists and to go forth as friends against the feudalistic-Cadet, officer bands of Denikin—Kolchak, and the Anglo-French aggressors. We demand the immediate proclamation of Soviet power in Mugan and the creation of a Red Army for its defense.

19. This mandate is issued to Comrade Otradnev by the Caucasian Regional Committee of the Workers' Communist party (RKP) in view of the fact that he has been delegated to go to Mugan with emergency powers in order to clarify the situation of party organizations and Soviet institutions. He is fully empowered to replace unsuitable or insufficiently trained personnel; if necessary, to reorganize existing institutions, dismiss them, or create new ones.

All party organizations, comrades, and Soviet institutions are obligated to treat him with complete trust and give him whatever aid and cooperation he requires in carrying out the tasks he had been assigned.

A. Mikoyan,
Member of the Bureau of the
Caucasian Regional Committee of the RKP.

20. One resolution, which was passed at the meeting of the union of apothecary workers, has been preserved.

> To be demanded from the government of Georgia:
> 1. An immediate declaration of war against the bands of the Tsarist general Denikin and his accomplices until such time as they be routed once and for all in the name of the triumph of the proletarian revolution.

2. The cessation of any negotiations whatsoever with Denikin's representatives, because every delay strengthens the counterrevolutionaries both at the front and here, in the rear.

3. For insuring the favorable outcome of the struggle, it is necessary to create a united worker-peasant front, which itself necessitates: immediate reelection of the Soviet of Workers', Peasants' and Soldiers' Deputies that was elected on the bases of proportional representation; and the convocation of a workers' congress as the only class organ capable of unifying revolutionary democracy around itself and taking concrete measures to insure the unification of all proletarian Transcaucasia.

21. I first became acquainted with Dzhugeli in March 1917 when he was still a Bolshevik. We were both in a group of agitators attached to the Tiflis Party Committee and occasionally I spoke alongside him at workers' meetings. I remember one meeting which took place at the Adelkhanov Shoe Factory. Because there were many Georgian and Armenian workers at this factory, Dzhugeli spoke in Georgian and I, in Armenian. I liked his speech on that occasion very much, and in general he made a good impression on us all at that time. But he turned out to be a real careerist at heart, and finally betrayed the Bolsheviks. When the split in the united party organization of Tiflis took place, he went over to the Menshevik side after insuring his position as leader of the so-called national guard, which became the Mensheviks' armed support; in actuality, it was an ordinary punitive detachment that had already helped suppress the peasant rebellions on more than one occasion. In a word, Dzhugeli was a consummate traitor.

22. A short time later this became still more obvious as we found out that the British command, desiring to bolster its influence in the Transcaucasian republics and using Denikin's dependence on the British for supplying his army and for political support, had established a so-called line of demarcation between Denikinite "domains" on one hand, and Georgia and Azerbaijan on the other. Apropos of this, a fascinating communiqué from the British command has been preserved in the archives. It says:

> General Denikin has been ordered not to allow his troops to move to the south of this line, and the Caucasian states must not move to the north of it. The Caucasian states must refrain from any aggressive actions against the volunteer army, and assist General Denikin at the very least by furnishing oil and other supplies for the Caspian Fleet, while simultaneously refraining from supplying the Bolshevik forces with these things.

The communiqué was signed by Major General G. N. Corey who was in command of the British forces in Transcaucasia.

23. In November 1919 I wrote an article, which was published at the time in *Pravda*, about the character of and driving force behind the mountaineers' rebellion in the Caucasus. The content of the article could still have some interest for contemporary readers. Therefore, I shall quote from it.

I pointed out that as a result of historical circumstances which had taken shape during the Tsarist period, the mountain laboring masses were still very far from participating in political life. Even the February Revolution had accomplished little in this regard.

Only after the October Socialist Revolution, and particularly in terms of the growth and strengthening of Soviet power in the northern Caucasus, did the broad masses of laboring mountaineers, Ingush, Chechens, Ossetians, Dagestanis, and other peoples of the Caucasus, mature politically, become conscious of their class and national interests, and become actively involved in the revolutionary struggle on the side of Soviet power, standing up for the defense of that power from the White Guard bands. It was namely the October Revolution that allowed them for the first time to feel free and equal to other nations. With the help of Soviet power they were saved from Cossack oppression and caprice, and were able to recover the lands which had been forcibly taken from them by tsarism and transferred to the well-to-do Cossacks.

In speaking about the invasion of the Denikinite bands, I wrote that at first the mountaineers "had still not sufficiently realized the (true) nature of the General-led counterrevolution which was attacking them: the question of whether Soviet rule should or should not exist had still not been understood by them as a question of their life or death."

This was why the mountaineers did not at first put everything into the struggle with "the volunteer" bands "that they could have, did not exert all their strength, and as a result, unwittingly facilitated the fall of Soviet power in the Caucasus."

They soon realized, however, on the basis of their own experience, (just) what the rule of the *oprichnik*-generals and "fine Cossack fellows" brought in its wake.

"At the very start Denikin revealed his diabolical countenance quite clearly, without any masks," the article said.

> His worthy comrade-in-arms, the gallant general Lyakhov who had been the bloodthirsty suppressor and executioner in the revolution in Persia, burned and completely destroyed the auls without mercy, and shot all men, women, children and old people he came across, consistently applying in this instance the principle of uncon-

ditional "equality for all." Thus, in January and February scores of Ingush and Chechen auls were wiped from the face of the earth. In April, and then in July in Dagestan and Ingushetiya the same thing was repeated in order to make a bigger impression. Having used such means for instilling fear and "respect" for their power, the Denikinites designated their own little princelings for each mountain people as the "rulers" of Dagestan, Chechnya, Ingushetiya, and Ossetia; they created punitive detachments made up of officers and mountaineers who sympathized with the counterrevolution, and levied unbearable heavy taxes on the mountaineers, the payment of which brought in its wake the complete ruin of the already poor mountain population, for in general the poor could not pay even a half of what was demanded from them . . .

After attaining fundamental mastery of the "object lesson" given them by the gentry-bourgeois counterrevolution, and after gathering their forces and carrying out the necessary preparatory work (and one must take into account that at that time the mountaineers still did not have strong political parties, newspapers, or cadres from the intelligentsia), the laboring masses of mountaineers, united by unanimity and a firm resolution to fight the impudent enemy, passed the following resolutions at their congresses: to give Denikin nothing, and to begin a struggle with him not for life, but to the death. The freedom-loving eagles of the Caucasian mountains rose up to the sacred struggle for their freedom against the general's black reaction. For more than three months the broad masses of all the mountain peoples were seized by the fire of general rebellion.

As I wrote in my article, it was very important for the sake of our party and Soviet power to clarify the nature of the mountaineers' movement. Elements of their social and class differentiation were clearly manifested in it. The mountaineers' rebellion was an expression of the struggle of the mountain poor not only against Denikin, but also against his agents from the mountain area.

If during the Denikinite bands' advance upon the Caucasus, the mountain poor, with a gnashing of teeth, offered them open or covert resistance, then, in contradistinction to them, the propertied strata of mountaineers, former civil servants, officers, and kulaks, not only helped Denikin to strengthen his position in the mountains in every possible way, but also begged and pleaded with him to set up his garrisons in the major auls in order to protect their property; they willingly entered the service of the executioner of their own people by occupying the posts of rulers (Denikin's "deputies"), district chiefs, and senior representatives, and by filling out the ranks of the detachments which carried out reprisals against the poor. Denikin cleverly used the mountain counterrevolutionaries to do all his dirty work of oppressing and pillaging the mountain people. Therefore, the poor people's boundless hatred of their enslavers extended not only to the Denikinites, but also to "their own" counterrevolutionary mountaineers . . .

There were still quite a few Russian Red Army men in the mountains. Many volunteered their services to help the insurgents. The Moslem mountaineers not only gave shelter to the Russian Bolsheviks, but received them with open arms, showing them special love and gratitude, going hand in hand with them to battle against the Russian generals and officers . . .

The situation in which the mountaineers found themselves enabled them to understand the nature of their movement in the correct way, and to seek allies. They turned not to the Azerbaijani, Moslem, bourgeois-gentry government, but to the committee of Bolsheviks, asking it for assistance and wanting to know whether or not the Bolsheviks and the Red Army would return to them in the near future.

The insurgent laboring masses of mountaineers, having resolved either to conquer or to die, turned their backs on their coreligionists and class-betrayers.

The partisans of the Caucasian mountains have extended their fraternal hand to the Red Workers-Peasant Army and will not retreat until they have smashed the counterrevolutionary wall which has been erected between the mountain peoples and socialist Russia by the Black Hundreds' bands.

24. A controversy about this arose unexpectedly in our party press. Our paper's editor, Lominadze, without the approval of the Baku Committee, published a long, confused editorial in which he affirmed that the removal of the British troops from Transcaucasia signified the downfall not only of British, but of world-wide, imperialism. Communists who were in prison at the time felt that the article was wrongheaded and ran directly counter to the party's policies. Two days later an article by Sarkis appeared, under the pseudonym "Danielson," which correctly took issue with Lominadze's article. The next day we read Lominadze's answer to the criticism. Thus, to everyone's surprise, a controversy had got started in the press. We wrote a sharply-worded letter from prison to the comrades on the outside in which we condemned the controversy as unnecessary and even harmful for our party at the time. We declared our agreement with Sarkis's point of view and disagreed with Lominadze's position, but we demanded that an end be put to the controversy immediately. The comrades themselves realized that a mistake had been made, and the debate was curtailed and the difference of opinion painlessly quelled.

25. "The older generation," I said,

> bears within itself a heavy burden of bourgeois nationalism. In many respects this both divides and weakens their ranks in the class struggle. Although nationalist sentiments are widespread among the young people as well, they are, nonetheless, more free of nationalism, and hence it is easier to draw them into the international revolutionary struggle . . . The Party Regional Committee attaches a great deal of importance to the first Transcaucasian Conference of Communist Youth. Despite the small number of delegates, the Conference has a great deal of significance because it represents all the principal districts of Transcaucasia. Your task is to strengthen in organizational terms the unity of all the Communist youth organizations of Transcaucasia as components of the All-Russian Union of Communist Youth. Some of the Russian Komsomol members are fighting valiantly on the fronts of the civil war, defending the achievements of Soviet power, and others have rolled up their sleeves and are devoting all their energy to the construction of a new socialist society. All of us here in Transcaucasia are working under very different conditions. Our task is to carry out an organizational campaign to rally the best young people of the working class and of all the working force of the urban and rural areas. We have to be active in our propagation of Marx's and Lenin's ideas, and we have to carry out extensive political propaganda and agitation. But that's not enough. The young revolutionary should not confine himself to imbibing the ideas of Marxism-Leninism, he should make these ideas come true, he should take an active part in the revolutionary struggle. At this time our contemporary youth has a wonderful opportunity to show its skills in this period of the revolutionary destruction of the old bourgeois order and the establishment of a new socialist state.

26. I went on to say that the modern-day youth of Transcaucasia were being given a wonderul opportunity to immerse themselves in the practical preparations for the proletarian revolution, and to imbibe the spirit of collectivism, since the proletarian revolution is an act of the highest order of proletarian collectivism, an act of fighting community togetherness, when the individual merges with the masses during the struggle.

> Certain members of the intelligentsia see an implicit tragedy in the contradictions between the interests of the individual and the collective. That's either fraud or delusion! In fact, there is no, nor can there be any, antagonistic contradiction between socialist society and the separate individual in this society: socialism blends the interests of the individual with those of society in a harmonious fashion.

I then discussed the Komsomol's work in the professional unions and the need to raise the backward members of the young working class to the level of the vanguard. I dwelled in detail on the principles on which the Komsomol organizations should be based and on their ties with the party organizations, and I spoke of the need for independence in all practical party work of the Komsomol, and of the unacceptability of any petty surveillance by the party organizations. Guided by the ideas of communism to help the party in every way possible to solve the problems facing it, the Komsomol must train fighters who are able to form a correct understanding of situations and of issues which arise, even the most complicated and unexpected of them; the Komsomol has to instill in its members a will for struggle and for victory, an ability to organize the masses and rally them to the cause of the Communist party. I spoke of the necessity of keeping working-class youth in the leading role in the Komsomol, again emphasizing the need for a struggle against the remnants of nationalistic sympathies.

27. Alluding to my report, he said this, in particular, in his telegram:

> Public opinion in the Caucasus is riveted on the rebellion of the mountain peoples of the Caucasus—the Dagestanis, the Ingushes, the Chechens, and the Kabardins (?)—that began at the end of August . . . Aside from a small group of traitors and betrayers—officers who sold out to Denikin, all segments of the mountain population, without any help from anyone, but driven to despair by Denikin's atrocities, have categorically refused to pay the indemnity imposed on them, and to provide the needed regiments for the struggle with Soviet power. With rifles and daggers alone, they have rushed into bloody combat with the bands of officers and Cossacks, having resolved to win a victory or to die. General enthusiasm, even to the point of fanaticism, have overcome even the women, children, and old men, who have the complex task of supplying the front and the rebel detachments, because all the men have been put under arms. The most unbattleworthy segment of the population was using horses and bullock carts to haul everything that they had in their villages (auls) to the front for the soldiers. All the recent victories have inspired the rebels who display miraculous acts of heroism, and the huge stock of war supplies which have been gained as booty helps to strengthen the detachments, providing them with arms, of which the mountaineers have very few. In a series of battles the Dagestanis alone seized more than three million cartridges, sixteen pieces of artillery, several dozen machine guns, and the garrisons of the mountain posts of Dagestan were completely destroyed, where up to three thousand men had been killed by the Cossacks alone . . . Azerbaijan and Georgia, having taken an interest in the victory of the mountaineers, who

serve as a strong buffer against Denikin, are playing the heinous role of an idle and indifferent onlooker, and are not giving the rebels any aid whatsoever . . .

S. M. Kirov. Articles, speeches, and documents, Vol. 1. 1912–1921, Moscow, 1936, pp. 1478.

28. In greeting the congress Ordzhonikidze's telegram said:

On behalf of the Ninth Kuban Army which has delivered a fatal blow to the Wrangel landing force in the Kuban and has destroyed it once and for all, we greet the first Congress of Peoples of the East. Although we have been kept from participating directly in the work of the congress, we are following enthusiastically the rising of the East against the yoke of western European imperialism. We are proud that our victorious Red Army, in fraternal unity with the Azerbaijani peasantry and workers, having overthrown the government of the beks and khans, has created a Red Azerbaijan, in the capital of which you are meeting today. The moment is not far off when the red dawn of liberation will light up the entire East . . .

Name Index

Abakidze, 448
Abilov, Ibragim, 223, 224, 227, 232, 234, 313, 374, 487, 564
Abov, Gevorg, 555
Abovian, Avetik, 338
Adamov, 450
Adelkhanov, 63
Adzhemov, 57
Agamali-ogly, Samed, 428
Agamirov, Suren, 158, 162, 174, 175–176, 177, 178, 493, 521
Agayev, 312, 353, 360, 449, 493
Agayev brothers, 158, 174, 324
Agayev, Vakhram, 158, 174, 305, 306, 325, 439
Agazade, 326, 392, 407
Aikuni, G., 481, 482, 503, 508, 509
Akhabadze, 379
Akhnarazian, 29
Akhundov, Rukhulla, 158, 342, 397, 409, 410, 411, 412
Akhundov, Tukhula, 106, 412
Akirtava, 63
Akushinsky, Hadji, 461, 462
Akopian, Akop, 25
Akopian, Suren, 554, 556
Alabian, Karo, 555
Alaniya, 186, 189, 190, 205, 217, 218, 316, 317
Alaverdian, 542
Alexander II, 254
Alikhanian, Gevorg, 30, 41, 42, 63, 85, 318, 451, 554, 556
Alimov, 540
Ali-ogly, Maadzhit, 461
Aliyev, Ashum, 326, 487, 488, 489
Alkhavi, 120
Alliluyev, Sergei, 432
Altunin, 225, 226
Amazasp, 124, 125, 129, 131, 132, 133, 140, 151, 154
Amfiteatrov, 222
Amirian, Arsen, 150, 187, 195, 199, 200
Amirian brothers, Tatevos and Alexander, 187
Amirkhanian, Shavarsh, 555, 556
Amirov, Armenak, 237, 238, 243, 244
Amirov, Tatevos, 127, 147, 149, 173, 174, 179, 181, 187, 191, 196, 205, 228, 236, 559
Amirova, Maria, 191, 217, 259
Amirov brothers, Armenak and Alexander, 217, 227, 231

Ananyev, V., 565
Anashkin, A. I., 207, 253, 310, 312, 322, 325, 326
Anashkin, Ivan, 307, 339, 360, 378, 379, 381, 392, 522, 565
Andranik, 34, 553
Andreassian, Napoleon, 29, 41, 554
Andzhiyevsky, 390, 468
Anvazian, Levon, 33
Aralov, 224
Arsenidze, 445
Arutiunian, Telemak, 555
Arveladze, 89, 90, 145
Asadulayev, 428
Askendarian, Tigran, 397, 399, 400, 401, 402, 403, 404, 405
Atarbekian, Gevorg, 48
Avakian, Bagdasar, 195
Avakian, Ensign, 75, 76, 187, 188, 189
Avanesov, 560
Avetisov, 141, 142, 143, 144, 149, 151
Avetissian, Suren ("Little Gevork"), 154, 155, 555
Avis, 325
Aydinbekov, 460
Ayollo, 77, 166, 170, 445–446, 447
Azimzade, 360
Azizbekov, Aziz and Aslan, 203
Azizbekov, Meshadi, 107, 110, 111, 112, 114, 136, 147, 154, 163, 164, 184, 187, 189, 195, 198, 200, 201, 202, 203, 220, 301

Babaev, Dzhafar, 158
Babenko, Ya., 565
Bablumian, Artem A., 555
Bagaturov, 161
Bailov, 322
Bairamov, 390, 391
Balian, Vagan, 554, 556
Bannayan, Ovaness, 556
Baranov, Alexander (Shura), 174, 175, 176, 177, 178
Baratov, 491, 492
Barkasho, 491
Basin, Meer, 164, 195
Batminov, 194
Bebel, 30
Beethoven, 517
Beker, 162, 164, 170, 175
Bekhaèddin-bay, 175, 177
Belenky, 568

Belinsky, V., 27, 28
Berg, 195, 200
Bertsinskaya, Shura, 397, 399, 403, 498
Bibinur, 550
Bicherakov, Lazar, 117, 119, 120, 121, 142, 146, 161, 178, 179, 186, 204
Blagochinny, I., 565
Bliumen, 161
Bogdanov, Anatolii, 195
Bogdanov, Solomon, 30, 104, 110, 164, 192, 195
Bohm-Bawerk, 47, 48
Bonch-Bruyevich, V. D., 219, 246
Borian, Armeniak, 195
Borian, Bograt, 31, 32, 90–91
Borian, Khoren, 399, 401
Borian, Zhenya, 31
Borkhart, 416
Brown, 362
Budyonny, 544
Buinaksky, 461
Buniatzade, 404, 451, 484, 485
Byron, 33
Bzhshkian, Gai, 556

Catherine II, 54
Chaikin, Vadim, 205, 206, 261, 262
Chairman, 311
Chekhov, 470
Chernomordik, 463, 465
Chernyshevsky, 28
Chicherin, 537
Chikarev, 432
Chikaryov, 312
Chukaryov, 322
Churayev, 262, 302, 306, 307, 312, 360, 370, 374, 376, 377, 378, 379, 380, 381, 392, 436, 560, 565, 567
Corey, Major General G. N., 570

Dadashev, Museib, 153
Dalakishvili, 89
Dalin, 162, 172, 173, 558
Danilian, Sarkis, 47
Darwin, 26
Denikin, 152, 250, 321, 348, 358, 399, 401, 406, 426, 429, 431, 432, 439, 440, 441, 442, 443, 446–450, 452, 459, 461, 462, 466, 487, 489, 491, 494, 500, 503, 506–507, 508, 512, 520, 547, 554, 563, 564, 568, 569, 570, 571, 574, 575
Denikinites, 368, 403, 414, 444, 454, 459, 461, 468, 471, 488, 501, 505, 520, 564, 572
Denikinite's ships, 438, 500, 501
Denikin's army, 365, 368, 403, 414, 426, 440, 444, 454, 460, 461, 462, 476, 506, 507
Densterville, 135, 181, 204
Dobroliubov, 27, 28
Dokhov, 209
Dostoevsky, 28

Dovlatov, Isai, 322, 347, 348, 429, 432, 437, 451, 486, 531
Druzhkin, Semyon, 206, 242, 316, 317
Dumas, Alexander, 28
Dumbadze, Lado, 64, 332, 367
Dutov, 511, 515
Dyganenko, F., 368A
Dzerzhinsky, F., 415, 416, 418, 423
Dzhabiyev, 531
Dzhaparidze, Elena and Liutsia (daughters), 141, 203, 257, 427
Dzhaparidze, Prokofii (Alyosha), 59, 62, 68, 69, 70, 72, 77, 78, 103, 107, 110, 111–114, 124, 136, 138, 147, 149, 153, 154, 160, 162, 163, 164, 167, 179, 181, 184, 186–187, 188, 189, 195, 196, 198–203, 220, 233, 246, 257, 301, 315, 316, 413, 427, 480, 559
Dzhaparidze, Varvara Mikhailovna (wife), 141, 162, 171, 179, 191, 192, 217, 257, 427, 559–560
Dzhevanshir, Beybut, 175–177, 311
Dzhugeli, 63, 445–447, 569
Dzneladze, 63, 85, 493
Dznuni, Daniel, 555

Efendiev, 111, 481
Egiazarian, Artavazd, 556
Egorov, Vasia, 486
Elbakidze, Arkadii (Agordiya), 491, 492
Eliava, 419
Ellis, 209, 210
Engels, F., 41, 46, 80, 107, 255
Enukidze, 219
Eremian, Vagan, 27, 30, 555, 556
Eshba, 465

Fatalibekov, 379, 380
Figatner, Yuri, 463, 468, 472–473
Fioletov, Ivan (Vanya), 103, 110, 114, 119, 154, 164, 184, 185, 187, 195, 199–201, 220, 257, 301, 559
Fioletova, Olga Bannikova (wife), 191–192, 217, 230
Forer, O., 565
Fourrier, 28
Frunze, M. V., 527, 529
Funtikov, Fyodor, 190, 191, 194, 195, 204–205, 208–210, 241, 242
Fyodorov, 416

Gabyshev, Ivan, 105, 142, 143, 144, 147, 196
Galstian, Tigran, 555
Gandiurin, I., 200, 565
Ganin, 105, 142, 143, 144, 147
Garagash, 40
Gardashyan, Garegin, 493, 555, 556
Garegin, Father, 48, 56
Garibdzhanian, 542
Garshin, 28
Gegechkori, 441
Gelovani, 470, 489, 490

George V, Catholicos, 52
George V, King, 259, 260
Gevondian, Gevorg, 555
Giandzhetsian, 55–56
Gigoyan, Emmanuel, 192, 217, 318
Gikalo, Nikolai F., 368, 503
Gogoberidze, Levan, 114, 174, 178, 303–304, 307–309, 312, 313, 320, 322, 325–326, 348, 360, 367, 370, 392, 393, 407, 442, 449, 451, 463–477, 487–489, 531
Goncharov, 28
Gorky, Aleksei Maksimovich, 426, 429–431
Gorodinskaya, Vera, 384, 565
Gotsinsky, Imam, 107, 108, 461
Grigorev, 563
Grigorian, Arutiun, 555
Gubanov, Fedia, 158, 347, 348, 360, 370, 378, 407, 443, 500, 503, 504, 565
Gudiyev, Captain, 488, 565
Gukasian, Gukas, 493, 542
Gukasov, 108
Guseynov, Mirza Davud, 305, 312, 325–326, 353, 395, 428, 451, 471, 484, 540

Hilferding, 117
Hugo, Victor, 28
Hurok, 226

Ibragimov, 360
Ibsen, 28
Ilin, 31, 360, 568
Iliukhin, S., 565
Ioanesian, Shiga, 346–347, 485
Ionessian, Ashot, 48, 56, 61
Isakin, I., 565
Israfilbekov, 111, 515
Ivanov-Kavkazsky, Vladimir, 525, 531
Ivanov-Razumnik, 54
Izmailov, Abkurakhman, 461

Jack, 483–484
Jouresse, Jean, 27

Kadyr, Fazma, 549
Kafarov, 387
Kakhiani, Mikhail, 527, 531
Kakhoyan, Asatur, 40, 82, 89, 367
Kalantadze, Fyodor, 63, 367
Kaledin, 113
Kaledinites, 104
Kameron, 207
Kandelaki, Samson, 191, 192, 217, 243–245, 248, 438
Kanevsky, 407, 436
Kapanakian, 28
Kaplan, L., 565
Karakhan, 543
Karayev, Ali Geydar, 305–306, 312, 325–326, 348, 353, 357, 370, 374, 376, 392, 395, 428, 449, 450, 467, 475, 476, 481, 484, 486, 493, 540, 564

Karinian, Artashes, 110, 203
Kasparov, Lev, 451, 453, 454
Kasparova, Grachiya, 451
Kasparova, Maria, 389, 450–453
Kasparova, Rosa, 451–454
Kasparova, Tatiana, 451, 452
Kasparov apartment, 367, 389, 450, 451, 463, 475
Kasumov, Mir-Bashir, 157, 207, 312, 322, 522, 565
Kasyan, Sarkis, 367, 368
Kautsky, 30, 46
Kavtaradze, Sergei, 64, 89, 207
Kazarov, 125, 126, 129, 131, 132
Kazbekov, 360, 461
Kazimamed, 460
Kediya, 459, 460
Kerensky, 76, 93–95, 97, 227, 302, 347
Khachiyev, Arshak, 342–344, 373
Khalimov, Saib Abdul, 461
Khankhoysky, 251
Khanoyan, Sergo, 40, 64
Khanum, Nadzhia, 550
Khoysky, F. Kh., 310–311
Khozhemiakin, 407–408
Khrenov, 194
Khununts, 41
Kibalchich, 254
Kirov, Sergei Mironovich, 204, 347, 407, 408, 410, 411, 412, 421, 422, 425, 426, 432, 451, 460, 461, 490, 497, 500, 502–504, 528, 529, 531, 540, 575
Klinger, 405
Knuniants, 70
Kobakhidze, 415, 416
Kochar, Ervand, 556
Koganov, Mark, 105, 195
Kolchak, 240, 439, 447, 494, 510, 512, 515, 547, 568
Kolesnikov, 110
Kolesnikova, Nadezhda Nikolaevna, 202, 203
Koliubiashkin, I., 368A
Kolomijtsev, Ivan, 435–438
Kolostov, 194
Kondakov, 183, 185
Korganov, Grigorii, 88, 90, 105, 110, 120, 138, 145, 154, 160, 162, 164, 184, 189, 191–192, 195, 199, 200, 211, 217, 559
Korneyeva, Olga, 325
Kostandian, Aram, 195
Kostanian, Alikaz, 554, 556
Koval, 306, 314, 378–379, 380–381, 384–386, 565
Kovalyov, P., 318, 385
Kramarenko, Artak, 158, 217, 235
Kramarenko, Marusia, 158, 217, 224, 225, 227, 232, 233
Krasin, L., 430
Krasnov, 140, 152, 240
Krivoi, T., 565
Krupskaya, Nadezda Konstantinovna, 203,

218, 413, 421
Kuibyshev, Valerian Vladimirovich, 423, 527
Kun, Bela, 189, 190, 204, 262, 546, 548
Kurginian, Shushanik, 25, 50
Kuzminsky, Arkady, 105
Kuznetsov, 98

Lakoba, Nestor, 465
Lelikhin, M., 565
Lenin, Vladimir Ilyich (Ulianov-Lenin), 25, 32, 37–38, 39, 46, 63, 66, 70–72, 77, 78, 79, 80, 89, 92, 94–96, 99, 100, 107, 108, 109, 110, 111, 112, 118–119, 123, 139, 152, 160, 190, 212–214, 218–219, 239, 246–247, 255, 397, 398, 399, 404, 405, 406, 409, 410, 413, 415–421, 423, 427, 428, 430, 431, 432, 455, 493, 497, 498, 503, 504, 505, 506, 507, 508–511, 512, 513, 514, 515, 524, 530, 539, 541, 545, 548, 550, 557, 573
Lessner, 108, 109
Levandovsky, Mikhail Karlovich, 426, 529, 530–532, 536–537, 540, 542
Liakhovsky, N., 565
Lidak, Otto, 253–254, 256, 325, 439
Liebknecht, Karl, 86
Lifshits, Lev, 411
Likhachev, 424
Lipartiya, 459–460
Lisinova, Liusia, 31
Lominadze, Beso, 334, 340, 367–369, 384, 394, 451, 487, 491, 493, 520, 521, 530, 565, 572
London, Jack, 28
Lordkipanidze, 534, 536
Lunacharsky, 493
Luxemburg, Rosa, 86

MacDonald, Ramsey, 260
Macdonnell, 204
Magoma-ogly, Abdul Barab Hadji, 461
Malleson, Jr., 193, 208
Malleson, Major General Sir Wilfred, 206–210, 240, 241
Magauzov, Suren, 452, 454–455
Malygin, Ivan, 105, 164, 195
Mahomed-Abdur-Rabe-Berk, 549
Makharadze, Gerasim, 442, 444–445, 491
Makharadze, Philip, 64, 77–78, 88–89, 91, 93, 98, 331, 335–337, 353, 367–369, 383, 442, 443, 450
Makhmudov, M., 565
Makintsian, 48
Mamed, Kazy, 73
Mamedyarov, Sarkis, 157, 312, 322, 325, 383, 451, 486, 521, 572
Mamontov, 432
Mandalian, Tatevos, 556
Mantashev, 28, 69, 72
Manukian, 47
Marat, 27
Markarian, Gurgen, 556
Markarian, Sedrak, 556

Martikian, Satenik, 191, 217, 257
Martikian, Sergo, 158, 171, 174, 179, 193, 232–233
Martikian, Tatul, Azat and Emma (children), 257
Marx, Karl, 30, 38, 40, 46, 47, 54, 55, 80, 107, 213, 255, 409, 410, 573
Maupassant, 222
Mayakovsky, Vladimir, 411, 412
Mayumedyarov, 322
Mazmanian, Mikael, 556
Mdivani, 333
Mekhanoshin, 529, 540
Melikian, Romanos, 16
Melkumov family, 257
Mendeleev, 26
Merkel, N., 565
Merkurov, 203
Meshcheriakov, A., 565
Meskhi, 191
Metaksa, Iraklii, 192, 196
Miakov, Andrea, 402
Miasnikian, Alexander, 48, 224
Mikhalchenko, 168–169, 171
Mikoyan, Anastas, 14, 31, 33, 42, 56, 70, 87, 124, 158, 160, 164, 170, 173, 337, 341, 360, 370, 381, 386, 388, 389, 390, 392, 400, 405, 410, 446, 450, 459, 464, 470, 474, 475, 498–499, 543, 557, 561, 565, 568
Mikoyan, Artem Ivanovich (Anushavan, brother), 21, 81, 247, 337, 457, 553
Mikoyan, Nerses (grandfather), 19
Mikoyan, Ovanes (father), 14, 84
Mikoyan, Vartiter (grandmother), 19
Mikoyan, Sergo (son), 210
Mikoyan family, 7, 8, 457
Miliukov, 94
Milne, 376
Minayev, Moriak, 362
Mirzoyan, Levon, 158, 174, 253, 306, 313, 322, 339, 367, 378, 379, 522
Mishne, Isay, 192, 196
Mogilevsky, 48
Molitozhko, 194
Mommsen, 27
More, Thomas, 28
Mravian, Askanz, 40, 64, 90, 329, 331, 367, 369
Mudryi, 191
Mukhtadir, 69–70
Mukhtarogly, Guseinn Kuli Kerbalai, 565
Musavekov, 531
Musayelian, 542
Musevi, Mirfatag, 326, 486, 487, 488, 489
Mzhavanadze, 340

Nadezhdina, N., 226
Nadzharova, Tamara, 453
Nakhichevansky, Dzhamshid, 248
Naneishvili, Victor, 78, 108, 112, 404, 484, 485

Narimanov, Nariman, 107, 110, 111, 114, 201–202, 224, 515, 539
Navasardian, 53–55, 60–61
Navasartian, 235
Nazaretian, Amayak, 62, 64, 332, 367
Nazariyan, 559–560
Negri, Ada, 25
Nerses, Catholicos, 45
Nicholas II, 233, 563
Nikolaev, 175
Nikolaishvili, Ivan, 192, 196
Nobel, 65, 313, 315, 561
Nogin, V. P., 323, 324, 480
Novgorodtseva, K. T., 510
Nuridzhanian, Avis, 156, 161, 463, 467–468

Ochkin, 555
Okinshevich, 260, 261
Okudzhava, Misha, 62, 64, 367
Oraz-Sardar, 242
Orakhelashvili, Mamiya, 329, 332, 367, 450–451
Ordzhonikidze, Sergo, 71, 152, 212, 247, 333, 405, 412, 413–418, 419, 420, 421–429, 431, 450, 516, 529–531, 540–546, 550, 575
Ordzhonikidze, Zinaida Gavrilovna (Zina, wife), 417, 427
Orlov, Count, 54
Orlov, P., 565
Osintsev, 169, 171
Otradnev, Timofey Ivanovich, 407–409, 437, 568
Ovsepian, Suren, 183, 195, 200
Owens, Robert, 28

Palladine, 26
Paronian, Akop, 25, 50
Parsadanian, Sergei, 555
Pastriusin, 32
Pepinov, 313, 392
Pereverdiev, 536
Perikhanian, Gevorg, 556
Perovskaya, Sofia, 254, 256
Petrosov, Dr., 194, 249
Petrov, Grigorii, 125, 130, 132, 138, 139, 142, 143, 148, 149–152, 160, 161, 167, 181, 182, 184, 185, 195, 199, 200, 211, 559
Pilsudsky, 432
Pirumov, 73, 340
Pisarev, Dmitri, 28
Platten, Fritz, 78
Plekhanov, 30, 38, 46, 54
Pleshakov, 157, 312
Plyashakov, 322
Pogossian, Ovanes, 48, 50, 555
Pokonov, Z., 565
Poltoratsky, 194, 339, 344, 360
Polukhin, Vladimir, 145, 155, 180, 195, 200
Pomerantsev, N., 565
Popoviants, S., 384, 388–389, 390, 392, 565

Potemkin (battleship), 302
Pushkin, 28

Raffi, 25
Ramishvili, Isidor, 79
Razhden, 459
Rebrukh, V., 377, 384, 392, 565
Reed, John, 546, 547–548
Robespierre, 27
Rodionov, E., 322, 565
Rogachik, Adolf, 232, 234
Rogov, N., 347, 348, 429, 565
Rokhlin, 260, 308, 320, 383, 567
Romanov, Tsar Nicholas, 416
Rossmer, 546, 549
Rothschild, 65, 175
Rozanov, 194, 493, 555
Rumiantsev, Kostya, 174, 175, 178, 452
Ryabushinski, 561
Ryskulov, 549

Sadovsky, 77, 103, 146, 156, 157, 159, 160, 162, 165, 170, 197
Safarov, 127, 131, 132, 154, 490, 491, 492
Safikiurdsky, 374, 375, 376, 564
Saint-Simon, 28
Saraikin, 347
Sardarov, Buniat, 348
Sarkisov, Semen, 556
Sarkissian, Mikhail, 556
Sedykh, 195, 241, 242
Semenisky, G., 565
Serafimovich, 419, 420
Shabanova, 550
Shakhbazov, Lyuba (wife), 532
Shakhbazov, Museib, 532
Shakhgaldian, Aram, 34, 41, 46, 55, 555
Shamrkhanian, 556
Shatunovskaya, Olga, 133, 141, 153, 158, 174, 176, 465, 498–499, 517, 521, 527, 531
Shaumian, Ekaterina Sergeevna (wife), 71–72, 141, 219, 510
Shaumian, Leva (son), 80, 81, 117, 163, 179, 190, 191, 194, 208, 209, 217–219, 225, 228, 229, 236, 243, 247–248, 261, 559–560
Shaumian, Manya and Serezha (children), 190
Shaumian, Stepan, 25, 37, 38, 59, 64, 65, 68, 70–73, 75, 77, 78, 80, 87, 89–93, 98, 99, 100, 103–105, 108, 109–112, 117–120, 122–124, 127, 131–133, 136, 137, 139–141, 143–144, 147, 149–150, 152, 153–154, 155, 156, 159–162, 164, 167, 171, 174, 175, 176, 179, 180, 182–186, 188–191, 193, 195–197, 200–202, 204, 207, 209, 214, 218–221, 224, 233, 246–247, 255, 257, 301, 309, 316, 319, 320, 369, 413, 427, 428, 435, 523, 554, 557, 559
Shaumian, Suren (son), 108, 163, 173, 175, 179, 190, 191, 200, 217, 218, 228, 237, 238, 242, 243, 246–248, 559–560

581

Shaumian family, 75, 162, 232, 261, 510
Shaverdian, Danush, 31–32, 37–38, 39, 40, 46, 62, 64, 70, 72, 88, 90, 331, 367
Schiller, 28
Sheboldaev, Boris, 105, 130, 131, 142, 145, 149, 440, 450, 453, 461, 466, 469, 473, 474, 475, 476, 477
Sherin, Ya., 384, 565
Shirvanzade, 25
Shishkin, 54
Shtern, 392
Shuisky, Judith, 226
Shuttleworth, 484
Simak, 41–42
Simonian, Martiros, 13, 14, 15
Skachko, 498–499
Slavinsky, 459
Smeliansky, 194
Smilga, 504
Smith, Adam, 47
Solntsev, Fyodor, 105, 145, 195
Spandarian, Suren (Timofei), 79
Stalin, Josef, 38, 139, 246, 333, 418, 419, 422, 560
Stamboltsian, Artak, 50, 53, 147, 155, 158, 217, 224, 225, 231–232, 236, 247, 249, 254, 256, 257, 261, 555, 556
Starozhuk, 407, 436, 539
Stasova, Elena Dmitrievna, 31, 32, 406, 428, 465, 497, 504–505, 510, 515–518, 531, 546
Stepaniants, Grisha, 386–390, 565
Stepanov-Skvortsov, 55
Steinecke, K., 226
Struve, 55
Sturua, Georgi, 89, 157, 164, 165, 167, 170, 334, 360, 367, 369, 370, 378, 379, 386, 388–390, 443, 446, 448, 450, 459, 460, 489, 490, 565
Sturua, Vano, 450
Sudenko, Fiodor, 565
Sultanov, Gamid, 74, 111, 404, 451, 460, 484–485, 540
Sumbat, 362
Sverdlov, Ya. M., 338, 510

Tagianosov, 108, 356
Tagiyev, 387, 470
Tamara, 454
Tarnovsky, Count, 222
Telliya, 194
Ter-Gabrielian, Laak, 124
Ter-Israyelian, 465, 470, 474, 475
Ter-Petrosian, Dzhavaira, 40
Ter-Petrosian, S. A. (Kamo), 40, 100, 421, 427, 429, 430, 431, 432, 451, 538
Ter-Saakian, Ashot, 145, 165
Ter-Simonian, Drastamat, 556
Tevosian, Ivan Fedorovich (Vanya), 174, 203, 424
Tevosian, Julia, 174
Thompson, General V. M., 206, 250–252, 262, 302, 311, 560–561
Tighe-Jones, Reginald, 205–207, 209, 210
Tigran, 498–499
Timiriazev, 25
Tiukhtenev, N., 368A
Toniants, Amalia, 174, 178
Topuridze, 240, 249, 250, 253, 254, 256, 261
Toroshelidze, Malakiya, 64, 91, 92, 122, 332–333, 367
Tovmassian, 47–48, 50
Trotsky, 123, 417
Tsarev, M., 565
Tseretelli, 93
Tsimon, K., 565
Tsintadze, Kote, 68–69, 72, 73, 74, 99
Tskeidze, 121
Tskhakaya, Mikha, 71, 77–79, 87, 89, 94, 95–96, 98, 333, 336, 448–449, 479, 493, 507
Tugan-Baranovsky, 47
Tukhachevsky, 530
Tumanian, Ashkhen (Mikoyan, wife), 23–24, 229–230, 335, 460
Tumanian, Gaik, 337, 338
Tumanian, Hovaness, 25, 559–560
Tumanian, Lazar, 13, 87, 335, 337, 338
Tumanian, Maro, 158, 191, 227, 247, 257
Tumanian, Verginia, 13, 15, 23, 87, 334–335, 337–338, 457
Tumanian family, 335, 336
Turgenev, 28
Tyukhtenev, 312

Uliantsev, T. I. Otradnev, 407
Urushadze, 442
Usubekov, 484
Uzan-Hadji, 461

Van, 547
Vasin, 162
Vazgen I, Catholicos, 52
Velunts, 167, 172
Veresaev, 28
Vezirov, 110, 138, 139, 164, 184, 195, 200
Voltaire, 54
Voroshilov, K. E., 421, 425, 517, 544
Voskanian, Gurgen, 41, 556
Voskanov, R., 261, 565
Voskresensky, 564
Voytinski, 445
Vrachev, 417–419

Walton, 364, 374, 375, 537
Welch, 545, 549
Wrangel, 545, 549, 550, 551, 575

Yakovlev, 205
Yakubov, 312, 322
Yefremov, Mikhail Grigorievich, 531–532, 534–538, 540
Yudenich, 512, 547

Yudenich's White Army, 510, 547
Yeremenko, Zina, 400
Yusifzade, A. B., 323, 480, 484

Zabeniagin, 424
Zalov, M., 565
Zarafian, 223
Zavriyev, 559–560
Zavyalova, 531
Zevin, L., 559
Zevin, Vladimir (son), 203
Zevin, Yakov, 110, 136–138, 164, 191–192, 195, 200, 203
Zheliabov, 254, 255
Zhitnikov, 194
Zhordaniya, Noi, 428
Zimin, 210, 261, 262
Zina, 178
Zinoviev, 422
Zorabov, Grisha, 400, 402
Zvonitsky, 67
Zubalov, 62, 94, 332
Zulfugar, 393
Zurabov, Arshak, 340–343, 447